Medieval Supposition Theory Revisited

Medieval Supposition Theory Revisited

Edited by

E.P. Bos

in Collaboration with

H.A.G. Braakhuis, W. Duba,
C.H. Kneepkens and C. Schabel

BRILL

LEIDEN • BOSTON
2013

Also published as Volume 51, No. 1-4 (2013) of Brill's journal *Vivarium*

Library of Congress Cataloging-in-Publication Data

Medieval supposition theory revisited/ edited by E.P. Bos in collaboration with H.A.G. Braakhuis, W. Duba, C.H. Kneepkens and C. Schabel.
pages cm
Includes index.
ISBN 978-90-04-25983-6 (pbk. : alk. paper) 1. Rijk, L. M. de, 1924-2012. 2. Rijk, L. M. de, 1924-2012. Logica modernorum. 3. Logic, Medieval. 4. Fallacies (Logic) I. Bos, Egbert P., 1947-

B4095.R554M43 2013
160.9'02—dc23

2013030368

This publication has been typeset in the multilingual "Brill" typeface. With over 5,100 characters covering Latin, IPA, Greek, and Cyrillic, this typeface is especially suitable for use in the humanities. For more information, please see www.brill.com/brill-typeface.

ISBN 978-90-04-25983-6 (paperback)
ISBN 978-90-04-26023-8 (e-book)

This book is printed on acid-free paper.

Contents

Preface 1

E.P. Bos and B.G. Sundholm, Introduction 3

Early Supposition Theory in General

L.M. de Rijk, Semantics and Ontology. An Assessment of Medieval Terminism 13

Sten Ebbesen, Early Supposition Theory II 60

Arabic Philosophy

Allan Bäck, Avicenna's Theory of Supposition 81

XIIth Century

Luisa Valente, Supposition Theory and Porretan Theology: *Summa Zwettlensis* and *Dialogus Ratii et Everardi* 119

XIIIth Century

Mary Sirridge, Supposition and the Fallacy of Figure of Speech in the *Abstractiones* 147

Julie Brumberg-Chaumont, The Role of Discrete Terms in the Theory of the Properties of Terms 169

Dafne Murè, Suppositum between Logic and Metaphysics: Simon of Faversham and his Contemporaries (1270-1290) 205

XIVth Century

Costantino Marmo, Scotus on Supposition 233

Simo Knuuttila, Supposition and Predication in Medieval Trinitarian Logic 260

Laurent Cesalli, Richard Brinkley on Supposition 275

Alessandro D. Conti, Semantic and Ontological Aspects of Wyclif's Theory of Supposition 304

FABRIZIO AMERINI, Thomas Aquinas and Some Italian Dominicans (Francis of Prato, Georgius Rovegnatinus and Girolamo Savonarola) on Signification and Supposition 327
CATARINA DUTILH NOVAES, The Role of 'Denotatur' in Ockham's Theory of Supposition 352
CLAUDE PANACCIO, Ockham and Buridan on Simple Supposition 371
E. JENNIFER ASHWORTH, Descent and Ascent from Ockham to Domingo de Soto: An Answer to Paul Spade 385
ERNESTO PERINI-SANTOS, When the Inference 'p is true, therefore p' Fails: John Buridan on the Evaluation of Propositions 411

XV-XVI-XVIITH CENTURIES
ANGEL D' ORS, Logic in Salamanca in the Fifteenth Century. The *Tractatus suppositionum terminorum* by Master Franquera 427
STEPHAN MEIER-OESER, The Hermeneutical Rehabilitation of Supposition Theory in Seventeenth-Century Protestant Logic 464

LOGIC: MEDIEVAL AND MODERN
SARA L. UCKELMAN, A Quantified Temporal Logic for Ampliation and Restriction 485
TERRY PARSONS, The Expressive Power of Medieval Logic 511

Index 523

Preface

From June 2nd to June 7th, 2008, the XVIIth European Symposium for Medieval Logic and Semantics was held at the University of Leiden, The Netherlands. The present volume contains the papers presented at the symposium by scholars from all over the world (the European Symposium extends beyond Europe in terms of its participants).

The general theme of the symposium was 'The Rise and Development of Supposition Theory'. It seemed appropriate that a symposium in Leiden, held about 50 years after the publication of L.M. de Rijk's *Logica Modernorum* (Assen 1962-1967), emeritus professor of the University of Leiden, should be devoted to medieval supposition theories. On June 30, 2012, De Rijk died at the age of 87 years.

All articles are preceded by an abstract and a list of keywords used. Indexes on places, names and things are added to this volume, as well as a list of manuscripts mentioned in the articles.

The organizer of the Symposium is very grateful to the Leids University Fund (LUF) and the Faculty of Philosophy at the University of Leiden (now Institute of Philosophy as one of the seven institutes of the Faculty of Humanities) for their financial and administrative support.

Special thanks are due to my colleague Dr. Joke Spruyt (University of Maastricht, The Netherlands) for her great help in correcting an earlier version of this edition.

E.P. Bos
University of Leiden

Introduction

E.P. Bos and B.G. Sundholm
University of Leiden

This volume contains the acts of the XVIIth European Symposium on Medieval Logic and Semantics, which was held at the University of Leiden from June 2nd till June 7th, 2008. In the first part of this introduction we shall indicate the subject-matter and describe the contents of these acts. What follows in the second part is the speech with which Prof. B.G. Sundholm, who holds the chair of logic and its history at the University of Leiden, opened the symposium.

I

In 1962-1967 Professor L.M. de Rijk published his *Logica Modernorum—A Contribution to the History of Early Terminist Logic.* It consists of two parts, divided over three volumes. The first part, entitled, *On the Twelfth Century Theories of Fallacy,* was published in 1962. The second part appeared in 1967 und the title *The Origin and the Early Development of the Theory of Supposition.* The first volume of the second part is a study of the early treatises, while the second volume contains texts and indices. De Rijk's *Logica Modernorum* provides the basis for the modern study of medieval theories of supposition.

Now, some 50 years later, scholars have made great progress in the study of the properties of terms. Editions and studies have been published. Some of the treatises, mostly those composed in the twelfth century, are anonymous. Others are composed by well-known medieval authors, notably those of the fourteenth century. For example, Ockham's (ca. 1285-1347) *Summa logicae* was edited as part of the edition of his *Opera Omnia*; similarly, logical treatises by Walter Burley (ca. 1275-1344/5), John Buridan (born between 1300 and 1305, if not earlier—died 1361) and Richard Brinkley (fl. third quarter of the fourteenth century) have been edited.

De Rijk's study was primarily about the *early* development of terminist logic, i.e., during the twelfth and thirteenth centuries. Scholars have also investigated later developments well into the sixteenth and seventeenth centuries. Not only logical texts, but also texts on grammar have been published. They have great bearing on the study of the properties of terms. Many of the scholars who have contributed to this development also participated in the symposium and now offer papers in this volume.

The medieval theory of supposition is part of the theory of the semantic properties of terms. The most prominent of these properties are *suppositio, ampliatio, appellatio* and *copulatio*. In treating these notions medieval authors attempt to analyze the presuppositions of natural language, and as a rule deal with the meaning of a term as it is used in a proposition. The theory is about what terms stand for and their relation to other terms. For instance, the sentence 'some animals are men' is constructed from what are traditionally called categorematic words ('animal', 'man'), which have a definite meaning of their own, and syncategorematic words ('some', 'are'), which only signify when joined to a categorematic word. In the analysis of sentences, both categorematic and syncategorematic terms play a part. These kinds of terms are discussed in logical treatises. Categorematic terms have supposition, but their supposition is constrained by (among other factors) the syncategorematic terms in the propositions where they are used.

This theory of the properties of terms is one of the important innovations of medieval semantics. It emerged in the twelfth century and developed well into the fourteenth century. Even in the post-medieval period, interesting though less creative traces can be found. The medievals themselves did not consider the theory as something new. In their view it only explained what was implicit in Aristotle's logic. However, from our modern perspective, the theory is original. Like, e.g., Paul Spade (2000) has elucidated, the medieval theory of the properties of terms is complicated and, unfortunately, it is not always clear what it involves.

De Rijk's *Logica Modernorum* is still the basis for the study of supposition. Many of De Rijk's conclusions still stand, but other scholars have made corrections and suggestions. New questions have arisen. For instance: What are the years of composition of the tracts? Is supposition theory a new medieval development? Did it replace the ancient logic, or did this logic still continue? What is the relation between the theory of fallacies and supposition theory? Did the latter originate from the former? Did it absorb the former, or not? What exactly are the different properties of terms? To which level of language (oral, written or mental) or to which kinds of terms do they belong? Do only subject terms possess properties, or do predicate terms as well? Does the theory of supposition form a single theory (is it a semantic, or a syntactic theory)? If it is not a single theory, was it at least initially a single theory? Can it be compared with modern semantics, in which formalization plays a part? If so, to what extent, and in the semantics of which authors? Does supposition theory produce an analysis leading to quantification on the basis of the equivalence of the analysing propositions and the proposition that is analysed? Does this apply to some authors but not to all? The articles collected here give answers to these and other questions.

Another development can be seen in the research on supposition theory, viz. its analysis with the help of modern logic. Formalizing medieval logical theories helps to appreciate their nature, and also to discover possible flaws. Since Ernest Moody's work in the 1930s, however, scholars have been well aware that modern quantification theory does not apply to medieval texts without problems. Jacobi and others have emphasized this feature. Unlike modern logicians, medievals analysed utterances that could be called scientific (like 'all men are mortal', 'thunder is a noise in the clouds'). They hoped to construct a theory which could help solve problems in everyday language. In contradistinction to modern logic, their logic was never purely formal.

In the present volume the reader finds altogether twenty contributions. The first eighteen investigate the theory of supposition in the long Middle Ages, more precisely from its origin in the early twelfth century well into the seventeenth century. They study the theory from what could be called an intrinsic, historical point of view. The last two studies explicitly draw upon tools from modern logic to elucidate medieval theories.

The overall history of the theory of the properties of terms has not yet been written, especially since many treatises still await an edition. Many historical and systematic questions such as the ones mentioned above still need to be answered.

II

The following remarks were made at the opening of the conference at the behest of the Dean of the Faculty of Philosophy.

It is a pleasure as well as a great honour to have this opportunity to offer a few remarks in order to inaugurate the conference. There are two customary requirements for being asked to perform such a service. First, the word of welcome should be spoken by a dignitary of some eminence—for instance, a Dean of a Faculty is held to be eminently suitable for the purpose—while, second, the chosen speaker should not be burdened by too much knowledge about the subject matter of the conference in question. In the present case, the Dean was disqualified on account of the second requirement, whence he thought of asking me. Of course I was happy to accept, the more so since the issue of how to speak about words is one that has long intrigued me. Thus, for instance, when my children were asked at school what their father did, I instructed them to tell their friends that he 'was thinking about the use of quotation marks'. Over the years, in spite of my early training as a modern mathematical logician, I have more and more come to appreciate the virtues of supposition theory as opposed to the use of quotation marks.

The standard account in terms of quotation marks we owe to Frege's *Grundgesetze* from 1893, where, as far as I have seen, they are used uniformly and with complete rigour. Their use was then further refined and given theoretical foundation in the seminal writings of Tarski (*Der Wahrheitsbegriff*) and Carnap (*Die Logische Syntax der Sprache*) during the thirties. The canonical reference here is Quine's *Mathematical Logic* from 1940 whose §4 bears the tell-all title *Use versus Mention*. This distinction is here made absolute with respect to expressions. Thus, for instance, Frege was an underpaid, overworked teacher of analytical geometry at Jena and 'Frege' was his name. Thus, I *use* the name (word) 'Frege' in order to *mention* the mathematician Frege. If, by any chance, I wish to mention his name rather than him, I have to use a name of his name, for instance, "Frege". Typical of this treatment is that words are treated as a special kind of (material) object. On this view, an expression is a thing, an object, among other objects, but charged with meaning. Meaningful expressions could be called 'things plus', shall we say, that is, a material substratum charged with meaning.[1] The relation of reference that holds between an expression and that for which it stands is very much a common or garden variety, or physical, relation that holds between ordinary things. Quine, in particular, has been especially tireless in singing the praises of reference when viewed as a thing-thing relation. In the words of his then colleague Nelson Goodman in the preface to *The Roots of Reference* from 1974, according to Quine, 'reference is an important relation of words to objects—or better, of words to other objects, some of which are not words—or even better, of objects some of which are words to objects some of which are not words'.

From the point of view of supposition theory, Quinean expressions really have *material* supposition, be it *discreta* or *communis*! This is a consequence of the view that an expression is a thing, and its meaning something additional to

1) Sometimes, e.g., by Tarski in *Der Wahrheitsbegriff*, however, expressions are treated not as *tokens*, but rather as *types* of equiform expressions.

it. To my mind the opposite view is more attractive: then the material *substratum*, that is, the expression in *material supposition*, is obtained by disregarding the (conceptual) content, the meaning of the expression, and looking instead only at that in which this content has been embodied.

Tarski, in his fine textbook *Introduction to Logic*, of which there is even a Dutch edition splendidly translated by E.W. Beth, explicitly rejected supposition theory in favour of quotation marks:

'Let us consider, for example, the following two words:

well, *Mary.*

Clearly, the first consists of four letters, and the second is a proper name. But let us imagine that we would express these thoughts, which are quite correct, in the following manner:

(I) *well consists of four letters;*
(II) *Mary is a proper name.*

We would then, in talking about words, be using the words themselves and not their names. And if we examine the expressions (I) and (II) more closely, we must admit that the first is not a sentence at all, since the subject can only be a noun and not an adverb, while the second might be considered a meaningful sentence, but, at any rate, a false one, since no woman is a proper name.

In order to avoid these difficulties we might assume that the words '*well*' and '*Mary*' occur in such contexts as (I) and (II) in a meaning distinct from the usual one, and that they here function as their own names. Generalizing this viewpoint, we should have to admit that any word may, at times, function as its own name; to use the terminology of medieval logic, we may say that in a case like this, the word is used in *suppositio materialis*, as opposed to its use in *suppositio formalis*, that is, in its ordinary meaning. As a consequence, every word of common or scientific language would possess at least two different meanings, and one would not have to look far for examples of situations in which serious doubts might arise as to which meaning was intended. We do not wish to resign ourselves to this consequence, and therefore we will make it a rule that every expression should differ (at least in writing) from its name.

The problem arises as to how we can set about forming names of words and expressions. There are various devices to this effect. The simplest one among them is based on the convention of forming a name of an expression by placing it between quotation marks. On the basis of this agreement, the thoughts expressed in (I) and (II) can now be stated correctly and without ambiguity, thus:

(I') *'well' consists of four letters;*
(II') *'Mary' is a proper name.'*

So Tarski's rejection of the supposition theory is based on fear of ambiguity: expressions have too many meanings. However, the Tarski-Carnap-Quine view has the opposite consequence. The semantic meaning-categories not only match the ontological categories. They actually *coincide*.

Tarski's example above brings this out very well. The expression ('name') 'Mary' stands for Mary, whence 'Mary' is an element of the semantic category of expressions and Mary is an element in the ontological category of (referential) objects. On the other hand, the expression ('name') ''Mary'' stands for the referential object 'Mary' that accordingly is an element of the ontological category of objects. Hence, the semantic category of expressions is a part of the ontological category of expressions. This is a bullet that I personally will not bite, and I feel much more comfortable accepting the Tarskian charge of ambiguity.

In his textbook cited above, Tarski further noted that quotation-mark names, such as ''Mary'', may be treated like single words of a language, and thus like syntactically simple expressions. The single constituents of these names—the quotation marks and the expressions standing between them—fulfill the same functions as the letters and complexes of successive letters in single words. Hence they can possess no independent meaning. Every quotation-mark name is then a constant individual name of a definite expression (the expression enclosed by the quotation-marks) and in fact a name of the same nature as the proper name of a man.

This may appear innocent enough, but as Miss Anscombe, following the Czech logician Karel Reach, has pointed out it has the remarkable consequence that it is impossible to be told the name of someone. Consider the famous Hungarian-American mathematician John von Neumann. Obviously, the European Union is not a semantic constituent of him. The occurrence of the letter combination EU in his name is purely accidental, but yields no contribution to meaning. In the same fashion, on the Tarski-Carnap-Quine view, the occurrence of the letter combination FREGE inside Frege's name 'Frege' is purely accidental and has no semantic function. When asked: What is the name of this logician? where a portrait of Frege is indicated, I must not answer: His name is Frege, because in using the name ‘Frege’ I name *him* and not his name. Accordingly I must use a *name* of (the object that is) his name, and say:

> His name is ‘Frege’,

where, indeed, the quotation-mark name ‘‘Frege’’ is used to refer to ‘Frege’, that is, the name of Frege. But in this name ‘‘Frege’’ (that is, the name of Frege's name) the string ‘Frege’, that is, the name of Frege, occurs only *accidently*, and there is no a priori link between the references of these names (recall the EU and John von Neumann!). In fact, I could as well use another name of his name [...] Of course, it is not very common to give names to expressions. The *Tetragrammaton* springs to mind as one of few examples. But in principle there would be nothing wrong with a formal baptism ceremony in which we decided to apply, say, ‘Kurt’ to Frege's name. (Note the quotation marks here: they constitute the starting point of an infinite regress [...]) Since Kurt = ‘Frege’, we could then rightly observe that *he* is Frege and his name is Kurt, but not Gottlob, since that is (part of a) name of *him* and not of his name.

EARLY SUPPOSITION THEORY IN GENERAL

Semantics and Ontology
An Assessment of Medieval Terminism

L.M. de Rijk†
University of Maastricht

Abstract

This paper aims to assess medieval terminism, particularly supposition theory, in the development of Aristotelian thought in the Latin West. The focus is on what the present author considers the gist of Aristotle's strategy of argument, to wit conceptual focalization and categorization. This argumentative strategy is more interesting as it can be compared to the modern tool known as 'scope distinction'.

Keywords

fallacies, focalization, categorization, epistemic procedure, ontology

The organizers of this year's symposium have made a happy choice in proposing the present theme. Every student of medieval thought knows that the ongoing expansion of supposition theory, and the doctrine of the properties of terms in general, can be followed like a thread through the development of medieval philosophy and theology. Being a drive and a device at the same time, terminism has both stimulated and governed doctrinal developments. The aim of this paper is to make a small contribution to evaluating the role of terminism in the broader perspective of Western philosophical thought.

The fons et origo *of Medieval Terminism*

Whoever is interested in the rise and development of supposition theory and terminism in the Middle Ages and has realised its unmistakable significance for the growth of medieval thought cannot help but wonder how and when the whole thing started. I still consider it indisputable that the terminist movement as such had its starting point in the vivid interest of eleventh- and twelfth-century scholars in linguistic analysis and the study of fallacies as indispensable instruments for solving serious (or less serious) problems. The doctrine of fallacy, including grammatical analysis as the apparatus *par excellence* for unmasking fallacious arguments, underlied terminist logic. This is not to say that, since, starting with the rise of the universities around 1200, medieval terminism did not progress far beyond the original scheme as outlined in the present author's history of early terminist logic.[1] Nor will anyone be inclined to identify terminism as just a series of vicissitudes from the debate on bad argument that followed along the lines of the well-known Aristotelian fallacies. Thanks to many of the attendees of the present symposium, research since the 1960s has made this perfectly clear, and current research continues to confirm it.

Thus fallacy theory was far from being an inherent hallmark or mould of later terminism. Nor was it a sort of innate virtue, the predominance of which, so to speak, could be compared to the impact of original sin on human nature. On the other hand, throughout the development of terminism, the scholastic predilection for its twin, to wit semantic analysis, was indeed a distinct characteristic of the progress of the terminist method.

To study the rise and development of supposition theory in the doctrinal context of the properties of terms requires more than merely spelling out the *methodological* aspects of the growth of terminism. It is true, philosophical progress owes quite a lot to methodology. Nonetheless, in basic matters of philosophy any *modus operandi* is itself determined by a certain initial, in a way aprioristic worldview. Likewise the terminist approach to philosophical and theological matters testifies to a basic philosophical attitude. Ultimately, it is the philosopher's primary approach to things that counts.

From this point of view, the anonymous author of a fifteenth-century treatise on the *parva logicalia* (of the so-called *Copulata* genre) had a point in his attempt to trace back these tracts of the *logica modernorum* to Aristotle the Philosopher. The man's noble effort is referred to by Philotheus Boehner in his short but pioneering outline of medieval logic. The author argues in favour of the completeness of Aristotle's logical writings by pointing out that the Stagirite established the true principles of logic and thus also invented the treatises of the *parva logicalia radicaliter et virtualiter*.[2] He tries to support this view

---

1) It should be underlined again that in my *Logica modernorum* only the origin and early development up to circa 1200 are discussed, without any presumption about its later development.

2) *Textus et copulata omnium tractatuum Petri Hispani* etc. in the Cologne edition of 1493 (p. 36), referred to by Boehner (1952), 16-18: '[...] quamvis Arestoteles (!) non invenit istam logicam que

with a rather detailed analysis of the aims of the various treatises. Boehner had good reason to condemn this endeavour as a crude and somewhat artificial derivation of the *parva logicalia* from Aristotelian writings, including the most profound of them, the *Metaphysics*.

However crude and strained this attempt at reconciliation may seem, we still are faced with a fascinating question concerning the continuity of Aristotelian thought in the Middle Ages. Doctrinal continuity would seem unlikely if the Medievals' methods and basic philosophical intentions were substantially different from Aristotle's. In fact, there was no such gap. Even though the terminist movement was to some extent a novelty, it was clearly in keeping with Aristotle's basic intentions.

1 Aristotle's Strategy of Argument

1.1 *Aristotle's Flexible Use of Language as His Semantic Method*

Aristotle's search for genuine knowledge runs along the following lines.[3] The ever-changing things of the outside world, he takes it, *are* in many ways—or they enjoy their 'being-ness' multifariously. Unlike Plato, Aristotle recognizes unchangeable ontic elements as *immanently* present in outside things. For him, philosophical research must focus on a thing's immanent essential elements and not take them as separated from the concrete individual. However, even though these essential, universal elements do not belong to a separate domain, we can take them apart conceptually, as long as we remember that they are actually immanent in outside things. This is where conceptualization and semantic behaviour in general come in.

Philosophers and linguists alike assume that there is a linguistic correlation between language and thought. As far as ancient and medieval thinkers are concerned, they were not so much interested in language itself. Rather they concentrated on the various ways in which correct linguistic expressions are representative of correct thinking and, by the same token, somehow disclose

hic traditur, in se et in propria forma istorum tractatuum, tamen invenit istos tractatus in suis principiis, quia posuit quedam principia ex quibus isti tractatus ulterius eliciuntur et fiunt. Etergo dicitur quodammodo, hocest radicaliter et virtualiter, istos tractatus invenisse. Unde patet quod magis est regratiandum Phylosopho quam Petro Hyspano, cum circa principia maior sit labor inventionis. Habitis enim principiis facile est addere et augere reliquum, ut inquit Phylosophus in secundo Elenchorum.'

3) This section is a distillation of De Rijk (2002), I, 12-16; 60-74.

the diverse features of extra-linguistic reality. In particular, Aristotle's attitude as a philosophical investigator is deeply rooted in his semantic views, namely his firm belief that the multifarious ways in which we use linguistic expressions are representative of what things really are.

More than once this attitude leads people to accuse Aristotle of imprecise or even inconsistent usage. The well-known historian of Greek philosophy W.K.C. Guthrie, for instance, blames Aristotle for 'an insouciant use of language which is full of traps for the unwary'.[4] But at the same time Guthrie acknowledges that the flexibility of Aristotle's language makes it a wonderful instrument compared to the resources of his predecessors, including Plato, because the aporias that baffled them dissolve and vanish in the face of Aristotle's famous 'in one sense [...] but in another' device. Unfortunately, Aristotle's critics fail to see that his use of this device is founded upon and justified by his view of the different formal ontic aspects that are constitutive of a thing's being. It is these aspects—the Medievals called them *rationes*, i.e., formal structures or ontic elements—that can and should be meticulously focused on, conceptualized, and categorized by whoever seeks the true nature of things.[5] What on the face of it may seem like ambiguous usage can turn out to be a fruitful employment of the ambivalence of linguistic expressions.

Aristotle's semantic practice may have led Guthrie (*ibid.*) to call Aristotle 'this astonishing man' (or 'our philosopher with his incorrigibly one-track-at-a-time mind'), but it is far from a haphazard procedure. A number of customary procedural features can be observed which have a strong impact upon his management of the phenomenon of 'meaning'. These features can even be listed as semantic rules, which are all concerned with the multiple use of certain expressions. Four main rules of Aristotelian semantics (and indeed ancient semantics in general) can be formulated, which are all intended to profit from, and at the same time to regulate, ambivalence.[6] To my mind, the constant use of these rules offers an answer to Guthrie's accusations.

The first is the rule of multiple applicability (RMA) of substantive and substantivated adjectives and articular participles. It concerns the threefold use of names (nouns), to wit (a) to denote things as self-contained, subsistent wholes, (b) to connote the special property or feature signified that these things possess, and (c) to refer to this property in an abstract manner, conceiving it apart from the things in which it is enmattered. Take for instance the various ways in which '*ousia*' is used to stand for either (a) a physical particular (primary

4) Guthrie (1981), 212 ff.

5) For this specific use of '*ratio*' see De Rijk (1994), 197-218.

6) See De Rijk (2002), I, 69-72; 162; 252; II, 154; 413.

substance), or (b) the characteristic eidos in virtue of which a thing possesses subsistence (like the secondary substance of the *Categories*), or (c) 'being-ness' in general, including the kind of being-ness invested in non-substantial ('non-subsistent') modes of being. Every student of Aristotle's *Metaphysics* is familiar with the pivotal role of similar uses of substantivated adjectives (e.g., *to leukon, to simon*)—which in fact convey a categorial modification of being(ness)—when Aristotle comes (namely in Books VII and VIII) to setting his own metaphysical position against that of Plato.[7]

The second rule concerns the absence of a clear-cut border between a linguistic expression (whether simple or compound) taken as a linguistic tool and its significate. A nice example is the well-known use of *logos* for, on the one hand, phrase, definiens (as linguistic tools) and, on the other, their respective contents (as what is expressed by these tools). Another example occurs in *Categoriae* V, 3 b 10-12, where, after the division of substance into primary and secondary substances, it is said to be indisputably true that primary substances *signify* a certain 'this' (*tode ti* or *hoc aliquid*). However, they *are* a *hoc aliquid.* An expression's autonymous use is often not clearly marked off from its significative use. Charles Kahn has rightly observed[8] that the Greeks rarely—and before Chrysippus (c. 280-207) never systematically—distinguished the word or sentence as linguistic expression, i.e., taken as a mere utterance, from the meaning or content it expresses. Regarding autonymous use as a sort of self-reference I have baptized this the 'Rule of indiscriminate reference' (RIR).

The third rule, which is closely connected with the second one, is about the double sense of the phrase 'the content of an expression', and can be labelled the 'Rule of the multiple significate' (RMS).[9] The phrase 'an expression's significate' can be used to stand for both (a) mental entities (the so-called 'affections of the soul' of *De interpretatione* I, 16 a 6-7), to wit things or states of affairs insofar as they are conceived of, and (b) the things (states of affairs) signified in their capacity as extra-linguistic entities. For instance, the expression 'Socrates is a white musician' both stands for the mental entity 'that-Socrates-is-a-white-musician' and the intended extra-linguistic entity, say, the real state of affairs of Socrates's being a white musician. This ambivalence does not imply, of course, that the Greeks made no distinction at all between extra-linguistic states of affairs as such and qua conceived of. Rather it underlines the ancient view that when a mental state of affairs is rightly claimed to apply to the

---

[7] See De Rijk (2002), II, 148-288.

[8] Kahn (1973), 366.

[9] Note that, like the foregoing and the fourth one and unlike the first, this rule concerns simple and compound expressions alike.

external world, the difference between the two states does not imply any real opposition.

A paradigmatic case of the concurrence of these main rules is found in the celebrated Aristotelian adage about the several senses of the phrase *to on legetai pollachôs* (in Latin less fortunately rendered without the article: *ens multipliciter dicitur*). This phrase indiscriminately and simultaneously means 'that which *is* is so named [or: is brought up] as be-ing in many ways' and 'the term 'being' has several senses'. The subject term indiscriminately stands for both the linguistic expression and its content (RIR); its content or significate can (or should) be taken in a twofold way, bearing, that is, on a mental entity *and* the corresponding extra-linguistic entity (RMS). In addition, our first rule (RMA) is in order: what is referred to by the phrase *to on* can (or should) be taken indiscriminately to stand for the *thing* which is and this thing *qua be-ing*. This rule in particular was to play a predominant role in the medieval debate on intentionalism.

A fourth main rule allows the simultaneous application of different senses or nuances of expressions, that is to say, their use on more than one semantic level at the same time. While the previous rules (RMA, RIR, and RMS) bear on the possibility of multiple application of expressions in terms of 'at one time this sense, at another that', the fourth rule is about the *simultaneous* application of their various nuances coming to the fore in the three other rules. Thus, with reference to RMA, the fourth rule allows the application of a noun to a thing's property and its possessor *at the same time*, and in fact inseparably. For instance, *to leukon* stands for the white thing, this thing's particular whiteness, and whiteness as such, and all this simultaneously. The same goes for the respective areas of RIR and RMS. Thus the fourth rule pertains to what is nowadays called 'double entendre'. With the analogy to music in mind I have called this rule 'the rule of semantic counterpoint' (RSC). In fact, RSC is preeminently the device by which the ambivalence of expressions is used to its very limits. Small wonder that—*n'en déplaise* Guthrie's displeasure—this rule in particular is Aristotle's favourite in support of his attacks on Plato's metaphysics of transcendent Being.

The close relationship between Aristotle's semantic behaviour and his basic manner of philosophizing of course most prominently comes to the fore in his *Metaphysics*. Let one single example suffice.[10] Aristotle's decisive argument in support of the enmattered form (*eidos*) as the true *ousia* entirely hinges

---

10) A wealth of evidence is found in De Rijk (2002), I, 61-69 and II, 135-288.

on his application of the fourth rule (RSC). His strategy of argument (from *Metaphysics* VII 6 onwards)[11] is to play with the ambivalence of the appellative noun signifying a thing's *eidos* (meaning both the form taken as such and qua enmattered in its hypokeimenon; rule RMA); he deliberately allows one of the two simultaneously applicable senses to dominate, depending on what he wishes to claim. The simultaneous use of appellative nouns for both an enmattered form (whether substantial or coincidental) and the composite endowed with it depends on the referential (or extensional) identity of the two. The assumption of this referential identity is *the* basic tenet of Aristotelian ontology (argued for against Plato): a form can only exist as enmattered in the outside world. Although the metaphysician is surely entitled to *conceive of* a form by formal abstraction apart from its hypokeimenon (in which it is immanent), the form thus conceived of is nothing but a mental entity.[12]

Before proceeding to Aristotle's general strategy of argument it may be useful to show that his linguistic practice, which I have tried to lay down in the above four main rules, is less peculiar than it might seem. Each of them finds its parallel in modern usage.[13] The use of RMA in Greek matches its application in modern languages (although the English idiom is less hospitable in this respect than Greek, Latin, and some other modern languages). For instance, supposing there is a white wall at the back of my classroom and I resentfully order: 'That whiteness over there should be removed', my intention is either to have the room enlarged, or to have the dazzling white colour replaced with a relaxing pastel shade. Our second rule (RIR) is not entirely alien to modern usage either. E.g., 'These lines are not easily decipherable and make clear that the emotional author is of quite a different opinion'. The same holds for the third rule (RMS). Take, e.g., this sentence: 'Hannibal's march across the Alps wrought terrible havoc there [sc. because of his elephants], and caused panic and fear in Rome'. A similar case occurs in 'The accused denied the accusations'. Finally, modern European parlance presents us with a nice example of the fourth rule of *double entendre* (RSC) practised, indiscriminately using the word 'Presidency' both as an abstract and a concrete noun, e.g., in saying now

---

11) See De Rijk (2002), II, 188 ff., and also the sections 1.2 and 2.4.1 below.

12) For applications of the four rules, particularly *RSC*, in other works see De Rijk (2002), I, 68 ff. In *Physics* IV, chs. 10-14, the intriguing problem concerning the proper nature of time is discussed along these semantic lines. See De Rijk (2002), II, 367-384. By the way, the quasi-problem of so-called 'prime matter' in Aristotle does not stand up to a clear-headed application of the four rules; see *ibid.* II, 384-395.

13) For the general theme see Vendler (1967), 131 ff. and De Rijk (1985), 36-47.

(May, 2008) 'The Slovenian Presidency, which only lasts six months, has to solve many tricky problems'.

Many examples of this kind of linguistic expression may sound a bit annoying because of their grammatical incongruity, but they are still perfectly intelligible. In any use of appellative terms there is a certain intermingling of the denotative and connotative aspects. Supposing a politician of good report has just arrived, and some journalist maliciously comments: 'The sly old fox is about to enter the premises', his statement about the man's arrival is true enough denotatively, but can be (rightly or wrongly) contradicted because of its connotation. As the problems surrounding the freedom of speech in modern Western society make patently clear, to address a person in an insulting manner is only possible—and accordingly liable to juridical quarrels—because any denotative use of appellative nouns is fused with connotation. As for Aristotle, as a philosopher he indeed made the most of the linguistic ambivalence of his mother tongue.

1.2 *Aristotle's Strategy of Argument: Focalization and Categorization*

Aristotle's strategy of argument pivots around finding the appropriate terms to assemble a proof, since, in his view, a question can only be answered insofar as the object under examination, including its properties, is brought up for discussion under the appropriate description.[14] This leads Aristotle to what I have labelled 'focalization and correct categorization'. To Aristotle, as a self-contained unity, each particular subsistent being is made up of various modes of being, to wit the subsistent mode accompanied by a number of coincidental, non-substantial modes. Aristotle's epistemonic procedure comes down to focusing on those ontic properties (whether essential or coincidental) present in the particular things under investigation that afford him the appropriate 'middle' or '*medium demonstrationis*' for proving his thesis. The investigator then categorizes the object accordingly, by using (most of the time) the qua-procedure.[15] For instance, when Callias's behaviour towards his slave Callicles is at issue, Callias is not brought up in his capacity as man or animal, nor as a musician or grammarian, but qua master of his slaves. Thus in the latter

---

[14] The epistemonic procedure as such is dealt with in *Posterior Analytics*; see De Rijk (2002), I, 594-749. The procedure of focalization and categorization is applied throughout Aristotle's works (see *ibid.* I, 133-189; 384-387; 449 ff.; 562-749; II, 23-27; 76-85; 264; 363; 384 ff.; 403-416).

[15] For this procedure see De Rijk (2002), I, 167-189 and II, 34-36.

capacity the relationship 'master-slave' is essential, whereas Callias's being human is coincidental.[16]

Time and again, Aristotle makes it clear to his audience that for any serious argument an appropriate quidditative appellation is required, even though this name on occasion may highlight one of the object's coincidental properties. For this reason, Aristotle's strategy of argument requires one first to focus on an object's specific property which is most appropriate to serve as a *medium demonstrationis*, and next to categorize the object precisely after this property. Every student of Aristotle's works (especially the biological works) is familiar with his use of the qua-locution as a favourite means for highlighting a certain ontic property.[17]

I have already mentioned Guthrie's worries about Aristotle's use of this 'in one sense [...] but in another' device. However successful it may be (even in Guthrie's opinion), it does lead to a serious question. Our speaking about outside things, focusing on their different modes of being, may make us wonder about the validity of this argumentative procedure. Is it well equipped to be a vehicle for attaining the truth about things adequately, that is, without jeopardizing their unitary nature?[18] To answer this question requires a clear understanding of what Aristotle means by 'speaking about things'. In line with one of the fundamental positions endorsed in my study of semantics and ontology in Aristotle, my claim is that the basic mental activity involved in his scientific procedure should be taken in terms of onomastics rather than apophantics. That is to say, naming and appellating things, rather than statement-making, play the pivotal role in Aristotle's epistemonic procedure. Focalization and categorization, he insists, do not affect the ontic integrity of the objects they refer to. Aristotle never comes up with a statement like 'Callias is essentially a master, period' or 'Callias is coincidentally a human being, period'. Such *statements* are out of the question and not even implicitly involved in name-giving either. To Aristotle, there *is* such a thing as 'what is essential or primary by nature in an absolute sense', to wit that which is given in answer to the *unqualified* question about a thing's quiddity (*ti estin haplôs*). Accidental or coincidental modes of being are only considered essential with respect to a specific discussion, for example, about the difference between master and slave, which

16) The famous passage *Categoriae* VII, 7 a 35-39 is often misunderstood in this respect, including Ammonius' correct interpretation of it. See De Rijk (2002), II, 406-410.

17) Cf. De Rijk (2002), I, 167-189; II, 31-36; 357 ff.; 388 ff.; 406 ff.

18) This question is understandably given full attention in Bäck (2000b), 269 ff. Cf. De Rijk (2002), II, 398-410.

is not affected by the fact that they both are essentially human.[19] Thus any ontological disturbance of a thing's substantial unity is out of the question.

Let us return to Guthrie's ambivalent judgement about Aristotle. Guthrie clarifies[20] the flexibility of Aristotle's language by pointing out the philosopher's way of treating some key metaphysical notions. With regard to, e.g., form (*eidos*), he remarks:

> [...] hardly surprisingly, specific form, the essence of individuals [...] is endowed in the *Metaphysics* with the titles reserved in the *Categories* and elsewhere for the true individuals—Socrates, Coriscus, this horse. [...] The title of 'a particular 'this' (*tode ti*)', elsewhere jealously reserved for the concrete object, is now transferred from the empirical to the scientific, or philosophical unit, the specific form, which as essence usurps also the title 'primary being'. [...] At the same time this astonishing man can identify *eidos* as subject of definition with *to katholou*! Seen in one light it is individual, in another universal.

However, appearances notwithstanding, Aristotle's flexible semantics is of a well-considered coherence, as can appear from a meticulous analysis of his arguments.[21] What is more, Aristotle also makes every effort to elucidate and justify his protocol language, as being the only means to adequately describe the outside world.

1.3 *On the Ambivalence Tool Matching Modern 'Scope Distinction'*

In his discussion of the lexical nuances and paraphrastic values of the Greek verb *einai*, Charles Kahn also speaks about the different nuances that may occur together in a single occurrence of a word without leading to ambiguity.[22] Dealing with the different uses of the Greek *logos*, he points out the intimate connection between the three different nuances within the semantic area of one single lexeme ('discourse', 'rational account or rational principle', and 'reason' or 'rationality'). Kahn recognizes the considerable philosophical advantage of a terminology that brings together the concepts of language and

---

19) Part, or rather the gist of the modern interpreters' problem is the wrong assumption of such a thing as 'essential vs. accidental predication' as already occurring in Aristotle. To explain statement-making in Aristotle in terms of the later view of predication is anachronistic; see De Rijk (2002), I, 75-100; II, 409-411.

20) Guthrie (1981), 216.

21) As I have tried to do in De Rijk (2002) by pointing out Aristotle's skilful use of the four semantic main rules.

22) Kahn (1973), 232 ff., 403. The ambivalence vs. ambiguity topic is discussed in De Rijk (2002), I, 69-72.

rationality as essentially related, in his words 'as it were two sides of the same coin'. This usage, he claims, 'contains the seed of an important philosophic insight'.

Notwithstanding his favourable attitude towards this type of multiple significance, however, Kahn keeps speaking of 'ambiguity'. To my mind, we should sharply distinguish between (advantageous) 'conceptual ambivalence' and (undesirable) 'conceptual ambiguity'. The following rule of thumb can perhaps make clear what I have in mind here. The user of ambiguous speech can be forced to clear up the situation by firmly making a choice between the different senses involved. Whoever uses ambivalent speech, however, needs the semantic ambivalence of certain key terms in order to bring about their intentions appropriately; to eliminate explicitly one of the alternative senses would result in one-sidedness and imbalance. Small wonder, then, that Kahn is, for instance, blind to Aristotle's skilfully taking the different nuances of the word *eidos* in terms of ambivalence to counter Plato's lore of the transcendent Substances. Kahn therefore fails to see how ingeniously Aristotle succeeds in avoiding Platonism by clearly opposing the substantial form to matter, on the one hand, but at the same time giving full weight to their intimate connection. Kahn[23] does not even recoil from speaking of 'the troubled course of *Metaphysics* Z'. In section 2.4.1 below I hope to corroborate my claim that it is in *Metaphysics* VII and VIII in particular that Aristotle and likewise an intelligent commentator like John Buridan make use of the ambivalence of linguistic expressions to underscore their metaphysical thought.

The use of the ambivalence of key philosophical terms—on occasion even using it to its very limits—is surely not an Aristotelian monopoly. Elsewhere I have investigated the fashion in which, e.g., Proclus copes with the intriguing transcendence-immanence antinomy, which is an unavoidable concomitant of the Platonic and Neoplatonic ideas of causation and participation.[24] As a faithful Neoplatonist Proclus tries to bridge the gap between the metaphysical Principle, the One or Good, and its productions. This gap is bound to occur in any metaphysical system that is based upon an unchangeable sublime Principle as opposed to the inferior ever-changing outside world. So how can the sublime Principle be the First Cause of such a trivial offspring as the outside world? Proclus's answer centres around the key notion of participation or sharing (*methexis*). To cause something is always co-ordinated with sharing, because any cause makes its product share in it. Sharing is in fact what

23) Kahn (1973), 461 f.

24) De Rijk (1992), 1-34.

guarantees the product's communion with its cause insofar as they are mutually connected in their kindred character. In Proclus's words

> [...] the participated bestows upon the participant communion in that which it participates. Well, surely it must be that what is caused (*to aitiaton*) participates in the cause as from whence it possesses its beingness.[25]

We have landed smack in the middle of the interpretive problems which have provoked so much irritation and confusion among modern scholars, who almost unanimously accuse Proclus of doctrinal and methodological obscurity and inconsistency.[26]

Two features in particular have provoked the confusion and irritation, namely Proclus's descriptions of the cause and of its counterpart, the participant. Proposition 98 says of the cause rather enigmatically:

> Every cause which is separated <from its effects> is at once everywhere and nowhere.

As to the participant, it is now referred to by *to metechomenon* (*litt.* 'what is participated'), now by *to metechon* (*litt.* 'what takes a share in'). The former phrase stands for the ontic element (*dynamis*, ontic power) communicated by the cause to the product, rather than the cause itself as shared in, the cause being designated by the phrase *to hou meteschen* (*litt.* 'that of which it [viz. the participant] has taken a part'). In other words, *to metechomenon* is the transmitted power deriving from the cause, rather than the cause itself. Grammatically speaking, the subject of the active verb *metechein* ('to participate') is the participant, while its indirect object (expressed by a partitive genitive) is the cause. On the other hand, the subject of the passive *metechesthai* is the power descending from the cause and immanently present in the participant. This means that the phrase *metechesthai hypo* should be rendered (or at least be understood) 'to be as a share present in', rather than 'to be shared in by'. Likewise the participle *metechomenon* should preferably be rendered 'being an immanent share'. This means that the word *metechomenon* is always used to stand for the ontic element or power that inheres in all that *is*, and, consequently, is found at every level of the universe.

---

25) Proclus, *Elementatio*, prop. 28, *expositio*, ed. Dodds (1963), 32, 19-21.

26) De Rijk (1992), 1 f.

The relationship between effect and cause is transitive; the one between causation and participation is not. In the intransitive relationship between causation and participation, the notion of *amethektos* (meaning at the same time 'unshareable' and 'unshared') is focal. Once the precise meaning of *amethektos* has been established, the participle *metechomenon* no longer poses a problem.[27] As something that transcends its effects, no principle (neither a secondary principle nor the First Principle, the One or the Good) is partaken of, nor can it possibly be something immanent; consequently, it is never said to *metechesthai* ('to be an immanent share'). As we have seen before, what is shared is not the cause itself, but an ontic power produced by a cause qua cause is a thing's share. So in the context of Neoplatonism, causation should be explained in terms of an ongoing procession out of the One; it is simply the One's overflow, due to its superabundance and unlimited perfection.

The entire procedure can be summarized as follows:

[a] Every product constitutes itself by contemplatively turning back to its immediate source.
[b] In the procession out of the One, oneness is *the* characteristic of what is produced on the lower levels, owing to the One's 'being divided'.
[c] From Iamblicus onwards, strong emphasis was laid on the principles as being separated from their products (emphasizing, that is, the opposition of transcendence to immanence).
[d] What is shared (the *participatum*) is the power deriving from the principle as a cause; the principle taken by itself, viz. preceding[28] the stage of procession, must be indivisible and, accordingly, not-shared and unshareable (*amethekton*).

Thus you have the intriguing triad, *amethekton, metechomenon, metechon. Amethekton* is any principle (whether the First or a secondary one) in the downward serial procession when considered apart from its being shared in; the *amethekton* par excellence is of course the One or the Good taken in its transcendence. *Metechomenon* is the principle qua divided or distributed and immanent in the product as an ontic power. Finally, *metechon* is the participant, the entity, that is, which possesses the ontic power, whereas insofar as it

27) De Rijk (1992), 16-17.

28) This term should not be taken in a chronological sense, the priority being simply a matter of logical entailment.

is considered from the ongoing downward procession, it is a secondary principle, being a *metechomenon* itself. This very feature is the nucleus of its basic role for continuity and reversion in the Universe. Whatever comes 'after' the One is not pure Being, but merely a mixture of 'one and not-one', and at the same time alike and unalike the One. Alikeness not only causes lower entities to be subsistent; it also makes them revert to their Principle, the One.

In the procession out of the One, the members of the triad *amethekton, metechomenon, metechon* all have their specific function. The *metechomenon* mediates between the cause qua unshareable and the effect qua participant, but it is certainly not something between two pre-existing entities. In its capacity of being representative of the superior cause it constitutes the effect and thus acts as the participant's immanent cause, to wit the ontic power deriving from the superior cause. Qua immanent share it is representative of the transcendent cause, which qua transcendent is unshareable (*amethekton*). At the same time, in its being 'alike to the cause' it binds itself and the participant to the superior cause and, ultimately, to the One, and in doing so it realizes Reversion in the Universe.[29]

As can be expected, Proclus is quite relaxed about statements that, as they stand, are flatly contradictory in simultaneously affirming a thesis and its antithesis. A notorious example is *Elementatio*, prop. 2:

> All that partakes of the One is both one and not-one.

In the exposition, Proclus declares that *being one* is meant as *being one according to participation* whereas *being not one* means *not being the One itself.* No doubt, this is a reasonable explanation, but one has to wonder why the proposition should be framed in such a provocative way. In the exposition of prop. 24, where the mutual relationships between the members of the famous triads are discussed, the *metechomenon* is described as what at the same time is 'one, yet not-one', and the participant as something 'not-one, yet one'.

Proclus's explanatory strategy actually comes very close to what we call 'scope distinction'. Now, scope distinction involves more than just considering one and the same object from different angles. It rather concerns an object now being observed from one aspect proper to it and now from another angle that is likewise representative, in order to obtain true knowledge of the object as a whole. A fine specimen of scope distinction in Proclus is found in prop. 99, which deals with the unshareable qua unshareable:

---

[29] De Rijk (1992), 26-29.

> Precisely in the respect in which an unshareable is unshareable, it does not obtain its subsistence from something else as its cause, but is itself the principle and cause of all its shares. And it is in this sense that in every series the principle is ungenerated.[30]

The notion 'unshareable' only applies within a certain scope, to wit by setting apart any relationship of sharing, both in the upward and in the downward direction, although at the same time these relationships are neither denied nor ignored.

What is more remarkable for the aim of my present section is that Proclus explicitly recommends[31] scope distinction as a cognitive device successfully applied by Plato's spokesman Parmenides. The old Parmenides, Proclus claims, takes (in the second hypothesis of the dialogue, 142B-155E) his starting-point from the Parmenidean One, i.e., oneness appearing *in* things, and considers it one time qua one, another time qua 'one and being', and still another time qua being that partakes of the One. Next different properties of the One are posited, and those are used now in the affirmative and now in the negative. The entire procedure is used to explore fourteen properties of the One, without splitting up the One into an inconsistent, self-contradictory whole. Thus Proclus offers a striking example of the use of the ambivalence of expressions to clarify intricate philosophical problems.

Other specimina of such scope distinctions occur in medieval speculative grammar, expounded in the well-known tracts *De modis significandi*. Basically, in this discipline the various relationships between modes of being (*modi essendi*) and modes of signifying (*modi significandi*) are discussed. The latter are derived from the former, but are not altogether representative of a thing's mode(s) of being. They fall short because of the actual abundance of an object's modes of being, whereas, on the other hand, not every mode of signifying matches a mode of being; an object can, for instance, be signified by a noun *per modum substantie*, without actually being a substance. From about 1300 onwards, the epistemological impact of the doctrine was a topic of lively interest. The lack of an unambiguous relationship of equivalence between *modi intelligendi* (annex *modi significandi*) and *modi essendi* forced the adherents of the doctrine to explore the proper nature of their mutual correspondence. The unclear area between thought and extramental being demanded the insertion of mediating elements to bridge the gap. Here is where the ambivalence device came in. The *modus significandi activus* was marked off from the *modus*

---

30) Proclus, *Elementatio*, prop. 99, *expositio*, ed. Dodds (1963), 88, 20-23.

31) Proclus, *In Parmenidem*, ed. Cousin (1864), VI, 1049, 37-1050, 25.

significandi passivus, the latter of which was regarded as a sort of aggregate of the mode of being and the active mode of signifying. This view entails that the active mode of signifying both formally and materially differs from the mode of being, whereas the passive mode of signifying, though being formally distinct from the mode of being, materially coincides with it. In addition, the active and the passive mode of signifying are formally the same, although they differ from one another when taken from the material point of view.[32] This multiplication of intermediary rationate entities (*entia rationis*) may be regarded as an analogue to the introduction of in-between, transitional entities in the Neoplatonic Universe, or to the insertion of intermediary mental products conjectured by, e.g., Hervaeus Natalis in his *De secundis intentionibus* (see section 2.4.2 below).

2 Aristotle and the Strategy of Terminist Argument

To evaluate the Medievals' lore of the property of terms in light of their basic adherence to Aristotelian doctrine, an important distinction should be made in advance. To acknowledge that medieval thinkers were familiar with Aristotelian thought and always found themselves on common ground with the Philosopher, even when criticizing him, is one thing, to claim that their strategy of argument was in perfect harmony with Aristotle's is another. Regarding the subject matter of the present paper we must therefore investigate whether there is also a methodological continuity between Aristotle and his followers.

2.1 *Onomastics vs. Apophantics*

If there is a basic similarity between Aristotle's linguistic analysis and the one involved in terminism, this is surely not obvious on the face of it. More specifically, there is a huge discrepancy, it seems, between the anatomy of the Aristotelian sentence (statement), centered around name-giving, and the special interest in the *propositio* found in medieval logic and grammar, particularly after Peter Abelard. Elsewhere I have extensively argued for what I consider Aristotle's deep structure analysis of statement-making, including

32) See, e.g., the discussion of these topics in Radulphus Brito, *Quaestiones super Priscianum minorem* I, q. 18 ('Utrum modi significandi et intelligendi et essendi sint idem'); q. 19 ('Utrum modi significandi activi et modi intelligendi activi sint idem'); q. 21 ('Utrum modi significandi accipiantur a modis essendi et proprietatibus rerum'); q. 22 ('Utrum modi significandi activi et passivi formaliter sint idem an differant'), referred to in Pinborg (1980), 73.

predication.[33] The following summary may suffice. Unlike the well-known dyadic 'S is P' construal, Aristotle's statement-making utterance (or assertion) is a monadic construct consisting of an assertoric operator (or functor) and an assertible complex which functions as its operand (or argument). The operator is the strong form of hyparctic 'be', whereas the operand (i.e., the assertible complex) is a nominal construct, which equals one of the following elements in English: gerundial phrase, infinitival phrase, or *that*-clause. The assertible is a compound consisting of a substratum (which is indicated by a noun or substantivated adjective) and its attributive determination (which is expressed by a rhêma, either adjectival or verbal). A fully-fledged assertion consists of an assertible and an assertoric operator ranging over it. For instance

'Is: [(a man&pale)'s be-ing]'

is the protocol sentence having the (indiscriminate) translation value of

'There is a white man'

or

'A man is white'.

Much evidence from several writings of Aristotle can be brought forward in support of the monadic ('copula-less') anatomy of the Aristotelian statement-making utterance and against the later 'S is P' construal.[34]

This dissimilarity is all the more important as the key notion of *suppositio* is its occurrence *within* the proposition (*in propositione*, definitely not *extra propositionem*), which I have termed the 'contextual approach'. This requirement could suggest that terminism should have the proposition as such in the focus of interest. However, the fact that terminism centres around the logical analysis of the proposition as the breeding ground for the diversification of a term's actual meaning does not make terminism a logic of propositions (let alone a system of propositional logic). In terminism—note the nomenclature

---

33) De Rijk (2002), I, 75-255.

34) De Rijk (2002), I, 87-93. One need only consider a purely linguistic feature: the putative copula (our assertoric functor) is said to be '*ad*iacens', as in the Greek texts there is also always talk of passive '*pros*tithesthai' and the like, never of the active notion of *sun*tithenai'. In addition, the key text in *De interpretatione* III, 16 b 24-25 reads: '*is* is a sign of some combination', rather than '*is* combines...'.

'terminism'—it is the propositional *analysis* that counts. Ernest Moody has made a nice comparison to elucidate the difference between the logic of terms and the logic of unanalyzed propositions. Supposing we were interested in making an exhaustive classification of the ways in which different kinds of houses could be built out of different kinds of materials, we could go about this in two ways. We might make a tour of inspection among all the houses that had been built out of such materials, noting the differences in their construction, and giving a name to each distinct type of house. Alternatively, we might try out the various combinations of materials, by building houses out of them, and, in addition to giving names to the different kinds of houses, also give names, or assign properties, to the materials, according to the way in which their presence in a house, in combination with the other materials, affects its construction. Moody rightly associates the latter method with the way in which terminism deals with propositions as logical and linguistic buildings.[35]

2.2 *The Strategy of Terminist Argument*

Let me put the foregoing observations in the broader framework of philosophical analysis. It is the philosopher's job to give a reasonable account of his viewpoints about reality. Any discussion on this score first and foremost requires that one should accurately determine the precise meaning of the words that make up the statements involved in the dispute. Your interlocutor's concern is to understand what precisely you are talking about. When it comes to arguing for profound (say, metaphysical) convictions in particular, it is indispensable to determine the meaning of words and all their connotations. Medieval philosophers and theologians never shirked this duty. Quite the contrary, their innate predilection for disputation more than once led them to dismember the meaning of words rather sophistically. Supposition theory, then (and more generally the doctrine of the properties of terms), was an attempt to formulate, on a metalinguistic level, descriptive rules concerning the referential function exercized by terms in a propositional context. Thus these rules concerned the various acceptances of meaningful words and expressions in various contexts. It was often by playing (whether or not in a sophistical manner) on the different shades of meaning that serious philosophical and theological controversies were disputed.

35) Moody (1935). However, Moody is entirely wrong in ascribing the former method to Aristotle.

Studying this matter one easily recognizes the application of the four Aristotelian semantic rules in the usual divisions of supposition. The main purposes of the doctrine find their fulfillment in the domain of these rules. The basic distinction between *significatio* and *suppositio*, to begin with, as well as that between connotation and denotation, including their various kinds, and in their wake the notions of simple, personal, common, discrete, determinate, distributive, and confused supposition can all in principle be explained in terms of the rule of a term's multiple applicability (RMA), the rule of indiscriminate reference (RIR), and the rule of the multiple significate (RMS). As for the fourth rule, the rule of semantic counterpoint (RSC), allowing the simultaneous application of various senses or nuances of single as well as compound expressions, or their use on more than one semantic level at the same time, this rule in particular is often applied in the domain of metaphysics in a manner similar to Aristotle's semantic behaviour. As will be clear, the medieval debate on intentionality is unthinkable without having the fourth rule in mind. In addition, this rule plays a central role in what I elsewhere called 'stratification semantics'.[36] In section 2.4.2 below I shall return to these applications in more detail.

2.3 *Some Additional Observations on the Vicissitudes of Aristotelian Semantics*

It may be useful to make three more observations.

[a] *Onomastics vs. apophantics*. The first observation is intended to prevent a misunderstanding about my claim that the doctrine of the properties of terms should be regarded in terms of onomastics, say, conceptualization and categorization, rather than in terms of apophantics, that is to say the lore concerning statement-making. It is important to be aware of the broader range of terminist logic, beyond, that is, the domain of the properties of terms. The noteworthy medieval doctrine of the so-called *Consequentiae*, for instance, does involve recognition of the logic of unanalyzed propositions as the primary part of inference theory, including a certain awareness of some theorems of modern propositional calculus. In this respect the achievements of medieval authors went beyond Aristotle's logic of terms, in a way comparable to how Stoic logicians left the confines of Aristotelian logic. This medieval advance over Aristotelian logic should not be underestimated.

---

36) De Rijk (1981), 48-52; (2000), 215-221; (2002), I, 74; II, 124; 199; 310; 409.

[b] *Problems surrounding the copula view.* My second observation is about the difference between the monadic anatomy of the Aristotelian statement-making utterance and the usual dyadic analysis of the medieval *propositio*. As I have tried to make clear before, this difference turns around the un-Aristotelian notion of '*copula*'. In the Middle Ages, the first occurrences of '*est*' as a copula I know of date from about 1100.[37] It was not until Peter Abelard that the copula comes to be regarded as an essential element for making an assertion (in fact acting as the (supposedly) indispensable binding agent of the statement). Hence Abelard in principle must reject the formation '*Socrates albus est*' as genuine, and instead requires that the copula '*est*' be put in between the subject and predicate terms: '*Socrates est albus*'.[38] Nonetheless, time and again medieval authors showed that they were aware of the Aristotelian analysis, particularly in their discussions of '*est*' in terms of its being a *substantive* verb.

One example from Ockham may suffice. In his *Reportatio in II Sent.*, q. 1, Ockham discusses the question whether the notion of copula implies an absolute conception of what the subject and predicate terms stand for, or is a concept distinct from and supervenient to these two. In the former case, Ockham argues, the notion of copula would be, so to speak, the *forma* of the entire compound 'S-P', in a way distinct from the components taken together or separately—like the notion Scotus adopts concerning the form of a whole in an outside thing; and that is precisely what you (Ockham says to his opponent)

---

37) See De Rijk (2002), I, 235-41. As late as in Boethius's monograph *De syllogismis categoricis* the author speaks of the verb '*est*' which is 'accommodated' to make up an assertion (as '*non est*' is added to produce a denial). This manner of expression has nothing to do with the copula idea and comes close to considering the '*est*' and '*non est*' assertoric operators to be added to an assertible, as is the case in Aristotelian statement-making.

38) In a twelfth-century *Perihermeneias* commentary (found in ms Orléans, *Bibliothèque municipale*, cod. 266), the anonymous author implicitly alludes to the Abelardian requirement in describing what he calls the process of the mental transposition needed for the congruous formation of a proposition. I quote this passage (cod. Orléans, *Bibliothèque municipale*, cod. 266, f. 261b) from Kneepkens (2003), 386, n. 74: 'Quando vero transpositionem [*sc.* facimus *De R.*], prius habemus omnes simplices intellectus, quibus habitis consideramus si ex eis eo ordine habitis posset convenienter totalis intellectus uniri. Quodsi ex eis tali ordine habitis non potest convenienter totalis componi, transmutat anima nostra illos simplices et alio ordine disponit, et ex eis alio congruo ordine dispositis totalis intellectus componitur. Verbi gratia, cum dico 'est homo albus', anima mea prius habet hos simplices intellectus, et postea considerat quod ex eis hoc ordine habitis, scilicet si intellectus 'est' sit in subiecto et intellectus 'homo' in predicato, non potest totalis [*sc.* intellectus *De R.*] convenienter componi. Transmutat ergo anima mea illos simplices intellectus, scilicet intellectus 'homo', qui erat ultimus, ponit primum, et intellectus 'est', qui erat primus, medium, et sic intellectum totalem componit.'

intend to avoid. In the latter case, the notion of copula does not seem to have anything to do with our notion of subject and predicate. Ockham's reply to the underlying question about the copula's proper nature is that the copula conveys a common concept that differs from the concepts of the two extremes (subject and predicate) in the same way that one rationate being (*ens rationis*) differs from another and from a real being. I claim, he continues, that if I only have the concept of copula without that of the extremes, I do not have the notions 'subject' and 'predicate' at all. Asked in what way the concept of the copula is a common one, Ockham answers that its commonness is a result of its being an agent mutually uniting the two extremes. However, in the case of a mental union there is no need (he insists) to indicate a real or rationate relationship [sc. between the extremes]. Next follows an intricate discussion of *pros* and *cons*, without a clear conclusion concerning the original question whether or not the notion of copula conveys something supervenient to the mere concepts of subject and predicate. Finally, Ockham claims that, after all, the copula '*est*' is only a syncategorematic concept, and therefore, even though it could signify a real relationship, it cannot stand for it (nor denote it, accordingly):

> Sed contra est quia: Numquam conceptus syncategorematicus potest supponere pro aliquo, quia tunc potest esse subiectum vel praedicatum, sicut nec dictio syncategorematica. Sed, sive dicat conceptus copulae absolutum sive respectivum, solum syncategorematicus est. Igitur, non obstante quod posset significare respectum realem, non tamen potest supponere nec praedicari de eo primo modo dicendi per se.[39]

In his *Quodlibet* VI, q. 29, the same question arises. Ockham once more argues for the position that the copula '*est*' is merely a syncategorematic concept: 'One can in an absolute sense know *man* and *animal* through a single act of cognition, but even so *man* is still not a subject nor *animal* a predicate. This is because of the absence of the syncategorematic concept 'is', whereas, once this has been added, *man* immediately becomes the subject and *animal* the predicate, without any other relationship [being posited between these two]; and there you have the complete proposition.'[40]

---

39) Guillelmi de Ockham *In librum secundum Sententiarum*, ed. Gál and Wood (1981), q. 1, p. 22, ll. 10-16.

40) Guillelmi de Ockham *Quodlibeta septem*, ed. Wey (1980), VI, q. 29, 696: '[...] illud absolutum in mente quod est subiectum vel praedicatum propositionis, potest esse et intelligi cognitione incomplexa, et tamen non unum erit subiectum nec aliud praedicatum. Sed hoc non est propter defectum alicuius respectus rationis, sed propter defectum conceptum absolutum copulae. Potest

Ockham's claim in both passages that the copula '*est*' conveys a syncategorematic concept is obviously intended to prevent us from thinking that the copula (which is a *substantive* verb, to be sure) should have any (absolute or respective) meaning other than consignification. By calling 'est' a syncategorematic term he puts its function on a par with the consignificative function commonly assigned to the rhêma by Aristotle,[41] where the author says of the rhêma (*litt.* 'what is said' or 'assignable') which is part of the statement-making utterance that 'it additionally signifies time'. This consignificative function is then explained by Aristotle:[42] 'By 'additionally signifying time' I mean this: 'health' (*hygieia*) is a name, but 'thrives' (*hygiainei*) a rhêma, because it additionally signifies something as obtaining *now*'. That is to say, as a name, *hygieia* refers to the entity *health*, but it does so in bringing it up qua form only, whereas the rhema *hygiainei* always refers to health as a form qua actually inhering in some substratum, hence as a form that is enmattered or actualized. And so the form involved in the verb manifests itself as actually[43] being invested in temporal conditions. In fact, the rhêma adds time-connotation (the notion, that is, of the thing's being realized in an actual case) to the semantic value it has in common with the corresponding name.[44] Accordingly, what is later called 'the substantive verb' (*rhêma hyparktikon*) '*est*' indicates that what is expressed by the assignable (e.g., 'healthy Socrates') is (or was, or will be) actually the case.

Ockham's assigning a consignificative function to the copula '*est*' should be explained along similar lines. Ockham's deep structure account of the copula '*est*' comes close to taking it, as did Aristotle, as an assertoric operator ranging over an assertible such as the compound 'healthy Socrates', rather than regarding it as the binding agent between two single concepts.[45]

Ockham's deep structure analysis of the role of the 'copula' '*est*' is all the more remarkable as he must try as hard as he can to play down the notion of

---

enim aliquis absolute cognitione incomplexa intelligere hominem et animal, et tamen nec homo erit subiectum nec animal praedicatum. Et hoc quia deficit iste conceptus syncategorematicus 'est', quo posito, sine omni alio respectu statim homo erit subiectum et animal praedicatum, et habetur tota propositio.'

41) Aristotle, *De interpretatione* III, 16 b 6-7.

42) Aristotle, *De interpretatione* III, 16 b 8-9.

43) It is extremely important to distinguish between the expressions 'actually' and 'factually'. Actuality is opposed to potentiality, and leaves factual existence out of consideration. So to Aristotle, factuality implies actuality, not the other way round. See De Rijk (1981), 38-40, and section 2.4.3 below.

44) Accordingly, '*est*' is going to mean 'it is the case that…'; see De Rijk (2002), I, 207 f.

45) For my interpretation of the hyparctic *estin* as an assertoric operator see *ibid.*, 248-255.

uniting (*unio*) which is almost inextricably connected with that of copula as a binding agent. Putting it straightforwardly, to call the mental union of subject and predicate a syncategorematic concept implies recognizing that the syncategorematic concept '*est*' ranges over a *preexisting* mental compound, just as syncategorematic terms such as 'all', 'every', 'some' and the like range over a preexisting conceptual formation merely to modify it in terms of quantification. It cannot be underlined enough in this context that, in spite of his usual adherence to the common Abelardian view of the superficial structure of the statement-making utterance, Ockham does not mistranslate Aristotle's plain language "be' additionally signifies some combination [...] etc.' with 'it *effects* the combination [...] etc.', meaning something like 'it *couples* subject and predicate':

> Primam particulam declarat, dicens quod verbum consignificat tempus, nam licet 'cursus' et 'currit' idem significent', quia tamen 'cursus' est nomen, ideo non consignificat tempus, sed 'currit' quia est verbum, consignificat tempus, nam denotat cursum nunc esse [...]. Nam sic dicendo 'currit' datur intelligi quod cursus nunc est.[46]

It is only at the end of the discussion[47] that the common interpretation seems to come to the fore, viz. that the copula is effective of the combination, although the *quasi* might suggest that Ockham's giving in to the common view is rather a matter of following convention.[48]

[c] *The 'conceptual approach' requirement.* Finally, an additional remark on the contextual approach requirement might be of importance. As I said before, this requirement concerns the difference between *significatio* and *suppositio*: the diverse ways in which significative terms can stand for (*supponere pro*) things only come about when their meaning is differentiated as a result of their being used in the context of a proposition. What are we to say now about

---

46) Guillelmi de Ockham *Expositio in librum Perihermeneias Aristotelis*, cap. 2 (*ad* 16 b 6-11), ed. Gambatese and Brown (1978), 383.

47) Cap. 3 *ad fin.*, 389.

48) Guillelmi de Ockham *Expositio in librum Perihermeneias Aristotelis*, ed. Gambatese and Brown (1978), cap. 2, 389, 14-19: 'Et tamen tale verbum significat compositionem quandam ex subiecto et praedicato, quae tamen compositio sine compositis, hocest sine subiecto et praedicato, non esse potest. Intelligendum quod propositio est quoddam compositum non tamquam per se unum, sed tamquam aggregatum ex subiecto et praedicato et copula, quae quasi unit subiectum cum praedicato.' Ockham's treatment of the famous *Perihermeneias* passage can profitably be contrasted to the perplexing problems Thomas Aquinas became entangled in as a result of his view of the copula ('*est' tertium adiacens*), to wit that it merely is the binding agent between subject and predicate. These problems are vividly described and evaluated in Bäck (2003).

my twofold claim that (a) supposition theory (and terminism in general) should be assessed in terms of Aristotelian focalization and categorization, which are key devices of a logic of terms—i.e., a system of logic primarily focusing on terms, and only secondarily, it seems, on propositions—, and (b) the terms' occurrence in propositions, rather than their being used outside statements is strictly required for supposition etc.? This problem is entirely a matter of appearance. In fact, the contextual approach requirement matches Aristotle's embedding of focalization and categorization in a situational context of discussion and dispute, without which focalization is pointless and inane. In a word, Aristotle's *argumentative* contextuality mirrors the medieval contextual approach.

2.4 *The Four Aristotelian Rules Multifariously Applied*

Here are some problem areas for whose solution Aristotle's strategy of argument, particularly his four main semantic rules, used to play a decisive role. This list is of course not exhaustive. It only aims to throw some more light upon the fact that they were commonly applied in medieval argument, although most of the time, they were applied implicitly. One general remark on the doctrinal influence of terminism as a philosophical method should be made in advance. The different directions taken by philosophical and theological discussions in the Middle Ages testify to the hospitability of terminism for realistic as well as nominalistic/conceptualistic positions of any fashion. One thing will be clear enough: the Medievals were successful in expanding and refining Aristotle's strategy of argument.

2.4.1 *The Debate on Metaphysical Matters. John Buridan*

In *Metaphysics* VII 6, Aristotle discusses the question whether a thing's quiddity as indicated by the definiens coincides with the particular thing itself. To his mind, the only condition for a quiddity to exist is to be embodied in this or that particular of the outside world. But what about the key notion *ousia* with its unmistakable flavour of universality? In this chapter, Aristotle tackles the relationship between a particular thing and its quiddity by investigating the semantic ambivalence of the word *ousia*. This happens from the angle of naming things. The onomastic approach comes to the fore from the frequent use of phrases like 'things *called up* after a coincidental mode of being' instead of the mere talk of 'accidental things'. Small wonder that the notions *ta kath'hauto legomena, ta kata symbebêkos legomena*, and *kat'allo legesthai* are in focus. Aristotle points out in particular the relationship of referential identity (see 1031 b 25-26: 'in one way they are the same') that exists between that to which the white attaches (viz. the *album* or 'white thing', Callias) and the particular form 'whiteness' inhering in him. This referential identity is based upon the fact that in Aristotle's view, *this* particular whiteness, which is a strictly individual form, is found nowhere else but in this person, Callias. At the same time Aristotle is not blind to the formal diversity between the white thing, Callias and his immanent form of whiteness, because this particular man is formally distinct from the particular instantiation of whiteness inhering in him. In a word, being a white man is not the same as being white.

In the third question of John Buridan's commentary on Aristotle's *Metaphysics*, which is about Book VII 6, 1031 a 19 ff., the author discusses the general problem posed by the Philosopher in VII 6 (running 'Utrum autem idem sit aut diversum *quod quid erat esse* (Greek, *to ti ên einai*) et unumquodque') insofar as coincidental terms are concerned, in which case Aristotle's answer is in the negative (a 19-28). Significantly, in order to have the opportunity of opposing *esse album* to *albedo* Buridan rephrases the *quod quid erat esse* formula (which stands for 'quiddity') with 'esse *ipsum*': 'Utrum in dictis secundum accidens sit idem *ipsum* et *esse ipsum*'.[49] In what he calls his 'metaphysical' solution to this question—which comes to envisaging for which *things* such terms supposit—Buridan points out that Aristotle gives a negative answer to the question 'whether in the case of things being said after a coincidental feature, the thing itself is the same as its quiddity [read 'being precisely such a thing']'. For Aristotle, the identity between the particular and its quiddity only applies to things designated by substantive terms. In what is designated as 'white thing' (e.g., a man or a stone), to speak of the thing itself does not coincide with speaking of its quiddity. Buridan explains what he thinks Aristotle intends to say. The term '*album*' only supposits for the subject-substrate in which whiteness inheres, and connotes the inhering whiteness. But, because, formally speaking, the thing in question is white owing to whiteness, the phrase '*esse album*' supposits either for this whiteness in virtue of which, formally speaking, there exists a case of being white, or it supposits for the aggregate consisting of the whiteness and the substrate in which the property of being white inheres. From this it clearly appears that *album* and *esse album* are not the same,

[49] For the general theme see Bakker (2001), 249 ff. and Tabarroni (2003). Buridan's rephrasing Aristotle's expression was rightly given special attention by Tabarroni, who, from the viewpoint of semantics, has extensively discussed Buridan's and Marsilius of Inghen's comments on *Metaphysica* VII, 3-5. Note that in *ipsum* and *esse ipsum*, the word '*ipsum*' is a dummy word used like our 'x', 'y', etc. just as '*unumquodque*' (Greek, *hekaston*) is in the Aristotelian formula.

because (a) a substrate and its inherent form differ from one another and (b) there is also a difference between a substrate and the aggregate of form and substrate.[50]

The twofold significative function of '*album*', viz. of denoting the substrate and connoting the inhering form, is common doctrine, based on the application of the first semantic main rule (RMA). But for the phrase '*esse album*' (let us call it a quidditative complex)[51] things are different. No mention is made of this quidditative complex's connotation, but only of its *suppositing* either the form whiteness or this form plus the substrate it inheres in. In the former case, it stands for the *particular* form in virtue of which the *particular* thing is white; in the latter for the particular taken as a self-contained thing composed of form and material substrate. This can be neatly explained as applications of the RMA rule. The twofold supposition involved in the latter case is presented in terms of an alternative, *either* the particular form alone *or* this form including its substrate. Buridan correctly thinks that a choice between this *either-or* can be left aside for the time being, because in either of the alternative cases the difference between the single term *album* and the quidditative complex *esse album* is sufficiently clear.

In his next question—which is about things that are designated according to their quiddity as self-contained thing (*'Utrum in dictis secundum se sit idem ipsum et esse ipsum'*) rather than according to one of their coincidental properties (which were under investigation in the preceding question)—Buridan once again comes to speak about the difference between *album* and *esse album*. First, he mentions Aristotle's position that, with regard to things designated according to their own quiddity (*in dictis secundum se*), the thing itself and its quiddity coincide, meaning that, e.g., *homo* and *esse hominem, lapis* and *esse lapidem* are the same, in a word, whenever substantive terms are involved. This leads him once again to consider the *esse album* issue and also to raise the question about the relation of identity between *albedo* and *esse albedinem*.

---

50) Johannes Buridanus, *Quaestiones in Aristotelis Metaphysicam* VII, q. iii (after Mss Paris, *BnF. lat.* 14.716, f. 154va and Carpentras, *Bibliothèque Inguimbertine*, cod. 292, f. 86ra): 'Et ideo ista questio potest magis metaphisice solvi, scilicet concipiendo pro quibus rebus tales termini supponunt. Et puto quod sit de intentione Aristotilis quod iste terminus 'album' supponit solum pro subiecto cui inheret albedo, et appellat vel connotat albedinem sibi inherentem; sed iste terminus 'esse album', quia res est alba formaliter per albedinem, supponit vel pro illa albedine secundum quam formaliter est *esse album*, vel supponit pro aggregato ex illa albedine et subiecto cui inheret. Et tunc statim manifestum est quod non est idem *album* et *esse album*, quia non est idem subiectum et forma sibi inherens, nec etiam est idem subiectum et aggregatum ex forma et subiecto.'

51) I would prefer this label to Tabarroni's (2003), 396 ff. 'quidditative term'.

Asked for the reason why *album* and *esse album* are not the same, Buridan points out that, unlike '*album*', which has the well-known supposition plus connotation, '*esse album*' does not supposit for just the white thing (e.g., stone), but for the aggregate of the white thing and its whiteness (that is, stone qua white stone).[52] This is to say that *esse album* does not convey the notion of the white thing qua *thing*, but rather qua *white thing*. Thus the quidditative complex *esse album* presents to our mind the white *thing* including the property by which it is a *white* thing.

Buridan already pointed out in his first question on *Metaphysics* VII that Aristotle thought that every coincidental term or concept connotes or implies a substantial concept, and what is more, that concrete *adjectival* terms such as '*album*', '*nigrum*', '*iustum*' etc. supposit for *substances* (the white, black etc. *thing*). To Aristotle, the same applies to abstract terms (such as '*albedo*'), that is to say, they convey—note that Buridan does not say '*supponit*'—a substantial substrate, because to Aristotle, *albedo* is nothing other than a particular's *esse album*. Now, the latter phrase conveys (in Buridan's view too, for that matter) the intellection that some *thing* is white, as it is unthinkable that there should be a case of *esse album* unless some *thing* is white. And obviously the term '*aliquid*' is a substantial term.[53] So far, so good, the metaphysician is ready to say. Unfortunately, however, the doctrine of the Eucharist prevents the Christian philosopher from assuming (together with Aristotle) that *albedo* should formally imply that some *thing* is white as actually affected by it and therefore cannot possibly exist separately from an underlying substrate. Consequently, as a Christian philosopher, Buridan has to reject Aristotle's formal identification of *albedo* and *esse album*.

What should he think, then, of quidditative complexes such as *esse albedinem*? Buridan has to maintain, against Aristotle's testimony, that *albedo* does

---

52) *Ibid.*, q. 4 (*P* 155ra; *C* 86va): 'Ideo credo esse dicendum quod protanto differt *album* et *esse album* quia hoc nomen 'album' sic diversimode plura significat quod pro uno illorum supponit et non pro altero, sed illud alterum connotat tamquam rem vel dispositionem adiacentem illi rei pro qua supponit. Tunc enim 'esse ipsum' non supponit pro ipso, sed pro aggregato. Verbi gratia, 'esse album' non supponit pro albo, sed pro aggregato ex albo et albedine sibi adiacente, scilicet per quam dicatur album.'

53) *Ibid.*, q. 1 (*P* 153va; *C* 84vb-85ra): 'Sed Aristotiles credidit quod omnis terminus sive conceptus accidentalis connotaret vel implicaret in se conceptum substantialem, ymo quod termini concreti (ut 'album', 'nigrum' etc.) supponunt pro substantiis. Et etiam termini abstracti secundum Aristotilem dant intelligere substantias, quia credidit Aristotiles quod non esset aliud *albedo* quam *esse album*, nec *figura* quam *esse figuratum*, et sic de aliis. Et tamen *esse album* dat intelligere quod aliquid est album. Non enim potest intelligi quod sit *esse album* nisi eo quod aliquid est album. Et iam iste terminus 'aliquid' est terminus substantialis.'

not formally include the notion of *esse album*. Consequently, he has to realize that, in accordance with his own, Christian view, the quidditative complex *esse albedinem* can only mean that there is an instance of whiteness, whether or not enmattered, that is to say, whether naturally enmattered or miraculously not-enmattered. From this point of view, it seems reasonable to claim that *albedo* (although it is not the same as *esse album*) is the same as *esse albedinem*. This surmise is supported by the conviction that '*albedo*' does not connote any disposition adjacent to whiteness, because such a disposition is not formally required for there to be *albedo*. Contrary to '*album*' (i.e., 'white *thing*'), indeed, which does not *supposit* for whiteness, but *connotes* it as an additional disposition of the white *thing* (required for a *white* thing to be), *albedo* needs nothing additional to be whiteness. Another claim could be made to the effect that if '*albedo*' were to *supposit* for the aggregate of whiteness plus substrate without any connotation of an additional disposition, then *albedo* and *esse albedinem* would coincide:

> Tunc, istis visis, esset[54] dicendum quod idem esset *albedo* et *esse albedinem*, quia quicquid Aristotiles diceret de hoc, tamen nos dicentes albedinem esse separabilem, diceremus quod hoc nomen 'albedo' non connotat dispositionem aliam adiacentem albedini secundum quam albedo formaliter dicatur albedo. Postea etiam dicendum esset quod si 'albedo' supponeret pro albedine et subiecto simul sine connotatione dispositionis addite, idem esset *a<lbedo>* et *esse a<lbedinem>*.[55]

The identification of *albedo* and *esse albedinem* puts *albedo* on par with other substantial terms pertinent to the *dicta secundum se*, such as '*homo*' and '*lapis*', leaving aside (for the time being) the usual distinction between absolute and connotative terms.[56] What now counts is the basic distinction between supposition and connotation. Buridan claims that if a substantial term like '*homo*'

---

[54] Buridan has a habit of using such subjunctives as *esset* as *modus potentialis*, even when it is preceded by '*si*' and could lead the reader to take it as a *modus irrealis*. This use of *modus potentialis* should be taken as a stylistic mode ('One could or might say . . .', 'I would like to say . . .').

[55] *P* and *C* as well as most other Mss here and in the next lines simply read *a.*, except for Erfurt, *Universitätsbibliothek, Amplon.* F 322, which each time has (f. 54ra-rb) *album*.

[56] It should be borne in mind that in Buridan's view, not-enmattered whitenes is not *naturally* subsistent or a substance. See *Quaestiones in Metaphysicam* IV, q. vi (*P* 131ra; *C* 61rb): '[. . .] omne illud est substantia quod naturaliter per se subsistit ita quod non inheret alteri, et omne illud etiam est substantia quod est pars talis per se naturaliter subsistentis; et omne illud est accidens quod sic non subsistit per se naturaliter nec est pars per se subsistentis, non obstante quod subsisteret per se miraculose. Et sic albedo, quamvis per se subsisteret, non diceretur substantia, quia non sic subsistit naturaliter, sed miraculose'.

signifies man, including body and soul, and also supposits for a man plus his body and soul, without formally connoting any disposition required for being a man, then *homo* and *esse hominem* are the same. Thus, from the viewpoint of referentiality, '*albedo*' will find itself in the company of other substantival terms, e.g., '*homo*', which does not *connote* soul and body as if they were something extrinsic to the quiddity *man*.[57] With regard to all these substantival terms devoid of connotation (irrespective of whether they are absolute or connotative terms) the identity of *x alone* (*ipsum*) and *being x* (*esse ipsum*) obtains:

> *Ibid.*: Ita similiter in proposito, si iste terminus 'homo' significat animam et corpus simul et supponit etiam pro illis simul indistincte et quod non connotet dispositionem aliquam per quam homo formaliter dicatur homo, tunc est idem *homo* et *esse hominem*. Modo ita intendit Aristotiles de omnibus terminis substantialibus supponentibus pro compositis ex materia et forma. Aristotiles enim intendit quod homo est formaliter homo per animam intellectivam et lapis est formaliter lapis per suam formam substantialem, et non per aliquam dispositionem sibi additam.

Concluding this section, Buridan returns to his statement about *albedo*, which gives him a fine opportunity to reject Aristotle's identification of *albedo* and *esse album* once again, because what may apply to *albedo* does not apply to *album*:

> *Ibid.*: Et sic oportet concludere quod est idem *homo* et *esse hominem* et *lapis* et *esse lapidem*, sicut dicebatur quod esset idem *a<lbedo>* et *esse a<lbedinem>*, si '*a<lbedo>*' supponeret pro albedine et subiecto. Sic autem non est de albo, quia 'album' non supponit nisi pro subiecto.

57) In this passage Buridan clearly takes 'connoting' as synonymous with 'conveying a notion that is extrinsic to the formal nature of the thing signified', so it can be viewed as concerning an additional (or adjacent) disposition of this thing. The inherence of whiteness in its subject can be expressed in terms of an additional disposition, which is natural, but can be taken away miraculously (as in the Eucharist). In his reply to an objection (*P* 154vb-155ra; *C* 86rb-va) concerning the separation of body and soul in death, Buridan answers (*P* 155rb; *C* 86vb-87ra) that if one regards the composition of body and soul as connoted by the term '*homo*', Aristotle would deny that it is an additional disposition. But if we wished to call it so, then, still, it is not owing to this composition that a man is called man, but to his soul. When someone, he goes on, wishes to maintain that the name '*homo*' connotes such a disposition beyond soul and body, and also that man is formally a man owing to this disposition, then that person must concede that *homo* and *esse hominem* are not the same. On the vital terminology *dispositio adiacens/addita* see De Rijk (1997), 407, and Bakker (2001), 255, n. 17, Zupko (1998), 588-599 and De Rijk (2008), LXIV-LXXIII. To my mind, *adiacens* conveys actual presence, whereas the use of *addita* refers to its status of being extrinsically added.

The gist of the entire discussion seems to be that, in Buridan's view, the real occurrence of whiteness in the Eucharist entails that the quidditative complex 'esse albedinem' can supposit for it (meaning that there is an instance of actual subsistent whiteness), but it does not supposit for there being some white *thing*, as '*esse album*' does.[58]

Buridan once again plays on the idea of referential identity in his solution to another question concerning Aristotle's basic ontology. In q. 12 ('*Utrum forma substantie materialis sit tota quiditas eius*') he has to comment upon Aristotle's position in *Metaph*ysics VII, 17-VIII, 1-6, where true *ousia* is finally identified as the enmattered form, that is to say, the compositum or aggregate of a thing's form and its material condition.[59] Buridan is of the opinion (as are many other commentators as well) that we should not be led astray by Aristotle's use of Platonic terminology. He summarizes: the Philosopher holds that a thing's quiddity is that which is signified by a quidditative predicate (designation) and which the quidditative predicate supposits for. Putting it briefly, he says, the quiddity of a horse or an ass is not its form, but the thing precisely as composed of matter and form. Next, he proves this by considering the supposition of the terms 'horse' and 'ass' when a particular is sensorially identified as (a particular instance of) horse or ass. The referential identity between the particular quiddity and the individual informed by it clearly comes to the fore through the use of the demonstrative noun (*hoc*):

> Sed Philosophus ponit quod quiditas rei est illud quod per predicatum quiditativum significatur et pro quo predicatum quiditativum supponit. Et sic breviter ego dico quod quiditas equi vel asini non est forma equi vel asini, sed est ipsum compositum ex materia et forma. Probatio quia: Nos concedimus quod iste equus singularis est compositus formaliter ex materia et forma (unde et Plato etiam hoc concessit). Modo ergo si queratur de isto equo 'Quid est?', convenienter respondemus 'Hoc est equus' vel 'Hoc est animal'. Ergo quiditas equi est illud pro quo supponit iste terminus 'equus' vel iste terminus 'animal'. Et tamen isti termini supponunt pro eodem pro quo supponebat illud pronomen 'hoc' quando demonstrabamus istum equum singularem, quia si non supponerent pro eodem, propositio non esset vera dicens quod hoc est equus, quia hoc est hoc et nichil aliud. Ideo si equus est aliud, non est verum quod hoc sit equus. Et tamen illud pronomen 'hoc' supponebat pro isto equo singulari, quem dicebamus esse compositum ex materia et forma. Ergo iste terminus 'equus' pro illo composito supponit. Et per consequens illud compositum est quiditas ipsius equi. Et hoc bene expressit Aristotiles dicens quod singuli *quod quid erat esse* est una substantia, et

[58] See the concluding sentence of the text quoted above from VII, q. 1, and De Rijk (1993), 45-47 and (1997), 407; Bakker (2001), 255, n. 17.
[59] See the extensive discussion in De Rijk (2002), II, 244-301.

singuli substantia non est aliud ab illo nec alteri conveniens, ymo cuiuslibet substantia et quiditas est ipsummet.[60]

The impact of supposition theory (along the lines of the main semantic rules 1 (RMA) and 4 (RSC)) appears from the frequent occurrence of the supposition vs. connotation issue, and the flexible way in which Buridan makes the most of the ambivalent meanings of linguistic expressions in order to solve problems of ontology. The Eucharist problem forces him to refine skillfully the notions of supposition and connotation.

2.4.2 *The Intentionality Debate*

The medieval intentionality debate is about how to evaluate the process of cognition.[61] It has both an epistemological and a psychological aspect. The Medievals' basic idea on this score is that there is some connection between any psycholinguistic tool or entity and the 'thing' expressed by it. Speaking from the viewpoint of semantics proper, the basic phenomenon in matters of intentionality is the referential force of any linguistic expression. Intentionality, then, is any psycholinguistic device's property of referring to something formally different from it.[62] The gist of the intentionality debate is about the benefit and indispensability (or inappropriateness, to other people's minds!) of the multifarious relationships (supposed to exist) between the various constituents that make up the semantic diagram of referentiality. In fact, the intricate problems concerning intentionality are all about the semantic topic of referentiality *annex* the epistemological issue of the representativeness and reliability of cognition, and, in the wake of it, the psychological issue of the role and proper nature of the mental entities that are (supposedly) involved in the process of cognition.

The diversity of the positions taken in the intentionality debate was a result of the fact that medieval scholars kept focusing on the specific nature of the various tools and devices (putatively) operating in the trajectory between the extramental object and the cognitional faculties, senses and intellect. They did so by examining, in particular, the cognitional roles of the two principal agents, the outside thing and the intellect, and by looking for the proper meanings of 'intellection' and 'being intellected', including the intentional relationships

---

60) John Buridan, *Quaestiones in Metaphysicam* IV, q. xii (*P* 159ra; *C* 90vb).

61) This section derives from the *Introduction* to my edition of Giraldus Odonis' treatise *De intentionibus* (2005), 113-357.

62) The focal meaning of the key word 'intention' is to convey a tendency to something else (*tendentia in quoddam alterum*).

between the intellect and the object of cognition. I will confine myself to some significant cases that enable us to see the impact of these different positions (both in terms of methodology and content).

As for the general notion of *intentio* itself, an intention can be taken as precisely a mental entity residing in the soul as its *habitus*, taken apart from its being significative ('abstractively', says, e.g., Simon of Faversham), or considered after its own mode of being (*pro esse intentionis*). In this case, e.g., the concept 'man' or 'stone' is taken as a psychic entity, apart from the significative force it has qua *intentio*. The concept can also (concretively) stand for the quidditative mode of being its significate possesses in its supposits (*pro esse quod habet in suppositis*). Every universal (or universal intention, corresponding to any appellative noun) signifies both the property of universality and the thing underlying the intention (*res subiecta intentioni*). Thus the intention *homo* conveys both manhood and the individual thing manhood inheres in. In an individual man his particular manhood ('being a man') can be distinguished from its actualization in the individual man, and the two different senses involved are both recognized as possible alternatives. In addition, their simultaneously obtaining is not excluded. Again we find ourselves on the familiar ground of our four semantic rules. Playing, in the wake of Aristotle, on the ambivalence (note that I do not speak of ambiguity),[63] the *intentionalistae* unanimously uphold the basic referential equivalence of the phrases 'intellect-in-its-actual-state-of-intellecting' and 'what-is-actually-intellected', as well as that of 'intellected thing' and 'thing intellected'. Some themes, however, particularly the identification of the intelligible species with the intellective act, were eagerly debated. The impact of terminism, including the application of the four rules (the fourth in particular) in these debates will be patently clear to anyone familiar with the intentionality literature.

A first or primary intention (*prima intentio*), then, is a primary intellection of an extramental thing, by which the soul apprehends it according to its quidditative nature and properties, for instance when a human being is grasped qua *man*, or *animal*, or *rational*. A second intention is the secondary intellection of the object in question, by which the soul apprehends it according to its being a genus or species or its acting as a *definitum* (qua opposed to *definiens*), or as a subject or a predicate, or its functioning as a compound intellection in sentence-making or discursive thought. Now, to grasp an outside object according to its being a genus or species, or its acting as a *definitum* is to grasp it in its

63) The salient distinction between ambivalence and ambiguity is highlighted in De Rijk (2002), I, 69-72; II, 154; 413; see also section 1.3 above.

generic, specific, or definitorial mode of being, which are all present in it qua so many ontic layers. Their distinction as well as their mutual relationships belong to the field of enquiry covered by the four rules. Obviously the quick-witted debates about the function (or the superfluity) of a (putative) host of intermediate 'things' existing (or supposed to exist) between the outside thing and the intellect reveal an adroit manipulation of the four semantic rules, which can serve both for recognizing the differences between the respective semantic layers and for arguing their referential equivalence.

To support his view of the proper nature of intentionality, Hervaeus Natalis proceeds to analyse accurately the many-sided relationships between being (universal and particular, and non-being as well) and the intellect. His focus (in his *De secundis intentionibus*) on the diverse relationships (*habitudines*) in particular leads to a real convolution of interconnections between the diverse cognitional tools and devices, both mutually and with respect to the intellect. The ensuing entanglement of interconnections between the agents and the patients of the process of cognition could not disguise even to Hervaeus' admirers its incoherence and shortcomings.[64] However this may be, the impact of the four rules is omnipresent as is the use of the doctrine of supposition/connotation as well.

2.4.3 *A Sophisticated Use of Semantic Stratification: Playing on Time Connotation*

As early as in Peter Abelard we are faced with a remarkable intermingling of the denotative and the connotative levels of appellative nouns by playing on the time factor. Appellative nouns (names) are taken to display a two-level stratification, to contain, that is, two significative layers. One concerns their job of denoting things (in accordance with their primary, natural function) as actually existent (or conceived of as actually existent) alone.[65] This function depends upon an appellative name's basic level, in virtue of which it refers to something merely being there as a particular self-contained entity, regardless of its specific mode of being (whether essential or coincidental), such as 'being-a-man' or 'being-a-stone' or 'being-something-white'. The other semantic layer concerns the name's appellative force to signify a thing's ontic modifications or specific modes of being.

[64] De Rijk (2005), 251-301; 350.

[65] It should be underlined that a thing's actuality is as such involved irrespective of its facticity in the present, past, or future time. For this important distinction see De Rijk (1981), 38-40.

Abelard's well-known example is the problematic sentence 'Hic senex erit puer'. To regard it as true, you have to focus exclusively on the subject term's basic layer designating merely the self-contained subject-substrate aspect, and to abstract from the term's appellative force, which only obtains at a time other than the moment in which the sentence is uttered. On this interpretive line, the sentence can be true, if said of a boy of, say, ten years of age with reference to the day or year after, meaning 'This entity, an old man at a later date, will be a boy'. This semantic move has everything to do with '*appellatio formae*'. An amusing case occurs in an anecdote about John Buridan. As a Parisian student he had a violent quarrel about a girl with a fellow-student of his, Pierre Roger, who later became Pope (Clemens VI). The story goes that when as a prominent Parisian master Buridan came to pay his respects to the new Pope at Avignon, and the Holy Father ironically asked him, 'Tu, quare percussisti Papam?', Buridan, playing on the *appellatio formae*, replied, 'Pater, papam percussi, sed non percussi papam' ('I didn't hit the Pope, but the Pope is the one I hit').

In her fine paper on the unity of semantics and ontology, Joke Spruyt discusses John Wyclif's interpretation of such problematic sentences, including the guidelines Wyclif explicitly presents and elucidates.[66] One of his examples runs 'Iste rex fuit genitus a muliere que numquam genuit istum regem' ('This king was begotten by a woman who never begot that king'). Wyclif explains:

> [...] quicumque terminus accidentalis, predicatus respectu verbi affirmativi de preterito vel de futuro, limitat ratione differentie temporis connotati suum significatum inesse subiecto pro conformi tempore connotato. Ut [...] si ista mulier genuit istum regem, tunc genuit ipsum pro instanti pro quo fuit rex.[67]

Note the position of '*iste rex*' at the beginning of the sentence. As is clear from the foregoing explanation, the sentence should be taken to mean that if this woman is the mother of this king, the preterite parturition concerns an entity that presently (i.e., many years after being born) is this king. The additional temporal connotation the name receives from the verb used should be accommodated

[66] Spruyt (2008), 24-58. Her discussion of the above type of problematic sentences is found *ibid.*, 32 ff.

[67] *Tractatus de logica*, ed. Dziewicki, I, ch. VII, 112 f. (corrected by Joke Spruyt after Assisi *Biblioteca Comunale*, cod. 662): 'A coincidental term of whatever kind, when it is predicated, qualified by an affirmative verb in the past or future tense, restricts, because of the difference of the time connoted, the term's significate to its inherence in the substrate obtaining for the fitting time (*tempore*) connoted. For instance, [...] if this woman has begotten this king, then for the portion of time (*pro instanti*) when he was king, <it obtains that> she has begotten him.'

in accordance with the name's present connotation. Thus the various times are to be conceived as precisely those portions that fit the time connoted.

Other examples are found in the syncategoreumata treatises and the so-called *Sophistaria*. For instance, in the treatment of *aliud* and *alter*, when he deals with the sophisma 'Sortes incipit esse alter istorum', Henry of Ghent introduces his solution by pointing out the term's ambiguity:

> Cum in hoc termino 'alter' duo sunt, scilicet suppositum ut sunt illa in quibus est alteritas, et accidens quod est alteritas [...] etc.[68]

In Matthew of Orleans' *Sophistaria*, the author also starts his solution to a sophisma concerning the meaning of the *dictio exclusiva* '*tantum*' (on account of '*Tantum homo albus currit*') by distinguishing between the two senses of '*aliud ab homine albo*':

> Sed quia 'aliud ab ipso' est dupliciter, scilicet in supposito et in forma accidentali, propter hoc potest fieri exclusio dupliciter: vel ratione suppositi vel ratione forme.

Some lines further Matthew explains:

> [...] hoc quod dico 'aliud' [...] non solum dicit diversitatem in substantia sed etiam in accidente.[69]

The above examples from Abelard and Wyclif could lead us to regard stratificational semantics as mere quibbling, or at most a matter of logical exercizing *pro acumine iuvenum*. However, the same Abelard uses stratificational semantics in his *Ethica*, when he tries to defend his fundamental thesis that sin is nothing but consent to evil. Against the Augustinians, who claimed that sin consists in bad will (*mala voluntas*), he introduces the famous example of the poor servant who kills his master in self-defence:

> Ecce enim aliquis est innocens in quem crudelis dominus suus per furorem adeo commotus est ut eum, evaginato ense, ad interimendum persequatur; quem ille diu fugiens et quantumcumque potest sui occisionem devitans, coactus tandem et nolens occidit eum ne occidatur ab eo.[70]

---

[68] Henrico de Gandavo adscripta *Syncategoremata*, ed. Braakhuis, Etzkorn and Wilson (2011), 58, 1772-1774.

[69] Matthew of Orleans, *Sophistaria* II 22 and 23, ed. Spruyt (2001).

[70] Petrus Abaelardus, *Ethica sive Scito teipsum*, ed. Luscombe (1971), 6, 24-28. 'For consider: there is an innocent whose cruel master is so burning with rage against him that he with a naked sword

Abelard points out that the sin the servant has committed does not consist in willing something bad, for the only thing the poor man wanted was to save his own life (which is something good). Properly speaking, he definitely did not want to kill his master, for in that case he would have wanted to endanger his own life (viz. in court, for he knew that the judges were going to sentence him to death).[71] Thus Abelard distinguishes in the material act of killing two different layers, one the deed intended ('saving one's own life'), the other the deed effected ('endangering one's own life'). Abelard's argument consists in identifying the deed of killing with the act of endangering oneself, separating it from the other layer, which cannot be regarded as bad willing (in the Abelardian line of thought, whoever passionately wants to smoke *eo ipso* passionately desires to die). For the sake of argument, Abelard arbitrarily chooses a secondary layer of a term's significative area, consciously ignoring its main layer. In this line of thought, sentences like 'This servant wanted to save his life' and 'This servant wanted to die' are both true, as (speaking of Mr. X, who passionately wants to remain a smoker for another 50 years) are 'Mr. X passionately wants to live' and 'Mr. X passionately wants to die'. Abelard's arguments are indeed far-fetched, but in his *Ethica* they are brought forward with serious intentions.

Incidentally, a similar use of stratificational semantics seems to underlie modern discussions concerning the metaphysics of modality. For instance, can we speak of identity through possible worlds in claiming that the term 'this man' should refer to one and the same person in all possible worlds? In other words. Is there such a thing as 'transworld identity' or are individuals 'worldbound'?[72] For those who are not amused by such far-fetched metaphysical speculations, there is a juridical phenomenon we are all familiar with, the preclusion of criminal proceedings by reason of lapse of time. Or putting it generally, am I the same person I was some 50 years ago? Forget it!, my children would say.

---

chases him for his life. For long that man flies and as far as he can he avoids his own murder; in the end and unwillingly he is forced to kill him lest he be killed by him.' (transl. Luscombe).

71) Note that, in accordance with the social convictions of his days, Abelard did not regard this event as a case of legitimate self-defence. See De Rijk (1986b), 8 ff.

72) See the still interesting reader on possible world semantics edited by Loux (1979), *passim*.

3 Conclusion

It may be profitable to preface the summary of the foregoing observations with some remarks on the general paradigm of medieval epistemology, the Medievals' confidence in human cognition in particular. This basic confidence never disappeared as a general attitude, yet was seriously undermined by doubts about the adequacy of the cognitional procedure and by various attempts to eliminate these doubts by restricting the modalities of knowledge. The development of terminism, including supposition and the other properties of terms, can also be assessed in this perspective.

3.1 *On the Medievals' View of the Reliability of Cognition*

The semantic views held by medieval thinkers were basically determined by a firm, twofold conviction to the effect that a) there is an extramental world around us, which possesses by itself—independently, that is, of the operation of the human intellect—certain ontic features, and b) in principle, our cognitional (sensitive and intellective) faculties provide us with effective access to extramental Reality. This conviction was grounded upon the assumption that there exists a fundamental parallellism between the various ontic articulations of the things of the outside world, on the one hand, and the various natural ways in which we are able to understand things, on the other. The latter idea went hand in hand with their optimistic conviction that it is up to the human mind to really *discover* the truth about things. Aristotle's doctrine of the ten categories in particular was regarded as a cognitional model that neatly runs parallel with Reality.

For this reason, the Ancients' and Medievals' conceptual engineering (including terministic approaches) as well as their trust in its efficacity and reliability were ultimately based on this parallellism paradigm. This is not to say of course that there was no debate about the ins and outs of the adequacy of language in general and linguistic referentiality in particular. Quite the contrary, increasing doubts were raised about people's ability to avoid the pitfalls of language with its ambiguous, ambivalent, or plainly misleading expressions. It is precisely the parallellism paradigm—which never entirely lost its influence throughout the Middle Ages—that could easily lead to an adoption of superfluous real entities as a result of the reification of mental tools and devices. The intentionality debate provides us with illustrative cases of such (putative) misapprehensions, like, e.g., the famous 'intelligible species'. We are all familiar with the (Aristotelian) reductive principle commonly going under the name

'Ockham's razor' ('Entities should not be multiplied beyond necessity'), which was intended as an antidote against such a proliferation of entities.

In order to assess medieval terminism as a cognitional device it may be useful to point out an interesting fourteenth-century criticism of the appropriateness of the Aristotelian categorial scheme for discovering how things really are. Whereas an Ockham rejects the appropriateness of most of the nine accidental categories, a contemporary, the Black Friar Crathorn goes a momentous step further in taking what we have called the flexibility of the focalization/categorization device more seriously. What I am trying to say is this. Ockham wielding his razor sliced away eight of the categories; only substance and quality came away unscathed. Why? Because Ockham considered them merely as aspects of substances and qualities, which could only function as various ways of talking about the real world. Obviously, he thought that, unlike the eight other categories, the categories of substance and quality *are* representative of ontic properties that distinctly exist outside the mind. Crathorn, on the other hand, held that even the distinction between substance and quality and the other categories is merely a matter of speech, so the substance/accident notions are interchangeable. It all depends upon our variable point of view.

Crathorn's extensive discussion of substance is found in q. 13 of the first book of his *Sentences* commentary (written in 1330-32).[73] The first thesis runs

> Nulla est substantia proprie loquendo.

The line of argument is based on the idea of subsistence conveyed by the noun 'substantia'. His main argument is to the effect that there is no reason why, speaking of a piece of wood and its properties, you should prefer maintaining that its so-called accidents by nature are dependent on the log's nature to saying that the log itself by nature is dependent upon the things that are actually united with it:

> Si dicitur quod natura ligni ab aliquibus rebus sibi unitis non dependet, immo [*ideo* Hoffmann] respectu illarum dicitur substantia, et illae res respectu illius dicuntur accidentia, contra: Sicut natura ligni ab aliquibus accidentibus non dependet sed potest naturaliter esse sine illis, ita aliquae res unitae naturae ligni quae sunt accidentia naturae ligni, possunt esse naturaliter sine aliquibus rebus quae sibi uniuntur, sicut illa accidentia sine quibus non possit esse naturaliter natura ligni. Igitur illae res possunt dici substantiae eadem ratione qua et natura ligni.[74]

---

73) My quotations are from the edition of Book I by Hoffmann (1988).

74) Crathorn, *In I Sententiarum*, ed. Hoffmann (1988), q. 13, 392, 16-23.

Our attachment of either the name 'substance' or 'accident' to things hinges on our observing a thing's effects. When we infer from a thing's effects that its subsistent nature is more perfect than its coincidental one, and that qua substance it is less dependent on its sensible qualities than the other way round, we call it a 'substance'. That is why names such as 'man', 'stone', or 'animal', which belong to the category of substance, are assigned to such things which are perceived as less dependent than the things united with it:

> Ideo videtur mihi quod nulla res corruptibilis potest proprie dici substantia, nec aliqua res cui alia non unitur. Quantocumque sic non dependens est proprie substantia, tamen quia ex effectibus arguimus aliquam rem esse unitam istis qualitatibus sensibilibus perfectiorem et minus dependentem a qualitatibus sensibilibus quam econverso, tali rei imponimus hoc nomen 'substantia'; et talibus rebus specie distinctis, quam distinctionem arguimus vel ex figura distincta accidentium vel ex distinctis accidentibus vel ex distinctis operationibus, imponimus talia nomina distincta 'terra', 'aer', 'ignis', 'aqua', 'lignum', 'lapis' et consimilia. Licet igitur nulla res corruptibilis possit proprie dici substantia considerando modum significandi istius termini 'substantia', imponitur tamen iste terminus ad supponendum pro rebus talibus. Est igitur substantia rerum sibi invicem unitarum res perfectior non dependens vel minus dependens a rebus sibi unitis quae ab illa naturaliter dependent. Et isto modo accipiendo istum terminum 'substantia' est unum decem praedicamentorum. Et omnes illi termini dicuntur esse in praedicamento substantiae qui sunt isto termino 'substantia' minus communes et de quibus significative acceptis primo modo dicendi per se iste terminus 'substantia' praedicatur vel potest praedicari. Per quales terminos convenienter respondetur ad quaestionem factam 'per quid' de substantia. Quales termini sunt isti: 'homo', 'lapis', 'animal' et similes.[75]

Three more theses argued for by Crathorn are devoted to the distinction 'substance-accident'. The fifth thesis explicitly claims that distinguishing between substance and accident is all a matter of perspective. One and numerically the same thing can be called substance or accident at one's own discretion. Once again, the subsistence criterium is decisive. Crathorn refers to the phenomenon of blazing iron (*ferrum ignitum*). Both iron and fire are substances and can be each other's substrate. So if you like, you can speak of *ignis ferreus*, meaning *iron fire*, i.e., 'fire enmattered in iron' (note that, unlike English, Latin has the adjective noun *ferreus*). Clearly, the substance fire is in the iron as its substrate,

---

[75] Crathorn, *In I Sententiarum*, ed. Hoffmann (1988), q. 13, 392, 24-393, 8. In his commentary on the *Categories*, Thomas Maulevelt entertained (probably in the 1320s and 1330s in Paris) a similar radical thesis about *substantia* to the effect indeed that we have no need to posit the real existence of substance, the only category required to describe the outside things being that of *qualitas*. See Andrews (2008).

and thus satisfies the definitional criterium for being an accident. Therefore properly speaking the terms 'substantia' and 'accidens' are interchangeable:

> Quinta conclusio est quod aliqua eadem res numero respectu diversarum rerum potest vere dici substantia et accidens, et quod una substantia potest vere dici subiectum alterius. Istud patet de ferro ignito, ubi ferrum est subiectum ignis et ignis accidens respectu ferri. Sed ferrum est substantia et ignis in ferro est substantia. Igitur una substantia potest dici accidens et vere esse accidens respectu alterius, et alia substantia illius subiectum. Quod ignis sit in ferro sicut in subiecto, probatio quia: Illud est in aliquo sicut in subiecto quod est in eo non sicut pars et impossibile est esse sine eo in quo est; et hoc naturaliter loquendo, ita quod naturaliter dependet ab eo in quo est, nec potest naturaliter sine eo esse, licet non econverso.[76] Sed sic est ignis in ferro. Igitur ignis est in ferro sicut in subiecto.[77]

Replying to a possible objection, Crathorn has the opportunity to display his atomistic ontology. If someone says that the fire is in the iron in the peculiar manner in which a form informs a substrate—as, e.g., Ockham still holds—, Crathorn replies that a sophisticated distinction between *esse in subiecto* and *informare subiectum* is pointless. The manner in which the adherents (like Ockham) themselves take 'being informative' as an extra condition over and above merely 'being present' inevitably evokes a contradiction, which can only be avoided by recognizing that it is an empty device:

> Si dicitur quod ignis non est in ferro informative [which would be required for its supposed subsistent nature], istud non satisfacit, quia ad intellectum illorum qui utuntur illo termino 'informative', ille terminus est [*read*, proves to be] signum fictum et nullum signatum sibi correspondet.[78] Credunt enim quod *albedinem informare parietem* sit aliquid aliud quam *albedinem esse in pariete*, et quod aliquid aliud requiratur in pariete ad hoc quod paries sit albus praeter parietem et albedinem et existentiam albedinis in pariete. Quod apparet esse falsum ex hoc quod implicat contradictionem albedinem esse in pariete et parietem non esse album.[79]

Contemporaneous philosophers and others before Crathorn regarded nature itself as organized along the lines of a categorial arrangement and, therefore,

---

76) As is clear from the good luck the so-called 'Three Youths' Shadrach, Meshach, and Abednego miraculously had, after having been cast on the command of King Nebuchadnezzar into the midst of the fiery furnace; see Daniel 3:19-25.

77) Crathorn, *In I Sententiarum*, ed. Hoffmann (1988), q. 13, 394, 4-9.

78) Meaning 'a fancy sign with no thing indicated'. Or should we read *significatum*? In the latter case these people's offence against the principle of parsimony is more conspicuous.

79) Crathorn, *In I Sententiarum*, ed. Hoffmann (1988), q. 13, 394, 18-26.

took our mind's ability to use it as a reliable means to obtain knowledge of the outside things just as they are in themselves, quite apart from our way of conceiving them. In full accordance with Aristotle, they used the categorial scheme flexibly enough, allowing them to focus upon the objects of investigation according to their essential or coincidental properties at their own discretion, and, on occasion (for the sake, that is, of the discussion at hand) to categorize subsistent things after one of their coincidental features. However, in the outside world, some things, factually and on their own, were substances, others accidents. Period.

Crathorn, on the other hand, abandons Aristotelian metaphysics by considering the substance-accident scheme no more than a linguistic convention evoked by sensorial perception. In fact, unlike Ockham, Crathorn is not so much engaged in a radical reduction of the number of the categories. Rather it is the categorial arrangement itself as representative of the (putatively) parallel categorial ordering of nature which has come under attack. Accordingly, Crathorn replaces the Aristotelian categorial arrangement of nature with an atomistic configuration, with which we become acquainted without any speculation about its mysterious character. Nature is surely accessible to the senses, and sensorial cognition can be reflected upon, but cognition should be shielded from interpretive overkill ('Hineininterpretieren') by the intellect. The substance-accident scheme has itself become a matter of ambivalence and scope distinction.

3.2 *The Aftermath*

What does this mean for the parallellism paradigm with its impact on the medieval cognitional procedure, including the terminist approach? The answer is predictable, I am afraid. Terminism and supposition theory lost their influence on mainstream Western philosophy, but were able to keep up their position in various (neo)scholastic systems, and could also maintain their influence in theological discussions. A special revival was the share of medieval logic in the past century. And in the wake of this revival, medieval fallacy theory too enjoyed some fresh interest.

A final remark on the continuity of mainstream philosophical thought. No doubt, in our circles, so seriously and successfully interested in medieval thought, there is no room for rude and glaring misunderstandings about its historical position. We medievalists are fully aware of the drastic epistemological turn which came about in the fourteenth and fifteenth centuries and, among other things, was conducive for the rise of empiricism. I recall once again Crathorn's highlighting sensation at the cost of rational speculation, such as investigating the form's informative activity beyond its pure presence in things.

There is an interesting parallel in the seventeenth century, when David Hume frequently displayed his radical dislike for speculation surrounding superfluous entities. It is found in *An Enquiry Concerning Human Understanding*, section 7. In sections 7 and 8[80] Hume discusses causation, focusing in particular on the cause's inherent power or force that his predecessors had assumed in order to adequately explain causality. Hume rejects this idea. We can discover causes by experience, he admits, but we do not experience any power in the agent. His favourite example for making this clear is the impact between two colliding billiard balls. As one ball rolls towards the other, we observe motion, then contact, hear a smack, and perceive the second ball speeding away from the first. However, our outward senses do not experience any internal power or force residing in the first ball. Likewise, that the motion of our limbs follows the will's command is a matter of common experience. But the power or energy by which this is effected is unknown and inconceivable, Hume claims (section 7, 14-15). I am sure that Crathorn's undermining of the 'substance-accident' paradigm would have been music to Hume's ears.

Elsewhere, in Section 6 of *A Treatise of Human Nature*, he asks those philosophers who imagine that we have clear ideas of substance and accident, whether the idea of substance is derived from the impressions of sensation or of reflection. If it is conveyed to us by the senses, he asks by which of them. If by the eyes, it must be a colour, if by the ears, a sound, and so on and so forth. Nobody will assert such things. On the other hand, the impressions of reflection resolve themselves into our passions and emotions, which cannot possibly represent a substance either. We have therefore no idea of substance, distinct from that of a collection of particular qualities, Hume concludes.

While not arguing that there is an obvious continuity of philosophical thought in this respect, let alone any familiarity on Hume's part with his fellow-countryman, Father Crathorn, the historian of philosophy can draw attention to medieval signs of a critical attitude towards the sensorial and intellectual faculties, long before the seventeenth-century breakthrough of empiricism with its abhorrence of unwarranted speculation. However, while its ongoing critical attitude towards unwarranted speculations somehow anticipated

80) Pp. 134-147 (*Of the Idea of Necessary Connexion*) and 148-164 (*Of Liberty and Necessity*) (1999).

modern tendencies, terminist methodology continues to serve some momentous post-medieval metaphysical systems.[81]

3.3 *A Summary*

The following statements can summarize our findings:

[1] The medieval theories of the properties of terms, supposition in particular, and medieval terminism in general developed around 1100 from a doctrine concerning the unmasking of fallacious arguments through grammatico-logical analysis.

[2] Unlike fallacies, as a root of terminism, grammatico-logical analysis continued to play a prominent role in its later developments.

[3] Like any philosophical method, the methodology of terminism had everything to do with a philosopher's basic attitude towards reality, and, accordingly, its diverse developmental directions mirrored diverse philosophical positions, varying from realism to nominalism/conceptualism, to each of which it was equally hospitable.

[4] The basic ancient and medieval philosophical positions were all governed by the paradigm of the parallellism (supposedly existing) between Reality and our cognitive faculties (for Aristotle and the Medievals, particularly the scheme of the ten categories).

[5] There are, consequently, unmistakable parallells between the Aristotelian and the medieval strategies of argument, which, however different they sometimes seemed, both served much the same philosophical positions.

[6] The nucleus of the Aristotelian strategy of argument, which consists in focalization and correct categorization, is also found in terminism, both of which are based on the conviction that any proper discussion requires highlighting the object's (whether essential or coincidental) ontic features that are suited to the investigation or discussion at hand.

[7] Taking advantage of the ambivalent semantic area of the key focal terms of the discussion *is* one of the main characteristics of terminism.

---

81) As a student of Aristotelico-Thomistic philosophy in the early 1940s at the archiepiscopal seminary of Utrecht I made my first acquaintance with supposition theory. In a few months the extensive treatment of the properties of terms in Jacques Maritain (1930) enabled me to make exercises on the most profound subtleties of the theory, including such a complex phenomenon as the '*supposition proprement dite, personelle, essentielle, universelle distributive, confuse mobile*'.

[8] The medieval debates on metaphysical issues and on intentionality, as well as sophisticated disputes about the significative force of words occurring in special contexts, gave rise to many semantic refinements, such as the issues of semantic stratification and time connotation.

[9] The entire development of terminism can also be assessed in the broader context of the post-medieval rise of empiricism, including an ongoing critical attitude towards rational speculation. As far as the Middle Ages are concerned, this attitude came to the fore in reducing the number of the categories regarded as ontologically representative (with Ockham among others), or even in giving up the idea of ontological representativeness altogether (with Crathorn).

Bibliography

Primary Literature

Anonymus, *Commentary on the Perihermeneias* in ms Orléans, *Bibliothèque Municipale*, cod. 266, ff. 257b-263b (unedited), cited in C.H. Kneepkens, 'Aristotle's Transposition and the Twelfth-Century Commentaries on *De interpretatione* 20 b 1-12. An Exploratory Study', in: H.A.G. Braakhuis and C.H. Kneepkens, eds., *Aristotle's* Perihermeneias *in the Latin Middle Ages. Essays on the Commentary Tradition* (Artistarium, Supplementa, X; Groningen-Haren 2003, 363-411)

——, *Textus et copulata omnium tractatuum Petri Hispani, etiam parvorum logicalium et tractatus syncathegorematum* [...], *iterum atque iterum diligentissime correcta secundum doctrinam irrefragabilem divi Thomae Aquinatis ac iuxta frequens exercitium magistrorum Coloniae infra sedecim domos in bursa Montis regentium in hunc unum librum congesta* (Coloniae 1493)

Guillelmus de Crathorn, *In I Sententiarum*: F. Hoffmann, *Quästionen zum ersten Sentenzenbuch. Einführung und Text* (Beiträge zur Geschichte der Philosophie und Theologie des Mittelalters. Texte und Untersuchungen. Neue Folge, 29; Münster 1988)

Guillelmus de Ockham, *Expositio in librum Perihermenias Aristotelis*, ed. A. Gambatese et S. Brown, in: *Guillelmi de Ockham expositionis in libros artis logicae prooemium et expositio in librum Porphyrii de praedicabilibus*, ed. E.A. Moody; *Expositio in librum Praedicamentorum Aristotelis*, ed. G. Gal; *Expositio in librum Perihermenias Aristotelis*, ed. A. Gambatese et S. Brown; *Tractatus de praedestinatione et de praescientia Dei respectu futurorum contingentium*, ed. Ph. Boehner, recognovit S. Brown (Guillelmi de Ockham Opera Philosophica, II; St. Bonaventure, N.Y. 1978, 341-504)

——, *Quodlibeta septem*, ed. J.C. Wey (Guillelmi de Ockham Opera Theologica, IX; St. Bonaventure, N.Y. 1980)

——, *In librum secundum Sententiarum, Reportatio*, ed. G. Gál and R. Wood (Guillelmi de Ockham Opera Theologica, V; St. Bonaventure, N.Y. 1981)

Henrico de Gandavo adscripta *Syncategoremata*, ed. H.A.G. Braakhuis, G. Etzkorn and G.A. Wilson (Ancient and Medieval Philosophy Series-2, no. 37; Leuven 2011)

Hume, David, *A Treatise of Human Nature*, ed. L.A. Selby-Bigge, revised by P.H. Nidditch (Oxford 1981)

——, *An Enquiry Concerning Human Understanding*, ed. T.L. Beauchamp (Oxford Philosophical Texts; Oxford 1999)

Johannes Buridanus, *In Metaphysicen Aristotelis. Quaestiones argutissimae Magistri Ioannis Buridani in ultima praelectione ab ipso recognitae* (Parisius 1518; repr. Frankfurt am Main 1964)

Johannes Wyclif, *Tractatus de logica*, ed. M.H. Dziewicki, 3 vols. (Wyclif's Latin Works 14, London 1893-1899)

Mattheus Aurelianensis: Matthew of Orleans, *Sophistaria sive Summa communium distinctionum circa sophismata accidentium* edited with an Introduction, Notes and Indices by J. Spruyt (Studien und Texte zur Geistesgeschichte des Mittelalters, LXXIV; Leiden 2001)

Petrus Abaelardus, *Ethica sive Scito teipsum*, in: D.E. Luscombe, *Peter Abelard's Ethics. An Edition with Introduction, English Translation and Notes* (Oxford Medieval Texts; Oxford 1971)

Proclus, *Elementatio theologica*: E.R. Dodds, *The Elements of Theology*. A Revised Text with Translation, Introduction and Commentary (Oxford [2]1963)

——, *In Parmenidem commentarius*, ed. V. Cousin (Paris 1864)

Secondary Literature

Andrews, R. (2008), 'Thomas Maulevelt's Denial of Substance', in: L.A. Newton, ed., *Medieval Commentaries on Aristotle's* Categories (Leiden 2008, 347-368)

Bäck, A.T. (2000), *Aristotle's Theory of Predication* (Philosophia antiqua. A Series of Studies on Ancient Philosophy, LXXXIV; Leiden 2000)

——, (2003), 'Aquinas on Predication', in: H.A.G. Braakhuis and C.H. Kneepkens, eds., *Aristotle's* Perihermeneias *in the Latin Middle Ages. Essays on the Commentary Tradition* (Artistarium, Supplementa, X; Groningen-Haren 2003, 321-338)

Bakker, P.J.J.M. (2001), 'Aristotelian Metaphysics and Eucharistic Theology: John Buridan and Marsilius of Inghen on the Ontological Status of Accidental Being', in: J.H.M.M. Thijssen and J. Zupko, eds., *The Metaphysics and Natural Philosophy of John Buridan* (Medieval and Early Modern Science, 2; Leiden-Boston 2001, 247-264)

Boehner, P. (1952), *Medieval Logic. An Outline of its Development from 1250 to c. 1400* (Manchester 1952)

Hoffmann, F. (1988), see *Guillelmus de Crathorn*

Kahn, C.H. (1973), *The Verb 'Be' in Ancient Greek* (Foundations of Language. Supplementary Series, 16; Dordrecht 1973)

Kneepkens, C.H. (2003), 'Aristotle's Transposition and the Twelfth-Century Commentaries on *De interpretatione* 20 b 1-12. An Exploratory Study', in: H.A.G. Braakhuis and C.H. Kneepkens, eds., *Aristotle's* Perihermeneias *in the Latin Middle Ages. Essays on the Commentary Tradition* (Artistarium, Supplementa, X; Groningen-Haren 2003, 363-411)

Loux, M.J., ed., (1979), *The Possible and the Actual. Readings in the Metaphysics of Modality* (Ithaca and London 1979)

Maritain, J. (1930), *Eléments de philosophie*. Fascicule I: *L'ordre des concepts. Logique* (Paris 1930)

Moody, E.A. (1935), *The Logic of William of Ockham* (New York 1935; anastatic repr. 1965)

Pinborg, J. (1980), 'Radulphus Brito on Universals', in: *Cahiers de l'institut du moyen âge grec et latin* 35 (1980), 56-142

Rijk, L.M. de (1962-1967), *Logica Modernorum. A Contribution to the History of Early Terminist Logic. I. On the Twelfth Century Theories of Fallacy*; II/1: *The Origin and the Early Development of the Theory of Supposition* II/2: *Texts and Indices* (Assen 1962-1967)

——, (1981), 'Abaelard's Semantic Views in Light of Later Developments', in: H.A.G. Braakhuis, C.H. Kneepkens and L.M. de Rijk, eds. (1981), *English Logic and Semantics from the End of the Twelfth Century to the Time of Ockham and Burleigh.* Acts of the 4th European Symposium on Medieval Logic and Semantics, Leiden-Nijmegen, 23-27 April 1977 (Artistarium, Supplementa, I, Nijmegen 1981, 1-58 [also in E.P. Bos, ed. (1989): L.M. de Rijk, *Through Language to Reality. Studies in Medieval Semantics and Metaphysics* (Variorum Reprints; Northampton 1989)

——, (1985), *La philosophie au moyen âge.* Traduit de néerlandais par P. Swiggers (Leiden 1985)

——, (1986a), 'Peter Abelard's Semantics and his Doctrine of Being', in: *Vivarium* 24 (1986), 85-127

——, (1986b), 'Abelard and Moral Philosophy', in: *Medioevo* XII (1986), 1-27

——, (1992), 'Causation and Participation in Proclus. The Pivotal Role of Scope Distinction', in: E.P. Bos and P.A. Meijer, eds., *On Proclus and his Influence in Medieval Philosophy* (Philosophia Antiqua. A Series of Studies on Ancient Philosophy, LIII, Leiden 1992, 1-34)

——, (1993), 'On Buridan's View of Accidental Being', in: E.P. Bos and H.A. Krop, eds. (1993), *John Buridan: A Master of Arts. Some Aspects of his Philosophy.* Acts of the Second Symposium Organized by the Dutch Society for Medieval Philosophy Medium Aevum (Nijmegen 1993, 41-51)

——, (1994), 'John Buridan on Man's Capability of Grasping the Truth', in: Andreas Speer, ed., Scientia *und* Ars *im Hoch- und Spätmittelalter* (Miscellanea Mediaevalia, 22; Berlin-New York 1994, 282-303)

——, (1997), 'Foi chrétienne et savoir humain. La lutte de Buridan contre les theologizantes', in: A. de Libera, A. Elamrani-Jamal, A. Galonnier, eds., *Langage et philosophie. Hommage à Jean Jolivet* (Etudes de philosophie médiévale, 74; Paris 1997, 393-409)

——, (2000), '*Logica Morelli.* Some Notes on the Semantics of a Fifteenth Century Spanish Logic', in: I. Angelelli and P. Perez-Ilzarbe, eds. (2000), *Medieval and Renaissance Logic in Spain.* Acts of the 12th European Symposium on Medieval Logic and Semantics, held at the University of Navarra, Pamplona, 26-30 May 1997 (Philosophische Texte und Studien, 54; Hildesheim-Zürich-New York 2000, 209-224)

——, (2002), *Aristotle. Semantics and Ontology,* I: *General Introduction. The Works on Logic*; vol. II: The *Metaphysics. Semantics in Aristotle's Strategy of Argument* (Philosophia Antiqua. A Series of Studies on Ancient Philosophy, XCI/II; Leiden 2002)

——, (2005), *Geraldi Odonis Opera Philosophica,* II: De intentionibus. *Critical Edition with a Study on the Medieval Intentionality Debate up to ca. 1350* (Studien und Texte zur Geistesgeschichte des Mittelalters, LXXXVI; Leiden 2003)

——, (2008), *Johannes Buridanus,* Lectura Erfordiensis in I-VI Metaphysicam *together with the 15th-century* Abbreviatio Caminensis. *Introduction, Critical Edition and Indexes* (Studia Artistarum. Etudes sur la Faculté des arts dans les Universités médiévales, 16; Turnhout 2008)

Spruyt, J. (2001), *Matthew of Orléans, Sophistaria sive Summa communium distinctionum circa sophismata accidentium,* Edited with an Introduction, Notes and Indices (Studien und Texte zur Geistesgeschichte des Mittelalters, LXXIV; Leiden 2001)

——, (2008), 'The Unity of Semantics and Ontology. Wyclif's Treatment of the *fallacia accidentis*', in: *Vivarium* 46 (2008), 24-58

Tabarroni, A. (2003), 'John Buridan and Marsilius of Inghen on the Meaning of Accidental Terms (*Quaestiones super Metaphysicam* VIII, 3-5)', in: *Documenti e Studi sulla tradizione filosofica medievale* XIV (2003), 264-408

Vendler, Z. (1967), *Linguistics in Philosophy* (Ithaca N.Y. 1967)

Zupko J. (1998), 'Sacred Doctrine, Secular Practice: Theology and Philosophy in the Faculty of Arts in Paris, 1325-1400', in: J.A. Aertsen and A. Speer, eds., *Was is Philosophie im Mittelalter?* (Miscellanea Medievalia, 26; Berlin 1998, 656-666)

——, (2001), 'On Certitude', in: Thijssen, J.M.M.H. and J. Zupko, eds., *The Metaphysics and Natural Philosophy of John Buridan* (Medieval and Early Modern Science, 2; Leiden-Boston 2001, 65-182)

Early Supposition Theory II

Sten Ebbesen
University of Copenhagen

Abstract

In 1981 I published an article called *Early Supposition Theory*. Then as now, the magisterial work on the subject was L.M. de Rijk's *Logica Modernorum*, and then as now any discussion of the topic would have to rely to a great extent on the texts published there. This means that many of the problems that existed then still remain, but a couple of important new studies and several new texts have been published in the meantime, so it may be time to try to take stock of the situation. I will first look at the origin of the term *suppositio* and then at the chronology of our source texts.

Keywords

supposition, appellation, *causa apparentiae, causa non existentiae*

1. Whence suppositio?

In 1981 I tried to weaken L.M. de Rijk's case for supposition in the logical sense being derived from Priscianic grammar, and more specifically his claim that in Priscian *suppositum* means 'grammatical subject'. I think I was reasonably successful on that score. I did not, however, deny that twelfth-century grammarians' use of *suppositum* was relevant, or that De Rijk's 'put as a subject' was a good translation of their *supponere*, but I suggested that a common idea underlay the grammatical and the logical use of *suppositum*, namely that the *suppositum* is or is claimed to be the bearer of a certain form: in the case of grammar the subject would then be called *suppositum* because it is claimed to be the bearer of the form indicated by the predicate; in the case of logic, the *supposita* of *homo* would be the individuals that bear the form of humanity, and which might be the bearers of whichever form is predicated of them in a sencence with *homo* for its subject. This idea of mine was not based on much hard evidence, but I continue to cherish it somehow.

In 1987, however, Kneepkens with his usual meticulous care argued forcefully for the view that the logicians' use of *supponere* was developed from the grammatical use of *supponere verbo* with an understood *personam*, and that ultimately the grammarians' usage should be traced back to their mullings over a passage in Priscian[1] containing the word *suppositum*.[2] But he also demonstrated that the *suppositum–appositum* analysis of sentences is not as old as we had previously thought. One of the key passages in Peter Helias turned out to be a later interpolation, and generally speaking, the *suppositum–appositum* analysis only becomes prominent some time after the middle of the twelfth century. There remained a couple of places in which Peter used *supponere* in a relevant way, and, following a suggestion of Pinborg's,[3] Kneepkens proposed that Peter had borrowed the terminology from Gilbert the Porretan. De Libera's paper for the 1987 symposium added more information about the Porretan trail, and more recently, Valente has further investigated that part of the history of supposition.[4]

While Kneepkens was not very keen on my idea that the key idea is that something is the bearer of a form, his suggested connection to the Porretans was, in fact, grist to my mill. To the Porretans the metaphysics of form and bearer is quite central, and predicates introduce a form—substantial, accidental or individual—for the subject to bear.[5]

1) Priscianus, *Institutiones grammaticee*, ed. Hertz (1855-1859), XVII, 3, 23.

2) See Kneepkens (1987), esp. pp. 341-342.

3) Pinborg (1968) and (1972), 47-49. See also Nielsen (1982), 105.

4) De Libera (1987), 455; Valente (2008), esp. 275 ff. See also Valente's contribution to this volume.

5) See, e.g., *Compendium logicae Porretanum* I, 23, ed. Ebbesen *et al.* (1983), 10-11. Also note the use of *suppositum* in I. 20, 9: 'Ratio cur dicatur demonstrationem cum nomine substantivo fungi loco proprii. Cum enim pronomen demonstrativum certum significet suppositum, ex vi demonstrationis determinat ipsum imitatione accidentium. Nomen vero substantivum adiunctum substantialem determinat proprietatem. Cum ergo sic discrete significat suppositum accidentialibus et substantialibus <proprietatibus> determinatum, quid amplius proprium nomen posset efficere?' See also III. 29, 52. Further Valente (2008a), 288 with footnote 41, in which she quotes a passage from Langton's commentary on the *Sentences*: 'QUIDAM TAMEN. Magister Gilebertus, quia omnis appositio formae est, et suppositio substantiae, et ideo haec vera 'tres personae sunt unus deus', i.e., unius deitatis. Ex parte vero suppositi vellet hoc nomen 'deus' supponere pro persona, et ideo hanc {*sc.* 'unus deus est tres personae'} dixit esse falsam sicut hanc 'una persona est tres personae'.' Note that when quoting Latin texts, edited and unedited alike, I impose my own orthography and punctuation.

As also stressed by Kneepkens, it is a striking feature of grammatical texts as well as of the logical ones from the twelfth and early thirteenth centuries that *supposita* are usually not linguistic items, but such extra-linguistic entities as the predication is about, usually signified and/or named by a grammatical subject term, but occasionally, as a *suppositum locutioni*, by other means, as for example the preposition *de* + the ablative in an utterance of the form *hoc dicitur de hominibus.* It is also a fact that,until well into the thirteenth century logicians—at least continental ones[6]—mainly speak about suppositing as something subject-terms do, and thus it can be a matter of discussion, sometimes, whether *homo supponit Ciceronem* or *supponit pro Cicerone* means 'the word *homo* stands for Cicero' or 'the word *homo* introduces Cicero as the subject of the verb of the sentence'. Usually it does not matter which interpretation you choose. To take an example from John Pagus:[7]

> Videtur quod terminus communis supponens verbo de praesenti non coartetur ad entia sed indifferenter supponit pro entibus et non entibus.

Here the common term *supponit verbo*, i.e., provides the verb with a subject, and the same term *supponit indifferenter pro entibus et non entibus.* The obvious translation in the context is:

> It seems that a common term which introduces something to function as the subject of a verb of the present tense is not restricted to existing things, but performs its subject-introduction on behalf of existing and non-existing things indiscriminately.

In another context, however, we might have rendered *supponit pro* as 'stands for' without bothering about the things being stood for having the role of subject. Only rarely are predicate terms said to supposit, and, it seems, mainly in England, but that seems to be a secondary extension of the use of *supponere*.[8]

---

6) Cf. De Libera (1982b), 176.

7) Johannes Pagus, *Appellationes*, ed. De Libera (1984), 224. I have changed the edition's *supponitur* into *supponit.*

8) For an example, see the anonymous mid-thirteenth-century commentator on the *Prior Analytics* in ms Cambridge, *Peterhouse* 206, f. 100va: 'Sed adhuc dubium est de suppositione subiecti et praedicati in talibus 'tantum homo currit'. Et potest dici quod ratione exponentis negativae stat subiectum confuse tantum et praedicatum confuse et distributive. In hac enim, sc. 'aliud quam homo currit' stat li 'homo' confuse et distributive per virtutem de li 'aliud', et praedicatum stat determinate, quia negatio de li 'aliud' semper sistit circa materialem compositionem, unde idem est dicere 'aliud quam homo' et 'aliquid quod non est homo', et ita negatio de le 'aliud' non

I would now like to point to the use of *supponere* in a couple of somewhat different contexts that have not been mentioned so far in the debate about the origin of supposition theory, at least not as far as I remember.

1. One of the earliest (ca. 1090) commentaries on Boethius' *De topicis differentiis*, the *'Primum oportet'*, talks about *voces suppositae homini* without specifying whether the author is speaking of tokens of 'this man' or of proper names. In view of later developments in medieval logic, his use of *suppositae* deserves to be noticed.[9] I am not sure which sort of relation he thought obtains between the general term and its subordinate words—perhaps one of containing, but it is not easy to assign a precise sense to the container-contained metaphor as used by this author (whose name may have been Arnulphus); *inter alia*, he claims that the thing signified by a proper name is contained under the name.[10]

2. The earliest Latin commentary on the *Prior Analytics* (by Anonymus Aurelianensis III) can with some probability be assigned a date between 1160 and 1180.[11] The work knows nothing like supposition theory, not even under the

---

attingit praedicatum. Praeposita igitur negatione sic 'nihil aliud quam homo currit' mutatur suppositio subiecti in confusam tantum et suppositio praedicati in confusam et distributivam.'

9) *Commentarium in Boethii De topicis differentiis*, ed. Hansen (2005), 67: " 'risibile' quidem aequale est homini tantundem quantum et 'homo' significando, sed tamen seiunctum est a ratione substantiae hominis, id est a proprietate significationis vocum suppositarum homini, per hoc videlicet quod illae significant in eo quod quid, 'risibile' autem in quale." Cf. p. 63: 'id est: hanc vocem quae est 'quaestio', quae dicitur principium quantum ad voces sibi suppositas', and p. 112: 'Nota quod totum ut genus ab integro differat toto, in eo videlicet quod se sibi suppositis omni modo tribuit, integrum vero totum non omnino se suis attribuit partibus. Potes enim dicere: 'Homo est animal' et 'Homo est substantia animata sensibilis', sed non recte dices: 'Paries est domus' nec 'Paries est constans ex pariete et tecto et fundamento'.'

10) *Commentarium in Boethii De topicis differentiis*, ed. Hansen (2005), 87-88: 'QUIA OMNE ARTIFICIUM DISSERENDI CONTINETUR QUATTUOR FACULTATIBUS. Quasi dicat: Ideo dicendum est quae argumenta admittant sibi suas facultates, quia quattuor tantum facultates comprehendunt omnem locum et omnem syllogismum. Vel ad illud potest esse causae redditio quod dixit: quae facultas quibus uti noverit argumentis. Per 'QUATTUOR FACULTATES' habes idem quod per 'dialecticam, rhetoricam, philosophicam, sophisticam doctrinam', per OMNE vero ARTIFICIUM omnem locum et omnes syllogismos, harum videlicet quattuor facultatum significata. Differt autem ars et artifex et artificium; nam ipsae doctrinae quibus aliqua docemur dicuntur ars, artifex vero qui per eas aliquid agit, artificium vero omnis argumentatio. Quod autem dicit omne artificium quattuor facultatibus contineri ita accipe ut significata in suis significantibus continentur; omne enim significans suum continet significatum; ut res significata ab hoc nomine 'Lungomarius' continetur infra idem nomen, sic et in aliis.'

11) For this commentary, see S. Ebbesen, 'Analyzing Syllogisms *or* Anonymus Aurelianensis III—the (presumably) Earliest Extant Latin Commentary on the *Prior Analytics*, and its Greek Model', in: *Cahiers de l'institut du moyen âge grec et latin* 37 (1981), 1-20 (rep. in Ebbesen (2008)). Yukio

guise of appellation, but the author does use *supponi* in a way that seems relevant to our topic, because he repeatedly uses it to say that some item is subsumed under or falls under another.[12]

In one place he wants to prove the validity of the syllogism:

> That every man is an animal is necessary,
> but that some body is not a man is necessary,
> therefore that some body is not an animal is necessary.

The way to prove it, he says, is so take something which is *suppositum minori extremo*, that is 'body', yet such that the middle term 'animal' can be universally denied of it. A stone fits the bill, and then you argue as follows:

> That every man is an animal is necessary,
> but that no stone is an animal is necessary,
> therefore that no stone is a man is necessary.

From which it follows that by necessity some body is not a man.

Iwakuma (Fukui Prefectural University) later did a preliminary transcription of the full text, and now Christina Thomsen Thörnqvist is preparing an edition. I am grateful to both of them for sharing their materials with me.

[12] Anonymus Aurelianensis III, *Commentarium in Analytica priora*, ms Orléans *Bibliothèque municipale* 283, 188A: 'SED NECESSE <30 a 9, *Aristoteles latinus* III, 20, 24-25>. Quasi: Non possunt probari per impossibile hii syllogismi de necessario, sed probantur per demonstrationem de exposito hoc modo: Proposito priori syllogismo de necesse in quarto modo secundae, causa expositionis ponatur aliquod *suppositum* minori extremitati a quo medium possit removeri universaliter, et fiat de illo *supposito* syllogismus in eadem figura et ceteris terminis eisdem, qui fiebat de minori extremo. Ut verbi gratia, cum hic sit quartus secundae 'Omnem hominem esse animal est necesse, sed quoddam corpus non esse animal est necesse, igitur quoddam corpus non esse hominem est necesse', probetur hoc modo: Ponatur causa manifestationis aliquod *suppositum* minori extremo quod est 'corpus', sed tale a quo medium possit universaliter removeri, velut 'lapis', et dicatur 'Omne animal esse hominem est necesse, et nullum lapidem esse animal est necesse, ergo nullum lapidem esse hominem <est necesse>', per quod demonstratur quoddam corpus non esse hominem necessario, quoniam hoc corpus, scil. lapis, ex necessitate non est homo.' Here, and in several similar cases, the author is clearly talking about what falls under a term. Less clear 184B: 'PER OSTENSIONEM <6.28 a 23>. Demonstratio per ostensionem dicitur cum ad probandum quod dixeras inducis *singularem suppositionem* rei quam primo per universale *supposueras*. Si ambigas ad probandum quod ex eis duabus 'omnis homo est animal' et 'omnis homo est risibilis' sequitur 'quoddam risibile est animal', inducas quod si omnis homo est animal, et omnis homo est risibilis, necessario hic homo Socrates erit simul et animal et risibile; ergo cum idem nunc sit hoc et illud, necessario quoddam animal est illud, et e converso. Quae demonstratio ostensio vocatur, quoniam quod dictum est generaliter *per suppositionem singularem* semper melius aperitur.'

This tallies very nicely with what Lambert of Lagny, Ligny or Auxerre says about the middle of the thirteenth century:[13]

> Dicuntur vero supposita quia supponuntur sive subiciuntur suis superioribus.

and is not very far from William of Sherwood's declaration that[14]

> Suppositio autem est ordinatio alicuius intellectus sub alio.

Although, as Kneepkens pointed out in 1987, medieval authors could, when necessary, keep various uses of technical terms apart, I am very tempted to think that the anonymous' use of *supponere* has to be added to the number of uses that influenced the use of the word in what became supposition theory.

At the same seventh European symposium at which Kneepkens presented his paper about supposition, I gave one about the theologian Stephen Langton,[15] who, I had recently discovered, had developed a fairly complex theological theory of supposition in the 1180s-90s, with a distinction between *suppositio essentialis* and *suppositio personalis* at its centre. I wondered aloud whether this meant that the logical distinction between *simplex* and *personalis* had its origin in theology. If this were so, the logical use of the notion of *suppositio* might be as late as the 90s, or possibly even later, depending on how many of De Rijk's early dates of logical treatises could be raised, and by how much. Of course, if *simplex* and *personalis* were artists' creations from the 70s or early 80s, Langton might have been inspired by the artists.

Langton's semantics and sentence analysis is very much influenced by that of the Porretans. I shall not catalogue the similarities, but just point to two important features of his theory, which both point back to the Porretans and forward to summulistic treatments of supposition. First, although he does not use the definition, for Langton supposition is definitely a *substantiva rei designatio*, as some logicians were to say. Only substantive nouns and substantivated adjectives supposit. When the talk is about created things, verbs predicate; but

---

13) Lambertus de Lagny, *De Appellatione*, ed. De Libera (1981), 254-255: 'Dicuntur autem appellata eo quod appellantur sive nominantur a suis superioribus. Superiora enim de suis inferioribus praedicantur secundum nomen et secundum rationem. [...] Dicuntur vero supposita quia supponuntur sive subiciuntur suis superioribus, et dicuntur singularia eo quod nominant aliquid discretum et individuum quod uni singulariter convenit.'

14) William of Sherwood, *Introductiones in Logicam* V, ed. Brands and Kann (1995), 132.

15) Ebbesen (1987). Pinborg (1968) had already pointed to Langton's pupil, Andrew Sunesen, without, however, knowing that Andrew was dependent on Langton.

when it is about God, they couple—this to avoid introducing Aristotelian categories in propositions about God.[16] However, this divine coupling is the divine analogue of attributing a form to the subject, which is the job of ordinary *copulatio* in logic books.[17]

As shown by De Rijk in 1967, *appellare* and *appellatio* competed to some extent with *supponere* and *suppositio* in the works of early logicians, *appellatio* probably being the older term. In fact, it now seems reasonably certain that there was a time when only *appellatio* was a fully developed technical term in logic. Unlike *supponere, appellare* cannot be used to say 'provide a subject for the verb', but it shares with *supponere* the ability to indicate descent to something within the range of a term's signification. In some 'classical' thirteenth-century authors it comes to be reserved for the relationship between a term and presently existing items signified by it, but the wider use was not soon forgotten. John Pagus' *Appellationes* from the 1230s[18] is about what we would call supposition, not about appellation in the narrow sense, and the same holds for Lambert's *De appellatione* from about the middle of the century. Notice his explanation of *appellata and supposita*:[19]

> Dicuntur autem appellata eo quod appellantur sive nominantur a suis superioribus. Superiora enim de suis inferioribus praedicantur secundum nomen et secundum rationem. [...] Dicuntur vero supposita quia supponuntur sive subiciuntur suis superioribus.

Both designations are explained in terms of a superior–subordinate relationship, and Lambert simply takes *appellata* and *supposita* to be extensionally

---

16) Actually, Langton is not consistent in avoiding *praedicare* when talking about the divine, whereas his pupil Andrew Sunesen is very consistent. See Ebbesen (1987). NB: Whereas verbs cannot supposit in Langton's and Sunesen's theory, nouns can both supposit and couple.

17) Already *Ars Meliduna*, ms Oxford, *Bodleian Library, Digby* 174, f. 217vb: 'Nos recipimus in his omnibus extensionem fieri appellationis, sicut et in nominibus illis quae substantiales vel naturales copulant proprietates.'

18) De Libera (1984), 193, follows Chenu in assigning a date of about 1230, but this presupposes that John's logical works were all written before he began to study theology. Heine Hansen, who is preparing an edition of John's commentary on the *Categories*, has pointed out to me that the commentary contains a number of references to theological authors, which suggests it was composed after John had commenced his study of theology. Assuming that he continued to teach the arts during his first years as a student of theology, we gain a wider span of time within which his logical works may have been written, roughly 1231-1241.

19) Lambert de Lagny, *De appellatione*, ed. De Libera (1981), 254-255. For the date, which is far from securely established, see De Libera (1981b).

equivalent. Only as an afterthought does he mention the newer, more restricted, use of *appellare*:[20]

> Properly speaking, however, only actually existing things are called appellata [...] and so it is correct when people say that appellation is supposition for existing things.

About 1240 Robert Kilwardby still calls the two rules that a verb of past tense ampliates the subject to past things and one of future tense to future things *regulae appellationum*, though he phrases them in *suppositio*-language. Thus the one about ampliation to the past runs:[21]

> Terminus communis supponens verbo de praeterito potest supponere pro hiis qui sunt vel pro hiis qui fuerunt.

Elsewhere, though, he refers to the same rules under the name of *regulae suppositionum*.[22]

A similar use of *regula appellationum* appears in the *Elenchi*-commentary of Anonymus Monacensis, which probably dates from the second quarter of the thirteenth century.[23] The indiscriminate use of *appellare* and *supponere* only seems to disappear after the middle of the thirteenth century.

---

20) Lambert de Lagny, *De appellatione*, ed. De Libera (1981), 255, continuation of the quotation above: 'Sciendum autem quod proprie loquendo non dicuntur appellata nisi sint actualiter existentia, appellatur enim proprie quod est et non quod non est, et ideo bene dicitur quod appellatio est pro existentibus suppositio.'

21) Robertus Kilwardby, *Commentum in Analytica Priora*, in: ms Cambridge, *Peterhouse* 205, ff. 88vb-89ra: 'Et potest dici quod duae priores instantiae multiplices sunt [secundum] per *regulas appellationum*. Haec enim 'nullus senex erit puer' multiplex est ex eo quod hoc subiectum 'senex' potest stare pro sene qui est vel qui erit. Si pro sene qui est, sic est sensus 'nullus senex qui est erit puer', sic est vera, et sic convertitur, et hoc modo est sensus 'nullus puer erit senex qui est'. Si pro sene qui erit, sic est sensus 'nullus senex qui erit erit puer', et sic est falsa et potest converti. [...] Similiter dicendum est de hac instantia 'nullus puer fuit senex' per *illam regulam appellationum*: Terminus communis *supponens* verbo de praeterito potest *supponere* pro hiis qui sunt vel pro hiis qui fuerunt.'

22) Robertus Kilwardby, *Commentum in Sophisticos Elenchos*, mss Cambridge, *Peterhouse* 205, f. 335rb and Paris, *BnF. lat.* 16619, f. 62vb: 'Quaeritur etiam *de duabus regulis suppositionum* quae iam positae sunt, sc. quod terminus communis non restrictus etc. *supponens* verbo de praeterito potest *supponere* pro hiis quae sunt vel pro hiis quae fuerunt, similiter terminus communis *supponens* verbo de futuro potest *supponere* pro hiis quae sunt vel {vel: et *CP*} erunt.'

23) Anonymus Monacensis, *Commentum in Sophisticos Elenchos*, mss Admont, *Stiftsbibliothek* 241, f. 17vb, and München, *Bayerische Staatsbibliothek, clm* 14246, f. 8rb: 'Sed contra. In appellationibus habemus regulam hanc quod terminus communis non habens vim ampliandi etc. supponens

2. Problems of Chronology

De Rijk in the 1960s tried to impose some chronological order on the mass of undated texts with which he was dealing. While some of his results stand, others do not. His methodology was all right, because it was and is the only one we have for such tasks. He relied to some extent on the date of manuscripts to establish *termini ante quos*—a text must have been composed no later than the time it was entered into an existing manuscript. The problem with this approach is, of course, that dating manuscripts is still a sub-scientific art. We are waiting for some method from the natural sciences that will allow us establish the year the animal was felled that provided the raw material for the parchment. That will give us a secure *terminus post quem* for the execution of the manuscript, and a probable *terminus ante quem*, since we may assume that most parchment was used within a decade of its production, I believe. Stocking such a precious commodity for years instead of buying just what you need here and now would appear to be bad economy.

Next, De Rijk tried to anchor his chronology by attributing particular works to particular persons whose careers were somewhat known. That yielded a few fixed points to be used in connection with the third part of his work.

The third task was to establish a relative chronology of the texts, based on the tacit assumption that there would be a linear development of doctrine. Again, he was perfectly aware that doctrinal development may not always be perfectly linear, if for nothing else, because even if the development did proceed linearly in each and every sub-branch of the big intellectual community, there might be a different pace in the several sub-branches. Toulouse, for example, might need a couple of decades to get abreast of new developments in Paris. But rarely was it possible for him to establish with certainty the place of origin of a relevant text.

People who are not trained as historians or philologists tend to brush aside the problems involved in dating, and simply accept what the most authoritative historian or philologist says. In this case it means that very few outside the circle of the European Symposia know how fragile the chronology is, and which of De Rijk's assumptions have been supported or undermined by later research.

---

verbo de praesenti non habenti vim ampliandi restringitur ad supponendum pro eis quae sunt sive ad praesentes; ergo cum dicitur 'laborans sanus est', ille terminus 'laborans' pro praesentibus solum supponit, et ita non habebit duo tempora.' See context in Ebbesen (1997), 149.

One of the anchors of the chronology was Guillelmus Arnaldi's commentary on Peter of Spain's *Summulae*, which, supposedly, could not be later than 1248, whence a date in the 1230s was a reasonable estimate for the composition of the *Summulae*, so much the more as De Rijk also found an anonymous commentary, the *'Cum a facilioribus'*, which seemed to be earlier than William's, and thus could hardly be later than ca. 1240.[24] I expressed my doubts about the date of Guillelmus Arnaldi in 1970, because the format of his commentary appeared to me too similar to works from the 1270s.[25] Nobody noticed a young scholar's squeak. Some more—but not all—noticed when some years later Gauthier demonstrated beyond reasonable doubt that the Arnaldi in question was not the one De Rijk had assumed, but someone who worked in the 1290s.[26] Unfortunately, that left us with no interesting *terminus ante quem* for Peter's *Summulae* and the commentaries on it, and we cannot even use any knowledge about Peter's life to date his work, for Angel d'Ors has taught us not to identify the author with Pope John XXI or some Portuguese scholar who may or may not have been identical with the pope. D'Ors has proposed other possible life-stories for our author, but none that carries total conviction.[27] Consequently, we simply have neither a *terminus ante quem* nor a *terminus post quem* for Peter of Spain's *Summulae.* The 1230s are still possible, but so are the 40s, even the 50s—or the 20s.

A second anchor for chronology was the theological *Fallacies* of Master William, and so far it stays in its anchor position. De Rijk's proposal to identify the author with Willelmus de Montibus has been accepted by Iwakuma, and, as such things go, it may be considered fairly safe. But this almost certainly means that the work was composed after 1186 when William began to teach theology in Lincoln.[28] As a mode of the fallacy of figure of speech William mentions univocation, which is *eiusdem dictionis in eadem significatione et terminatione varia appellatio*, and he also mentions that the appellation may be restricted or ampliated. His theological examples do not offer themselves very easily to analysis by means of standard supposition rules, but there can be little doubt that he knew a secular logic that called variation of appellation 'univocation' and put it under figure of speech. William's definition of univocation is very

[24] De Rijk (1970), 17-18.

[25] Ebbesen and Pinborg (1970), 44 n.

[26] Thomae de Aquino, *Expositio libri Peryermeneias* (1989), 69*-72*.

[27] D'Ors (1997), (2001), (2003).

[28] Iwakuma (1993), 1-4. In ms Cambridge, *Jesus College* Q.B.17, William's *Fallaciae* occurs together with theological works by Willelmus de Montibus.

close to that found in *Tractatus de univocatione Monacensis*,[29] which, of course, is nice for the relative chronologist.

A third anchor is *Ars Meliduna* because of a reference to King Louis of France and an unnamed bad king of England. Louis, de Rijk realized, must be Louis VII, who unfortunately reigned for an intolerably long time (1137-1180). The uncomplimentary reference to the king of England, however, suggests a date after the early 1150s.[30] Anyway, we have 1180 as a reasonably certain *terminus ante quem*. Now, the author of the *Ars* does not speak of *suppositio*, but of *appellatio*, and he does have rules about restriction and ampliation of appellation, and indicates that there was some discussion about the matter, so that he cannot have been the first to introduce the subject. However, he only has a very rudimentary terminology for types of appellation: thus a term may be put or taken *confuse* or *discrete*, and occasionally the notion of appellation is introduced in that context, as when he says:[31]

> ibi ponitur nomen confuse, id est non pro aliquo suorum appellatorum.

On one occasion, at least, de Rijk put the *Ars Meliduna* as early as the middle of the twelfth century,[32] but I believe most scholars would now agree that ca. 1175 is a safer guess.

Other of de Rijk's suggested attributions of works to definite persons may be considered obsolete. Anonymus Digbeianus' mutilated commentary on the *Elenchi* cannot plausibly be attributed to Edmund of Abingdon, who, according to Roger Bacon, was the first to lecture on the book in Oxford, and the *Abstractiones* of master Richard cannot plausibly be attributed to Richard Fishacre.[33] Nor can *Summae Metenses* be considered the work of an early

---

29) Iwakuma (1993), 3.

30) De Rijk (1962-1967), II/1, 280-281.

31) *Ars Meliduna*, ms Oxford, *Bodleian Library, Digby* 174, f. 218va. Cf. f. 225ra: 'Ad id etiam improbandum sufficit quod iste terminus 'coloratum hac albedine' nihil discrete supponit, unde potius quoddam commune significat quam singulare.' *Poni* or *accipi confuse* vs. *infinite* occurs in several places. F. 227va: 'Quae vero unum terminorum sumit discrete, alterum communiter, a communi denominabitur, ut 'Socrates vel asinus currit' indefinita est.'

32) De Rijk (1982), 165.

33) As done by De Rijk in his *Logica modernorum* (1962-1967), II/1, 72-74. The two identifications were linked to each other. Ms Oxford, *Bodleian Library, Digby* 24 contains both texts. Having identified Richard as Richard Fishacre, De Rijk proposed to identify Anonymus Digbeianus = SE59 in Ebbesen (1993) with Edmund, because he and Fishacre had been in contact. I believe Anonymus Digbeianus' commentary is no earlier than the middle of the thirteenth century.

thirteenth-century Nicholas of Metz; it is much more likely by a mid-century Nicholas of Paris.[34]

Though it could not be linked to any definite person, the *Dialectica Monacensis* became another important anchor for the chronology of supposition, because it has a fairly well-developed account of the matter, and it was placed in the 1170s by de Rijk.[35] So, the more primitive stages of supposition- or appellation theory probably lay in the preceding decades. However, Braakhuis, myself and Iwakuma soon came to suspect that the date was too early by some decades. Braakhuis pushed it towards the end of the twelfth or very early thirteenth century. I myself inclined towards a date close to 1220, and Iwakuma in 1993 took a sort of middle position: '1190s if not later'.[36]

Among my reasons for wanting a late date is the occurrence in the treatise on fallacies of the doctrine of causes of appearance and non-existence or falsity, which does not occur in any work surely dated to the twelfth century, but became a standard item in thirteenth-century theory of fallacies.[37]

It is notable that the same manuscript that transmits the *Dialectica Monacensis* contains another set of treatises, which I shall call *Tractatus Monacenses*, written in the same hand, and sharing numerous traits of formulation and doctrine with the *Dialectica*, both in the field of supposition and in having the notions of *causa apparentiae et non existentiae* for the fallacies.[38] The two works must come from the same environment and be approximately contemporaneous. Interestingly, and disconcertingly, the *Tractatus Monacenses* uses the river Elbe instead of the standard Parisian Seine in the example *Albea currit, ergo habet pedes*.[39] De Rijk had claimed the *Dialectica Monacensis* for an Englishman with contacts in Chartres and Paris, but on extremely slender grounds.[40] The occurrence of the Elbe in a related text suggests that we should consider the northern part of the Holy Roman Empire as a possible place of origin.

---

34) De Rijk (1962-1967), II/1, 450-452; Braakhuis (1979), I, 317-328.

35) De Rijk (1962-1967), II/1, 414.

36) Braakhuis (1979), 1, 427, n. 12; Iwakuma (1993), 4, n. 16.

37) A thirteenth-century date also makes the quotation of *Liber de Sex Principiis* mentioned by De Rijk (1962-1967), II/1, 410 unproblematic. For the use of *causa apparentiae & non-existentiae*, see the table in the appendix.

38) *Tractatus Monacensis* occupy ff. 121r-141r of ms München, *Bayerische Staatsbibliothek, clm* 14763, in which *Dialectica Monacensis* occupies ff. 89r-121r.

39) Anonymus, *Tractatus Monacensis, Fallaciae*, ms München, *Bayerische Staatsbibliothek, clm* 14763, f. 123vb.

40) De Rijk (1962-1967), II/1, 414.

Moving the *Dialectica Monacensis* up in time has serious consequences for the birthdate of supposition theory. It was supposed to be a very early instance of a fairly developed lore of supposition. A late date not only saves us from believing that its author was precocious, it also makes it slightly less of a mystery why the *Summa 'In omni doctrina'*[41] totally ignores supposition, although it would seem to belong somewhere in the first half of the thirteenth century.[42]

Conclusion

A host of questions concerning the dates of the relevant texts remain unresolved, but this is what I think the available evidence points to at this moment:

The main outlines of the story about supposition remain as in 1967, but the dates change. First, the birth of supposition theory took place in the very late twelfth century. The first signs of what was to come appear in the 1170s, but in logic centered round the notion of appellation, while supposition was becoming a key notion in theology.[43] A stage with a fairly developed terminology for types of supposition is not reached till about the 1190s, when also *suppositio* begins to outmanœuvre *appellatio*, though this was to be a slow process. The majority of our early texts that teach or employ supposition, English and continental alike, were composed in the thirteenth century.

41) Anonymus, *Summa 'In omni doctrina'*, ed. Bos (2001).

42) Bos (2001), 6, proposes a date between 1200 and 1220, but I am afraid that is too early. There are references to the *Posterior Analytics* in II/1.1.1, 95, and II/1.1.5, 97; and to *Physics* II in III. 0, p. 134. In II/1.1, 85, we find 'Nullus enim artifex probat sua principia', which seems to indicate a date when both *Posterior Analytics* and *Physics* I were commonly read. Cf. Thomas Aquinas, *In Sententiarum* I, q. 1, a. 3: 'sicut nec aliquis artifex potest probare sua principia', Boethius of Dacia, *Quaestiones super libros Physicorum* ed. Sajó (1954). I. 12, 152-154 'Quaeritur utrum aliquis artifex possit probare sua principia si sibi negantur. [...] Item, nullus artifex potest probare aliquid contra illum qui nihil sibi concedit'. The debate in Boethius (and others from the second half of the century) is linked to Averroes' discussion in his commentary on *Physics* I, comm. 8.

43) I am inclined to think that the *Summa Zwettlensis* is a work from about the 1170s. Häring's date 'before 1150' rests on his very doubtful attribution of the work to one Peter of Poitiers/Vienna. See Valente (2008a), 25. If I am right, the *Summa* is approximately contemporary with Peter of Poitiers' *Sententiae*, in which *supponere* is used in a relevant way, but without any developed system of types of supposition.

Finally, I think that although some authors may have had very clear ideas of which of the many uses of *supponere* was relevant in each particular context, they would generally be influenced both by the grammatical 'putting as a subject'-tradition, the logical one of saying that what may be subsumed under a term *supponitur* under it, and the metaphysical thesis that bearers of forms *supponuntur* under their forms.

Appendix

The following table lists a number of commentaries on the *Sophistici Elenchi* and treatises on fallacies, whether separate of parts of *summulae*. Column 2 gives the number the work has in the list of texts on fallacies in Ebbesen (1993). Column 3 offers my best guess at a date. Column 4 registers whether the work uses either of the terms *appellatio* and *suppositio* in the technical sense. Column 5 whether the text lists univocation as a type of the fallacy of figure of speech (*figura dictionis*). Column 6 whether, in the description of figure of speech, specific types of supposition, such as *confuse* and *determinate*, are referred to. Column 7 whether the text assigns a *causa apparentiae* (= *principium motivum)* and a *causa non-existentiae* (= *causa* or *principium falsitatis* or *defectus*) to the several fallacies.

Among other things, the table shows that having univocation as a type of figure of speech is restricted to a very tiny group of texts, which may, therefore, be assumed to be roughly contemporary.

A	= *appellatio*
S	= *suppositio*
S/A	= both *suppositio* and *appellatio* used
(A)	= a single relevant use of *appellatio* occurs
c. a. / non-e. & fals.	= *causa apparentiae* and both *causa non-existentiae* and *causa falsitatis* occur
p. mot.	= *principium motivum*

1. Name	2. SE	3. Date	4. *suppositio / appellatio*	5. *univocatio under f.d.*	6. *confuse & sim. in f.d.*	7. *causa aparentia[e] / non -existentia[e]*
Glosae SE	5	1140-60	—	—	—	—
Summa SE	6	1140-60	—	—	—	—
Anonymus Parisiensis	8	1140-70	— (*A*)	—	—	—
Anon. Aur. I	13	1160-80	—	—	—	—
Anon. Aur. II	14	1160-80	—	—	—	—
Anon. Cantabr.	15	1160-90	—	—	—	—
Fallaciae Vindobonenses	16	1160-90	— (*A*)	—	—	—
Introductiones Parisienses[44]	19	1190-1210	*S*	—	—	—
Fallaciae M. Willelmi	20	1186-1200	*A*	+	—	—
Fallaciae Parvipontanae	17	1190-1210[45]	*A /S*	(—)[46]	—	—
Fallaciae Londinenses	18	1190-1210	*S*	—	+	—
Fallaciae Lemovicenses	23	1190-1210	*S/A*	+	+	—
Dialectica Monac.	27	1200-20	*S*	+	+	*c. a. / non-e & fals.*
Tractatus Monac.	28	1200-20	*S*	+	+	*c. a./ non-e & fals.*
Summa 'In omni doctrina'	29	?	—	—	—	*c. a. / fals.*

[44] Dated ca. 1170 by De Rijk (1962-1967), II/1, 447, but on the slenderest of grounds (including an invalid argument from the way 'Socrates' is abbreviated ms Paris, *BnF. lat.* 15170). There is a fairly developed system of types of supposition, which is distinguished from appellation in the way that many thirteenth-century authors do. The fallacy of figure of speech 'provenit ex variata suppositione vel ex variato modo supponendi vel copulandi', which is close to the formulations used by *Fallaciae Lemovicenses* and *Dialectica Monacensis* (see Ebbesen and Iwakuma (1993), 28 with references in footnote).

[45] De Rijk (1962-1967), I, 152 says 'Internal evidence makes me date this work in the last decades of the twelfth century.' He does not, however, say what the internal evidence is.

[46] Mentioned but rejected.

Table (*cont.*)

1. Name	2. SE	3. Date	4. *suppositio / appellatio*	5. *univocatio under f.d.*	6. *confuse & sim. in f.d.*	7. *causa aparentiae / non -existentiae*
Introductiones Antiquae	30	1220-60	*S*[47]	—	—	*c. a. / fals.*
Petrus Hisp., *Summulae*	32	1220-50	*S*	—	(—)[48]	*p. mot. / defectus*
Anon. Monac., *Comm. SE*	34	1230-50	*S/(A)*	—	+	*p. mot.*
Grosseteste (?) *Comm. SE*	31	1230s?	— *(?)*	—	—	*c. a. / non-e.*
Fallaciae ad modum Oxoniae	33	1220-60	*S*	—	+	*c. a. / non-e.*
Sherwood, *Introd.*	36	1230s?	*S*	—	—	*c. a./non-e.*
Kilwardby, *Comm. SE*	35	ca. 1240	*S*	—	+	*c. a. / non-e.*
Nicolaus Parisiensis, *Summae Metenses*	42	1240-60	*S*	+	+	*c.a./non-e & fals.*
Nicolaus Parisiensis, *Comm. SE*	41	1240-60	*S*	—	—	*p. mot. & c. a. / defectus*
Ripoll Compendium	40	1240-60	*S*	—	—	*p. mot.*
Bacon, *Summulae*	43	1250-55	*S*	—	+	*c. a./non-e.*
Robertus, *Comm. SE*	45	1250-70	*S*	—	+	*c.a.*
Robertus de e Aucumpno, *Comm. SE*	48	1250-70	*S*	—	+	*c. a. / non-e.*

47) Not in section on fallacies.

48) Mentioned but rejected.

Bibliography

Primary Literature

Anonymus, *Ars Meliduna*, in: ms. Oxford, *Bodleian Library, Digby* 174, ff. 211ra-241vb

——, *Commentary on the Prior Analytic*, in: ms. Cambridge, *Peterhouse* 206, ff. 13va-21rb

——, *Tractatus Monacenses* in: ms München, *Bayerische Staatsbibliothek, clm* 14763, ff. 121r-141r, ed. L.M. de Rijk (1962-1967), II, 2, 453-638

——, *Dialectica Monacensis*, in: ms München, *Bayerische Staatsbibliothek, clm* 14763, ff. 89r-121r, ed. L.M. de Rijk (1962-1967), II, 2, 453-638

——, *Summa 'In omni doctrina'*, ed. E.P. Bos, in: id., *The Summa 'In omni doctrina'* (Philosophes Médiévaux, 43; Louvain-la-Neuve and Louvain-Paris 2001)

——, *Commentarium in Boethii De topicis differentiis*, ed. H. Hansen (2005), in: *Cahiers de l'institut du moyen âge grec et latin* 76 (2005), 45-130

——, *Compendium logicae Porretanum*, ed. S. Ebbesen, K.M. Fredborg and L.O. Nielsen, in: id., '*Compendium logicae Porretanum* ex codice Oxoniensi Collegii Corporis Christi 250. A Manual of Porretan Doctrine by a Pupil of Gilbert's', in: *Cahiers de l'institut du moyen âge grec et latin* 46 (1983), iii-xviii + 1-113

Anonymus Monacensis, *Commentum in Sophisticos Elenchos*, mss Admont, *Stiftsbibliothek* 241, ff. 1ra-96rb, and München, *Bayerische Staatsbibliothek, clm* 14246, ff. 1ra-48vb

Anonymus Aurelianensis III, ed. S. Ebbesen (1981), in id., 'Analyzing Syllogisms *or* Anonymus Aurelianensis III—the (presumably) Earliest Extant Latin Commentary on the *Prior Analytics*, and its Greek Model', in: *Cahiers de l'institut du moyen âge grec et latin* 37 (1981), 1-20 (repr. in Ebbesen (2008))

Boethius of Dacia, *Quaestiones super libros Physicorum*. ed. G. Sajó, in: id., *Boethii Daci Opera. Quaestiones de Generatione et corruptione—Quaestiones super libros Physicorum*, ed. G. Sajó, V, pars II. *Quaestiones super libros Physicorum continens* (Corpus Philosophorum Danicorum Medii Aevi, 5,2; Hauniae 1974)

Guillelmus de Shireswood: William of Sherwood, *Introductiones in Logicam. Einführung in die Logik*, ed. H. Brands and C. Kann (Hamburg 1995)

Johannes Pagus, *Appellationes*, ed. A. de Libera (1984), in: id., 'Les *Appellationes* de Jean le Page', in: *Archives d' histoire doctrinale et littéraire du moyen-âge* 51 (1984), 193-255

Lambertus Autissiodorensis: Lambert of Lagny, *De appellatione*. ed. A. de Libera, in: id. 'Le *Tractatus de appellatione* de Lambert de Lagny (*Summa Lamberti* VIII)', in: *Archives d'histoire doctrinale et littéraire du moyen âge* 48 (1981), 227-285

Priscianus, *Institutiones grammaticae*, ed. M. Hertz, in: *Prisciani grammatici Caesariensis Institutionum grammaticorum libri XVIII*, 1-II (Grammatici Latini ex recensione Henrici Keilii II-III; Lipsiae 1855-1859, repr. Hildesheim 1961)

Robertus Kilwardby, *Commentum in Sophisticos Elenchos*, mss Cambridge, *Peterhouse* 205, ff. 277ra-344vb and Paris, *BnF. lat.* 16619, f. 1ra-72vb

——, *Commentum in Analytica Priora*, in: ms Cambridge, *Peterhouse* 205, ff. 85ra-135rb

Thomas Aquinas, *Scriptum super Sententiis*, ed. P. Mandonnet and M.F. Moos (Paris 1929-1956)

——, *Expositio libri Peryhermeneias*, ed. Leonina (Rome-Paris 1989)

Secondary Literature

Braakhuis, H.A.G. (1979), *De 13de eeuwse tractaten over syncategorematische termen: Inleidende studie en uitgave van Nicolaas van Parijs' Sincategoreumata*. 2 vols. (Dissertation Leiden, Nijmegen 1979)

Ebbesen, S. (1981), 'Early Supposition Theory (12th-13th cent.)', in: *Histoire épistémologie langage* 3, 35-48 (repr. in Ebbesen (2009))

——, (1987), *The Semantics of the Trinity according to Stephen Langton and Andrew Sunesen*, in: J. Jolivet and A. de Libera, eds., *Gilbert de Poitiers et ses contemporains aux origines de la* Logica Modernorum. Actes du septième symposium européen d'histoire de la logique et de la sémantique médiévales, Poitiers 17-22 Juin 1985 (Napoli 1987, 401-435)

——, (1993b), 'Medieval Latin Glosses and Commentaries on Aristotelian Logical Texts of the Twelfth and Thirteenth Centuries', in: *Glosses and Commentaries on Aristoteian Logical Texts. The Syriac, Arabic and medieval Latin Traditions*, edited by Ch. Burnett (The Warburg Institute, University of London, London 1993, 129-177)

——, (1997), 'Texts on Equivocation', in: *Cahiers de l'institut du moyen âge grec et latin* 67 (1997), 127-199

——, (2008), *Greek-Latin Philosophical Interaction: Collected Essays of Sten Ebbesen*, I (Aldershot 2008)

——, (2009), *Topics in Latin Philosophy from the 12th-14th centuries: Collected Essays of Sten Ebbesen*, II (Aldershot 2009)

Ebbesen, S. and J. Pinborg (1970), 'Studies in the Logical Writings Attributed to Boethius de Dacia', in: *Cahiers de l'institut du moye-âge grec et latin* 3 (1970), 1-54

Ebbesen, S. and Y. Iwakuma (1993), 'Fallaciae Lemovicenses', in *Cahiers de l'institut du moyen âge grec et latin* 63 (1993), 3-42

Iwakuma, Y. (1993), 'The *Fallaciae* and *Loci* of William de Montibus. An Edition', in: *Journal of Fukui Prefectural University* 2 (1993), 1-44

Kneepkens, C.H. (1987), '*Suppositio* and *Supponere* in 12th-Century Grammar', in J. Jolivet et A. de Libera, eds., *Gilbert de Poitiers et ses contemporains aux origines de la* Logica Modernorum. Actes du septième symposium européen d'histoire de la logique et de la sémantique médiévales, Poitiers 17-22 Juin 1985 (Napoli 1987, 325-351)

Libera, A. de (1982), 'The Oxford and Paris Traditions in Logic', in: N. Kretzmann, A. Kenny and J. Pinborg, eds., E. Stump, ass. ed., *The Cambridge History of Later Medieval Philosophy. From the Rediscovery of Aristotle to the Disintegration of Scholasticism 1100-1600* (Cambridge-London-New York-New Rochelle-Melbourne-Sydney 1982, 174-187)

——, (1987), 'Logique et théologie dans la *Summa Quoniam homines* d'Alain de Lille', in: J. Jolivet and A. de Libera, eds., *Gilbert de Poitiers et ses contemporains aux origines de la* Logica Modernorum. Actes du septième symposium européen d'histoire de la logique et de la sémantique médiévales, Poitiers 17-22 Juin 1985 (Napoli 1987, 437-469)

Nielsen, L.O. (1982), *Theology and Philosophy in the Twelfth Century. A Study of Gilbert Porreta's Thinking and the Theological Expositions of the Doctrine of the Incarnation during the Period 1130 1180* (Acta Theologica Danica, 15; Leiden 1982)

Ors, A. d' (1997), 'Petrus Hispanus O.P., Auctor Summularum', in: *Vivarium* 35 (1997), 21-71

——, (2001), 'Petrus Hispanus O.P., Auctor Summularum (II): Further Documents and Problems', in: *Vivarium* 39 (2001), 209-254

——, (2003), 'Petrus Hispanus, O.P., Auctor Summularum (III): "Petrus Alfonsi" or "Petrus Ferrandi"?', in: *Vivarium* 41 (2003), 249-303

Pinborg, J. (1968), 'Review of L.M. de Rijk, *Logica Modernorum* II', in: *Vivarium* 6 (1968), 155-158

——, (1972), *Logik und Semantik im Mittelalter. Ein Überblick*, mit einem Nachwort von H. Kohlenberger (Stuttgart-Bad Cannstatt 1972)

Rijk, L.M. de (1962-1967), *Logica Modernorum. A Contribution to the History of Early Terminist Logic*. I. On the Twelfth Century Theories of Fallacy; II/1: *The Origin and the Early Development of the Theory of Supposition* II/2: *Texts and Indices* (Assen 1962-1967)

——, (1970), 'On the Genuine Text of Peter of Spain's *Summule*, part V. Some Anonymous Commentaries on The *Summule* Dating From The Thirteenth Century', in: *Vivarium* 8 (1970), 10-55

——, (1982), 'The Origins of the Theory of the Properties of Terms', in: A. Kenny, N. Kretzmann, I. Pinborg, eds., E. Stump, ass. ed., *The Cambridge History of Later Medieval Philosophy. From the Rediscovery of Aristotle to the Desintegration of Scholasticism 1100-1600* (Cambridge-London-New York-New Rochelle-Melbourne-Sydney 1982, 161-173)

Valente, L. (2008), *Logique et théologie. Les écoles parisiennes entre 1150 et 1220* (Sic et non; Paris 2008)

Postscript

After this article was handed in for publication I have discovered evidence that Anonymus Cantabrigiensis' commentary on *Sophistici Elenchi* must have been composed no earlier than 1204, not betwen 1160 and 1190, as proposed in the table on p. 74.

The edition of Pagus on the *Categories* referred to as forthcoming in footnote 18 has now appeared. See H. Hansen, *John Pagus on Aristotle's Categories. A Study and Edition of the Rationes super Praedicamenta Aristotelis*, Ancient and Medieval Philosophy, De Wulf-Mansion Centre, Series I, XIV, Leuven University Press 1912.

ARABIC PHILOSOPHY

Avicenna's Theory of Supposition

Allan Bäck
Kutztown University

Abstract
Although he does not have an explicit theory of supposition as is found in the works of Latin medieval philosophers, Avicenna (Ibn Sīnā) has two doctrines giving something equivalent: the threefold distinction of quiddity (*triplex status naturae*), corresponding to a division of simple, personal and material supposition, and his analyses of truth conditions for categorical propositions, where sentential context determines in part the reference of their terms. While he does address which individuals are being referred to by the universal terms used there, Avicenna concentrates more on the varied temporal durations of the predications being made. In Western terms, he has incorporated ampliation and restriction into the theory of supposition itself.

Keywords
threefold distinction of quiddity, *triplex status naturae*, paronymy, *dictum*, intention, ampliation, restriction, *status*, predication, supposition: simple, personal, material

I claim that Avicenna (Ibn Sīnā) has what amounts to a theory of supposition.[1] I am not claiming that this theory is the very one to be found in the Latin medieval West, nor that it was ever called that. Yet it does have strong similarities to it, indeed an isomorphism with it—not so surprisingly given their common roots in Greek philosophy. To be sure, these clusters of doctrines, Arabic and Latin, have their differences in content and in emphasis. Yet, so too, Latin supposition theory proper has its different versions with different logicians concentrating on different problems.

In short, the basic distinctions of Latin medieval supposition theory—the material and the formal, with the latter divided into the simple and the personal—correspond to Avicenna's threefold distinction of quiddity. The

[1] This differs from a common current view. E.g., Street (2004), 249: '[...] no doctrine like supposition was developed [...]'.

divisions of personal supposition, into the confused, determinate etc., have parallels in Avicenna's theory of the various ways in which the simple assertion signifies. However, Avicenna has quite a different emphasis. The divisions of personal supposition tend to focus on the descent to singulars: how universal terms in a given propositional context refer to individual objects. To be sure, Avicenna recognizes such a descent to singulars but does not discuss it much. Perhaps he finds it just obvious—or, as it is an inference, something to be dealt with in his syllogistic and dialectic, as he does somewhat. In contrast, he concentrates on the temporal duration of the existence of things signified by the terms in a proposition. In Latin medieval terms, he is focusing on ampliation and restriction—but not in the usual way. For he locates, quite plausibly, different terms signifying various temporal durations of things even within statements of present tense. We moderns can construe this doctrine in terms of reference: Avicenna is dealing with how the subject and predicate terms refer to various temporal stages or time-slices of the substances or objects being described.

Avicenna's theory differs from the Latin medieval theories. On the one hand, I am not urging a return to Carl Prantl's doctrine that Latin medieval logic had no originality but merely transmitted Greek and Arabic doctrine.[2] Latin supposition theory did not arise merely via translation from the Arabic. To start with, the Latin philosophers had no access to the bulk of Avicenna's logical writings, as far as I know; the *Avicenna Latinus* has as its *Logica* only Avicenna's commentary of Porphyry's *Eisagoge*. On the other hand, I am not urging a return to the classicist doctrine, that Arabic philosophy, including its logic, has no originality but only copied and transmitted inaccurately and slavishly received Greek doctrines.[3] What I am urging lies elsewhere and mostly in the details, to which I shall now turn.

Standard Supposition Theory

Talking of a standard Latin medieval theory of supposition invites ready refutation. Supposition theory has its versions. Given that it concerns what terms

---

[2] Prantl (1855), 263 ff.

[3] Elamrani-Jamal (1983), 10-11, notes (and protests against) this view that there is no Islamic logic but only a slavish repetition of the Greek corpus by a few Muslims. So says, e.g., De Boer (1967), 23.

'stand for' or 'suppose for', it is not ontologically neutral.[4] Thus nominalists like Buridan and Ockham tend to reject, or parse away, simple supposition, as it posits real things other than singulars. Realists like Burley support it.

Still, we have the textbooks, like those of William of Sherwood, Lambert of Auxerre, and Peter of Spain (non-papa).[5] They in turn were using and revising the distinctions and doctrines of supposition in earlier anonymous treatises like the *Dialectica Monacensis*. From these works there came to be a fairly standard, basic tradition of the types and features of supposition. Even those who differed from it tended to start from it, comment upon it, and critique it. For my purposes, to start with, it suffices to use just the basic distinctions taught and commented upon in the universities.[6]

My textbook division of supposition follows William of Sherwood most closely.[7] He divides supposition into material and formal, and then formal into simple and personal. A term has material supposition when it stands for the term itself and formal supposition when it stands for what it signifies. A term has simple supposition when it stands just for what it signifies and personal supposition when it stands for 'its inferiors', namely the things falling under what it signifies.[8] As the *Dialectica Monacensis* said earlier, a term has personal supposition when it is taken for its inferiors, and has simple supposition when the term is taken for something common and not for any of its inferiors, in

[4] I am aware that Dutilh Novaes (2007), 7, 18ff., claims that supposition theory is not a theory of reference, in a modern, technical sense: giving syntactic rules for how signs refer to objects. Here, though, I am taking supposition theory to be a theory of reference in her general sense, p. 29: how 'words can stand for things. [...]' (Why not just take supposition to concern Frege's UF and signification Frege's UO relation?—but I won't press the issue here). Cf. Spade (1988a), 212; Spade (2002), 243-245; Read (2008).

[5] Cf. Peter of Spain, *Tractatus*, ed. De Rijk (1972), lxxviii-lxxix.

[6] On the other hand, if I will use a textbook tradition for the Latin medievals, shouldn't I use the textbooks of Avicenna, or the *Shamsiyya* by al-Kātibī, used for teaching in *Sunnī madrasa*, the schools attached to the mosques? Cf. Street (2004), 254-255.

[7] See Peter of Spain, *Tractatus*, ed. De Rijk (1972), lxxvii-lxxxiii for useful charts giving their divisions. These authors in turn were using and revising the distinctions and doctrines of supposition in anonymous treatises like the *Dialectica Monacensis*. Cf. ed. De Rijk, *Logica Modernorum* (1962-1967), II/2, 409; 446-448; 456; 584-586.

[8] William of Sherwood, *Introductiones in Logicam*, ed. Brands and Kann (1995), 270, n. 174, says that the difference between personal and material supposition is rather like use and mention, but differs: no mentioned expression can stand for itself, but not so for a term in material supposition. One finds an important contribution on Sherwood's theory of supposition in Braakhuis (1977). So too Walter Burley, *De puritate artis logicae tractatus longior*, transl. Spade, I.1.1 (9): proper supposition is material or formal, and (11) the formal is either simple or personal.

order to make the proposition true.[9] For instance, 'goat' has material supposition in 'goat is a noun', simple supposition in 'goat is a species', and personal supposition in 'Britney has a goat'.

Personal supposition has various types, according to which individuals falling under the term are being referred to. Personal supposition is divided into determinate for singular expressions having a unique reference and common for universal terms that do not.[10] The latter have different sorts of supposition: determinate, for some definite singular in particular: exactly one of this one or that one..., and confused; which subdivides into merely confused and confused and distributive. In 'Britney has a goat', both 'Britney' and 'a goat' have determinate supposition: respectively, to that-there pop star and to just a single goat: to this goat or [exclusive] to that goat or.... In 'every goat is an animal', 'goat' has confused and distributive supposition, as a descent to each individual goat can be made; it stands for 'this goat and that goat and...' 'Animal' has merely confused supposition, as a descent to each individual animal cannot be made; still it stands for this animal or that animal or...[11]

Individual authors make further refinements and modifications of these divisions.[12] Still, these will suffice for me to make a broad comparison.

Avicenna's Theory

Avicenna discusses much of the doctrine of supposition theory under two main headings: his threefold distinction of quiddity (*triplex status naturae*) and his analysis of truth conditions for the categorical proposition. This bifurcation need not disqualify him from having an ersatz theory of supposition. Often the Latin theory of supposition is said to have such a bifurcation too, by combining semantic rules of reference with syntactic rules of instantiations for quantifiers: the main divisions into material, personal and simple being

9) Anonymus, *Dialectica Monacensis* VI, ed. De Rijk, in: id., *Logica Modernorum* (1967), II/2, 608, 14-19.

10) William of Sherwood, *Introductiones in Logicam*, ed. Brands and Kann (1995), 136-138 (59-65).

11) William of Ockham, *Summa logicae*, ed. Boehner *et al.* (1974), I, 70. Perhaps the 'or' is inclusive when the original subject, 'every goat', or the whole conjunction of individuals is retained. Still, once descent to one singular is made, the 'or' becomes exclusive.

12) For instance Walter Burley, *De puritate artis logicae tractatus longior*, transl. Spade, I.1.4. (75); I.4 (82); I.1.4 (100), offers a different scheme.

semantic, with the divisions of personal supposition being syntactic.[13] Here I summarize his doctrines.[14]

The *Triplex Status Naturae*

Avicenna has a famous doctrine that quiddities or essences have three respects: in themselves, in individuals, and in the mind. The doctrine extends to the terms signifying such quiddities. (Like Aristotle, Avicenna tends to assume an isomorphism between real things and a technical, protocol language describing them.) Typically such terms are universal—unless one holds that there are individual essences as well as universal ones, as Avicenna himself seems to do at times. This doctrine, of *the threefold distinction of quiddity*, thus describes what terms signifying essences stand for—that is, how they refer or 'suppose'.

The most explicit formulation of Avicenna's threefold distinction of quiddity available to the Latin West appears in his commentary on Porphyry's *Isagoge* (known as the *Logica*):

> The quiddities of things may be in individual things, and they may be in the mind. So they have three respects: the respect of quiddity inasmuch as that quiddity is not related to one of the two [modes of] existence, or to what is attached to the quiddity, insofar as it is in this respect. Also quiddity has a respect insofar as it is in individuals. There, accidents, which make particular its existence in-that, are attached to it. Also it has a respect insofar as it is in the mind. There-accidents that make particular its existence in that are attached to it; like being a subject and being a predicate, and universality and particularity in predication. [...][15]

Avicenna is saying that quiddities have three respects: in themselves, in things, and in the mind. Quiddities in themselves have no accidents, whereas quiddities in individuals and those in the mind each have accidents proper to them.

---

13) Scott (1966), 30; Kaufmann (1994), 118-120; Spade (1988a), 189, claims that this bifurcation of supposition theory holds for the fourteenth century but not for the thirteenth, before the period of the Modists. Spade (1988b) and (2002) has attacked the theory of the modes of personal supposition as it was presented by Ockham and Buridan. Spade believes that the theory has no function, partly because it fails to be a theory of truth conditions as captured by a certain kind of quantificational analysis, and partly because no medieval logician that he knows of said what the theory was trying to do.

14) The summaries are taken from Bäck (1996b) and (2004).

15) Avicenna, *Al-Madkhal*, ed. Al-Ahwānī (1952), 15, 1-6.

Quiddities in individuals and quiddities in the mind 'exist', in different ways, while quiddities in themselves do not exist, yet they have 'being' (*kuwn*).

The doctrine of the threefold distinction of quiddity claims that a quiddity has three modes, not that there are three distinct types of things comprising that quiddity. For then the universal term naming that quiddity would have three distinct referents. If there were three referents, the universal term would just be ambiguous and name three different things. In contrast, here the same thing is being talked about somehow, yet in three different ways or respects. This suggests that in such cases universal terms have the same signification but vary in supposition; in Avicenna's terms, in all cases the universal term signifies the quiddity or essence, which varies in what mode of being it has.

Avicenna seems to view the connection between these different senses of quiddity as similar to the relation between the different senses of 'healthy' or 'medical' that Aristotle discusses in *Metaphysics* IV.2.[16] As Aristotle says, these terms have different uses and definitions, but a focal meaning *(pros ti)*. So too, he argues, 'being', though said in many ways, has a focal meaning. In effect, Avicenna is now extending this doctrine to essences in general in his threefold distinction of quiddity.

Thus Avicenna claims that essences have three respects, in virtue of which they may be spoken about. Quiddities in these three respects serve as truth-makers to ground our assertions.

Next I consider each of the three respects of quiddity in more detail.

The Quiddity in Itself

To speak of the quiddity in itself (*in se*) is to stipulate that the universal term is to be considered solely with respect to its definition.[17] 'Man', taken to represent a quiddity in itself, stands for only what being a man is. To be a man is, let's suppose, to be a rational animal, and hence also to be what it is to be an animal. The quiddity in itself will thus include the definition of the universal term, and the definition of the parts of that definition.[18]

Because the quiddity in itself considers only what the quiddity is in its definition (the *ti esti*), Avicenna often refers to the quiddity in itself by an abstract term; instead of 'being human', 'being animal', he says 'humanity', 'animality'. Thus humanity is a quiddity in itself, and it is rationality plus animality, that is,

---

16) This is suggested by his theory of homonymy. See Bäck (2008a).

17) Avicenna, *Al-Ilāhiyyāt*, ed. Anawati (1960), 201, 7-13; 205, 2; 207, 5-12.

18) Avicenna, *Al-Madkhal*, ed. Al-Ahwānī (1952), 36, 8; 48, 15.

rational mobile animate corporeal substantiality.[19] He refers to such quiddities also by *qua* complexes: 'man *qua* man'; 'horse *qua* horse'.

Avicenna is quite explicit that even those attributes that are necessarily inseparable from the definition—the *propria* or *per se* accidents—do not constitute the quiddity in itself. So, for example, humanity is a quiddity in itself, and being a body belongs to it, since, when the parts of the definition of 'humanity' are given, and then defined in turn, being a body is predicated of humanity. The definition of 'humanity' is 'rational animality', and that of 'animality' is 'mobile life', and of 'life' 'being an animate body'. Hence being a body is an element constitutive of the quiddity in itself humanity. However, neither corporeity, which Avicenna takes to be mere extension in space, nor three-dimensionality is such an element, although Avicenna holds that if something is a body, it must be three-dimensional and occupy space. Rather, corporeity and three-dimensionality are *propria* of humanity, which come to be attached to it necessarily when it comes to exist via combination with matter.[20] Again, neither will a *proprium* like risibility belong to humanity or man *qua* man, as it not in the definition.

The Quiddity in the Individual

Quiddities in individuals are individual material objects. These things *have* quiddities in themselves. For instance, an individual, say, Socrates, may have the quiddities of humanity, risibility, justice, whiteness, snub-nosedness, fatherhood. Thus, this man, Socrates, is a man, is risible, is white, is just, is snub-nosed, is a father. There being individuals with many different attributes is made possible through a substratum that is able to receive and link up many compatible quiddities in themselves—humanity, whiteness, justice.[21] Quiddities are said to exist *in* individuals because they are linked together in this matter.[22]

For Avicenna, matter is in general the notion of serving as a foundation or substrate for the reception of a type of accidents.[23] He recognizes various types

19) Avicenna, *Al-Ilāhiyyāt*, ed. Anawati (1960), 236, 6-8; 241, 5-16; Avicenna, *Al-Madkhal*, ed. Al-Ahwānī, etc. (1952), 28, 13-29, 6.

20) Bäck (1989).

21) Avicenna, *Al-Ilāhiyyāt*, ed. Anawati (1960), 202, 3-8; 204, 16-205.

22) Avicenna, *Al-Ilāhiyyāt*, ed. Anawati (1960), 208, 5-9; Avicenna, *Al-Madkhal*, ed. Al-Ahwānī (1952), 74, 11-75, 21; Avicenna, *Al-Ṭabiyyāt*, ed. Zāyid (1965-1983), vol. 2.1, 13, 1-12.

23) Avicenna, *Al-Najāt*, ed. Al-Kurdi (1938) 451; Avicenna, *Al-Ishārāt*, ed. Dunya (1971), 1010-1012; cf. Goichon (1937), 468-473.

of matter. The type of matter here, 'real' matter, enables quiddities in themselves to become associated with certain quiddities that are not contained in their definitions, namely the sort of quiddities classified in the ten categories. In this way, 'accidents happen' to quiddities in themselves. Some of these accidents necessarily accompany the instantiation of a quiddity in itself; these are the *propria* of the substance convertible with constituents of the definition. The individual is just the totality, a structured whole, of all of these quiddities associated materially.

The complex of essential constituents and *propria*, in association in the material substratum, constitutes an individual thing's *nature*. The persistence of that nature is necessary and sufficient for the persistence of that individual. Quiddities in themselves other than those constituting that nature may come to be and cease to be attached to that nature: these are the common accidents. Accidents of this type received by these individuals are those of first intention, namely, those appearing in the Aristotelian categories.

Quiddities in individuals are normally signified by concrete terms used concretely. Thus, 'man' in general is not a quiddity in individuals. Rather, 'this man', 'a man' is. Such singular terms indicate an instantiation of the associated quiddity in itself, humanity. That individual instance has the quiddity in itself humanity, and also other quiddities in themselves: essential constituents of humanity, material accidents inseparable from the presence of the substantial quiddity in itself in matter—the *propria*, material accidents that are not inseparable—and contingent accidents.

Although some quiddities appear substantial and essential to the individual, and others accidental, still, with respect to quiddities in individuals, there is no such distinction. A thing has all of its attributes, and cannot be divorced from any of them, so long as it has them. The differences between the quiddities in themselves that serve as the secondary substance of the individual, the quiddities that are inseparable from these, and the quiddities that are separable from these, lie in (1) how long and permanently the individual has them: always or not always, and (2) under what conditions the individual has them. The first distinguishes the necessary quiddities, those constituting a thing's nature ('nature' here includes both the essential constituents and the *propria*) from contingent ones. The second distinguishes the essential constituents from the *propria*. Yet these distinctions of substance, nature, necessity, contingency, etc. do not lie on the level of quiddities in individuals. Like the notion of having common attributes, the distinction of essence and accident is made only in another respect, namely, that of quiddities in the mind.

The Quiddity in the Mind

Quiddities in the mind are concepts (*ma'nan*). These quiddities are abstracted, via sense perception, imagination, and the intellect, from quiddities in individuals.[24] Being abstractions, quiddities in the mind have a dual nature. On the one hand, they represent individual as well as common features of quiddities in individuals. Thus, two real individuals, Socrates and Plato, are both human, and each may be called man, where 'man' signifies the nature that they have in common, sc. all the constituents of the quiddity humanity and its *propria*. Likewise they have many other common attributes: being male, Greek, philosophical, in the forum. All these attributes—man, Greek, in the forum, etc.—are concepts, quiddities in the mind.

On the other hand, quiddities in the mind, being abstractions, also have certain formal features not found in the things from which the abstractions arose. Thus man, a quiddity in the mind, is a common notion, is essential to things that are men, and indeed is a species. Being common, being essential, and being a species are formal attributes or accidents that apply to these abstractions but not to the things.[25]

Quiddities in the mind are strictly signified by concrete terms used abstractly. Thus, 'man is a species', 'man is animal', and 'Socrates is man' (as distinct from: 'Socrates is a man') are true claims about quiddities in the mind. On account of quiddities in the mind having a distinct type of accidents, formal accidents, given that each such type has a distinctive sort of matter and existence, Avicenna says that quiddities exist in the mind in a distinctive way.

The Structure of the Categorical Assertion

For Avicenna, a normal predication always includes an assertion of existence. He has what I have called *an aspect theory of predication*. For him a simple, categorical affirmative proposition normally asserts the existence of its subject explicitly.[26]

On the aspect theory of predication a statement of *tertium adiacens*, 'S is P', has the logical form 'S is (existent) as a P'. So, for example, 'Socrates is (a) man'

---

24) Avicenna, *Al-Nafs*, ed. Anawati and Zayid (1975), 148, 14-15; 184, 9-10; 208, 3-209, 7.

25) Bäck (1996a), n. 27.

26) Bäck (1987).

is to be read as 'Socrates is existent as a man'; 'Socrates is just' as 'Socrates is existent as just'; 'every man is an animal' as 'every man is existent as an animal'; 'man is animal', taken as a predication of genus of species, as 'man is existent as animal'. On such a reading, even a seemingly simple predication will have compound truth conditions: e.g., the truth of 'Socrates is existent as a man' requires both that Socrates be existent and that Socrates be a man (i.e., that 'man' signifies one of the attributes of Socrates). A simple categorical affirmation then makes two claims: 1) S is existent, 2) 'P' is predicated of S'.

Avicenna recognized explicitly in his treatment of the square of opposition that the contradictories of simple affirmative predications, understood in this way, will be implicit disjunctions, and so have disjoint truth conditions, each of which suffices for the truth of the contradictory.[27] So, 'Socrates is not a man', taken to be the contradictory of the simple affirmation, 'Socrates is a man', is equivalent to 'it is not the case that Socrates is existent as a man', and hence to 'it is not the case that Socrates is existent and Socrates is a man'. Hence, for the truth of 'Socrates is not a man', either 'Socrates does not exist' or 'man is not predicated of Socrates' suffices.[28]

In working out the details of this aspect theory, the question would naturally arise: for what time duration does the proposition assert the predication of the subject to hold? Is the predicate being claimed to belong to the subject so long as the subject exists? This seems to be so when the predicate belongs necessarily and gives a constituent or necessary concomitant of the nature of the subject S. Then the predication is essential. However, when the predication is accidental, the analysis of the proposition becomes more complicated: first, because the predicate signifies an accident of the subject S, and so need not belong to the subject always; second, because accidental terms, being paronymous, have an ambiguous reference. Such factors complicate the reading of particular examples considerably. Avicenna offers an elaborate account, part of which I summarize below.[29]

---

[27] Avicenna, *Al-ʿIbāra*, ed. Al-Ḳuḍayrī (1970) 77, 8-84, 13.

[28] Avicenna, *Al-ʿIbāra*, ed. Al-Ḳuḍayrī(1970) 84, 6-17.

[29] Street (2000) has an analysis of the material in the Avicenna, *Ishārāt*. Yet that and like books of Avicenna are summaries of complex doctrines, with most of the complexities omitted.

Ambiguities of Paronymous Terms

A paronymous term 'R', derived from 'R-ness', has both the ambiguity of (1) signifying an accident of its subject and that of (2) signifying a paronym. It can be taken to refer to its subject, the substance, or to refer to a different subject. (1) Signifying an accident, 'R' need not be predicated truly of its substance S so long as S exists, although it may, just as 'living in Europe' held of Socrates always. Being paronymous, 'R', derived from one of the accidental categories, can be understood as 'the thing that is R'. This 'thing' here refers to S but need not refer to S at all times. It may signify the substance S that serves as a subject so long as S exists, through having S-ness, or it may signify that substance S only so long as it exists and has R-ness. That is, for the paronymous expression 'R', we may focus more on the R-ness or on the S-ness. (2) 'R' may also be taken to refer just to what has R-ness, apart from its being in a subject. Then it is taken to refer to a subject term in its own right: perhaps just the quiddity in itself R-ness or perhaps the accident R *in re*, apart from its being in the substance S.

As in the Aristotelian tradition all accidents are paronyms, Avicenna tends to run these two considerations, of accident and paronym, together:

> The sense of 'every white' is not: everything that is white insofar as it is white only, but rather everything that is characterized as being white, and everything of which white is said, whether that thing be white in being the nature [*nafs*] of white insofar as it is white, or whether it be something characterized as being white, while having another reality, like a man or log characterized by whiteness. Also it is necessary for us to know that the sense of our saying 'every white' is not: everything that is characterized as being white always. So our saying 'every white' is more common than our saying 'every white always'. 'The white' is more common than 'white at some time' and than 'white always'. The sense of 'every white' is: each thing that is characterized as being white always or not always, whether it be a subject for the white characterized by it, or whether it be the nature [*nafs*] of the white. This characteristic is not the characteristic of possibility and soundness. So it is not understood at all [by saying 'every white'] that it is everything of which it is sound [to say now] that it is white, but rather that it is everything that is characterized in act as being white at some time, be it indefinite or definite, or always, after it comes to be in actuality.[30]

These distinctions come from Aristotle. In discussing paronymy in the *Categories*, Aristotle also distinguishes as two paronyms two objects, for instance,

30) Avicenna, *Al-Qiyās*, ed. Zāyid (1964), 20, 11-21, 5.

whiteness and the thing having whiteness.[31] Whiteness is in the category of quality, while the white, the thing having whiteness, is an individual substance, this horse, that happens to be white, that is, that has the quality of whiteness.[32] In his *Metaphysics*, Aristotle goes on to say that 'the white' too can signify in two ways.[33] Primarily, it signifies the individual substance that happens to be white. Secondarily, it signifies the complex, the individual substance's having that whiteness. Avicenna makes use of all these distinctions. Furthermore, an abstract term like 'whiteness', which Aristotle takes to signify the quality, now signifies the quiddity in itself. The two senses of the concrete term ('white') become the two senses distinguished here.

In Aristotelian science, paronymous terms assume great importance. There the terms standing for universal accidents, like 'white' and 'rational', do not strictly speaking signify items in the categories. Rather, terms paronymous to them do, like 'whiteness' and 'rationality'. So, for example, the horse is white, although the horse is not whiteness, and likewise for 'rational', as opposed to 'rationality', and 'risible' as opposed to 'risibility'. That is, even the *propria* and *differentiae*, the *per se* accidents, appear in categories other than substance.[34] Accordingly their paronyms too will be predicated of items in the category of substance.

Now Aristotelian demonstration also requires such doctrines concerning paronymy like Avicenna's. For, when a paronymous term like 'white' or 'rational' appears in a syllogism as a predicate, it will often end up as the subject term via conversion steps in the proof. To be sure, Aristotle does protest against terms like 'white' serving as subject.[35] Yet, at the same time, he uses such terms regularly in his syllogistic.[36] I have suggested that the conflict can be resolved if we make a distinction between a concrete term like 'white' being used concretely and its being used abstractly in the way that Avicenna will do.[37] E.g., it follows from 'every swan is white' that some white *thing* is a swan—not that some white (*simpliciter*) be a swan. Accordingly, we can see why a working Aristotelian like Avicenna would have interest in the details of how to construe paronymous terms when they are used as subjects of propositions.[38]

---

31) Aristotle, *Categories* I, 1 a 12-15.

32) Aristotle, *Categories* X, 10 a 27-b 11.

33) Aristotle, *Metaphysics* VII, 4 and 6.

34) See Aristotle, *Categories* V, 3 a 21-24; *Topics* VI, vi, 144 a 20-1; Bäck (2000b), 151-158.

35) Aristotle, *Posterior Analytics* I, xxii.

36) E.g., Aristotle, *Prior Analytics* I, vi, 29 a 4-10.

37) Bäck (2000b), 185-195.

38) So too Ross (1949), 40, worries about *Prior Analytics* I, ix, 30 b 5-6: why does Aristotle say that 'no animal is moving' is not necessary, and that 'some white thing is an animal' is necessary?

Discussions of paronymous terms assume great importance in Latin medieval logic for the same reasons, with some similarity of doctrine.[39]

The Predicational Versus the Quantified Proposition

Either the connection between subject and predicate or the 'quantifier' may determine the truth conditions concerning the duration of the statement. The statement may be taken as a proposition (a *dictum*), or it may be considered as asserting an attribute to belong to an existing subject:

> If the existence of the subject is not considered but rather the truth of the proposition is considered, be the subject existent or non-existent, so that the simple [proposition] is that in which the judgment through its quantifier is true at some time, whether the subject be existent or non-existent [...][40]

We have in this distinction an analogue of the compound and divided senses of the Latin medievals.[41]

Avicenna's theory allows for the universal statements used in science to have two grounds or causes for their truth. First, Avicenna finds the basis for necessary truth and falsity on the level of quiddities in themselves. Because of the subordinate relation of caninity to animality, it is necessary that every dog be an animal. Because of there not being a subordinate relation of being tailed to caninity, it is not necessary that every dog have a tail, even if all dogs *in re* always have tails. Thus Avicenna advocates considering statements used in scientific demonstration with respect to the predication and without reference to time.[42] For here we seek essential connections. Essential connections are based on the connections of quiddities in themselves. Here, then, only possible existence is required for the predication to hold.

---

39) For instance, take Peter Abelard, *Glossae super Peri Hermenias*, ed. Geyer (1927), 360, 23-34; *Glossae super Porphyrium*, ed. Geyer (1919), 17, 12-28. *Glossae super Praedicamenta*, ed. Geyer (1919), 122, 29; *Dialectica*, ed. De Rijk (1970), 65, 24-31. See too Jacobi (1986), 149-155; Wilks (1998), 369, n. 4; Marenbon (1997), 138-9; 142; Bäck (2000b), 285.

40) Avicenna, *Al-Qiyās*, ed. Zayid (1964), 84, 16-85, 2; Avicenna, *Al-Ishārāt*, ed. Dunya (1971), III. 2.1, 271, 8-12; Avicenna, *Manṭiq al-Mašriqiyyīn*, ed. Al-Ḵaṭīb *et al.* (Cairo 1910, 64, 2-4).

41) Not in the senses of 'compound' and 'divided' given by Street (2000), 46-47 in discussing Avicenna's theory!

42) Avicenna, *Kitāb al-Burhān*, ed. 'A. 'Afifi and I. Madjur (1956), 71, 15. Avicenna, *Al-Qiyās*, ed. Zayid (1964), 83, 10-11 'So what they have supposed of the categorical, that it is necessary that the judgement in it is of existents at a time, has turned out to be absurd.'

In effect, Avicenna is advocating considering the statement taken as the predication to express a *dictum*, that the predicate is predicated of the subject. The *dictum* as a whole can have necessary truth; a predication of the subject, which is being asserted to have a contingent, *a posteriori* existence, cannot.

Second, we may ground the truth of universal propositions on the relations between the concepts signified by the statement that exist *in intellectu*, as abstracted from the sense perceptions of individuals existing *in re*. In this way, we may abstract the concept of dog from our observations of individual dogs. Avicenna grounds the truth of 'Homer is a poet' in the same way. Our past experiences of Homer produce a phantasm of Homer in the mind. This phantasm can exist *in intellectu* even after Homer has died. It is true of that phantasm to say that it is a poet, even though it is no longer true to say that Homer, a human being existing *in re*, exists as a poet.[43]

This way of grounding the truth of statements in respect of the quantifier can justify only contingent, *a posteriori* facts. For it just so happens that, in human experience up to this point or even always, all dogs are animals. Likewise, it might be that, in our actual experience, every dog has a tail. Yet it is necessary that every dog be an animal, while it is not necessary that every dog have a tail. How does one separate the necessary from the sempiternal (*kata pantos*) truths? To get at necessary truth, we need another foundation: the first one.

Given that science seeks necessary, universal knowledge, Avicenna prefers the former ground, where the relations between the quiddities in themselves provide the truth makers for modal propositions. In contrast, the latter sort, based on quiddities in the mind, abstractions from sense perceptions of individuals existing *in re*, provides the truth-makers for categorical propositions that may or may not be necessary.

Of course, the relation between the two levels of truth-makers, that of the quiddities in the mind and that of the quiddities in themselves, itself has complexities. For the existence of the perceptible individuals, the quiddities *in re*, is based upon the quiddities in themselves and their interweavings. On the

---

43) Avicenna, *Al-ʿIbāra*, ed. Al-Kuḍayrī (1970), 109, 2-110, 1. Cf. Aristotle, *On Interpretation* X, 21 a 25-28, and Al-Farabi's *Commentary* 160, 23-27 (transl. Zimmermann (1981), p. 155); Avicenna's discussion of the griffin, 110, 2ff. and 82, 16-18; Avicenna, *Kitāb al-Najāt*, ed. Al-Kurdi (1938), 6, 2-3. On the relation between phantasms and concepts, see Avicenna, *An-Nafs*, ed. Anawati and Zāyid (1975), 32,7 ff.; 147,1 ff. In general, Avicenna, *Al-ʿIbāra*, ed. Al-Kuḍayrī (1970), 110, 7-14, holds that phantasms are based on particular experience, whereas concepts are of universals. On predication of non-existent objects, cf. Ammonius, *In Aristotelis De Interpretatione*, ed. Busse (1897), 186, 15.

other hand, for us to have knowledge of quiddities in themselves, they must exist *in intellectu*.[44] Moreover, ordinarily, science, especially natural science, will concern things existing in the world of *nature*, namely, in this world. In this case, a proposition can have both grounds of truth, and the two levels will run together.

A common feature of both ways of taking the statement is that there is no descent to singulars existing presently *in re*. They differ in that the second sort requires instances *in re* at some time, while the former does not require instances *in re* at any time. Avicenna allows for possible things, and possible types of things, that are never actualized. The causes of truth for modal claims differ from those for (simple) categorical claims. The predicational propositions have a modal foundation of truth, while the categorical ones have an actual one.

That is, in Latin medieval terms, neither sort requires appellation (in the early sense of present existence). However, the second sort requires personal supposition, strictly speaking, personal supposition ampliated to all times (as in later supposition theory). As Avicenna says, normally, 'the sense of 'every C is B' is that everything characterized as being C in the past and the present is characterized as being B.'[45]

Yet, it can be said, the second sort does require appellation in some ways. For 1) it requires the concepts to exist as unified 'individuals' *in intellectu* now, 2) it requires the individual minds having them to exist *in re* now, and (3) it requires individual instances of those concepts to have existed *in re* in the present or past. These requirements have some logical merit: we speak of eclipses and a person being halfway to Baghdad in the present tense, even when there is no eclipse or anyone at that halfway point.

In contrast, the former sort requires no descent to singulars or existence in any sense. Terms in statements expressing the relations between quiddities in themselves might then be said to have a sort of 'simple' supposition, as I discuss below. Strictly, these terms should be abstract, as in 'caninity is animality'. Complications arise when concrete terms are used. For, in 'it is necessary that every dog is an animal', the use of the concrete terms normally suggests that the subject be construed to exist *in re*. Sometimes the subject does not have any present instances or even any instances at all: 'it is necessary that every eclipse has one celestial body obscured by another'; 'it is necessary that every

[44] —or at any rate our intellects must gain access to those quiddities. See Bäck (2005).

[45] Avicenna, *Al-Qiyās*, ed. Zayid (1964), 82, 15-83, 1.

heptagonal house has more than five sides.'[46] Certain contexts—for Avicenna, especially the modal contexts—disambiguate the statements.

On the Side of the Subject or the Predicate

The time duration during which the predicate is asserted to belong to the subject may be fixed by various features of the subject or predicate term.

> The sense of 'every B is A' is: every one that is characterized and determined to be B in actuality, always or not always, is characterized also by being A, without paying attention to when that is, and in any of the divisions that there might be.[47]

Avicenna then gives various ways to understand the temporal claims being made in a statement of the form 'B is A':

> One of them is: B is A always, and the second is: as long as it is characterized as being B, and the third is: as long as it is characterized as being A.[48]

If we think of the aspect theory of predication, we can see how Avicenna arrived at this distinction. Consider that statement: 'S exists as being a P'. That is, something named by 'S' exists also while being named by 'P'. Both the subject term and the predicate term describe the same thing in a true proposition. Sometimes which description is used will change what is being claimed about the temporal duration of the thing claimed to exist, the logical subject. Each description of this thing can have different time determinations: e.g., whatever is named 'S' now; whatever is named 'S' as long as it is existent as S; whatever is named 'S' as long as it is existent at all; whatever is named 'S' at all times. Likewise for 'P'. Sometimes a description of a thing in terms of 'S' and a description in terms of 'P' cause the reference to differ. These differences arise when the terms signify accidents. When all the terms of a proposition are essential, these differences disappear; when they signify features of the quiddities in themselves, there will not be any differences in modal contexts either.

---

46) On the latter example, see Avicenna, *Al-Ilāhiyyāt*, ed. Anawati (1960), V.1.

47) Avicenna, *Al-Qiyās*, ed. Zayid (1964), 26, 18-27, 2.

48) Avicenna, *Al-Qiyās*, ed. Zayid (1964), 27, 4-5; Avicenna, *Manṭiq al-Mašriqiyyīn*, ed. Al-Ḵaṭīb *et al.* (1910), 71, 26-72, 6. Cf. Johannes Philoponus, *In Analytica Priora*, ed. Wallies (1905), 43, 8-18.

Ad-verbial Qualifiers

The assertion of existence may be qualified in various ways by what might generally be described as 'ad-verbial qualifiers' (sc. the original, general sense of 'mode') including tense markers, modal ('necessary' etc.) and other operators (e.g., 'it is dubious that'), hypothetical and other subordinate clauses ('while traveling to Baghdad'), and other qualifications (e.g., 'not really'; 'non-existent').

Avicenna uses these distinctions in analyzing particular examples and paralogisms and in his syllogistic. He is also willing to recognize non-standard uses and special contexts. Understandably, Avicenna has no systematic doctrine for such qualifiers, for they amount to a hodge-podge. Like Aristotle, he deals with different ones in different places. He discusses the logical modalities in his modal logic. Ones like 'dead' or 'non-existent' he discusses in dealing with fallacious inferences. In his *Topics* he discusses further varieties.

The Greek Background

To gain some perspective on Avicenna's doctrines, let me compare his doctrines briefly to Aristotle's. Detailing ancient accounts of theories of reference, even just those elements pertinent to later theories of supposition, would be quite a large project in itself. I shall confine myself to making a few observations, hopefully not too controversial ones.

Both the Arabic and the Latin philosophers had Greek philosophy as a common base. Greek philosophers also worried about how our words described our world. They wanted 'their words to match their deeds'—and the things in the world.[49] Plato advocates a philosophy that chops reality up at its joints.[50] In order to continue this project, Aristotle regiments his language by rejecting some ways of talking as illegitimate and making up others when his Greek had none. In effect, he constructs a protocol language.[51] More to the point here, he worries about how expressions signify and about the many ways in which they are said.

Aristotle himself asserts that certain expressions signify items in the various categories. Some expressions signify substances.[52] Proper names and other

49) Plato, *Laches* 188 D.

50) Plato, *Phaedrus* 265 E.

51) Bäck (2000b), 137-147.

52) Irwin (1982), 242-6. See Wheeler (1999) for a review and critique of the various interpretations of Aristotle's concept of signification. I generally agree with his views, pp. 194; 209-211, except for his attributing a modern concept of sense to Aristotle.

singular expressions signify individual substances: 'Britney'; 'this goat'. Some general terms signify species and genera: 'goat'; 'animal'. Other expressions signify accidents. *Pace* G.E.L. Owen, Aristotle recognizes individual accidents.[53] At any rate, so the Aristotelian tradition has it. So then 'this white', 'the sitting of Socrates now', and 'this one's being at that corner now' would signify individual accidents. Some general terms signify universal accidents. Here the English may mislead. It uses abstract terms, like 'whiteness', 'sitting', and 'being on a corner', to signify universal accidents. Yet Aristotle hardly ever uses abstract terms thus. Like Plato, he uses concrete terms taken abstractly, like 'the white' (*to leukon*) and likewise for the essences of substances: humanity, horseness.

Avicenna has concrete individual terms taken concretely that signify substances or accidents referring directly to quiddities or individuals *in re*. Concrete universal terms taken abstractly refer directly to quiddities in the mind or concepts taken materially. As these intend or refer to quiddities in themselves insofar as they are in more than a single individual, the universal terms refer indirectly to instantiations of quiddities in themselves. Similarly the concrete individual terms can be taken to refer to quiddities in themselves insofar as they are in just a single individual.[54]

Aristotle says that strictly it is not a *quale* like white, but a quality like whiteness that is in the category of quality.[55] Evidently he means this point to apply to all categories of accidents. Whiteness and white are paronyms. A term like 'white' generally refers to the thing, ultimately the individual substance, that is white. Thus 'white' does not signify a simple item in a category, since it signifies something complex: the quality-in-the-subject.[56] As abstract paronyms like whiteness do not exist except as in a subject, they are not the things that exist *in re*; rather, their concrete paronyms like the white are the things that exist *in re*. 'The white' is not the same as its essence. For the essence, say, of the white (thing), which happens to be a goat, is the essence of its substance: being a goat.

Avicenna lays heavy stress on this doctrine of paronymy.[57] Following Aristotle, he distinguishes 'the white' into 'the thing that is white so long as it is white'

---

53) Owen (1965).

54) Here the concrete terms taken concretely are individual and the ones taken abstractly are universal. Yet the other options are possible, although not relevant here: the former taken universally might be Platonic Forms, and the latter taken individually might be individual essences.

55) Aristotle, *Categories* VIII.

56) Aristotle, *Metaphysics* VII, 6.

57) Bäck (2000b), 147-57; 85-95.

and 'the thing that is white so long as it exists', namely so long as the subject exists. He then considers various ways in which its being white can be construed, mostly in terms of temporal duration. Likewise, in his syllogistic Aristotle allows for subjects like 'white'. In his modal syllogistic he admits two readings of 'it is possible that every S is P': everything that is S can be P, or everything that can be S can be P.[58] Aristotle prefers propositions holding of every instance at all times (*simpliciter*) but does allow for those holding of every instance only at a particular time (*ut nunc*).[59]

Aristotle claims that a substance is the same as its essence in formula (*logos*).[60] 'Goat' and 'being goat' signify the same thing: there are no essences existing over and above the individuals having them. At the same time, 'S' and 'the essence of S' are not interchangeable. Otherwise Aristotle could not distinguish the two in his theory. Thus abstract terms, both those signifying accidents and those signifying the essence of substances, will have formal features different from their concrete correlates. Still, they need not refer to different real objects.

Likewise, Avicenna does not have abstract and concrete terms referring to different things existing *in re*; he is no Platonist. He distinguishes the ground of truth of a real definition of a quiddity, namely it in itself, from the existence of that quiddity *in re*.[61] Unlike Aristotle, he seems willing to admit 'real' definitions of essences even when they have no instances: having possible instances suffices.

Aristotle often uses terms like 'matter', 'form', 'genus' and 'potentiality'.[62] He locates such expressions themselves in no category. Yet surely he uses them prominently in his theory about existing things. For Avicenna such terms are second intentions, quiddities existing *in intellectu*. Here the concepts are being taken as subjects in their own right, having distinctive, formal attributes.

Aristotle allows that things recognized to exist in theory may differ from those recognized to exist *in re*. He says that the mathematician may treat figures and numbers *as if* they existed on their own, independently of their substances—although, in fact, they cannot.[63] Note that the figures and numbers

58) Aristotle, *Prior Analytics* I, xiii, 32 b 24-32.

59) Aristotle, *Prior Analytics* I, xv, 34 a 34-b 6.

60) Aristotle, *Metaphysics* VII, vi, 1032 a 1.

61) As in *Posterior Analytics* II, 1-2.

62) Linguistic items like names and verbs, when used, are in the category of quantity (*Categories* VI). What about them when they are mentioned, as in 'walks is a verb'? Aristotle locates statements (*logoi*) in the categories too (*Categories* V).

63) Aristotle, *Metaphysics* XIII, iii, 1077 b 31-1078 a 5.

of geometry and arithmetic are not substances; strictly speaking, they are *qualia* and *quanta*, the qualities and quantities existing in substances. In science they are treated abstractly, that is, as if they were *in re* yet apart from being in substances. So too we might consider 'the white' to signify a simple thing, the mere *quale*, apart from its being in a substance.

Avicenna offers an account of how such sciences have an objective basis. Aristotle has a somewhat embarrassing position if it is not augmented or qualified: his most precise sciences then seem to deal with *ficta*, fictitious objects. For Avicenna, in contrast, ultimately they deal with quiddities in themselves. These still do not exist, except insofar as they are in individual things or in individual minds. Quiddities in themselves also provide the basis for distinguishing the components of real definitions from their necessarily concomitant *propria*. The concepts in the mind, gained by induction from sense perceptions, cannot make this differentiation.[64]

Aristotle discusses the descent from universals to singulars somewhat in his syllogistic and the square of opposition. In his doctrine of exposition Aristotle allows for a substitution of a singular, given by sense perception, for its universal species.[65] This gives a descent for a universal term when it is the subject: if every S is P and R is one of the S's, then R is P. Given conversion, where the original predicate becomes the subject (as in 'every S is P; therefore some P is S'), we get a descent for the predicate terms too. Avicenna likewise deals with the inference relations between particulars and universals in his discussions of the squares of opposition, the syllogistic, and to a lesser extent in his *Topics*.

The later Aristotelians inherited Aristotle's doctrines. Already in the late Greek period, the commentators would speak of things 'serving as subjects (*hupokeisthai*)' and 'serving as predicates (*katēgoriesthai*)'.[66] They had to deal with the notion that some terms, particularly the paronymous or derivative ones, might make reference to their ultimate subjects, the substances in which they inhere, or to a more proximate subject, like the accidents themselves or to some other parts or aspects of those substances. Given Aristotle's insistence that individual substances are primary—if they did not exist, nothing else would—they had a motive to refer the universal terms to individuals. Still, as

---

64) Ardeshir (2008) lays more stress on mathematical objects existing in the mind.

65) As Aristotle uses it in his syllogistic, 'exposition' (*ekthesis*) involves 'putting forward' an instance of a universal claim. Cf. Alexander of Aphrodisias, *In Aristotelis Analytica Priora*, ed. Wallies (1883), 32, 32-33, 15; 99, 19-100, 26.

66) E.g., ps. Simplicius, *In libros Aristotelis De Anima*, ed. Hayduck (1882), 127, 26-32; 278, 20-29; so too Avicenna: see n. 15; Avicenna, *Al-'Ibāra*, ed. Al-Kuḍayrī (1970), 15; Avicenna, *Al-Qiyās*, ed. Zayid (1964), 76, 2-4.

Aristotle says that universal substances and accidents belong in the categories of beings, they had to explain how species and genera are real too. All this provided a common basis for medieval theories of reference, both Latin and Arabic.

Comparisons with Latin Supposition Theory

How do these doctrines of Avicenna match up with Latin medieval doctrines? We shall see—sketchily—that they constitute what I call, by baptism, an *Arabic supposition theory*. Moreover, like Latin supposition theory, Avicenna's views are not silly by modern standards. I shall compare his theory mostly to Ockham's, not only to keep my project manageable but also on account of the two theories having great similarities.

Recent historians of supposition theory give the following account: In the thirteenth century the standard, 'textbook' theory was developed. Then, in the 1270s the *modistae* came to have a different emphasis on the signification of terms through acts of imposition, more so in Paris than in England. Then in the fourteenth century we get the later supposition theories of those like Burley, Ockham and Buridan.[67] Avicenna had a strong influence while the textbook theory was developing, partly through his *Logica* (his commentary on Porphyry) and *Metaphysica*, and partly via Averroes' critiques in his commentaries translated into Latin. As a result of those critiques, Ebbesen says, he had less influence on the Modists.[68] Later, if we judge by his being cited often by Ockham *et al.*, he regained his influence in the fourteenth century—if he had ever lost it at all in other areas of philosophy, notably in metaphysics. (At all these times, the bulk of Avicenna's logical writings on reference were not available to the Latin West.)

I shall now compare Avicenna's theory to supposition theory, in both its earlier and later versions.

The Triplex Status

Terms having simple supposition, like 'goat' in 'goat is a species', refer to quiddities in the mind. Avicenna's theory differs from the textbook supposition theory, where terms in simple supposition refer to things outside the mind.

67) Spade (2002); Read (2008).

68) Ebbesen (1988), 114. On Averroes' critique: see *Al-'Ibāra*, ed. Al-Ḳuḍayrī (1970), 99, 1-100, 7.

Terms having material supposition, like 'goat' in 'goat is a name', present more difficulty. When 'name' here refers to an element in the universal mental language, it refers to a quiddity in the mind. When 'name' here refers to a verbal or to a written expression in a particular language, it seems to refer to a sound or to a mark in some particular language.[69] Then the reference amounts to a type of standard personal supposition, to an external object. The sound or mark being referred to may be taken particularly, so as to stand only for the one indicated in that particular linguistic act, or generally, so as to stand for any instance similar to the one being indicated. Still, it will not refer to the type itself; that would make the supposition simple. As with Aristotle, a written expression refers to a spoken expression and that to a concept in the universal mental language of the mind.[70] So indirectly a written or spoken name refers to a quiddity in the mind.

Avicenna does recognize, in passing, some different types of quiddities in the mind, like names and species, but he does not offer much of a systematic theory. He does of course distinguish first from second intentions. He does not distinguish referring to verbal or written signs and referring to concepts in the mind much, probably because like Aristotle he views concepts as constituents in a universal mental language.[71] In short, Avicenna does make some distinctions about quiddities in the mind, but nothing resembling a full-blown theory of types. But then the same can be said for Ockham *et al.*[72]

Expressions having personal supposition, concrete terms used concretely, like '(a) goat' in 'Britney has a goat' or 'every goat is tragic', refer to quiddities in individuals. As Avicenna thinks that only individuals exist *in re*, all universal terms having such a reference will allow and indeed require a descent to singulars. Usually the sentential context—having the term quantified or at least the sort of predication being made—will make the universal term refer to singulars directly. What Stephen Read says about Ockham seems to apply equally well to Avicenna:

---

69) Avicenna, *Al-'Ibāra*, ed. Al-Ḵuḍayrī (1970) I, 9, 16 ff.; Avicenna, *Al-Ilāhiyyāt*, ed. Anawati (1960), 29, 5-6; 31, 5 (*Metaphysica* I. 5).

70) Avicenna, *Al-'Ibāra*, ed. Al-Ḵuḍayrī (1970) 103, 5.

71) Following the lead of Aristotle, *De interpretatione* I, 16 a 3-8 in *Al-'Ibāra* part I, ch. 2.

72) In his *Summa logicae*, ed. Boehner *et al.* (1974), I. 13 William of Ockham says that a first intention signifies something that is not an intention of the soul, and that a second intention is an intention of the soul which is a sign of first intention. He has no theory of types here either.

'Man', he says, signifies Plato and Socrates and all men equally. Once signification is treated extensionally in this way, its only difference from supposition lies in its priority: a general term signifies all those things of which it can be truly predicated.[73]

Avicenna takes abstract terms of first intention, like 'humanity' and 'horseness', to stand for quiddities in themselves. Those of second intention, like 'universality' and 'being a predicate', stand for quiddities in the mind. Like other such expressions, these terms can be taken particularly, so as to stand only for that particular instance, or generally, so as to stand for any instance of that type. These terms can refer to more than one instance, as Avicenna, like Aristotle, recognizes intelligible matter and needs to multiply instances, so as to have 'animality' appear in the definitions of 'humanity', 'horseness' etc., and so as to have 'universality' apply to different concepts in the mind.

Abstract terms of first intention, like 'horseness' and 'whiteness', have the complication that they do not stand for anything that actually exists. It seems too weak to say that for Avicenna therefore they do not have supposition but only signification.[74] Instead, these terms seem to fit the textbook definition of simple supposition: 'A term has simple supposition when it stands just for what it signifies.' Perhaps it is better to say that such terms have supposition and signification but not appellation. The abstract terms of second intention can likewise be said to have simple supposition, just like 'species' in the Latin textbook theory.

However, Avicenna's theory resembles more Ockham's theory than the textbook doctrine for concrete terms taken either concretely or abstractly. On both theories the terms stand for existing individuals. Ockham holds that 'goat' in 'goat is a species' as well as in 'Britney is a goat' has personal supposition. He takes such simple supposition as a type of personal supposition, reference to the *intentiones* themselves, which may be first or second ones.[75] For him a term in personal supposition may stand for something outside the soul, a *vox*, an *intentio animae*, or a *scriptum*.[76] As with Avicenna, simple supposition then becomes a special type of personal supposition, where the referent is a concept.

---

73) Read (2008), William of Ockham, *Summa logicae*, ed. Boehner *et al.* (1974), I, 33.

74) —unless by 'signify' one means the original Aristotelian sense 'is a sign for'? See above. Then this amounts to the later 'supposition'.

75) In contrast to Walter Burley, *De puritate artis logicae tractatus longior*, ed. Boehner (1955), I.1.3 on Ockham's rejection of simple supposition. See also Kaufmann (1994), 76; Schulthess (1992).

76) William of Ockham, *Summa logicae*, ed. Boehner *et al.* (1974), I, 64, p. 195, 5-6.

Ockham takes material supposition to occur when a term supposes for a *vox* or *scriptum* but does not hold significatively.[77] Material supposition could thus be understood here to have to do with sounds or marks, with no further condition of intentionality being required. The marks and sounds are things *in re*, and so material supposition would be a special type of personal supposition. Particularly for nominalists like Ockham and Buridan, material supposition amounts to referring to the spoken or written signs of the *intentiones*, the concepts in the mental language, and not to the *intentiones* themselves. Then why not treat material supposition as a case of personal supposition, as suggested above? Perhaps, rather, we need to distinguish mere marks from linguistically significative marks. If so, the intentionality of a language user would be required. Then, for Avicenna, if not for Ockham, the sounds or marks would refer to quiddities in the mind. This conclusion does not seem inconsistent with Ockham's conception of mental acts, but he does not state or at least emphasize it in his account of material supposition. Buridan, however, has some remarks along these lines.[78]

In contrast to the Latin medieval nominalists, Avicenna takes abstract terms and concrete terms taken concretely to differ in their logical type of reference. Generally, abstract terms of first intention refer to quiddities *in se*, and the concrete ones to quiddities *in re*. Perhaps this corresponds to formal supposition, with its division into terms having simple supposition allowing no descent to singulars, and those having personal supposition allowing a descent to singulars. We could equate the latter with personal supposition. However, the former does not amount to the Latin sort of simple supposition: just look at examples like 'goat is a species'. For Avicenna 'goat' there has to refer to a quiddity in the mind and not to quiddities *in se*.

Moreover, there is a problem with saying that 'animality' in 'humanity is animality' does not allow descent to singulars. As mentioned above, Avicenna seems to allow for more than a single instance of a quiddity in itself. He seems to have to in order to account for differences in the subordinate genera and species. For instance, humanity is animality plus rationality, and horseness is animality plus neighability. Horseness is opposed to rationality. To allow for horseness to appear in these mutually incompatible definitional complexes seems to require some sort of multiplication of animality, perhaps in intelligible matter.

---

77) William of Ockham, *Summa logicae*, ed. Boehner *et al.* (1974), I, 64, p. 195, 38-39.

78) John Buridan, *Summulae. De suppositionibus*, ed. Van der Lecq (1998), 4.1.2. [transl. Klima (2000), 223].

If so, we can construe Avicenna as holding that both abstract and concrete terms have personal supposition, with their being two logical types of referents. Once again, this does not conflict with some versions of supposition theory: Again, simple supposition along the lines of Peter of Spain or William of Sherwood looks a bit like reference to quiddities in themselves.

Likewise, Ockham takes any term, including an abstract one, to have personal supposition when it refers to something extra-mental—indeed when it refers to anything other than itself.[79] So far he agrees. Still, Avicenna seems to have quiddities in themselves as being formally or really distinct from the individuals having them.[80] Here he seems much more like Scotus, who has been said to have an *augustinisme avicennant*.[81] Ockham has universals being only concepts in the mind.

I do not wish to belittle this difference: Avicenna is more realist than Ockham. Nevertheless, they do not differ much in the details of supposition theory here. Consider how Ockham deals with abstract terms:

> For there are certain abstract nouns, or they can be made up (*ad placitum instituentium*), which include equivalently some syncategorematic terms or some adverbial determinations, or something else, such that the abstract term in signifying is equivalent to a concrete or another term taken with some syncategorematic term or some other expression or expressions. [...] For if that abstract (noun) 'humanity' is equivalent in signifying to the whole, 'a man insofar as he is a man' or 'a man in virtue of the fact that he is a man', 'a man runs' would be true, and 'humanity runs' false, just as 'a man insofar as he is a man runs' is false. Similarly, if 'humanity' is equivalent to the whole, 'a man by necessity', so that the expression 'humanity' is substituted for the whole, 'a man by necessity', 'humanity is a man' would be false, just as 'a man is a man by necessity' is false, for no man is a man by necessity but only contingently, and in the same way 'humanity is white' would be false, just as 'a man is white by necessity' is false. And in such a way it can be established, whenever it is wanted, that a concrete term and an abstract term do not signify distinct things nor suppose for distinct things, and still the predication of one of the other is false without qualification, and what is predicated of one is not what is predicated of the other.[82]

Ockham proposes that 'S-ness is P' be understood as 'S qua S is P' or as 'S is P by necessity'. He seems to favor the first proposal, namely to analyze propositions containing abstract terms reduplicatively. His reference to Avicenna

79) William of Ockham, *Summa logicae*, ed. Boehner *et al.* (1974), I, 64, pp. 195-196, ll. 4-37.

80) Kaufmann (1994), 23

81) Gilson (1927), 171-172; 181; Bäck (2000b).

82) William of Ockham, *Summa logicae*, ed. Boehner *et al.* (1974), I, 8, p. 196, ll. 8-32.

bears this out.[83] Avicenna uses 'S qua S' as equivalent to 'S-ness'.[84] Ockham takes the explicit step, which is perhaps implicit in Avicenna, of reducing talk about S-ness to talk about S qua S. 'S qua S', or 'S *per se*', taken reduplicatively, has the same reference as 'S'.[85]

Moody summarizes Ockham's analysis of abstract terms into reduplicative complexes:

> What Ockham insists on, is that in such propositions we are not making a statement about an abstract form that is distinct from individuals, or, in the instance adduced, from the man called Socrates; all we are doing, says Ockham, is to affirm that something is true of Socrates which is also true of every man that exists, or of any man that existed in the past, or that might exist in the future, with the further indication that Socrates is a man.[86]

So, in short, it is Ockham's view that the reference of the subject in 'an S is P' is an individual S. The reference of the subject in an 'S-ness is P', i.e., 'an S qua S is P', proposition is all S's, past, present, and future.

Although Avicenna might accept this analysis, to the extent that it specifies to what *existents* the abstract terms refers, still, he would hold to the need for quiddities in themselves and hence to the need for abstract terms to have a distinctive reference. Otherwise real definitions could not be distinguished from ones composed of *propria* or even from merely contingent descriptions holding of all the individuals that have happened to exist.

Is then Avicenna a nominalist like Ockham? Well, he is no Platonist, despite all his talk of essences. However, his instance of quiddities in themselves removes him from nominalism. Ockham does not accept Avicenna's having individuals in all the categories nor his stress on quiddities in themselves.[87] Still Avicenna's theory of reference is closer to Ockham's than it might appear.

---

83) William of Ockham, *Summa logicae* ed. Boehner *et al.* (1974), I, 8, 55 ff.

84) Avicenna, *Al-Ilāhiyyāt*, ed. Anawati (1960), 196, 6-197, 5; 202, 2-203, 8 (*Metaphysica* V, 1). The connection of reduplication to a modality like necessity is strongly suggested by the tradition which characterizes a mode as determining, or commenting upon, the relation of the predicate to the subject, since a qua phrase does this too. Cf. Ammonius, *In Aristotelis De Interpretatione*, ed. Busse (1897), 214, 25; 215, 14-16.

85) Bäck (1996a), 328-330.

86) Moody (1935), 203-204.

87) It should be noted that Ockham admits individual items only in the categories of substance and quantity. Cf. Kaufmann (1994), 75-90.

The Categorical Assertion

What about the distinctions of personal supposition? In his treatment of categorical propositions with concrete terms taken concretely, Avicenna does have something like a theory of personal supposition but with quite a different emphasis: on the temporal duration during which the things being referred to in a proposition have the quiddities signified by its terms. He focuses not so much on the singular instances of the terms being used but on their temporal parts or segments. In Latin medieval jargon, this amounts to the supposition of the subject term being *ampliated* or *restricted* by features of the sentential context, particularly when it contains paronymous terms. But, like Ockham, Avicenna finds ambiguity in what certain propositions, especially those containing paronymous terms, assert. Once the meaning of such propositions is clarified, the normal theory of reference applies. So he has no need for ampliation. What Stephen Read says about Ockham again holds well for Avicenna:

> It is an interesting fact that almost alone among terminist logicians, Ockham does not speak of ampliation and restriction. The reason appears to be that he disagrees with the truth-condition given above for 'A white thing was black', and similar cases. This proposition, he says, is ambiguous. Rather than meaning that what is or was white was black, it equivocates between 'What is white was black' and 'What was black was white'.[88]

Avicenna devotes a lot of attention to paronyms. The paronymous subject ('R') can have a history different from that of the substantial subject ('S') in which it inheres. Moreover, the attachment of 'R' to 'S' can then dictate the periods of time of the history of S being referred to by that statement, especially when the predication is accidental. A term like 'awake' or 'white' can be taken as a subject in its own right as well as an attributive of an implicit, substantial subject. Indeed, it has to be taken as a subject in its own right in syllogistic reasoning where a proposition is converted.[89] Latin medievals too worried a lot about this problem in their commentaries on the *Prior Analytics*.[90] The *modistae* worried about it in their treatment of intentions.[91] Yet neither group related their discussions to supposition theory proper much.

---

88) Read (2008); Priest and Read (1981).

89) I suggest that the medieval restriction of *suppositio* to substantive terms is based upon the conversion of propositions in syllogistic. Peter of Spain, *Tractatus*, ed. De Rijk (1972), VI. 8-9; VI. 17; Anonymus, *Dialectica Monacensis* VI, ed. De Rijk (1962-1967), II/2, 606, 23-26.

90) Ebbesen (1988), 152.

91) Knudsen (1982), 486.

Also following Aristotle's doctrines on paronymy, Latin supposition theory worries about ambiguities about paronymous, or derivative, terms. As these are adjectives, they are discussed mostly with appellation.[92] Thus Buridan says that 'white' supposits for the thing that is white and appellates whiteness.[93] Still they do not stress as much as Avicenna how the details of the context changes the temporal reference of the terms, especially when one or more of the terms in the propositions are paronymous.

In focusing on paronymous terms, Avicenna has an emphasis more like that of the Latin Modists than that of the logicians in their supposition theory. Thus Peter of Auvergne worries about them, as in the example, '*album potest esse nigrum*' ('[something] white can be black').[94] Avicenna would not agree with the modist contention, that 'both the concrete and the abstract accidental terms have a single *significatum*, an accidental form.'[95] At the very least, Avicenna has a more complex theory than this.

Unlike the Modists however, Avicenna does not separate off paronymous terms for special attention.[96] For him a term need not be paronymous to change the time reference of a term, as with 'eclipse' and 'moon'. Ockham has a similar position about 'Judas is reprobate'. For this means that Judas will be judged and dammed in the future, but not at present.[97] Hence 'Judas' in this context—namely, in the scope of 'reprobate'—refers to the future Judas, and the statement expresses not a simple categorical but a future contingent proposition. The predicate 'reprobate' has changed the period of time in which the subject term is intended to refer to Judas. However, Avicenna manages to make similar claims about much more mundane statements, like 'the scribe is awake'.

For Avicenna quantified propositions allow for their terms to have a descent to singulars existing *in re*. So these terms have personal supposition of various sorts. The predicational propositions do not allow for such a descent. If I had to

92) Originally appellation, like predication, consisted in an assertion of present existence, while supposition, or being a subject, concerned the reference of the subject term, at whatever time it existed. Later on appellation came to mean just the predication of attributes, while supposition concerned reference to the present time only. Cf. Walter Burley, *On the Purity of the Art of Logic: The Shorter and the Longer Treatises*, ed. and transl. Spade, I. 2 (204).

93) Buridan, *Summulae*, ed. Van der Lecq (1998), 4.1.4; 4.5.1.

94) Ebbesen (1988), 108.

95) Ebbesen (1988), 121.

96) This may be due more to the disorganization in his writings than to anything else.

97) William of Ockham, *Tractatus de praedestinatione et de praescientia Dei respectu futurorum contingentium*. ed. Boehner-Brown (1978), q. 1, p. 508, l. 42. Cf. Spade (2002), 321.

name a type of supposition for the predicational sort, I would call it 'simple [*mu-laq*] supposition', sc., a supposition where the terms are taken absolutely or simply, apart from the existence of any instances at some time. In this sense—not the usual, Latin sense of 'simple supposition'—we can say that the lion, the (real) phoenix, and even the stegosaurus are animals. As Avicenna notes, such expressions appear commonly in science. These days we call them 'generic descriptions'. They do not describe formal features, like being species or nouns, but rather essential, necessary attributes of the objects being studied.

I find it unclear what sort of descent to singulars Avicenna would allow in predicational affirmations. On the one hand, judging by what Avicenna says about the goat-stag and the griffon, such terms cannot serve as subjects of true predicational affirmations. After all, Avicenna has an explicit existential import assumption for them. On the other hand, Avicenna allows for there to be more possible things than the ones actually existing at some time. He speaks of there being a heptagonal house being a real universal even if there had never been a heptagonal house. God can actualize more species of things than those that actually exist.[98] So perhaps he would allow for true predicational assertions and even science about goat-stags and griffons having no actual instances—if not for us, in the mind of God.[99] If we allow Avicenna to have a domain of all possible beings, then all propositions will have existential import and all universal terms a descent to singulars. I am inclined to this solution, but do not find the text decisive.[100] Latin medievals are clearer: the chimera has no personal supposition since it has no instances.

Again Avicenna has a different focus when he considers the descent from a universal to singulars than standard medieval Latin supposition theory. He focuses on the time segments, the temporal parts of the individuals *in re* having that universal quiddity. In contrast, the usual discussion of medieval supposition focuses on the individuals being referred to: whether all taken conjunctively or disjunctively, or whether only one taken definitely or indefinitely. Avicenna, perhaps finding it obvious, largely skips this stage of the analysis.

---

98) Avicenna, *Al-Ilāhiyyāt*, ed. Anawati (1960), V.1, 195, 8; Bäck (2001).

99) Avicenna, *Al-Qiyās*, ed. Zayid (1964), 83, 10-11: 'So what they have supposed of the absolute/categorical, that it is necessary that the judgement in it be of existents at some time, has turned out to be absurd.'

100) Thom (1996), 344-349 claims that Aristotle himself has the position in his modal syllogistic.

In Latin terms, Avicenna focuses more on appellation (in the earlier sense), where the reference is at the present time only, than on supposition, where reference is at some time. So the early *Dialectica Monacensis* says that 'man' supposes for Caesar and the Antichrist but does not appellate them as they do not exist now.[101] Yet 'now' has its obscurities, if we do not restrict it to the mere instant of Aristotle's 'moving now'. A proposition in the present tense covers a period of time, a specious present. Later medievals seem to recognize this and deemphasize appellation.

The Latin medievals do get around to Avicenna's concerns somewhat in their doctrines of ampliation and restriction.[102] But they don't do as well. For instance, Buridan speaks of the *status* of the term without ampliation or restriction, 'when the term supposits for or appellates all of its *significata* at the present time.'[103] A term with such a normal *status* has its usual supposition. When the term is ampliated or restricted in various ways, its supposition changes. Peter of Spain and Buridan speak of changing the tense of the verb, putting the term in the scope of a modal or intensional operator, and other grammatical variants, like using adjectives such as 'future' and adverbs like 'necessarily'. Such ways amount to Avicenna's talk of adverbial qualification.

In contrast, Avicenna concentrates on problems coming from how the sentential context varies the reference of the terms even in a statement in the present tense. In Latin medieval terms, he is locating ambiguities within the *status*. These ambiguities have various causes: 1) the signification of the terms specifies, sometimes ambiguously, the time period for which the present-tense statement is meant to hold, 2) the present tense covers more than the present instant: a specious present, and 3) scientific propositions are stated in the present tense, and yet are intended to hold for more than the present time but usually for all time.

I submit that Avicenna does better with these details concerning the *status*. For him propositions dealing with quiddities *in re*, namely, those having concrete terms taken concretely, do refer to individual things, just as with personal supposition. However, in contrast to the Latin medievals, he has a descent to singulars primarily of periods of time during which those things persists. As we *moderni*—or *post-moderni*—might put it, the descent is to time slices of space-time worms. This doctrine complements Latin medieval supposition theory.

---

101) *Dialectica Monacensis* VI, ed. De Rijk, in: id., *Logica modernorum* (1962-1967), II/2, 616, 20-30.

102) Already the *Dialectica Monacensis* VI has a long discussion of *restrictio*.

103) John Buridan, *Summulae*, ed. Van der Lecq (1998), 4.6.1. Cf. Klima (2001), n. 95.

Also it fits better with the ontology of modern physics than the usual Aristotelian metaphysics of substance.

Conclusions

Avicenna seems a lot more concerned about contextual ambiguities of temporal duration than most Latin authors on supposition. His treatment of paronymous terms looks more like the theories of the Modists, who in fact seem to have been somewhat influenced by Avicenna. On the other hand, the Latin medievals discuss their simple and material supposition, as well as the descent to singulars, in much more detail than Avicenna does.

Both Avicenna's theory and Latin supposition theory agree that a term has its reference fixed only within a sentential context.[104] Likewise, today, Walter Kintsch, in constructing an algorithm for determining the meaning of a predicate, says, 'In N-VP sentences, the precise meaning of the verb phrase depends on the noun it is combined with.'[105] Avicenna focuses on nouns having paronymous features and on the determination of the temporal meaning of the verb phrase. In contrast, in supposition theory the Latin medievals focus more on whether and in what ways the referents of the subject and predicate, taken nominally, are universal or particular. Avicenna may have a theory more congenial with a modern ontology of space-time slices to go along with his interest in infinitesimals.[106]

Bibliography

Primary Literature

Alexander of Aphrodisias, *In Aristotelis Analytica Priora Commentaria*, ed. M. Wallies (Commentaria in Aristotelem Graeca, II,1; Berlin 1883)

Al-Fārābī, *Commentary and Short Treatise on Aristotle's* De Interpretatione. Translated with an Introduction and Notes by F.W. Zimmermann (Classical and Medieval Logic Texts, III; London 1981; repr. 1982, repr. with corrections 1991)

——, *Greater Commentary on De Interpretatione*, ed. W. Kutsch and S. Morrow (Beirut 1960)

104) Avicenna, *Al-'Ibāra*, ed. al-Ḵuḍayrī (1970), 23,1; William of Ockham, *Summa logicae*, ed. Boehner *et al.* (1974), I, 65

105) Kintsch, 'Predication', *Cognitive Science* (in press).

106) Rashed (2000).

Ammonius, *In Aristotelis De Interpretatione Commentarius*, ed. A. Busse (*Commentaria in Aristotelem Graeca*, IV.5; Berlin 1897)
Aristotle, *Prior and Posterior Analytics*, ed., transl. and comm. by W.D. Ross (Oxford 1949)
Avicenna (Ibn Sīnā), *Al-ʿIbāra*, ed. M. al-Ḵuḍayrī, *aš-Šifāʾ* (Cairo 1970)
——, *Manṭiq al-Mašriqiyyīn*, ed. Al-Ḵaṭīb *et al.* (Cairo 1910)
——, *Kitāb al-Najāt*, ed. A. Al-Kurdi (Beirut 1938)
——, *Al-Madkhal*, ed. A. al-Ahwānī, G. Anawati and M. Al-Ḵuḍayrī, *aš-Šifāʾ* (Cairo 1952)
——, *Kitāb al-Burhān*, ed. ʿA. ʿAfīfī, *aš-Šifāʾ* (Cairo 1956)
——, *Al-Ilāhiyyāt*, ed. G. Anawati, *aš-Šifāʾ* (Cairo 1960)
——, *Al-Qiyās*, ed. S. Zayid, *aš-Šifāʾ* (Cairo 1964)
——, *Kitāb aṭ-Ṭabīʿiyyāt*, ed. S. Zāyid, *aš-Šifāʾ* (Cairo 1965-1983)
——, *Al-Ishārāt wa'lTanbīhāt*, ed. S. Dunya (Cairo 1971)
——, *Al-Nafs*, ed. G.C. Anawati and S. Zayid, *aš-Šifāʾ* (Cairo 1975)
Gualterus Burlaeus, in: P. Kunze, *Walter Burleigh. Von der Reinheit der Kunst der Logik. Erster Traktat: Von den Eigenschaften der Termini* (lat.—dt.) (Hamburg 1988)
——, *De Puritate Artis Logicae Tractatus Longior, with a Revised Edition of the Tractatus Brevior*, ed. by Ph. Boehner O.F.M. (Louvain-Paderborn 1955)
Gualterus Burlaeus: Walter Burley, *On the Purity of the Art of Logic: The Shorter and the Longer Treatises*, ed. and transl. P.V. Spade (Yale Library of Medieval Philosophy; New Haven 2000)
Guillelmus de Ockham, *Summa logicae*, ed. P. Boehner, G. Gál and S. Brown (Guillelmi de Ockham Opera Philosophica I. St. Bonaventure, N.Y., 1974)
——, *Tractatus de praedestinatione et de praescientia Dei respectu futurorum contingentium*, ed. Ph. Boehner, rec. S. Brown (Guillelmi de Ockham Opera Theologica, II; St. Bonaventure, N.Y. 1978)
Guillelmus de Shireswood: William of Sherwood, *Introductiones in Logicam. Einführung in die Logik*, ed. H. Brands and C. Kann (Hamburg 1995)
Johannes Buridanus, *Summulae. De suppositionibus*, ed. R. van der Lecq (Artistarium, 10-4; Nijmegen 1998)
Johannes Buridanus: John Buridan, *Summulae. De dialectica*, transl. G. Klima (New Haven 2001)
——, *Sophisms on Meaning and Truth*, transl. T.K. Scott (New York 1966)
Johannes Philoponus, *In Aristotelis Analytica Priora Commentaria*, ed. M. Wallies (Commentaria in Aristotelis Graeca 13.2; Berlin 1905)
Petrus Abaelardus, *Dialectica*. First Complete Edition of the Parisian Manuscript, with an Introduction, by L.M. de Rijk (Philosophical Texts and Studies, I; Assen 1972)
——, *'Logica Ingredientibus'*, part 1: Glossae super Porphyrium, in: *Peter Abaelards Philosophische Schriften. I. Die Logica 'Ingredientibus. 1: Die Glossen zu Porphyrius*, ed. Bernhard Geyer (Beiträge zur Geschichte der Philosophie des Mittelalters 21.1; Münster 1919)
——, *'Logica Ingredientibus'*, part 2: *Glossae super Praedicamenta Aristotelis*, in: *Peter Abaelards Philosophische Schriften. I. Die Logica 'Ingredientibus'. 2: Die Glossen zu den Kategorien*, ed. Bernhard Geyer (Beiträge zur Geschichte der Philosophie des Mittelalters, 21,2; Münster 1921)
——, *'Logica Ingredientibus'*, part 3: *Glossae super Peri Ermenias*, in: *Peter Abaelards Philosophische Schriften. I. Die Logica 'Ingredientibus'. 3: Die Glossen zu Peri hermenias*, ed. Bernhard Geyer (Beiträge zur Geschichte der Philosophie des Mittelalters, 21,3; Münster 1927)
Petrus Hispanus: Peter of Spain, *Tractatus, called afterwards Summulae logicales*. First Critical Edition from the Manuscripts with an Introduction, by L.M. de Rijk (Philosophical Texts and Studies, XXII; Assen 1972)

Ps.-Simplicius, *In libros Aristotelis De anima commentaria*, ed. M. Hayduck (Commentaria in Aristotelis Graeca, 110; Berlin 1882)
Walter Burley *vide s.n.* Gualterus Burlaeus
William of Ockham, *vide s.n.* Guillelmus de Ockham
William of Sherwood: *vide s.n.* Guillelmus de Shireswood

Secondary Literature

Ardeshir, M. (2008), 'Ibn Sīnā's Philosophy of Mathematics', in: S. Rahman, T. Street, and H. Tahiri. Berlin, eds., *The Unity of Science in the Islamic Tradition* (Berlin: 2008, 43-61)
Ashworth, E.J. (1978), *The Tradition of Medieval Logik and Speculative Grammar* (Toronto 1978)
Bäck, A. (1987), 'Avicenna and Existence', in: *Journal of the History of Philosophy* 25 (1987), 351-367
——, (1989), 'Ibn Sina on the Individuation of Perceptible Substance', in: *Proceedings of the PMR Conference*, 14 (1989), 23-42
——, (1996a), *On Reduplication* (Leiden 1996)
——, (1996b), 'The *Triplex Status Naturae* and its Justification', in: I. Angelelli and M. Cerrezo, eds., *Studies in the History of Logic* (Berlin 1996, 133-153)
——, (1996c), 'Avicenna and Descartes on the Wax Example', in: S. Brown, ed., *Proceedings of the S.I.E.P.M. Conference* (Turnhout 1996, 163-178)
——, (2000a), 'The Formal Structure of Scotus' Formal Distinction', in: I. Angelelli and P. Perez-Ilzarbe, eds., *Medieval and Renaissance Logic in Spain*. Acts of the 12th European Symposium on Medieval Logic and Semantics, held at the University of Navarra, Pamplona, 26-30 May 1997 (Philosophische Texte und Studien, 54; Hildesheim-Zürich-New York 2000, 411-438)
——, (2000b), *Aristotle's Theory of Predication* (Philosophia antiqua. A Series of Studies on Ancient Philosophy, LXXXIV; Leiden 2000)
——, (2001), 'Avicenna and Averroes: Modality and Theology', in: T. Buchheim, C. Kneepkens and K. Lorenz, eds., *Potentialität und Possibilität* (Stuttgart and Bad Canstatt 2001, 125-145)
——, (2004), 'Avicenna on the Categorical Assertion', in: A. Maierù and L. Valente, eds., *Medieval Theories on Assertive and Non-assertive Language* (Florence 2004, 141-162)
——, (2005), 'Imagination in Avicenna and Kant', in: *Topícos* 29 (2005), 101-130
——, (2008a), 'Avicenna the Commentator', in: L.A. Newton, *Medieval Commentaries on Aristotle's Categories* (Leiden 2008, 31-72)
——, (2008b), '*Islamic Logic*?', in: S. Rahman, T. Street and H. Tahiri, eds., *The Unity of Science in the Islamic Tradition* (Berlin 2008, 255-279)
Biard, J. (1989), *Logique et théorie du signe au XIV*[e] *siècle* (Paris 1989)
Boer, R.J. de (1967), *The History of Philosophy in Islam*. Transl. E.R. Jones (London 1903, repr. New York 1967)
Braakhuis, H.A.G. (1977), 'The Views of William of Sherwood on Some Semantical Topics and Their Relation to Those of Roger Bacon', in: *Vivarium* 15 (1977), 111-142
Brands, H. and C. Kann (1995), *vide s.n.* Guillelmus de Shireswood
Dutilh Novaes, C. (2007), *Formalizing Medieval Logical Theories*—Suppositiones, Consequentia *and* Obligationes (Berlin 2007)
Ebbesen, S. (1988), 'Concrete Accidental Terms: Late-Thirteenth-Century Debates about Problems Relating to such terms as "album" ', in: N. Kretzmann, ed., *Meaning and Inference in Medieval Philosophy. Studies in Memory of Jan Pinborg* (Dordrecht-Boston-London 1988, 107-161)

——, (1998), 'The Paris Arts Faculty: Siger of Brabant, Boethius of Dacia, Radulphus Brito', in: John Marenbon, ed., *Medieval Philosophy. The Routledge History of Philosophy III* (London-New York 1998, 269-290)

Elamrani-Jamal, A. (1983), *Logique Aristotélicienne et Grammaire Arabe* (Paris 1983)

Gilson, E. (1927), 'Avicenne et le point de départ de Duns Scotus', in: *Archives d'histoire doctrinale et littéraire du moyen âge* 2 (1927), 402-408

Goichon, A.-M., *La distinction de l'essence et de l'existence d'après Ibn Sina* (Brugge 1937)

Irwin, T.H. (1982), 'Aristotle's Concept of Signification', in: M. Schofield and M. Nussbaum, eds., *Language and Logos* (Cambridge: 1982, 241-266)

Jacobi, K. (1986), 'Peter Abelard's Investigations into the Meaning and Functions of the Speech Sign "Est" ', in: S. Knuutila and J. Hintikka, eds., *The Logic of Being. Historical Stdues* (Dordrecht 1986, 145-180)

Kaufmann, M. (1994), *Begriffe. Sätze, Dinge: Referenz und Wahrheit bei Wilhelm von Ockham* (Leiden 1994)

Kintsch, W. (2001), 'Predication', in: *Cognitive Science*, 25 (2001), 173-202

Knudsen, C. (1982), 'Intentions and Impositions', in: N. Kretzmann, A. Kenny, J. Pinborg, eds., E. Stump, ass. ed., *The Cambridge History of Later Medieval Philosophy. From the Rediscovery of Aristotle to the Desintegration of Scholasticism 1100-1600* (Cambridge-London-New York-New Rochelle-Melbourne-Sydney 1982, 479-495)

Leff, G. (1975), *William of Ockham* (Lanham, Maryland 1975)

Madkour, I. (1966), 'Duns Scotus entre Avicenne et Averroës', in: *Studia Scholastica Scotistica* 1 (1966), 169-182

Marenbon, J. (1997), *The Philosophy of Peter Abelard* (Cambridge 1997)

Moody, E.A. (1935), *The Logic of William of Ockham* (New York 1935; anastatic repr. 1965)

Owen, G.E.L. (1965), 'Inherence', in: *Phronesis* 10 (1965), 97-105

Pironet, F. (1997), *The Tradition of Medieval Logic and Speculative Grammar. A Bibliography (1977-1994)* (Turnhout 1997)

Prantl, C. (1855), *Geschichte der Logik in Abendlande* (Leipzig 1855)

Priest, G., and S. Read (1981), 'Ockham's Rejection of Ampliation?', in: *Mind* 90 (1981), 274-279

Rashed, R. (2000), *Les Mathématiques infinitésimales du IXe au XIe siècle*, 3 (London 2000)

Read, S. (2008), 'Medieval Theories of Properties of Terms', in: Edward N. Zalta, ed., *The Stanford Encyclopedia of Philosophy (Fall 2008 Edition)*, URL = <http://plato.stanford.edu/archives/fall2008/entries/medieval-terms/>

Rijk, L.M. de (1962-1967), *Logica Modernorum. A Contribution to the History of Early Terminist Logic*. I. *On the Twelfth Century Theories of Fallacy*; II/1: *The Origin and the Early Development of the Theory of Supposition;* II/2: *Texts and Indices* (Assen 1962-1967)

Rosier-Catach, I. (1983), *La grammaire spéculative des modistes* (Paris 1983)

——, (1995), '*Res significata et modus significandi*: Les implications d'une distinction médiévale', in: S. Ebbesen, ed., *Sprachtheorien in Spätantike und Mittelalter* (Tübingen 1995, 135-168)

Ross, W.D. (1949), *Aristotle, Prior and Posterior Analytics*. A Revised Text with Introduction and Commentary (Oxford 1949)

Schulthess, P. (1992), *Sein, Signification und Erkenntnis bei Wilhelm von Ockham* (Berlin 1992)

Scott, T.K. (1966), *John Buridan: Sophisms on Meaning and Truth*. Translation and Introduction (New York 1966)

Spade, P.V. (1988a), 'The Logic of the Categorical: The Medieval Theory of Descent and Ascent' in: N. Kretzmann, ed., *Meaning and Inference in Medieval Philosophy* ed. (Synthese Historical Library, 32; Dordrecht-Boston-London 1988, 187-224)

——, (1988b), *Lies, Language and Logic in the Late Middle Ages* (London 1988)
——, (2002), *Thoughts, Words and Things: An Introduction to Late Mediaeval Logic and Semantic Theory.* Version 1.1. August 9, 2002. http://www.pvspade.com/Logic
Street, T. (2000), 'Avicenna and Ṭūsī on the Contradiction and Conversion of the Absolute,' in: *History and Philosophy of Logic* 21.1 (2000), 45-56
——, (2004), 'Arabic Logic', in: Peter Adamson and Richard C. Taylor, eds., *Cambridge Companion to Arabic Philosophy* (Cambridge 2004, 247-265)
Thom, P. (1996), *The Logic of Essentialism* (Dordrecht 1996)
Weinberg, J. (1967), *A Short History of Medieval Philosophy* (Princeton 1967)
Wheeler, M. (1999), 'Semantics in Aristotle's Organon', in: *Journal of the History of Philosophy* 37/2 (1999), 191-226
Wilks, I. (1998), 'Peter Abelard and the Metaphysics of Essential Predication', in: *Journal of the History of Philosophy* 36 (1998), 356-385

XIITH CENTURY

Supposition Theory and Porretan Theology: *Summa Zwettlensis* and *Dialogus Ratii et Everardi**

Luisa Valente
University of Rome

Abstract

The article investigates how the problem of (linguistic) reference is treated in the theology of two pupils of Gilbert of Poitiers by means of *suppo** terms (*supponere*; *suppositus,-a,-um*; *suppositio*). Supposition is for Gilbert an action performed by a speaker, not a property of terms, and he considers language as a system for communication between human beings: key notions are the 'sense in the author's mind' and the 'interpreter's understanding'. In contrast, the two Porretans tend to objectify language as a formal system of terms. *Suppositio* becomes in the *Summa Zwettlensis* the name itself as subject term in a proposition, and is divided into many kinds; formal rules are described which govern the influence of the predicate on the subject term's denotation. In Everard of Ypres' *Dialogus Ratii et Everardi, supponere* is a function (*officium*) of the name, and 'human is a species of individuals' is, as in some logical treatises and differently from Gilbert, a case of rhetorical transfer.

Keywords

Gilbert of Poitiers, *Summa Zwettlensis*, Peter of Poitiers, Everard of Ypres, *suppositio discreta, suppositio communis, suppositio confusa, suppositio simplex*, contextual approach

Introduction

Gilbert of Poitiers (d. 1154) develops in his commentaries on Boethius' *Theological treatises* a particularly fine philosophical system which also contains a theory of language in its many possible forms. Gilbert's reflections on language are deeply connected with his ontology, which has been meaningfully

*) I would like to express my sincere gratitude to Chris Martin for his help both in correcting the English and in commenting on the content of this paper.

defined as a 'metaphysics of the concrete', and with his epistemology.[1] According to him the concrete name (the basic type of name, e.g., 'human') signifies the subsisting thing (this is the substance of the name: *substantia nominis*, e.g., a particular human being) and also one of its forms (this is the quality of the name: *qualitas nominis*, e.g., its own humanity). The basic type of proposition expresses the inherence of the form signified as *qualitas nominis* by the predicate term in the subsistent signified as *substantia nominis* by the subject term: in '[a] human is white' we declare that a form, whiteness, is inherent in a particular human being. The linguistic fact that we can predicate more predicates of the same name used as subject terms in a proposition reflects the ontological fact that 'each subsistent is 'one' but also 'many'', since it is and is what it is because of its many forms: 'Quapropter cum multa predicantur de uno, quodam modo illud unum est multa quoniam scilicet est multis'.[2] Nevertheless, language does not always reflect reality. There are propositions which have the form subject + predicate, but do not express any inherence of forms in subsistents. This happens mainly in logic, mathematics and theology, where the relations between subject terms, predicates, subsistents and forms are different than in basic language, the language used for speaking about natural subsistents. Mathematics (*mathematica* or *disciplinalis speculatio*) is intended by Gilbert as a science which investigates forms as abstracted from things, their internal structure if they are composed and not simple forms, and their connections to other forms. Thus the basic type of name in the language of mathematics is the abstract name, as in 'albedo est color'. Mathematics is fundamental for natural philosophy or physics (*rationalis speculatio*), for natural philosophy cannot understand the subsistent in its many formal features if mathematics does not beforehand investigate what these formal features are in themselves and how they relate to each other: 'Neque enim rationalis speculatio perfecte id quod est esse aliquid capit, nisi disciplinalis quoque id unde

[1] For secondary literature about Gilbert's philosophy see Valente (2008b), n. 3 and 4, and Valente (2011). The following studies concern in particular Gilbert's semantics: Maioli (1979); Nielsen (1976) and (1982), 103-114; De Rijk (1987), (1988) and (1989); Jolivet (1987) and (1990); Kneepkens (1987) and (2000); Jacobi (1995a) and (1995b); Valente (2008a), 123-149. The expression 'metaphysics of the concrete' is by Maioli. For Gilbert's epistemology, see, besides the studies mentioned above (in particular Nielsen (1982), 87-95; Maioli (1979), 131-143; Jacobi (1995b), also Haas (1987); Jolivet (1990); Marenbon (2002); Catalani (2005).

[2] Gilbert of Poitiers, *Expositio super librum Boecii De trinitate*, ed. Häring (1966), 176, 17-19; in the rest of the article the quotations from Gilbert's commentaries on Boethius are taken from this edition.

illud est quid est firmiter teneat'.[3] Natural philosophy, mathematics and logic are made up of some propositions which are true in their proper meaning and other propositions which are the result of a transfer of meaning (*transumptio, denominatio*) and must therefore be interpreted in order to disclose their true sense.[4] In theology, due to the radical transcendence of its object, which cannot be grasped by the intellect and consequently cannot be properly expressed in words, proper propositions do not exist, but we always have a semantic and formal transfer from the language of philosophy of nature. Thus, theological discourse is always improper. Nevertheless, this transfer is not arbitrary but 'proportional' (*proportionalis transumptio*), and this means that to a certain extent we are allowed to analyse and check the validity of theological propositions (i.e., their truth as well as their orthodoxy) using as a model the language of natural philosophy.

The aim of this contribution is to investigate—within the context described above—how two of Gilbert's pupils use in theological contexts the complex *suppo**—that is the verb *supponere*, the participle *suppositus,-a,-um*, the substantive *suppositio*—as technical terms in semantics. As has been already noticed by many scholars, Gilbert and his pupils in their theological writings have something interesting to say about the reference of subject terms in the propositional context, and their theories on this theme on some points anticipate later developments in the field of logic centered on the notion of *suppositio* as one of the 'properties of terms'.[5] Among the different Porretan theological texts, I have chosen, by reason of their particular speculative finesse, the *Summa Zwettlensis* and Everard of Ypres' *Dialogus Ratii et Everardi*. While for the *Dialogus* we have a quite certain date between 1190 and 1198,[6] the date of *Zwettlensis* is much more uncertain, falling somewhere between the end of the 1140s and the 1280s.[7]

---

3) *Ibid.* 84: 74-76. Cf. Maioli (1979), 138.

4) Valente (2008a), 123-149.

5) See Pinborg (1968), 156 and (1972), 48-49; Nielsen (1976), 43 and (1982), 105; Maioli (1979), 66 and 101; Kneepkens (1987), 337; De Rijk (1987), 170; De Libera and Rosier (1992), 117 and 124-126. De Libera (1987), 455 f., writes about *suppo** terminology in the Porretan theologian Alain de Lille, and Ebbesen (1987), 419-424, about *suppositio* in Steven Langton, who was very much influenced by Porretan theology; more generally on *suppo** terminology in theological works of the second half of the 12th century, both Porretan and non-Porretan, see Valente (2008a), ch. III.

6) Everard of Ypres, *Dialogus Ratii et Everardi*, ed. Häring (1953), 143-147.

7) Häring, who edited the text, dates it before 1150 since he identifies the author, named *Petrus Pictavensis* in one of the two manuscripts, with Peter of Vienna, who left France in 1150, and the *Summa* seems to be written in France (*Summa Zwettlensis*, ed. Häring (1977), 9). Sten Ebbesen tends to consider the *Summa* later, around the 1170s, because of its developed semantic

My inquiry follows, as is obvious, two lines: first, how the *suppo** terminology was used by the two Porretans; second, whether and how these masters reflected on problems connected to the reference of subject terms within the proposition, whether using this terminology or not. The main thesis of this contribution is that the two Porretan authors, while sharing with Gilbert the same ontology which consists of considering the subsistents as 'bearers of forms', approach the problem of linguistic reference in a manner that is quite distinct from Gilbert's. In Gilbert's Commentaries on Boethius' *Opuscula sacra*, *suppo** terms mainly concern the act of a speaker (or of the author of a written text) that consists of referring—by choosing a name as subject term in a proposition—to one or more subsistent things as what the speech act (or the text) is about. Supposition is for Gilbert an action performed by a speaker, not a property of terms,[8] and his 'contextual approach' has a pragmatic touch: 'we do not predicate something in order to 'supposit' as much as we 'supposit' in order to predicate'.[9] Language is considered by Gilbert as a system for communication between human beings; key notions are the 'sense in the author's mind'[10] and the 'reader's attention'.[11] The phenomenon of 'disciplinal' discourse ('human is a species of individuals') is treated by means of these hermeneutic

terminology (see his article in this volume). For my part, I think that this *Summa* uses the semantic terminology in such a peculiar way—if compared both to Gilbert and to the other theologians, Porretans—and non-Porretans—, that we cannot deduce from it any conclusion about its date other than that it clearly depends on Gilbert's *Commentaries*.

8) Cf., e.g., Gilbert of Poitiers, *Expositio in Contra Euticen*, ed. Häring (1966), 301, 77-82: 'si quis de Platone loquatur—siue unum siue multa de ipso affirmet uel neget—uerbo singulari hoc faciet. Non enim dicet 'Plato legunt', sed 'legit'. Nec 'Plato sunt' sed "est homo albus astrologus' et huiusmodi alia. Si uero Platonem et Ciceronem supponat, non dicet 'Plato et Cicero est' sed 'sunt' hec uel illa'; Gilbert of Poitiers, *Expositio in Contra Euticen*, ed. Häring (1966), 345, 40—346: 47: 'Nam et cum homo sit anima et corpus, hoc eius i.e., hominis nomine quod est 'anima' uel 'rationale' eodem supposito, poterunt uere predicari non modo illa que sunt propria animarum uerum etiam illa que sunt propria corporum: ut *octo animae intraverunt in archam*. Hic enim hoc plurali nomine, quod est 'animae', non animas hominum sed ipsos homines auctor supposuit et quod non animabus sed solis corporibus conuenit, 'intrare' uidelicet, predicauit.'

9) Gilbert of Poitiers, *Expositio in Contra Euticen*, ed. Häring (1966), 349, 50-51: '[...] in predicatiuis enuntiationibus non tam suppositurі aliquid predicamus quam predicaturi supponimus.'

10) Cf. Gilbert of Poitiers, *Expositio super librum Boecii De trinitate*, ed. Häring (1966), 67, 50-54, where the 'intellectus quem scripta faciunt' is opposed to the 'intellectus ex quo <scripta> facta sunt', which must be the real goal of the interpreter. On Gilbert's hermeneutics, see Jolivet (1990) and Valente (2004).

11) Cf., e.g., Gilbert of Poitiers, *Expositio in Contra Euticen*, ed. Häring (1966), 296, 31-33: 'Illud enim lectoris uigilantia debet attendere, acceptis dictionum significationibus, quibus significatorum propositi conueniat ratio et de quibus interpres id quod dictum est, intelligendum explanet.'

notions and not of a special kind of supposition.[12] With his pupils, things begin to change on these points.

SUMMA ZWETTLENSIS: A Quasi-formal System of Trinitarian Predications

In the first book of the *Summa Zwettlensis* the author—a master called Peter of Poitiers, but who is not the Peter of Poitiers, author of the *Quinque libri sententiarum*[13]—analyzes different kinds of Trinitarian propositions and tries to build up a sort of formal system aimed at deducing from the different kinds of predicates the objects referred to by the subject terms. The Trinitarian system of this part of the *Summa* is based on the following elements: three persons individuated by some properties, and different kinds of names.[14]

From the text, we can infer that the author considers, in this part of his work, three types of names:

- *discretiva <nomina>*, or *propriae appellationes* may refer to just one of the three persons, e.g., *Pater, genitor, Filius, genitus, Spiritus Sanctus*; these names are similar to proper names in ordinary language;
- *communia discretiva <nomina>* or *appellationes communes <discretivae>* can refer to two of the three persons, e.g., *emittens, procedens*, and are similar to 'dual' names in ordinary language;
- *communia <nomina>* or *communes appellationes* can refer to the three persons, and are similar to appellative names in ordinary language; some of these names can be said of the three persons singularly and in common (*singillatim et simul de omnibus*), e.g., *Deus, potens, dominus*; some may be said of the three persons in common but not singularly (*simul de omnibus sed non singillatim*), e.g., *trinus*.

While most theologians of the second half of the twelfth century introduce terminology connected with questions of reference within the context of the classification of divine names, this author builds up his theory of Trinitarian language on the basis of different types of propositions, described as different combinations of different kinds of subjects and predicates. In this sense, the

12) On Gilbert's use of *suppo** terms and his approach to the problem of linguistic reference, see Valente (2011).

13) His philosophy is in fact clearly influenced by that of Gilbert of Poitiers, and his style and terminology are very different from those of Peter of Poitiers.

14) On the semantics in this treatise, see Kneepkens (2000), 260-261 and 271-273.

author of this *Summa* has more in common with Gilbert of Poitiers than with theologians such as Alain of Lille, Simon of Tournai and the other masters of the last decades of the twelfth century.

The aim of this part of the *Summa* is to locate 'formal' rules which would enable one to avoid the Sabellian confusion between divine nature, persons, and personal properties by interpreting all theological propositions ('omnia que de Deo dicuntur') as referring to one, two or all three persons:

> Abhorring the Sabellian confusion of the persons, a deep enemy of the Trinity, we believe that all the things which are said of God have to be understood in the sense that they are said correctly either of one of the single persons, or of two of them, or of all three of them.[15]

This corresponds to the Gilbertinian and Porretan view, which states that what subject terms refer to in proper predications are always and only subsistents, that is, the *substantiae subiectae*: these are, in the Trinitarian system, the persons. The divine nature and what is pertinent to it may only be predicated of the three persons:

> The divine nature and everything that is understood to pertain to it are understood to be correctly predicated of all the three persons, these being 'supposited for' singularly or all together.[16]

When we use as subject or predicate terms names which signify the nature, the object of the proposition (*subiectum, id de quo est sermo, id de quo agitur*) is always the persons:

> Using a noun which signifies the nature and not a person we sometimes talk of a person. We say for example: 'Divine nature is Father', or 'Divinity is Father', and vice versa. For in these propositions the discourse concerns only the Father.[17]

---

15) *Summa Zwettlensis*, ed. Häring (1977), 62, 127: 'Nos autem Sabellianam personarum confusionem Trinitati penitus inimicam abhominantes credimus quoniam omnia que de deo dicuntur ita intelligenda sunt ut uel de singulis personis discretiue uel de duabus communiter uel de tribus recte intelligantur dici.'

16) *Summa Zwettlensis*, ed. Häring (1977), 63, 132: 'Ipsa [sc. natura] et quicquid ad ipsam intelligitur pertinere de singulis personis et singillatim et communiter suppositis recte praedicari intelligitur.'

17) *Summa Zwettlensis*, ed. Häring (1977), 67, 149: 'Nomine itaque quo natura non persona significatur agimus non nunquam de persona. Sic enim dicimus: Diuina natura est Pater uel deitas est Pater et e conuerso. In hiis enim sermo est de solo Patre.'

The location of which of the persons is (are) referred to by the subject term(s) depends on the kind of terms used as subject terms and predicates.

The author distinguishes three main sorts of Trinitarian propositions plus some subsets:

1. Propositions with *suppositiones discretivae* and *communia <nomina>* as predicates, e.g., 'Pater est Deus': the objects (*subiecta*) of these propositions are determined by properties distinguished by the subject terms ('Qualia sunt subiecta earum discretive proprietatum certificant suppositiones'): distinguishing the property of fatherhood, the *propria appellatio 'Pater'* refers only to the Father.[18]
2. Propositions with *suppositiones communes discretivae* and *communia <nomina>* as predicates, e.g., 'Emittens est Deus': the objects of these propositions are whichever of the persons for which the *commune discretivum <nomen>* used as subject term 'supposits' ('de qualibet illarum agimus personarum quarum siquidem sunt supponentes appellationes communes'): the *appellatio communis 'emittens'* 'supposits' for the Father and the Son; thus the proposition is indifferent about the Father or the Son.[19]
3. Propositions with *suppositiones communes* and predicates of whatever kind: the objects of these propositions depend always on the predicates ('qualia sint subiecta, praedicata certificant').[20]
   - 3.1. When the subject terms are *communia <nomina>* which may be predicated of their own *supposita* both singularly and in common, e.g., *Deus*, and
     - 3.1.1. we have *discretiva <nomina>* as predicates, e.g., 'Deus est Pater', 'Deus est Filius', 'Deus est Spiritus Sanctus', then the objects of the propositions (*subiectum*) are chosen on the basis of the property predicated;[21]
     - 3.1.2. when the subject terms are *communia <nomina>* which may be predicated of their own *supposita* both singularly and in common, e.g., *Deus*, and we have *communia <nomina>* as predicates, such that they may be predicated of their *personae*

[18] *Summa Zwettlensis*, ed. Häring (1977), 62, 128.

[19] *Summa Zwettlensis*, ed. Häring (1977), 62, 129.

[20] *Summa Zwettlensis*, ed. Häring (1977), 62, 130.

[21] *Summa Zwettlensis*, ed. Häring (1977). 64, 134: 'Sic enim dicimus: Deus est Pater, de solo Patre loquentes cuius est proprietas predicata.'

suppositae both singularly and in common, e.g., 'Deus est dominus', 'Deus est potens', 'Deus est persona', then the objects of the propositions may be each one of the persons;[22]

3.2. when the subject terms are *communia <nomina>* which may be predicated of the three persons only in common, not singularly, e.g., *trinus*.[23]

3.2.1. we cannot use *discretiva <nomina>* as predicates: 'Trinus est Pater' is impossible;[24]

3.2.2. but if we use as predicates *communia <nomina>* of any sort, the objects of the propositions are all the three persons: 'Trinus est Deus'; the same holds if we invert the two kinds of *communia <nomina>*: 'Deus est trinus'.[25]

Here are the texts concerning points 1 and 2 (for point 3 see the next section):

> It has to be remarked that, when we use as subject terms discrete names and as predicates common names, the discourse concerns only that person who is named by the proper name, as when we say: Father or begetter or unbegotten or who is not from anyone else is God. In fact, the discourse concerns only the Father. [...] In these sayings, the subject terms discern how the subjects of the properties are.[26]
>
> In a similar way, also every time we use as subject terms common names which are proper to two persons, and as predicates common names, we will talk about each one of the two persons for which the common names [used as subject terms] 'supposit'.[27]

22) *Summa Zwettlensis*, ed. Häring (1977). 64, 135: 'Intelligimus quamlibet illarum subici praedicato ad quas pertinet predicatum nomen.'

23) *Summa Zwettlensis*, ed. Häring (1977), 62, 136.

24) *Summa Zwettlensis*, ed. Häring (1977), 62, 136.

25) *Summa Zwettlensis*, ed. Häring (1977), 62, 137.

26) *Summa Zwettlensis*, ed. Häring (1977), 62, 128: 'In quo quidem notandum quod quando discretiuis utimur suppositionibus communia predicaturi, de illa sola persona sermo est cuius utimur propria appellatione ut cum dicimus: Pater uel genitor uel ingenitus uel qui est a nullo est deus. De solo enim Patre sermo est. Filius quoque uel genitus uel natus uel qui a solo Patre est, est deus: sermo de solo Filio est. Spiritus sanctus uel donum Patris et Filii uel amor eorum uel procedens a Patre et Filio est deus: de solo Spiritu sancto est sermo. In hiis siquidem locutionibus qualia sunt subiecta earum discretiue proprietatum certificant suppositiones.' For a Porretan analysis of the meaning of the predicate terms in these propositions, see Everardus Yprensis, *Dialogus Ratii et Everardi* (1953), 273 f., passage quoted here n. 34.

27) *Summa Zwettlensis*, ed. Häring (1977), 62, 129: 'Similiter etiam quotiens communibus duarum personarum discretiuis utimur suppositionibus trium communia predicaturi de qualibet illarum agimus personarum quarum siquidem sunt supponentes appellationes communes'.

The author doesn't say it, but it is clear that in this case too the object of the proposition will be determined by means of its subject term.

It seems that by *suppositio* here the *Summa* means the name itself as used as subject term in a proposition, by *appellatio* the name considered independently from its being used within the proposition, and by *supponere* the action performed by names when used as subject terms in propositions—and not by speakers or authors—and consisting in referring to some objects (*subiecta*).

These uses of *suppo** terminology in the *Summa Zwettlensis* only partly recall those of Gilbert. As I have tried to show elsewhere,[28] for Gilbert, the agent of *supponere* is the author—at least most of the time; here, it is the noun. Moreover, the *Summa* distinguishes different kinds of *suppositiones*, something which Gilbert does not. Finally, both certainly use *suppositio* in connection with the act of referring to things by means of a subject term *in a proposition*; in other words, both share a 'contextual approach' to language. In Gilbert this was expressed by the rule of *suppositio* 'non tam suppondituri aliquid predicamus quam predicaturi supponimus' (see *supra*), and the *Zwettlensis* has similar 'contextual' rules. But though contextual, the *Zwettlensis* approach to language is different from that of Gilbert, since it is not 'pragmatic' as his is.

Summa Zwettlensis' 'Contextual Approach'

Particularly interesting from this point of view is situation 3, described above with its subtypes. This concerns propositions which have as subject terms nouns which apply to all three persons (*suppositiones communes*):

> When we use as subject terms names which are common to the three persons, independently of which name we are going to use as a predicate, which ones are the objects depends on the predicates. But the different kind of 'community' of the names used as subject has to be considered. In fact among the common names some are predicated both singularly of each of the [things] which are 'supposited for' and of all of them in common, others not singularly but only in common. 'God' is a common name which can be said both of each of the three persons singularly 'supposited for' and of all of them 'supposited for' together. 'Trinus' on the contrary or 'Trinitas' are common names for the three persons which cannot be said of each one of them singularly.[29]

---

[28] Valente (2011).

[29] *Summa Zwettlensis*, ed. Häring (1977), 63, 130-131: 'Quod et si communibus trium personarum utimur suppositionibus quidlibet predicaturi qualia sint subiecta predicata certificant. In quo quidem diuidenda erit ratio communitatis. Conmunium enim appellationum alie singillatim de singulis, quibus sunt communes, predicantur et de omnibus simul suppositis: alie quidem non de singulis singillatim sed de omnibus simul suppositis. Deus enim ita trium communis est

The propositions 3.1.—i.e., the ones which have as subject term a name which may be predicated of its *supposita* both singularly and in common, e.g., *Deus*—are in turn divided in two other sub-sub-types, depending (3.1.1) on their having as predicate terms discretive names, e.g., 'Deus est Pater', 'Deus est Filius', 'Deus est Spiritus Sanctus'; in these cases the object of the proposition (*subiectum*) is chosen on the basis of the predicate ('secundum predicata eligimus subiectum'):

> Pay attention: if we use as subject terms those common names which can be predicated singularly of the *supposita*, and we predicate proper names, we will choose the object of the discourse according to the predicates. In fact we say: 'God is father', and talk only of the Father to whom the property which we predicate belongs.[30]

But the propositions with appellative subject terms may also have appellative terms as predicates (3.1.2.), and the author considers here only appellative terms which may be predicated of their 'supposited persons' (*personae suppositae*) both singularly and in common, like *dominus*. In these cases, the

appellatio ut et de singulis personis singillatim suppositis dicatur et de omnibus simul suppositis. Trinus uero uel Trinitas ita communia sunt tribus personis ut de nulla illarum singillatim supposita dici queant.' Cf. Gilbert of Poitiers, *Expositio in Utrum Pater*, 166: 78 ff.: 'quicquid de Deo substancialiter predicatur, id et de Patre et de Filio et de Spiritu sancto et diuisim de quolibet et simul de omnibus dicitur.' *Ibid.*, 17: 37 f.: '(Trinitas) [...] non de unoquoque illorum, i.e., Patre et Filio et Spiritu Sancto diuisim dicitur.'

30) *Summa Zwettlensis*, ed. Häring (1977), 63, 134: 'Age ergo si hiis communibus suppositionibus que de singillatim suppositis predicantur utimur, predicaturi discretiua, secundum predicata eligimus subiectum. Sic enim dicimus: Deus est Pater, de solo Patre loquentes cuius est proprietas predicata. Sic quoque dicimus: Deus est Filius, de solo Filio agentes cuius est proprietas predicata. Sic quoque dicimus: Deus est Spiritus sanctus, de solo Spiritu agentes cuius proprietas est quam predicamus.' In the language of philosophy of nature, a proposition corresponding to 'Deus est Pater' would be 'homo est Socrates'. It has to be remarked that the same *discretivus* name *pater* is said to predicate a property if it is used as predicate term, and to 'supposit for' a person if it is used as subject term. So the proposition 'Deus est pater' is read by the Porretans in the sense that the name *pater* predicates a property which is discrete for one person, and this person is the subject of the discourse, what it is about (*agere de*). Cf. *Dialogus Ratii et Everardi*, ed. Häring (1953), 270: 28. Differently, Peter the Lombard (*Sent.* I, XXVII, chap. 2, §4) affirms that *pater* in this proposition does not just predicate the property, but also represents the person in whom this property inheres—*pater* then must be intended in a substantive sense, not in an adjectival sense. Correspondingly the subject *Deus* does not represent only one person but the whole divine nature (see on this Valente (2008a), 301 f.). The passage quoted here has in the edition *praedicamenta* instead of *praedicata* ('Secundum *praedicamenta* eligimus subiectum'), but one of the two known manuscripts has *praedicata* and this is in my opinion the right reading. Cf. Valente (2003).

object will be any one of the 'supposited persons' since the predicate term is 'pertinent' to each of them:

> As when we use these common names and predicate proper names, we are referred to persons, so also every time we predicate common names—those common names which can be said singularly of the persons who are 'supposited for'—, we understand every person to be subjected to the predicate, to which the predicate name is pertinent. In fact these propositions are true of the single persons: 'God is the Lord', 'God is powerful', 'God is a person'. Thus if it is used as subject term, each common predicate can refer singularly to each of the persons.[31]

It should be remarked that the *Summa* speaks of *personae suppositae* by the predicated term. Nevertheless, it doesn't seem to me that we can see here an anticipation of a theory of the *suppositio* of the predicate term. In a Porretan context, the predicate term always expresses a form as inherent in a *res* 'supposited for' by the subject term. Here the author is making a distinction among common predicates. Some, taken as independent nouns, can refer to all three persons taken both singularly and in common, like *dominus*; others, only to the three persons taken in common but not singularly, as *trinus*. We also note here an uncommon use of the participle *suppositus,-a,-um*. This participle seems here to be used for an object to which a certain noun can refer, but does not necessarily refer to. If this is true, *suppositio* and *suppositus, -a, -um* seem to be used in a different way: while *suppositio* is the name within the propositional situation, *supposita* (plural) or *personae suppositae* are all the objects to which that noun can refer. It is clear, I think, that according to this author the *supposita* are related to a noun depending on its imposition, and not on its being used as subject or predicate in a given proposition. In order to indicate the actual reference of the noun considered as subject term inside a given propositional context, the author uses the terms *subiectum* and *subici*.

After having treated some particular names like *trinus* (3.2), *persona, individuus, singularis, trinitas*, the *Summa* comes to a general conclusion which could be described as the manifesto of the Porretan contextual approach. This is paragraph n. 145:

---

[31] *Summa Zwettlensis*, ed. Häring (1977), 64, 135: 'Sicut ergo quando hiis communibus utimur suppositionibus, discretiua predicantes, mittimur ad certas personas, ita quoque quotiens communia predicamus—talia siquidem communia que singillatim suppositis reddantur personis—, intelligimus quamlibet illarum subici predicato ad quas pertinet predicatum nomen. De singulis enim uerum est: Deus est dominus, deus est potens, deus est persona. Quolibet igitur talium supponente, predicatum commune potest singillatim reddi cuilibet illarum personarum quarum est commune.'

> With no doubt we understand that in the divine discourses one must consider with caution not only what is said but also about what [something is said]. And we must pay attention not only to the cause of what is said, but also to that of which it is said and the cause of saying. We will then understand that in theology the predication of everything which is said—be it the nature or what belongs to it, or the property or what belongs to it—must be submitted either one person or two or all the three persons.[32]

In divinis locutionibus, we are told, one must consider carefully both the predicate and the subject terms (*quid dicatur* and *de quo*): both the cause of what we say (*causa dicendi*)—that is, it is stated at the beginning of the text, the property or form which is predicated by the predicate term[33]—and 'that of which both what is said and its cause pertain' ('id cuius est et dictum et causa dicendi')—that is, the thing(s) about which we are talking when formulating the proposition. Once we have considered all these elements, says the *Summa*, we will be able to understand that in a given theological proposition the predicates may signify the nature or the properties, while only the persons are therefore what we are talking about.

In other words, our master is using in theology the same basic propositional analysis which Gilbert of Poitiers theorized for ordinary language: in non-figurative language, the predicate always expresses either the nature-form or a property-form, while the subject term always represents the substance in which those forms are inherent.[34]

---

32) *Summa Zwettlensis*, ed. Häring (1977), 66, 145: 'Hiis igitur et hiis similibus indubitanter intelligitur in diuinis locutionibus fideliter considerandum esse non solum quid dicatur sed etiam de quo. Nec tantum attendendam esse causam dicti sed etiam id cuius est dictum et causa dicendi, ut ex hiis intelligamus quicquid in diuinis dicitur, siue natura siue quod est nature siue proprietas siue quod est proprietatis, supponi debere predicationi omnium istorum uel unam personam uel duas uel tres.' In the lines which follows this passage the author proactively defends himself from the accusations of Sabellian heresy which could be provoked by the assertion that it is not possible to talk about the divine nature but only about the persons.

33) *Summa Zwettlensis*, ed. Häring (1977), 30, 18: '[...] II. Aliud itaque dictum est, aliud dicendi causa: ueluti cum hoc corpus dicatur album, causa dicendi albedo est. [...] IV. Item diuersa est causa dicendi a re cuius est et dictum et causa. Aliud namque est color, aliud coloratum. [...] V. Item omnis dicendi causa rei inest cuius est dictum. Color namque corpori inest quod dicitur eo coloratum.' The notion of *causa dicendi* plays also an important role in twelfth-century hermeneutics as related to that of the speaker's or the author's intention (*mentis / auctoris intentio / mens*); see Rosier (1998).

34) Cf. Gilbert of Poitiers, *Expositio in Contra Euticen*, ed. Häring (1966), 293, 49 f.: 'Numquam enim id quod est predicatur, sed esse et quod illi adest predicabile est, et sine tropo nonnisi de eo quod est.'; *Compendium logicae Porretanum* (1983), V, I.7 and I.8. On the theory of predication in Gilbert, see in particular Nielsen (1976) and (1982), 111-114, Maioli (1979), 79-101, De Rijk (1989), Kneepkens (2000). A clear exposition of this theory of predication is Everard of Ypres,

Summa Zwettlensis's Semantic Terminology

Let us now try to draw some conclusions about the meaning of the terminology used in the *Summa*:

- *Appellatio*: name; *propria appellatio*: proper name; *communis appellatio*: appellative name;
- *Suppositio*: subject term in a proposition; *discretiva suppositio*: proper name used as subject term in a proposition; *communis suppositio*: appellative name used as subject term in a proposition;
- *Supponere*: usually, the ability of a name to represent things if used as subject term in a proposition;
- *Suppositus, -a, -um*: the extra-linguistic objects which may be represented by a name if considered *in se*, outside the given proposition. Usually *suppositus* is used as an adjective connected to the word *persona*. It seems to be used in accordance with the Boethian and Priscianic tradition.
- *Subiectum-subiecta*: corresponds almost always to the real object(s) of the proposition, 'what we are talking about' ('[...] de illa sola persona sermo est [...]'), 'what we are treating of' ('[...] de Patre et Filio indifferenter agentes').

The *subiecta* may be either identical with the potential *supposita* of the subject term or a subset of them, depending on the relative range of the reference of the subject and the predicate terms. The *subiecta* will be identical with the

---

Dialogus Ratii et Everardi, ed. Häring (1953), 273 s.: 'Ex calumnia hujusmodi verborum magistri verba Boethii exponentis manifestum est arguentes magistrum Gillebertum in artibus non fuisse exercitatos. Nam si naturalem facultatem novissent, inter substantiam subjectam et substantiam subjecti discernere scivissent, i.e., inter subsistens et subsistentiam; si moralem, cum moralis facultatis sit pars theologia, scirent utique, cum dicitur 'Pater est Deus', quod in hac propositione praedicatur hoc termine 'Deus' essentia, i.e., substanta, i.e., usia divina, non persona, non substantia, i.e., non subsistens. [...] Nam sicut nomen habet duplicem significationem, scilicet substantiae subjectae, i.e., rei quae est substantia et qua non est substantia, et substantiae subjecti, i.e., substantialis formae, i.e., substantiae qua est homo et quae non est homo, ita hic terminus 'homo' habet duplex officium, i.e., subjicendi rem locutioni et praedicandi rem quae ostenditur inhaerere rei de qua est sermo. Eodem modo cum dicitur 'Pater est Deus, Filius est Deus, Spiritus sanctus est Deus', hoc nomen 'Deus' refertur ibi ad substantiam, i.e., ibi ponitur ad significandam divinam essentiam, i.e., usiam quae Latine hoc nomine 'substantia' significatur, qua usia quilibet illorum trium est substantia, i.e., Pater, Filius, Spiritus sanctus. Sed non refertur ibi ad substantiam quae est Deus, i.e., non ponitur ibi ad significandam substantiam quae est Deus, i.e., ad significandam personam, sed essentiam, quia ponitur ibi ad significandam substantiam non subsistens, appositum non suppositum, formam non materiam, usiam non personam.'

supposita of the subject term when the predicate term has a wider acontextual reference than the subject term, that is, the proposition has as subject term a *discretivum* <*nomen*> and as predicate term a *commune* <*nomen*> (kinds of proposition 1 and 2: 'Pater est Deus', 'Emittens est Deus'), and when the predicate term has the same reference as the subject term, but the subject term or the predicate term is a *commune* <*nomen*> which may be predicated of the three persons only in common, not singularly (3.2.2.: 'Trinus est Deus', 'Deus est trinus'). The *subiecta* will be a subset of the *supposita* of the subject term when the predicate term has an acontextual reference which is narrower than the one of the subject term, that is, when the subject term is a *commune* <*nomen*> and the predicate term a *discretivum* one (3.1.1.: 'Deus est Pater'); and it can be a subset of the *supposita* of the subject when the predicate term has the same reference as the subject term and they are *communia* <*nomina*> which may be predicated of their *supposita* both singularly and in common (3.1.2.: 'Deus est dominus').

General Remarks on Suppo* *Terminology in the Summa Zwettlensis*

In conclusion, we find in this *Summa* a notion of the reference of names for extra-linguistic objects as clearly differentiated from their connotation for the nature or property signified; we also find the differentiation between the extra-propositional (potential) reference of a name (*supposita*) and its intra-propositional (effective) reference of it (*subiectum, subiecta*), as well as a terminology which, on the one hand, is based on grammar (distinction between common and proper names, notion of *suppositum* for the extra-linguistic thing or person), and which, on the other, is to some extent similar to the later logical terminology, as in the case of the terms *suppositio discretiva* / *suppositio communis*.[35]

It is clear how the use of *suppo** terminology here is peculiar and different, both in its details and in its general approach, from Gilbert's uses as much as from the uses of other theologians and of the terministic logicians. The difference in the general approach is probably the most interesting: the author of the *Summa* approaches the problematic of Trinitarian language as if he had in front of him a formal system of propositions built up by combining names which already have their meanings and their possible referents in relation to a

[35] *Suppositio communis* and *suppositio discreta* are subdivisions of the *suppositio personalis* in the *Logica 'Cum sit nostra'* (ed. De Rijk, in: *Logica Modernorum* (1962-1967), II/2, 447); in the *Summae Metenses* (ed. De Rijk, in: *Logica Modernorum* (1962-1967), I, 455) the *suppositio communis* is divided in *personalis* and *simplex*.

certain number of entities (persons, nature, properties). There is nothing about the speaker and the interpreter, their *electio* or *vigilantia*, which are so important in Gilbert's eye.[36] It is then probably not by chance that here the action of *supponere* is attributed to names, and not to the author of the text, as it is in Gilbert, and thus we find a distinction between different kinds of *suppositiones*, the *suppositio discretiva* and *communis*. These do not exist in Gilbert but will be very important in the logical tradition. Terministic logic will share with this *Summa* a formal and quasi mechanical approach to language, putting aside the pragmatic and hermeneutic approach of Gilbert. It is no surprise, then, that we find in the *Summa*, even if it is not described by means of a specific terminology, the idea of a *restrictio* of the *suppositio* as brought about by the context: the limitation of the reference of the subject term effected by the predicate.[37] There are thus many elements which are common to the *Summa* and to the terministic logic concerning the theory of reference, even if *suppositio* is here a name put as subject term, and not one of its functions or properties.

Even if he follows Gilbert's ontology concerning the use of *suppo** terminology and the general approach to language, the author of the *Summa Zwettlensis* is going his own way. His use of *suppo** terminology will not be successful, just as his description of the principles of natural worlds in terms of *causae dicendi* is not going to be successful, except for the fact that this description will be introduced by Alain de Lille in his *Rules of theology*.[38] But his attitude toward language as a quasi-formal system with no real role attributed to the author and the interpreter is the same as that which underlies terministic logic, and we can find it also in an other interesting Porretan theological text, the *Dialogus Ratii et Everardi* by Everard of Ypres.

DIALOGUS RATII ET EVERARDI. Suppositum, Subiectum and Substantia Nominis

In the *Dialogus Ratii et Everardi*, Everard of Ypres tries to solve the question of why one is not allowed to say 'Deus est essentia', 'Deus est sapientia' and the

---

36) See Jolivet (1990), Jacobi (1995a), Valente (2004), Valente (2011).

37) This is hinted at by the adjective *solus* ('Sic enim dicimus: Deus est Pater, de *solo* Patre loquentes cuius est proprietas predicata', *Summa Zwettlensis*, ed. Häring (1977), 64, 134, or by the passive *mitti* ('Sicut ergo quando hiis communibus utimur suppositionibus discretiua predicantes *mittimur* ad certas personas. [...]', *ibid.* 64: 135).

38) *Summa Zwettlensis*, ed. Häring (1977), 30-33; see also Alain of Lille, *Regulae caelestis iuris*, ed. Häring (1981), 217-226.

like, comparing these theological propositions with similar non-theological ones, which he explicitly analyzes working with both the grammatical as well as the logical instruments.[39] The terms *suppositum* and *supponere* are introduced in the context of explaining the semantics of concrete appellative names, considered both as independent terms and as subject terms in a proposition.

The point of departure is the grammatical theory of the double signification of nouns, *pro substantia* and *pro qualitate*: a name (*homo*) is imposed to a thing ('rei quae est homo impositum est'), since it expresses a property which that thing possesses (*ex humanitate hominis*) in order to produce in the hearer the concept of it (*intellectum de homine*). The *res* to which the name is imposed is called *substantia nominis* as well as *suppositum locutioni*.

> A name is imposed on something from something else and because of something else. On something, i.e., on a corporeal or incorporeal thing. E.g., the name 'human' is imposed on the thing which is a human being from something else, since it is imposed from the humanity of the human being, because of something else, i.e., in order to produce the concept of a human, by means of his humanity. This is why Priscian says: 'It is proper of the name to signify substance with quality'. You have the same in the description of the name: 'The name is that part of speech which distributes a proper or common quality to each of the subjected bodies or things'. And note that on whatever thing a name is imposed, this thing is called the 'substance' of the name. And it is not called just substance, it is also called 'suppositum of the discourse'. 'Quality' of a name is said to be each form or property from which the name is imposed or by means of which a thing is conceived in the mind, be it a real quality or not [...].[40]

It seems from this passage that the *suppositum* is—as in the *Summa Zwettlensis*—the referent of a name if it is considered from the point of view

39) Everard of Ypres, *Dialogus Ratii et Everardi*, ed. Häring (1953), 252: 'Sciendum solutionem hujus quaestionis accubare prae foribus grammaticae et logicae'. On the semantics in the *Dialogus*, see Nielsen (1976), Schweiss (1987), Jacobi (1999), 252-256 and Kneepkens (2000): 266-267, 271-273.

40) Everard of Ypres, *Dialogus Ratii et Everardi*, ed. Häring (1953), 253. '[...] Nomen alii et ex alio et propter aliud impositum est: Alii, i.e., rei corporeae vel incorporeae ut hoc nomen 'homo' rei, quae homo est impositum est; ex alio, quia ex humanitate hominis; propter aliud, i.e., propter intellectum de homine, sua humanitate mediante, constituendum. Unde Priscianus: Proprium est nominis significare substantiam cum qualitate. Idem habes in nominis descriptione: Nomen est pars orationis, quae unicuique subjectorum corporum seu rerum propriam vel communem distribuit qualitatem. Et nota quod cuicumque aliquod nomen impositum est, illud 'substantia nominis' dicitur. Nec tamen simpliciter 'substantia' dicitur. Dicitur et 'suppositum locutioni'. 'Qualitas' autem nominis vocatur omnis forma vel proprietas, a qua nomen imponitur quave mediante res mente concipitur, sive simpliciter sit qualitas sive non. [...]' Everard is trying to answer the question why it is not correct to say 'Deus est essentia, sapientia' and so on.

of its imposition, that is, independently of a given propositional context. But some lines later we are told that *substantia nominis, persona* or *suppositum* is what a name 'properly situated in a proposition' signifies, and this is also 'what the proposition is about'. We are also explicitly told that the proper function of a name is its being used as subject term in a proposition and that the term *suppositum* is a grammatical term, since in logic we would say *subiectum*:

> Notandum quod nomen in oratione proprie positum significat id de quo est sermo. Quod in grammatica dicitur substantia nominis vel persona vel suppositum. Verbum vero ex officio significat appositum. In logica vero nomen dicitur significare subjectum ex officio. Unde nomen est subjectus terminus propositionis et verbum praedicatus. Nam significat appositum, i.e., rem praedicatam.[41]

Apparently, for the author of the *Dialogus* the terms *suppositum* and *subiectum* mean the same, that is, the referent of the subject term within a proper proposition, but *suppositum* belongs to the terminology of grammar while *subiectum* to that of logic. He doesn't use the distinction between *suppositum* and *subiectum* in order to convey the difference between potential and contextual references of names, as the *Summa Zwettlensis* seems to do, even if he is very well aware of the difference between the two.

Nomina Concretiva and Confuse Significare

The name *Deus* is analysed as a sort of concrete appellative name—it was such a name *ab origine*, when it was instituted in the context of natural philosophy. The fact that this name was later 'transferred' in theology and consequently changed its meaning doesn't seem to have any influence on the description of its role inside propositions (it has the same *modus significandi* in theology and *in naturalibus*).[42] Everard accepts the Gilbertinian thesis which states that concrete appellative names signify the *substantia subiecta* (*persona, subsistens*) when they are used as subject terms in propositions, and *substantia subiecti* (*forma, qualitas, subsistentia*) when they are used as predicates. From the double signification of names derives their double function: that of referring to (*supponendi*) the *subsistens, suppositum* or *persona* when the name is placed as subject term in a proposition, and that of predicating (*apponendi*) the form or quality when it is used as predicate term:

---

[41] Everard of Ypres, *Dialogus Ratii et Everardi*, ed. Häring (1953), 254.

[42] Everard of Ypres, *Dialogus Ratii et Everardi*, ed. Häring (1953), 270.

> I recall that we have said before that this name 'God' in the natural faculty from its first institution is a concrete name and this name 'divinity' a mathematical name. Both names, once they have been transferred to theology, lose their first signification. In fact, since in God there is no true concretion, there is also no true abstraction of a form. But since this name 'God' signifies the thing with which the discourse is concerned as a substance, i.e., as a person, and deity, i.e., essence, as a quality, also this name 'deity' signifies deity as a substance and its effect as a quality. But, as we have anticipated, from its double signification the name draws its double function, i.e., the function of *supponere* and *apponere*, and its double place in the proposition, i.e., as subject term or as predicate term. Then when we say '[a] human is human' and 'God is God', the name 'human' as subject term submits to the discourse a thing which is a human. And as predicate term it predicates and apposes the thing thanks to which it is a human, i.e., humanity. It happens similarly with the name 'God'.[43]

Particularly interesting, even if not completely clear, is the description of how concrete appellative names used as subject terms in propositions signify and represent things. The text seems to me to say that they signify confusedly—*significare confuse*—and therefore represent a person in an indefinite way—*subicere indefinitive personam*—as long as we consider them as potential terms of a proposition—*ratione proponendi*—, but that they represent a person in a clear and definitive way if we consider the actual content of the proposition—*determinate* and *finite ratione propositi*. In any case, the *confusion* may be avoided if we add to the subject name a qualification which can help to identify the referent:

> Concretivo enim nomine hoc, scilicet 'Deus', pro supposito significatur persona, ut cum dicitur 'Deus est Deus', pro qualitate essentia, quae ibi praedicatur. Eodem modo 'Deus est Pater', 'Deus est Filius', 'Deus est Spiritus sanctus'. Talia enim in his propositionibus sunt subiecta qualia praedicata admittunt.[44] Paternitas enim et filiatio, quae ibi praedicantur, de eodem et secundum idem praedicari non possunt, ut ille sit illius Pater cujus est Filius, quod esset si Deus esset Pater Filii et ejusdem Patris idem Deus Filius esset. Cum igitur hoc

[43] Everard of Ypres, *Dialogus Ratii et Everardi*, ed. Häring (1953), 269: 'Recolo dictum superius quod hoc nomen 'Deus' in naturali facultate a prima institutione est concretivum et hoc nomen 'deitas' mathematicum. Utrumque igitur, translatum ad theologiam, cadit a sua significatione prima. Nam cum in Deo non sit vera concretio, nec ibi vera est formae abstractio. Sed cum hoc nomen 'Deus' rem, de qua est sermo, significet pro substantia, i.e., pro persona, et deitatem pro qualitate, i.e., essentiam, et hoc nomen 'deitas' deitatem significat pro substantia et effectum ejus pro qualitate. At, sicut praemissum est, nomen ex duplici significatione duplex sortitur officium, scilicet supponendi et apponendi, et duplicem in propositione locum, scilicet ut sit subjectus terminus et praedicatus. Cum ergo dicitur 'homo est homo' et 'Deus est Deus', in eo quod hoc nomen 'homo' est subjectus terminus subjicit locutioni rem, quae est homo. Et in eo quod est praedicatus terminus praedicat et apponit rem, qua est homo, i.e., humanitatem. Sic et hoc nomen 'Deus'.'

[44] See *supra* for this principle, pp. 116 and 118.

nomen 'Deus' ibi positum confuse significet et ita ibi indefinitive personam subiciat, sed determinate et finite ratione propositi, licet non proponendi,[45] ideo nomini confuso characterica adjiciuntur nomina ut dicatur 'Deus generans est Pater, Deus genitus est Filius, Deus procedens ex utroque est utriusque Spiritus sanctus'.[46]

Without using the terminology of the *restrictio suppositionis* carried out by the predicate or by an adjective or participle, or that of the *suppositio confusa*,[47] we find clear reflections which go in these directions.

Impropriissima Usurpatio and *Mathematica Abstractio*

Analogously, we have something similar to what logicians would call *logica transumptio* or *suppositio simplex* treated as *impropria* and *impropriissima usurpatio*. The author here analyses two different sort of *usurpationes*, opposing to 'homo est animal' not the proposition 'homo est species', as is usually the case in logic,[48] but the two propositions 'Homo est species animalis' and 'homo est species individuorum'.[49]

The starting point is the explanation given by Gilbert of Poitiers, in his Commentary on Boethius' *De trinitate*, concerning the three different ways to proceed in the three speculative disciplines: 'in naturalibus igitur rationaliter, in mathematicis disciplinaliter, in divinis intellectualiter versari oportebit'.[50] The development of the arguments is also clearly connected with the logical notion of the relation between the subaltern species to the genus, on the one hand,

---

45) At p. 254 we read: 'Et praedicatur in prima propositione (sc. 'Petrus est homo') humanitas Petri, i.e., propositi et non proponendi'. According to this interpretation of the distinction between *ratio proponendi* and *ratio propositi* we should probably have here '*ratione*' instead of '*i.e.*' About *ratio propositi* see also Gilbert of Poitiers (1966), 296: 32 (see *supra*, n. 11).

46) Everard of Ypres, *Dialogus Ratii et Everardi*, ed. Häring (1953), 270.

47) See, e.g., *Fallacie Parvipontane*, ed. De Rijk, in: *Logica Modernorum* (1962-1967), I, 565, 569.

48) See for 'homo est species', De Rijk, *Logica Modernorum* (1962-1967), I, 55 (a case of *translatio dialectica* for Peter Abelard); 114, 116 (example of *univocatio*); 139 (a case of *dialectica transumptio* in the *Ars disserendi* of Adam of Petit Pont); 147 (*Fallacie Parvipontane*); 287 (a case of *univocatio* in the *Summa Sophisticorum Elencorum*); 294 ('homo' appellat improprie species'); 357, 384; 562, 594 (a case of *transumptio dialecticorum* in the *Fallacie Parvipontane*); 614 ('homo' as proper name in 'homo est species'); De Rijk, *Logica Modernorum* (1962-1967), II/2, 448 (example of *suppositio discreta* in *Logica Cum sit nostra*).

49) Cf. Marenbon (2002). At the foundation of this analysis is certainly Gilbert's *Expositio in Contra Euticen*, ed. Häring (1966), 296, 31—297, 66 and his *Expositio super librum Boecii De trinitate*, ed. Häring (1966), 86, 31-37.

50) Boethius 2000, 169: 78-80.

and that of the *species specialissima* to the individuals on the other. In accordance with the Boethian and Gilbertinian character of this text, the author interprets these relations in a realistic way.

This seems be the nucleus of the argument, which is not completely clear to me in all its passages.

- a proposition is proper when the subject term represents the *suppositum* which it signifies *pro substantia*, and the predicate term is taken as signifying the property or form which it signifies *pro qualitate*; this proposition expresses the inherence of the form in the *suppositum*;
- a concrete appellative name (*homo*) signifies *pro substantia* the *res* (a singular human being), which it signifies properly when it is used as subject term, and *pro qualitate* the property (*humanitas*), which it signifies properly when it is used as predicate term;
- an abstract name (*humanitas*) signifies *pro substantia* the form (humanity), which it signifies properly when it is used as subject term, and *pro qualitate* its effect, which it signifies properly when it is used as predicate term. This effect consists in the fact that the form 'produces the individual' ('humanitas facit hominem'). To be more precise: each individual substance is what it is due to its complete form, which is constituted by many singular forms: e.g., the individual form of Socrates, called *socrateitas*, is constituted by many forms, one of which is its own *humanitas*. Thus, the form *humanitas* causes the man Socrates since it causes his individual form *socrateitas*.

Then, we have three different (kinds of) propositions: 'homo est animal', 'homo est species animalis', 'homo est species individuorum'.

In the proposition 'homo est animal' we are speaking, and properly, says the author, about a singular *res* which is a human being. The proposition is a proper one since *homo* is used as subject term for signifying its *suppositum*, a human being, and *animal* signifies one of his substantial forms, his *animalitas*. The proposition is proper and true, and it belongs to the *naturalis facultas*.

In the proposition 'Homo est species animalis' we are not talking about a singular human being, but about the form *humanitas* signified *pro qualitate* by the name *homo*, using, improperly, the concrete name *homo* instead of the abstract name *humanitas*: it is an improper use since we use a concrete name as a subject term to refer to the *qualitas* ('ipsa qualitas supponitur'), while properly concrete names signify qualities only when they are used as predicate terms ('quo ipsa apponenda significatur').[51] Even if this is an improper proposition, it belongs to natural philosophy since it makes explicit one of the forms which divide the individuals belonging to the genus animal into different species. The corresponding proper proposition would be 'animalium aliud est homo, aliud non homo'.

In 'Homo est species individuorum' we have an 'impropriissima' *usurpatio*, since the predicate makes the subject term stand neither for the singular human being (the proper meaning of *'homo' pro substantia*), nor for the form *humanitas* as such (signified by the noun *'homo' pro qualitate*), but for the effect of the *humanitas*, that is, its being a part of the complete forms of an individual human being which allows it to be that particular human being (Socrates' *humanitas* as part of the *socrateitas*). Now, this effect is not signified by the name *homo* but only, *pro qualitate*, by the name *humanitas* when it is used as predicate term; but *humanitas* doesn't appear at all in the proposition 'Homo est species individuorum', thus the proposition is not just improper but most improper (*improprissima*). The proposition 'Homo est species individuorum' is an example of *mathematica abstractio*, since it is proper to mathematics to predicate the function of forms, that is, their informing individual substances. The corresponding proper propositions would be 'socrateitas est humanitas', 'platonitas est humanitas' etc.:

> Cum in naturali facultate hoc nomine 'homo' concretive significetur res, quae homo est, et forma, qua homo est, secundum naturalis concretionis proprietatem homo est species animalis quia homo est animal. At cum dicitur 'homo est species animalis', fit sermo de qualitate hujus nominis 'homo', i.e., de humanitate concrete significata. Sed de ejus effectu fit sermo, cum dicitur quod homo est species individuorum eodem nomine quo superius, impropriissima usurpatione, cum improprie etiam ipsa qualitas supponatur nomine quo ipsa apponenda significatur. Multo improprius effectus ejus qualitatis, qui eo nomine nullo modo significatur, sed hoc nomine 'humanitas'. Igitur humanitas, significata hoc nomine 'homo' pro qualitate, est species generis, i.e., animalis, quae ut significatur hoc nomine dividit hoc genus animalis, cum dicitur 'animalium aliud homo, aliud non-homo'. Sed non eadem humanitas, immo effectus ejus, qui pro qualitate significatur hoc nomine 'humanitas', est species individuorum et praedicatur de eis, cum dicitur 'socratitas est humanitas, platonitas est humanitas' et sic de singulis, quae individua ibi subjiciuntur mediantibus effectibus suis. Planum est ergo, quomodo haec propositio 'homo est species animalis' exemplum faciat

---

51) Cf. Everard of Ypres, *Dialogus Ratii et Everardi*, ed. Haring (1953), 254: 'Attende quod nomen quandoque significat id cui impositum est, et hoc proprie, ut 'homo est animal'; quandoque id ex quo impositum est, et hoc improprie, ut de eo fiat sermo ut 'homo est species', 'homo est assumptus a Verbo', i.e., humanitas.'

> naturalis speculationis et quomodo constituat exemplum concretionis, quia ista locutione 'homo est species animalis' hoc genus animal de hac specie 'homo' praedicari invenitur et sic rebus ipsius speciei composita intelligitur. Hac vero locutione 'homo est species individuorum' non datur intelligi, quid cui componatur, sed quae forma cui effectui supponatur. Et sic exemplum est mathematicae abstractionis, quia proprietas mathemaseos est non genus praedicari sed generis genus, i.e., non id quod est genus sed id quo est genus, non de eo quod est species sed de eo quo est species, i.e., effectum de effectu assignare ut hic 'humanitas est animalitas'.[52]

The problems here are similar to those addressed by the notion of *suppositio simplex* in logic, but the theoretical context is completely different and this explains why what a logician would probably call *logica transumptio* is described as *mathematica abstractio*. The epistemological framework is not the one constituted by natural science as opposed to logic, but the Boethio-Gilbertinian division of the speculative sciences *naturalis*, *mathematica*, and *theologica*, with the connected realistic ontological background as well as the theory of universals as collections of singular forms (causes) informing individuals substances. It has to be remarked that Everard takes a position which is totally different from Gilbert's: Gilbert in fact does not use the notion of *transumptio* or *improprietas* while analysing the propositions of logic or of mathematical disciplines, but the idea of the attentive *intelligentia lectoris*.[53]

General Conclusions

The Porretan theologians are important for those who are interested in medieval logic and semantics since they located problems, proposed solutions, and created terminology. Their theological motivation didn't prevent them from being genuinely fascinated by purely linguistic problems. In the end, theological language was necessarily transferred and improper as far as the meaning of the words is concerned, but virtually consistent in its formal structure, and to extend the boundaries of this consistency was for them a philosophical challenge. Supposition was one of the instruments they used—I would dare to say: partly invented—to do it, but each author had his own attitude toward language and his own uses of technical terminology, which could be similar but never identical to those of others. Thus, concerning the history of supposition theory, they have surely much to say, though it is very difficult to reconstruct how their speculations may have influenced those of contemporary or later

---

52) Everard of Ypres, *Dialogus Ratii et Everardi*, ed. Häring (1953), 257.

53) See on this difference Kneepkens (2000), 256-257.

logicians. In any case, the basic onto-linguistic tenet on which all the Porretan uses of *suppo** terminology are based is the Gilbertinian idea of the things as concretions of many different forms and of the concrete name as signifying both the whole thing and one of its forms. Particular to Gilbert is a pragmatic attitude toward language, while his followers share with terministic logic a more formal and mechanical approach to language. Gilbert's pupils transferred the *suppo** terminology, the main agent of which was in their Master, as in Boethius, the speaker or the author, into language itself considered as an objective system, and started to speak about *suppositio* and *supponere* as actions performed by the names themselves. The path to the theory of *suppositio* as a *proprietas terminorum* was paved.

Bibliography

Primary Literature

Alain de Lille, *Regulae caelestis iuris*, ed. N.M. Häring, in: *Archives d' histoire doctrinale et littéraire du moyen-âge* 48 (1981), 97-226

Anonymus, *Compendium logicae Porretanum*, ed. S. Ebbesen, K.M. Fredborg and L.O. Nielsen, in: id., '*Compendium logicae Porretanum* ex codice Oxoniensi Collegii Corporis Christi 250. A Manual of Porretan Doctrine by a Pupil of Gilbert's', in: *Cahiers de l'institut du moyen âge grec et latin* 46 (1983), iii-xviii + 1-113

——, *Summa Zwettlensis*, ed. N.M. Häring (Beiträge zur Geistesgeschichte der Philosophie und Theologie des Mittelalters, Neue Folge, 15; Münster in W. 1977)

Boethius, *Opuscula sacra*, ed. C. Moreschini (Bibliotheca Teubneriana; Monachi and Lipsiae 2000, 163-241)

Everard of Ypres, *Dialogus Ratii et Everardi*, ed. N.M. Häring, in: id., 'A Latin Dialogue on the Doctrine of Gilbert of Poitiers', in: *Mediaeval Studies* 15 (1953), 243-289

Gilbertus Porretanus (Gilbert of Poitiers), *Expositio in Boecii librum De trinitate; Expositio in Boecii librum De bonorum hebdomade; Expositio in Boecii librum Contra Eutycen et Nestorium*, ed. N.M. Häring, in: id., *The Commentaries on Boethius by Gilbert of Poitiers* (Studies and Texts, 13; Toronto 1966)

Petrus Lombardus: Peter Lombard, *Sententiae in IV libris distinctae*, editio III, ed. I. Brady, (Spicilegium Bonaventurianum, 4-5; Grottaferrata 1971-1981)

Secondary Literature

Catalani, L. (2005), 'Modelli di conoscenza tra Gilberto di Poitiers e Alano di Lille', in: J.-L. Solère, A. Vasiliu and A. Galonnier, eds. (2005), *Alain de Lille, le docteur universel. Philosophie, théologie et littérature au XIIe siècle*. Actes du XIe Colloque internationale de la Société internationale pour l'étude de la philosophie médiévale. Paris, 23-25 octobre 2003 (Rencontres de Philosophie Médiévale, 12; Turnhout 2005, 217-245)

——. (2009), *I Porretani. Una scuola di pensiero, tra alto e basso medioevo*, (Nutrix, 2; Turnhout 2009)

Ebbesen, S. (1981), 'Early Supposition Theory (12th-13th cent.)', in: *Histoire épistémologie langage* 3, 35-48 (repr. in Ebbesen (2009))

——. (1987), 'The Semantics of the Trinity according to Stephen Langton and Andrew Sunesen', in: J. Jolivet and A. de Libera, eds., *Gilbert de Poitiers et ses contemporains aux origines de la* Logica Modernorum. Actes du septième symposium européen d'histoire de la logique et de la sémantique médiévales, Poitiers, 17-22 Juin 1985 (Napoli 1987, 401-435) (repr. in Ebbesen, 2009)

——. (2004), 'Porretaneans on Propositions', in: A. Maierù and L. Valente, eds., (2004), *Medieval Theories on Assertive and Non-Assertive Language* (2004), 129-139

Fredborg, K.M. (1973), 'The Dependence of Petrus Helias' *Summa super Priscianum* on William of Conches' *Glosulae super Priscianum*', in: *Cahiers de l'institut du moyen-âge grec et latin* 11 (1973), 1-57

Haas, M. (1987), 'Die Wissenschaftsklassifikation des Gilbert von Poitiers', in: J. Jolivet and A. de Libera, eds., *Gilbert de Poitiers et ses contemporains aux origines de la* Logica Modernorum. Actes du septième symposium européen d'histoire de la logique et de la sémantique médiévales, Poitiers 17-22 Juin 1985 (Napoli 1987, 279-295)

Häring, N.M. (1955), 'The Cistercian Everard of Ypres and his Appraisal in the Conflict between St. Bernard and Gilbert of Poitiers', in: *Mediaeval Studies* 57 (1955), 143-172

——. (1957a), 'Petrus Lombardus und die Sprachlogik in der Trinitätslehre der Porretanerschule', in: *Miscellanea Lombardiana* (Novara 1957, 113-127)

——. (1957b), 'Sprachlogik und philosophische Voraussetzungen zum Verständnis der Christologie Gilberts von Poitiers', in: *Scholastik* 32 (1957), 373-398

——. (1981), 'Die theologische Sprachlogik der Schule von Chartres im zwölften Jahrhundert', in: W. Kluxen, ed. (1981), *Sprache und Erkenntnis im Mittelalter*. Akten des VI Internationalen Kongresses für mittelalterliche Philosophie der Société internationale pour l'étude de la philosophie médiévale, 29. August.-3. September 1977 in Bonn, 2 vols. (Miscellanea Mediaevalia, 13; Berlin-New York 1981, 930-936)

Jacobi, K. (1995a), 'Sprache und Wirklichkeit: Theorienbildung über Sprache im frühen 12. Jahrhundert', in: S. Ebbesen, ed., *Geschichte der Sprachtheorie 3. Sprachtheorien in Spätantike und Mittelalter* (Tübingen 1995, 77-108)

——. (1995b), 'Natürliches Sprechen-Theoriesprache-Theologische Rede. Die Wissenshaftslehre des Gilbert von Poitiers', in: *Zeitschrift für philosophische Forschung* 49 (1995), 511-528

——. (1999), 'Dialogus Ratii et Everardi. *Gründliche Gedanken unterhaltsam in Gesprächsform gebracht*', in: K. Jacobi, ed., *Gespräche lesen. Philosophische Dialoge im Mittelalter* (ScriptOralia 115; Tübingen 1999, 242-261)

——. (2002), 'Philosophische und theologische Weisheit. Gilbert von Poitiers' Interpretation der 'Regeln' des Boethius (*De hebdomadibus*)', in: R. Berndt, M. Lutz-Bachmann and R.M.W. Stammberger, eds., *'Scientia' und 'Disciplina'. Wissenstheorie und Wissenschaftspraxis im 12. und 13. Jahrhundert* (Berlin 2002)

Jolivet, J. (1987), 'Rhéthorique et théologie dans une page de Gilbert de Poitiers', in: J. Jolivet and A. de Libera, eds., *Gilbert de Poitiers et ses contemporains aux origines de la* Logica Modernorum. Actes du septième symposium européen d'histoire de la logique et de la sémantique médiévales, Poitiers 17-22 Juin 1985 (Napoli 1987, 183-198) (also in id., *Aspects de la pensée médiévale: Abélard. Doctrines du langage* (Paris 1987, 293-310))

——. (1990), 'Le jeu des sciences théorétiques selon Gilbert de Poitiers', in: S. Knuuttila, R. Työrinoja, S. Ebbesen, eds., *Knowledge and the Sciences in Medieval Philosophy.* Proceedings of the Eighth International Congress of Medieval Philosophy (SIEPM), II (Helsinki 1990, 71-88)

——. (1992), 'Trois versions médiévales sur l'universel et l'individu: Roscelin, Abélard, Gilbert de la Porrée', in: *Revue de métaphysique et de morale* 97 (1992), 111-155

Kardaun, M., and J. Spruyt, eds. (2000), *The Winged Chariot: Collected Essays on Plato and Platonism in Honour of L.M. de Rijk* (Brill's Studies in Intellectual History, 100; Leiden 2000)

Kluxen, W., ed. (1981), *Sprache und Erkenntnis im Mittelalter. Akten des VI Internationalen Kongresses für mittelalterliche Philosophie der Société internationale pour l'étude de la philosophie médiévale, 29. August.-3. September 1977 in Bonn,* 2 vols. (Miscellanea Mediaevalia, 13; Berlin—New York 1981)

Kneepkens, C.H. (1987), '*Suppositio* and *Supponere* in 12th-Century Grammar', in J. Jolivet et A. de Libera, eds., J. Jolivet and A. de Libera, eds., *Gilbert de Poitiers et ses contemporains aux origines de la* Logica Modernorum. Actes du septième symposium européen d'histoire de la logique et de la sémantique médiévales, Poitiers 17-22 Juin 1985 (Napoli 1987, 325-351)

——. (2000), 'Grammar and Semantics in the Twelfth Century: Petrus Helias and Gilbert de la Porrée on the Substantive Verb', in: M. Kardaun and J. Spruyt, eds. (2000), *The Winged Chariot: Collected Essays on Plato and Platonism in Honour of L.M. de Rijk* (Brill's Studies in Intellectual History, 100; Leiden 2000, 237-275)

Legendre, P., ed. (1998), *Du pouvoir de diviser les mots et les choses* (Bruxelles 1998)

Libera, A. de (1987), 'Logique et théologie dans la *Summa Quoniam homines* d'Alain de Lille', in: J. Jolivet and A. de Libera, eds., *Gilbert de Poitiers et ses contemporains aux origines de la* Logica Modernorum. Actes du septième symposium européen d'histoire de la logique et de la sémantique médiévales, Poitiers 17-22 Juin 1985 (Napoli 1987, 437-469)

Libera, A. de, and I. Rosier (1992), 'La pensée linguistique médiévale', in : S. Auroux, ed., *Histoire des idées linguistiques. Tome 2* (Liège 1992, 115-186)

Maierù, A., and L. Valente, eds. (2004), *Medieval Theories on Assertive and Non-Assertive Language.* Acts of the 14th European Symposium on Medieval Logic and Semantics, Rome, June 11th-15th, 2002 (Firenze: Olschki 2004)

Maioli, B. (1979), *Gilberto Porretano. Dalla grammatica speculativa alla metafisica del concreto* (Rome 1979)

Marenbon, J. (2002), 'Gilbert of Poitiers and the Porretans on Mathematics in the Division of the Sciences', in: R. Berndt, M. Lutz-Bachmann and R.M.W. Stammberger, eds., *'Scientia' und 'Disciplina'. Wissenstheorie und Wissenschaftspraxis im 12. und 13. Jahrhundert* (Berlin 2002, 37-78)

Nielsen, L.O. (1976), 'On the Doctrine of Logic and Language of Gilbert Porreta and His Followers', in: *Cahiers de l'institut du moyen-âge grec et latin* 17 (1976), 40-69

——. (1982), *Theology and Philosophy in the Twelfth Century. A Study of Gilbert Porreta's Thinking and the Theological Expositions of the Doctrine of the Incarnation during the Period 1130-1180* (Acta Theologica Danica, 15; Leiden 1982)

——. (2001), *'Peter Abelard and Gilbert of Poitiers',* in: G.R. Evans, ed., *The Medieval Theologians: An Introduction to Theology in the Medieval Period* (Malden, Mass. 2001, 102-128)

Pinborg, J. (1968), 'Review of L.M. de Rijk, *Logica Modernorum* II', in: *Vivarium* 6 (1968), 155-158

——. (1972), *Logik und Semantik im Mittelalter. Ein Überblick,* mit einem Nachwort von H. Kohlenberger (Stuttgart-Bad Cannstatt 1972)

Rijk, L.M. de (1962-1967), *Logica Modernorum. A Contribution to the History of Early Terminist Logic*. I. *On the Twelfth Century Theories of Fallacy*; II/1: *The Origin and the Early Development of the Theory of Supposition* II/2: *Texts and Indices* (Assen 1962-1967)

——. (1987), 'Gilbert de Poitiers, ses vues sémantiques et métaphysiques', in: J. Jolivet and Alain de Libera, eds., *Gilbert de Poitiers et ses contemporains aux origines de la* Logica Modernorum. Actes du septième symposium européen d'histoire de la logique et de la sémantique médiévales, Poitiers 17-22 Juin 1985 (Napoli 1987, 147-171)

——. (1988) and (1989), 'Semantics and Metaphysics in Gilbert of Poitiers. A Chapter of Twelfth Century Platonism', part I, in: *Vivarium* 26 (1988), 73-113, and part II, in: *Vivarium* 27 (1989), 1-35

Rosier-Catach, I. (1987), 'Les acceptions du terme *substantia* chez Pierre Helie', in: J. Jolivet and A. de Libera, eds., *Gilbert de Poitiers et ses contemporains aux origines de la* Logica Modernorum. Actes du septième symposium européen d'histoire de la logique et de la sémantique médiévales, Poitiers 17-22 Juin 1985 (Napoli 1987, 299-324)

——. (1994), 'L'introduction des notions de sujet et prédicat dans la grammaire médiévale', in: *Archives et documents de la SHESL*, 2e série, no. 1 (1994), 81-119

——. (1998), 'Les mots, les choses et l'intention: autour de maximes d'Hilaire et de Grégoire', in: P. Legendre, ed., *Du pouvoir de diviser les mots et les choses* (Bruxelles 1998, 39-56)

Schweiss, K. (1987), 'Logik und Theologie im *Dialogus Ratii et Everardi*', in J. Jolivet and A. de Libera, eds., *Gilbert de Poitiers et ses contemporains aux origines de la* Logica Modernorum. Actes du septième symposium européen d'histoire de la logique et de la sémantique médiévales, Poitiers 17-22 Juin 1985 (Napoli 1987, 219-228)

Valente, L. (2003), '*Talia sunt subiecta qualia praedicata permittunt*. Le principe de l'approche contextuelle et sa genèse dans la théologie du XIIe siècle', in: J. Biard and I. Rosier-Catach eds., *La tradition médiévale des catégories (XIIe-XVe siècle). Actes du XIIIe Symposium européen de logique et de sémantique médiévales (Avignon, 6-10 juin 2000)* (Louvain-Paris 2003, 289-311)

——. (2004), '*Virtus significationis, violentia usus*. Porretan views on Theological Hermeneutics', in: A. Maierù and L. Valente, eds., *Medieval Theories on Assertive and Non-Assertive Language*. Acts of the 14th European Symposium on Medieval Logic and Semantics, Rome, June 11th-15th, 2002 (Firenze 2004, 163-184)

——. (2008a), *Logique et théologie. Les écoles parisiennes entre 1150 et 1220* (Sic et non; Paris 2008)

——. (2008b), 'Un realismo singolare: forme e universali in Gilberto di Poitiers e nella Scuola Porretana', in: *Documenti e studi sulla tradizione filosofica medievale* 19 (2008), 191-246

——. (2011), 'Gilbert of Poitiers', in: H. Lagerlund, ed., *Encyclopedia of Medieval Philosophy, Philosophy between 500 and 1500* (Dordrecht-Heidelberg-London-New York 2011, 409-417)

XIIITH CENTURY

Supposition and the Fallacy of Figure of Speech in the *Abstractiones*

Mary Sirridge
Louisiana State University

Abstract

I undertake to examine the practice of Richard, Master of Abstractions, with respect to supposition in his dealing with the fallacy of figure of speech. His practice turns out to support the 'single theory' account of the theory of personal supposition, as does his treatment of a functional equivalent of simple supposition, but his practice of proposing additional solutions points to changing attitudes with respect to species as separate entities. Questions having to do with material supposition and the like are completely absent, even in discussions where we might expect to find them.

Keywords

fallacy of figure of speech (*fallacia figurae dictionis* = ffd), *similis figuratio, Abstractiones*, Master of Abstractions, *sophismata*, supposition, simple supposition, material supposition

What has the *Abstractiones* got to tell us about supposition? At first glance, not much. The *Abstractiones* is not a treatise *De Suppositionibus*; indeed, it is not a treatise at all, but a collection of some 305 *sophismata*, attributed to one Richard the Sophist, 'Master of Abstractions', 'Flower of Logicians'.[1] It is, moreover, not very explicitly theoretical. The most theorizing we get from Richard is the occasional short section, like the one on *consequentiae* in which he offers a division of the various kinds of conditionals that he proceeds to use in showing

[1] Ms B = Brugge, *Stedelijke Bibliotheek*, 497; ms C = Oxford, *Corpus Christi College*, E 293B; ms D = Oxford, *Bodleian Library, Digby* 24; ms K = København, *Det Kongelige Bibliotek*, Fragm. 1075; ms O = Oxford, *Bodleian Library, Digby* 2; ms P = Paris, *BnF. lat.*, 14069; ms R = London, *British Library, Royal* 12. F. xix. I follow the numbering of sophismata in the *Abstractiones* in Streveler (1993), 154-167, since it has become standard in the literature.

that the premises of arguments involving *'si'* and the like often have more than one interpretation, depending on how the conditional is interpreted.

Thus from *Abstractiones* we are not going to get a presentation of the theory of supposition *per se*. Nonetheless, since in the *Abstractiones* supposition and its near relatives like 'distribution' and 'descent' have taken over as tools of analysis, we are offered a good opportunity to watch the theory in action. In the end, I think, it can even tell us something about whether the theory of supposition is actually two theories, as Paul Spade[2] has maintained, or whether it is a single theory, as Gareth Matthews[3] and Catarina Dutilh Novaes[4] have argued.

We can perhaps also do a bit of historical bootstrapping. The *Abstractiones* as we have it is almost certainly from the last quarter of the thirteenth century, undoubtedly in the English style. Ockham's *Summa Logicae* gives us a very conservative *terminus ad quem* for the work, as he and his contemporaries refer to the 'Master of Abstractions,' which must mean that by his time the work was well established. Even if we do not succeed in finding out which Richard our Richard was, the more cogently we can assign Richard a spot in the developmental story about supposition leading up to and beyond Ockham, the steadier we can be about the span of time in which to place *Abstractiones*.

There is a close connection between the theory of supposition and the study of fallacies. The fallacy of figure of speech (*fallacia figurae dictionis* = ffd) is a significantly recurrent tool of logical analysis in the *Abstractiones*. In fourteen of Richard's own solutions his favored analysis is defended by the explicit allegation that the fallacy of figure of speech is present in a purported proof or disproof or in some related argument he considers germane. In a number of other cases the absence of an explicit label seems to be a mere oversight, since the analysis fits the usual pattern of his ffd discussions. Finally, and perhaps most interestingly, there are several cases in which Richard says that someone might try to use this pattern of analysis, and explains how a solution of this kind would go—then says that the approach is incomplete or just wrong.

A large number of arguments in which Richard detects a problem having to do with supposition fall under the fallacy of figure of speech. To be sure, as Richard understands it, an inference diagnosed as committing ffd must meet a syntactic condition: there is a 'similis figuratio' between two occurrences of the same or nearly the same sentence or sentence-part—they 'look the same

2) Spade (1988), 189-190.

3) Matthews (1977), 13-24.

4) Dutilh Novaes (2007), 7-76; (2008), 391-393.

or similar'. In fact, what is normally required for *similis figuratio* is the syntactic identity of the sentences or—more rarely—of the crucial sentence-fragments or expressions, i.e., the exact same wording.[5] The only variations regularly allowed are cases of transposition, cases in which there is 'identity of the expression with itself as transposed'.[6] But the question of *similis figuratio* is normally disposed of in a sentence or so, and the main work of logical analysis of fallacious inferences falls to a second, semantic criterion: there is some sort of slippage with respect to the supposition of terms. (i) In one kind of case, purportedly 'similar' expressions have different semantic significance in one premise and in the other, or in premise and conclusion, rendering the argument invalid; in such cases the fallacy is explicit in the argument. (ii) In the second kind of case the difficulty with an inference turns on the fact that there is some proposition that for some reason 'looks the same as' one of the premises (frequently an occurrence of the sophismatic proposition),[7] but which does not 'signify the same thing' and has a different truth value, the difference in truth value being traceable to a difference in supposition. In such a case, one occurrence of the 'similar expressions' is in fact outside the argument. The 'outsider' may be another interpretation of the sophismatic proposition, as when that proposition has a composite and a divided sense. In a few cases, though, it is just some other proposition that Richard thinks someone who proposes this argument must have in mind instead of the one that actually occurs in the inference, with the result that the person is fooled into thinking the inference valid, or fooled into thinking that a false premise is true. Cases of transposition are frequently of this second sort.

Given the number and variety of discussions that involve the ffd and the fact that it essentially involves slippage of supposition, there is some promise that an examination of these cases can shed some light on Richard's understanding

---

5) Neither apparent logical equivalence of expressions nor apparent entailment relationships are sufficient to qualify as *similis figuratio*, even if an incorrect inference turns on them. In the solution of <22> TU SCIS QUIDLIBET VEL NIHIL, Richard rejects the inference from 'Tu dubitas aliquid' to 'Tu non scis aliquid', which involves slippage in mode of supposition, but does not diagnose the ffd, apparently because 'dubitas aliquid' and 'non scis aliquid' do not 'look enough alike' to count as *similis figuratio*.

6) Why is *transpositio* singled out? For one thing, exactly the same words are used. In addition, Aristotle's *De Interpretatione* X, 20 b 1-12, makes an issue of the fact that transposition sometimes does, but frequently does not preserve truth value; cf. Kneepkens (2003), 363-411.

7) We will refer to sophismatic *propositions*, as Richard does, despite the inconvenience that these propositions often have no determinate truth value, but have multiple meanings (*multiplex*), and are said to be true or false according to which meaning (*sensus*) is selected, and *a fortiori* according to what is asserted (*denotatur*) or signified (*significatur*) in this sense.

of supposition and closely related notions like 'descent' and 'distribution'. To anticipate, an examination of cases in which ffd is involved turns up a variety of semantic shifts that seem to qualify in *Abstractiones* as a change in supposition; in the majority of cases there is slippage with respect to the mode of supposition (*modus supponendi*), and with a few exceptions, these are associated with a difference in what is signified (*significatur*) or asserted (*denotatur*). But Richard never uses the terminology of simple or material supposition in the analysis of instances of the ffd, even when it would seem natural to do so. In the case of simple supposition, there are several cautious uses of manoeuvres that are functionally equivalent to simple supposition. But so far as I can see, once a couple of apparent counterexamples are explained away, his solutions do not ever involve a functional equivalent of material supposition.

Let me remark somewhat parenthetically that investigating through the lens of a particular fallacy in the *Abstractiones* imposes some limitations on the significance of the result for an understanding of Richard's overall approach. Richard is interested in inferences, particularly in those that appear to be valid and to have true premises and false or crazy conclusions; having detected a fallacy in an argument of this kind, he will often not return to another argument that is just as bad and involves the same logical misstep, apparently because a false or crazy conclusion from obviously false premises is ordinarily of no interest. Moreover, Richard pays some attention to Aristotle's description of the ffd[8] and to his taxonomy of the various fallacies in *Sophistici Elenchi*. Aristotle's description of problems due to misleading surface grammar is extended by Richard to cover cases in which an expression functions as a noun in one proposition and thus signifies *hoc aliquid*, but as an adjective in another, thus signifying *quale quid*. Richard will ordinarily not explain an argument as ffd if he thinks it turns more obviously on an equivocation or amphiboly, even if the double-meaning results in slippage of supposition. For Richard the *fallacia accidentis* and the ffd do not overlap; as a result some inferences that we would like to see discussed as instances of the ffd fall under the fallacy of accident instead, though <56> THE WHITE WAS GOING TO BE DISPUTED ABOUT (*ALBUM FUIT DISPUTATURUM*) and <288> THE NECESSARY CAN BE FALSE (*NECESSARIUM POTEST ESSE FALSUM*) are discussed under the ffd.[9]

---

8) Aristotle, *Sophistici Elenchi* IV, 166 b 10-19.

9) Except for the odd classification of some examples like 'Album fuit disputaturum', then, *Abstractiones* is not in line with the 'tendency to link figure of speech with the fallacy of accident (that is to say, with a fallacy *extra dictionem*)' described by Tabarroni (1994), 15-24. Cf. also Ebbesen (1988), 107-115; Huelsen (1988), 175-185.

By contrast the fallacy of composition and division has dissolved into a very common way of explaining the difference in meaning that attaches to sophismatic propositions that 'must be distinguished,' including those that involve the ffd; and the *fallacia consequentis* is diagnosed whenever Richard detects an incorrect inference from q to p, when it is the inference from p to q that is valid, regardless of whether another fallacy is adduced to explain why 'q→p' is incorrect.[10] Overall, Richard's choice of a fallacy seems to be more a matter of what he sees as the principal logical problem than a result of distinguishing sharply between fallacies *in dictione* and fallacies *extra dictionem.*

1. The Fallacy of Figure of Speech

Let us first examine two basic patterns of analysis for arguments having to do with common personal supposition that are explicitly brought under the ffd, and then turn to examples involving something like simple supposition. Finally, we will turn to the more complicated question of Richard's attitude toward material supposition.

The first pattern of analysis relies on determining what 'descent' is allowable. Richard says <262> FROM SOCRATES ALONE DIFFERS WHATEVER IS NOT SOCRATES (A SOLO SORTE DIFFERT QUICQUID NON EST SORTES)[11] is to be conceded as true. The purported disproof of <262> is a fairly standard example of an argument said to commit the ffd on account of a mistake about 'descent'.

(1) From Sortes alone differs whatever is not Sortes <262>
(2) Plato is not Socrates
(3) From Sortes alone Plato differs

(1) is our sophismatic proposition. From (1) and the obviously true (2) (3) should follow, but (3) is crazy, since Plato differs from lots of people other than Socrates. And so <262> must be false.

But this counterargument against <262> does not work. People are fooled into thinking that (3) follows from <262> and an obviously true added premise,

10) See the discussion of <288> below.

11) In fact the initial statement of <262> adds 'NEC PARS SORTIS', but B never repeats this formula in its discussion, having eliminated parts of Socrates at the start of the discussion. D, which does not contain the exclusion, makes the addition several times.

showing <262> to be false, because the true <262> is easily confused with the false 'outsider' <262*> 'Whatever is not S differs from S alone'.

Here we have a case of *similis figuratio* because the two sentences have the same expressions as components, only in transposed word order. <262*> allows 'descent' to Plato, so that (3) follows and the argument is valid; but <262*> is false. <262> is true, but does not allow descent to (2), and so the inference to (3) is not valid. In this case, <262*> is an 'outsider,' and not a legitimate reading of <262>. The supposition that <262> and <262*> have the same meaning because they look so much alike, differing only in word order, is the reason people are fooled. Given the 'similis figuratio' of <262> and <262*>, and the difference with respect to the possibility of descent, we have ffd.

Here Richard does not, as he often does, go on to further explain this sort of difference of meaning as a difference in distribution or in the signification of the propositions or syncategorematic expressions involved.[12] Even in the absence of a detailed discussion of the respective meanings of <262> and <262*>, however, it is fairly clear what logical distinction is marked by the difference in word order.[13] In <262>, which is true, 'whatever-is-not-Socrates' designates the group or collection of things that are not S taken as a whole; it says that as a whole, the group of things defined by excluding only S excludes only S. In <262*> , 'whatever is not Socrates' refers to everyone who is not Sortes individually; <262*> is, of course, false, since the members of the group taken individually differ from each other as well as from Socrates.

The discussion of <288> THE NECESSARY CAN BE FALSE (*NECESSARIUM POTEST ESSE FALSUM*) gives us the second variant on the analysis, which appeals to a difference in mode of supposing (*modus supponendi*) within the argument.[14] Here Richard says that the sophismatic proposition is true, and the disproof fails. He analyzes the argument:

---

12) In <215> OMNIA DECEM PRAETER UNUM SUNT NOVEM and <216> OMNIUM DUORUM FRATRUM UTERQUE PRAETER UNUM EST ALBUS, Richard attributes the difference in descent to different distribution of the subject term, but in the case of <215> he also points to a difference in what is signified by 'omnis'.

13) Whether or not the distinction holds for non-logical Latin, Richard marks the difference between these two easily confused claims by the transposed word order of <262> and <262*>. Angel D'Ors (1993), 390, refers to attention given by logicians to a 'highly normalized language in which the function that the terms carried out in propositions and the mode in which they are affected by the syncategoremata—on which their species of supposition depends—are narrowly linked to the position which the terms occupy in the proposition.'

14) There are only four sophismata, all involving a disjunctive subject term, in which the analysis in terms of ffd cites only a difference of supposita, and not a difference in mode of supposing, as the source of the invalidity. Of these, strictly speaking, only <17> OMNE RATIONALE

(4) the necessary is able to be false <288>
(5) the necessary is not necessary

(5) is a contradiction, and so if the argument is sound, the sophismatic proposition is false. In this case it is a syncategorematic expression that has 'similis figuratio.' The *similis figuratio* of 'necessarium' 'to itself' in (4) and (5) fools people into thinking it has the same mode of supposition in both, which is wrong, since in (4) 'necessarium' supposits for present entities, but also for future entities, given the modal 'is able'. (4) is actually equivalent to (4') something that is necessary or that will be necessary can be false

It *will be necessary*, for example, that the Antichrist existed—at some point after the Antichrist has started to exist—though it is now false that the Antichrist existed; and so (4) is true because something that will be necessary can be false. But it is not possible for what is now necessary at the same time not to be necessary, which is what (5) says, since in (5) 'necessary' supposits for present entities only.[15] The problem with the inference is semantic slippage, a change of *mode of supposition* caused by the modal operator.

2. Simple and Material Supposition

No sophismata like MAN IS SPECIES (*HOMO EST SPECIES*) are included in the *Abstractiones*, and there is no explicit mention of simple supposition where we might most expect it, namely in the discussion of <9> EVERY ANIMAL IS WELL (*OMNE ANIMAL EST SANUM*) (there being of each species one well animal and one sick one); <10> EVERY ANIMAL WAS IN NOAH'S ARC (*OMNE ANIMAL FUIT IN ARCA NOE*); and <11> EVERY COLORED IS (*OMNE COLORATUM EST*)[16] (there being only one white thing, one black thing, and only one of

VEL IRRATIONALE EST SANUM and <18> OMNIS PROPOSITIO VEL EIUS CONTRADICTORIA EST VERA are said to commit the ffd, though in the solution of <17> it is said that the preceding ones <16> QUICQUID EST VEL NON EST EST and <13> OMNE BONUM VEL NON BONUM EST ELIGENDUM 'had the same problem.' It may be significant that <13> is introduced as 'this old sophism,' perhaps indicating that these are well-known examples, whose likewise well-known solutions are here recorded.

15) And so in addition to its other flaws, the disproof commits the fallacy of affirming the consequent.

16) If there is an idiomatic English translation that captures the ambiguity of the Latin, I have failed to find it.

in-between color).[17] We do, however, find in each case an appeal to the distinction that precedes the entire section:

> There can be division or distribution over singulars of a genus vs. genera of singulars or with respect to proximate parts and remote parts, or with respect to parts divided by species vs. parts numerically divided.[18]

In each case the sophismatic proposition is said to be true if it is about species as a whole and false if it is about individuals. And in each case, the disproof is said to fail if there is distribution over proximate parts because there is a switch between predication *quale* and predication *quid*. An example is the purported disproof of <11>:

(6) Every colored is (*omne coloratum est*) <11>
(7) every white (thing) is colored (*omne album est coloratum*)
(8) every white (thing) is (*omne album est*)

Here, I think, (6) is said to be *quale quid* because it says that every kind of color or way of being colored exists; there is a 'division' or 'distribution' over kinds of things, 'genera of singulars,' not individuals of these kinds, 'singulars of genera.' (7) is about the individuals that are white; it says that every white thing is colored. All of this seems to be equivalent to invoking a distinction between personal and simple supposition, understood as a distinction between referring to individuals of a kind vs. species of individuals, even if the terminology of simple supposition is not used; and I think it is.[19]

But in his discussion of <11> Richard goes on to offer a second solution that does not rely on this distinction (*respondendo vero non per distinctionem*), but instead on an alleged equivocation on 'esse'. If no reference to species is allowed, Richard says, <11> comes out simply true, and both disproofs fail. We have seen the first disproof; it is now to be rejected on the grounds that being is referred to equivocally, as 'operational being' in (6), the being of that which

17) In fact, only <9> is explicitly said to involve the ffd, but <10> and <11> are said to be 'the same,' and involve a pattern of analysis identical to that of <9>.

18) [B 74rb] 'Quandocumque signum divisive tenetur distinctio est communis quia potest fieri divisio vel distributio pro singulis generum vel pro generibus singulorum, sive pro partibus propinquis vel pro partibus remotis, sive pro partibus secundum speciem vel pro partibus secundum numerum. Sed sic non distinguitur nisi signo addito termino in quo est differentia huiusmodi partium, ut patet in hoc sophismate <10> OMNE ANIMAL FUIT IN ARCA NOE et in similibus.'

19) But cf. Spade (1977), for a different understanding of simple supposition.

is (= all these colored things exist), and in (7) as 'consequential or dispositional being' (= if something is white, it is colored). The second disproof is rejected as well; it says that, since a universal affirmative proposition is true only if the subject term designates at least three things (appellata), the following are all false: 'every white thing is', 'every black thing is', 'every intermediate thing is'; and so their opposites are true, i.e., 'it is not the case that every white thing is', etc.; and these imply 'it is not the case that every colored thing is', which is the contradictory of <11>, and so the sophismatic proposition is false. This disproof is wrong, says Richard, because 'every colored thing is' <11> is true (since there are three colored things), and does not in this case imply 'every white thing is' and the rest of them, all of which are false (since in each case there is only one thing for the universally quantified subject term to refer to). A true proposition cannot entail a false one. But then the negation of the consequent does not imply the negation of the antecedent, and so there is no disproof of the sophismatic proposition.[20] In <9> and <10> the distinction between proximate and remote parts is in fact attributed to 'certain people' (*distinguitur a quibusdam*); and for <10> as for <11> a second solution is put forward that does not allow for the sophismatic proposition to be about species taken as a whole.

It is clear, I think, that there is a kind of workaday (ho-hum) realism here that is consistent with Richard's casual reference to 'a unity that is one according to species grounded in things that differ numerically' in the discussion of <118> SOME THINGS INSOFAR AS THEY AGREE DIFFER (*ALIQUA INQUANTUM CONVENIUNT DIFFERUNT*). It does appear that at some point, perhaps when *Abstractiones* was first written, or possibly later on, alternative solutions were added for <10> and <11> that do not require anything like simple supposition. Still, even if the solutions involving a functional equivalent to simple supposition are not the favored ones, they are not disallowed.

As for material supposition, including quotation and self-reference of linguistic entities by self-instantiation, perhaps not surprisingly, it is not used explicitly in the resolution of problems in the *Abstractiones*, despite plenty of occurrences of expressions introduced by '*li*' and the like. We do not find the likes of MAN HAS THREE SYLLABLES on the list of sophismata. But the question of whether some functional equivalent of material supposition is being used is more complicated, and it looks initially like such an equivalent does occur. In the case of <303> IT IS POSSIBLE FOR SORTES TO KNOW WHATEVER

---

[20] This sophism and the next <12> OMNE PHOENIX EST are the only ones in the whole collection to invoke the requirement that a universal affirmative proposition needs three *appellata* for its truth, or indeed to make any reference to *appellata*.

PLATO KNOWS (*POSSIBILE EST SORTEM SCIRE QUICQUID PLATO SCIT*), it is supposed that S and P both know three propositions; P also knows a fourth, *a*; and P immediately forgets *a*, and it is impossible for S to know *a*. At issue is the first inference of the disproof:

(9) It is possible for S to know whatever P knows <303>
(10) P knows *a*
(11) It is possible for S to know *a*

<303> has more than one meaning (*est multiplex*), Richard says, and so it needs to be distinguished. In the divided sense it means that (*significat illud quod*)

(9a) S can know whatever P knows.

(9a) is false. Richard does not say why, but obviously the modal expression 'can' allows 'whatever P knows' to range over whatever P knows now or in the future; this includes *a*, which he now knows. In the composite sense <303> means

(9b) This proposition 'S knows whatever P knows' is possible

i.e., that it can be true. And, this proposition can be true, since it will in fact be true as soon as P forgets *a;* and so (9b) is true. (9a) and (9b) have *similis figuratio* on the basis of their wording (*secundum dictionem*) because they are both written as <303>; this fools people into thinking that descent to (10) is possible from the true (9b), as it is from the false (9a), which it is not.

Why not? We might be inclined to answer: because (9b) is about a proposition which is named or quoted in it, so that quantifying-in or instantiating-out is blocked by virtue of the fact that 'whatever P knows' is just part of the self-referential name of the proposition that is said to be possible.[21]

This is not Richard's answer. He is not inclined to say that the inference fails because 'whatever P knows' has material supposition—or in any event occurs

[21] This would amount to no more than what Quine (2004), 379, calls 'the first grade of modal involvement.' 'Quotation is the referentially opaque context par excellence. Intuitively, what occurs inside a referentially opaque context may be looked upon as an orthographic accident, without logical status, like the occurrence of 'cat' in 'cattle'. The quotational context '9>5' of the statement '9>5' has, perhaps, unlike the context 'cattle' of 'cat,' a deceptively systematic air which tempts us to think of its parts as somehow logically germane,' *ibid.*, 380-381. This is the interpretation of Read and Priest (1981), 275-277.

within a quotation context. The inference from (9b) to (11) fails, Richard says, because (10), which says P knows *a*, is false if 'S knows whatever P knows' is true (and so (9), which says it is possible that S knows whatever P knows, is false in this case), given that by the *casus* S does not and cannot know *a*.[22] Whatever it means, the occurrence of 'this proposition' in (9) does not take us to the meta-language. This is not because Richard is unfamiliar with the language/meta-language distinction; he will often introduce the self-referential name of an expression by 'li' and there are plenty of references to 'this proposition' or 'that expression' in the *Abstractiones.* But apparently that is not the logical issue here. This is not a context rendered fully opaque by quotation. It is just that 'whatever P knows' in (6b) is not in a *fully* transparent position and so you cannot freely substitute on or instantiate from this expression.[23]

But if this is no quotation context, why not? In the discussion of <303> Richard gives no reason for calling (9b) 'the composite sense'; but it turns out that 'composite/divided sense' is not just technical terminology for difference in scope of some kind or other. In several nearby sophismata, he says about this same pattern of analysis that in the composite sense, the modal expression 'determines the verb with respect to *the combination of* subject and predicate' (*ratione compositionis*), while in sentences like (9a) it determines the verb with respect to what is signified by the verb (*ratione rei verbi*). (9a) therefore means that whatever P knows, which includes *a,* S *possibly-knows* it, perhaps that S *has the power or potential* to know it.[24] I think we must read (9b) as saying that whatever exactly is required to make exactly 'S knows whatever P knows' come out true can obtain, i.e., that at some point *it can happen* that S knows whatever P happens to know at that point. Richard does not appeal to material supposition here to block the inference and it is clear that he does not need to. His point is that in (9b) it is the kind of connection between specific predicates that is at issue. The reference to 'this proposition' serves here to signal that

---

22) In fact, Richard is here rejecting the disproof, which proceeded obligatio-style as follows: [B 95rb] 'Et posito Sortem scire quicquid Plato sciat, negandum est Platonem scire *a* eo quod repugnant; cum sequatur: si Sortes scit quicquid Plato scit, et Sortes non scit *a*, Plato non scit *a*.'

23) A similar point is made in Ebbesen and Pinborg (1983), 12-13, about the way to understand the function of 'officiable' expressions. 'S believes that a is F' is to be analyzed as: S believes 'a is F', whose precise meaning is that a is F.

24) Letting *P* represent possibility and *K* a two-place predicate *x knows y*, p = Plato and s = Socrates, we can represent (6a) as: (x)(Kpx → PKsx). Whether we understand *PK* as a different predicate from *K*, something like *can+know*, or just as a modal operator on the claim that S knows x (9a) expresses something very like Quine's third grade of modal involvement; and that means that in *PKsx*, *K* ranges over everything P knows at any time. Cf. Priest and Read (1981), 275.

descent or instantiating-out is blocked on 'whatever P knows', not to introduce a quotation context.[25]

We find the same pattern of reasoning in Richard's discussion of <289> EVERY MAN OF NECESSITY IS ANIMAL (*OMNIS HOMO DE NECESSITATE EST ANIMAL*). The customary view, he says, is that <289> is true; and the disproof fails by ffd:

(12) Every man of necessity is an animal <289>
(13) Socrates is man
(14) Socrates of necessity is an animal

The problem, it is said, is that in (12) the supposition of 'homo' is ampliated universally, so that it refers to or supposits indeterminately for all men, while in (13) 'homo' supposits determinately. Thus we have the fallacy of figure of speech.[26]

As we have seen, Richard is willing to appeal to an equivocation on 'esse', and as we have seen in his discussion of <288>, he has nothing against the idea that the mode of supposition of a common noun is affected by its being conjoined with a modal or temporal expression or by the tense of a verb. And if there is *similis figuratio*, this sort of slippage makes for the ffd. But in this case, he says, this customary solution misses a more fundamental problem: <289> has more than one meaning: the composite sense,

(12a) This proposition is necessary: 'omnis homo est animal'

and the divided sense,

(12b) Every man necessarily-is animal.

From (12b), in which the modal expression attaches to what the verb signifies, it follows that this man, that man, and all the rest of them are-necessarily animal. In this divided sense, the universal <289> is false. The conclusion of the disproof (14) is an instance of (12b), all of which, Richard adds, are false too.

---

25) Thus we have not got just Quine's first grade of modal involvement.

26) It is also 'sometimes said' that the disproof fails because it depends on equivocation, on the grounds that in (12) = <289> it is habitual or consequential being that is referred to, and in (13) the 'operation of being.'

No man has necessary existence. (12a) is true, since it says that there is a necessary connection between being man and being animal.

The proposed proof for <289> depends on the definition of 'necessary *per se*', i.e., 'this is true, and it always was true and always will be true', but the proof offered for <289> does not support (12b), and would not support (12b), even if we added a further premise (15+) to get the argument:

(15) this is true: 'every man is animal' and always was true and always will be true
(15+) and it cannot not be true
(12b) every man of necessity is animal

For the true (12a) the situation is reversed. The proof does work, probably even without (15+), given the definition of necessity *per se*. The disproof, as we would expect, fails because the 'this proposition' locution signifies that descent or instantiation-out is blocked.

In this case Richard does not say, as he does about the 'customary solution' to the preceding sophism, that the customary solution is fine as far as it goes. This is because the 'customary solution' is in this case pretty much wrong on all counts. <289> does have a true reading, i.e., (12a), and the disproof does fail because of ffd. But the fallacy has nothing to do with ampliation; rather it has to do with the opacity of 'homo' in (12a), which is due to the fact that (12a) is about a necessary connection between being a man and being an animal. The false (12b) does, but the true (12a) does not, allow descent to Socrates as an instance. On the other hand, the proof fails for (12b), which is false, as Richard points out. Again, though Richard does not say so, the problem again is the fallacy of figure of speech, for the less modally involved (12a) is what really follows from (15).

As an indication of Richard's attitude toward material supposition these cases are inconclusive. Since Richard does not really need to appeal to self-reference or a quotation context if he can appeal to his theory about the composite and divided attachment of modal operators, it is not all that significant that he doesn't. More significant is the discussion of <92> IF SOMEONE SAYS YOU ARE AN ANIMAL HE SPEAKS THE TRUTH (*SI ALIQUIS DICIT TE ESSE ANIMAL DICIT VERUM*), for here it seems that Richard really might appeal to something very closely related to material supposition as one interpretation of the sophismatic proposition. 'Some people' he says, say that <92> is true and reject the disproof:

(16) if someone says that you are an ass he says that you are an animal
(17) if someone says that you are an animal he speaks the truth <92>
(18) if someone says that you are an ass he speaks the truth

Intuitively speaking, it is easy to see why these 'some people' would take this approach. Even if the accusative + infinitive construction is normally transparent, (17) is an instance of indirect discourse. It is obviously true, it seems, only in the opaque sense, only if it is tied to a verbal formula or utterance, only if it says (17a) that if someone pronounces the sentence 'you are an animal' (or one whose 'precise meaning' is that you are an animal) he speaks the truth, though (16) would be false if we read it as being about utterances, i.e., as saying (16a) that if someone pronounces the sentence 'you are an ass' he pronounces the sentence 'you are an animal'.

But Richard does not 'distinguish' <92>. Our hypothetical (17a) is an outsider, and not a legitimate reading of <92>. <92> is false, he says; and there is nothing wrong with the disproof. (16) is true; saying somebody is an ass does entail saying that somebody is an animal, which is a legitimate entailment from inferior to superior; and from (16) and (17) (18) follows. (18) is obviously false, which entails that (17), which is our sophismatic proposition <92>, is not true. And indeed, (17) is obviously not true, he adds; it is not generally true that someone who says you are an animal speaks the truth. If, for example, what the speaker in fact says is that you are an ass or a goat when you are a human being, or if the speaker says that you are an animal, but you do not exist, the speaker does not say anything true. In fact, Richard says, it is arguments like the proof of <92> that are faulty because they commit the ffd; one mode of signifying is changed into another:

(19) that you are an animal is true
(20) everyone who says that you are an animal speaks the truth →(17)[27] <92>

Which mode of signifying is changed into which? Richard does not tell us or identify the *similis figuratio*. But in explaining why there is no set of sophismata for 'true and 'false' as there are for 'possible' and necessary', Richard says that this is because '*p* is true' and '*p*' have the same truth conditions.[28] (19) would

27) <92>, and thus (17) of the disproof, was technically 'si *aliquis* dicit te esse animal dicit verum'.

28) [B 93rb] 'Quia 'esse' et 'esse verum' convertuntur et 'non esse' et 'esse falsum' convertuntur, non sunt sophismata specialia cum istis 'verum' et 'falsum'; sed sunt specialia cum modis aliis'.

be unproblematically true, provided that you exist, if you were an animal, i.e., if you were this human being or that human being, or an ass or the Chihuahua over there, and so on for all the rest of the animals; (19) is in fact true because it is true that you are one of these, this human being. But given what Richard has been saying, (20) can easily be false because 'animal' is in transparent position. As Richard expresses it, the problem is that what someone might actually say is that you are an ass or a goat, which is false—and so in such a case, (20) is false. (16), we remember, was true according to Richard, because 'saying that you are an ass' is one of the 'inferiors' of 'saying that you are an animal'.

There are several things wrong with the proof of (20). With respect to 'verum' it looks like we go from *quale* to *quid*. There is another problem with 'verum' as well, for in connection with the next sophism <93> IF I SAY YOU ARE AN ASS I SPEAK THE TRUTH (SI DICO TE ESSE ASINUM DICO VERUM) Richard says that from 'I say nothing true' and 'it is true that you are animal' it does not follow that I do not say that you are an animal, but only that I do not say you are an animal *insofar as it is true*, that is that I do not say you are any kind of animal you actually are—from which it does not follow that I do not say you are an ass. It may be, then, that all that Richard thinks that follows from (19) is that anyone, or someone who says that you are an animal *insofar as it is true* speaks the truth. But given what Richard has explicitly said, his main objection will be that in (19) 'animal' has determinate supposition—(19) is true if you an animal of some kind or other, and that is true if you are some animal or other. In (20) 'animal' is not distributed in this way. From what Richard says, for (20) or (17) to be true, it would have to be true that you are a man **and** are an ass **and** so on for all the subspecies of animal, so that there was no way of saying that you were an animal without saying something true. Matters would not improve if we were allowed to descend to propositions involving singular terms like 'this Chihuahua'.

We could spend quite a while working out exactly what is going with (17) and the like, but the point I want to make is rather simpler, which is that Richard goes to great lengths to preserve some kind of transparent reading of (17), even accepting the simple falsity of <92>, when he could very plausibly have distinguished between senses and blocked the disproof by allowing that (17) has a true reading, on which it is about what somebody actually says, so that 'animal' is not in a fully transparent position, and thus has something like material supposition. He does not, probably because the transparent reading is the usual one for his 'highly normalized Latin'.[29] (But unless Richard has spun

[29] D' Ors (1993), 390.

the solution he rejects out of thin air so as to have a foil for his own proposed solution, there was someone who allowed the alternative reading.)

3. Conclusion

Paul Spade has put forward the idea that the theory of supposition, having started out as a theory of reference chiefly for common nouns and the like in propositional contexts, developed into two theories, one a theory about how to determine the reference of common nouns in context; the distinction between personal, simple and material supposition is part of this theory. The other, the theory of modes of personal supposition, involving the notions of 'descent' and 'ascent' and 'ampliation,' served as a way of explicating the inferential relationships between quantified propositions like 'Every man is an animal' and its 'singulars,' like 'Plato is an animal' or 'Socrates is this animal or that one, etc.'[30] Priest and Read had described the theory of the modes of personal supposition as a kind of quantification theory that was supposed to determine a truth through a functionally or logically equivalent analysis of quantified propositions.[31] Spade rejects this interpretation of the theory; its weaknesses would have been glaringly obvious, and it would in addition have been redundant.[32] But that leaves this second theory, once separated from the first theory, to have really been about nothing. The unstable situation developed gradually without anyone really noticing.[33] Spade also observes that although the theory of personal supposition was at the outset about the subject and predicate terms of quantified propositions, it came to be applied to all referring expressions in a proposition, and he raises the question of how these very complex propositions could have been dealt with in terms of the theoretically meager techniques of the theory of ascent and descent.[34] Presumably he has in mind such cases as *Abstractiones* <72> *UTERQUE ISTORUM VEL RELIQUUS ISTORUM QUORUM NEUTER DIFFERT AB HOMINE EST ASINUS.*

Gareth Matthews has argued that this dual-theory approach is unpromising. He made no grandiose claim about what the whole theory was about, but

---

30) Spade (1988), 188-190. Scott (1966), 30, whose basic insight is endorsed by Spade, had described the theory of the modes of personal supposition as a system of quantification, though Spade does not endorse the description.

31) Read and Priest (1981), 274-279.

32) Spade (1988), 206-207.

33) Spade (1988), 188-190, 212.

34) Spade (1988), 198.

he did observe that the *modes of personal supposition* were called that for a reason, i.e., that they indicated *how* the expression in question refers to the individuals the expression was imposed to signify, e.g., with the subject term referring to all of them so that the predicate had to apply to each and every one of them for the proposition to be true. He showed pretty successfully that the theory of common personal supposition worked coherently, once certain demands were dropped, e.g., that the pattern of descent yields an analysis, or a logically equivalent set of propositions.[35] Recently, in *Formalizing Medieval Logic*, Catarina Dutilh Novaes has given the theory a grandiose name: a system of 'algorithmetic hermeneutics',[36] which means that the theory gives us a systematic way of analyzing propositions so as to determine their possible readings, what they can be used to assert. She has also gone beyond what Matthews claimed to propose that simple and material supposition are part of the same unified system.[37]

I am not sure whether Matthews and Dutilh Novaes are saying quite the same thing. Matthews sees the theory of personal supposition primarily as a theory of reference,[38] and Dutilh Novaes argues that this is at least the wrong emphasis, at least partially because she is interested in a unified treatment of personal, material and simple supposition, and partly because she thinks that dealing with problems like <72> was always part of the purpose of the theory of supposition as a whole. Still, it seems to me that they are on the same right track. The question is whether we can get Richard, Master of Abstractions, to support this approach. I think that to a great extent we can. His use of the theory of personal supposition substantially 'tracks' the approach described by Matthews and laid out in much greater detail by Dutilh Novaes. Moreover, we have clearly got a functional equivalent of simple supposition, at least understood as reference to species, which is part of Richard's system; and that is just what we should expect from someone who is a ho-hum realist, who thinks that species *are* something in addition to the individuals of the species. There is no hint, though, in *Abstractiones* of simple supposition understood as reference to what a term signifies or of Ockham's understanding of simple supposition as non-significative.

What of the complete absence, so far as I can see, of material supposition or any functional equivalent? The answer, I think, is that whereas for the ho-hum

35) Matthews (1997), 35-40.

36) Dutilh Novaes (2007), 8; 30-31.

37) Dutilh Novaes (2007), 26-28.

38) Matthews (1997), 36.

realist simple supposition is of a piece with personal supposition,[39] material supposition, in which a term 'supposits for itself or its like,'[40] is weird. Nearly everything about this definition is problematic, and variants that attempt to improve on it show why it is perplexing.

There is something strange about an expression suppositing for itself. If linguistic signs are supposed to refer conventionally, and nearly everyone holds that they are, there is something wrong with reference that depends on resemblance as these instances of material supposition do—'god' does not supposit materially for the inscription or the utterance 'homo'. Ockham will simply deny that expressions used in this way supposit significatively, which comes down to saying that they refer without having any relevant meaning.[41]

Moreover, expressions with material supposition do not always refer to themselves, strictly speaking.[42] What about 'nomen est vox' (a noun is a vocal sound), which is false if 'nomen' refers to itself, it being an inscription? And what precisely counts as *itself*? This inscription? That works for 'homo is written oddly', but not for 'homo is disyllabic', where 'homo' can easily refer to a linguistic type, as when someone inspecting a dictionary exclaims, 'Ah! 'homo' is disyllabic.' And it is not just any case of self-supposition that counts as material supposition, since 'vox' supposits for itself in 'omnis vox profertur' (every vocal sound is uttered).[43] And finally, the item referred to need not even be spelled the same way, but only in a relevantly similar way. In 'omnem hominem currere est verum' (that every man runs is true), it is really 'omnis homo currit' that is referred to and said to be true; and in 'animal predicatur de homine' (animal is predicated of man) 'homine' supposits for 'homo'.

This is not to say that the theory cannot be extended to cover such cases. In his article, 'How is Material Supposition Possible?' Stephen Read sketches just such a development in the generations after Ockham, in which a theory

39) In fact, Burley would probably consider the references to species endorsed by Richard to fall under personal supposition. Cf. Spade (1997), 8.

40) This is a paraphrase of the definition of Marsilius of Inghen, quoted in Read (1999), 3. It covers one sort of case of material supposition. For Burley (Read (1999), 2), the definition is broader: 'when a spoken word supposits for itself as spoken or for itself written or also for some other word which is not an inferior of this word taken in this way.'

41) William of Ockham, *Summa logicae*, ed. Boehner *et al.* (1974), I, 74, 196. Cf. Read (1999), 2, and Normore (1997), 27-30.

42) Most of the examples that follow are from Ockham's *Summa logicae* I and are discussed in Normore (1997), 2-9.

43) The example is found in Read (1999), 6.

of 'the non-ultimate concept, the concept of the term itself,'[44] evolves. But a workable theory of this kind does not develop naturally as an extension of a theory of reference to non-linguistic entities. It presupposes a well-developed theory about how to get concepts of linguistic things, so that expressions with material supposition can signify these, concepts that are ordinarily indifferent to which inflected form is being referred to by which. Moreover, in order to cover both spoken and written types and tokens of expressions, these concepts presuppose an ontology of types and tokens and need to function as spelling rules of a sort, which requires in turn a sophisticated understanding of what it is to be a letter.

Read sees the development of such theories as a result of increased perception that a term cannot supposit for what it does not signify in some way;[45] and this seems plausible. But it may also be significant that the development Read describes is on the Continent, where, as Spade points out, 'modism prevailed until the 1320s.'[46] And Modists, and the generation of Aristotelian-style grammarians which preceded them, could provide the philosophical foundation for such theories because the grammarians, particularly the *modistae*, have a well-developed theory of orthography and pronunciation. John of Denmark, for instance, gives specific consideration to whether 'letter' is used univocally for spoken and written letters.[47] The *modistae* also have an elaborate syntactic theory which includes the important idea that inflection and surface syntactic transformations are accidental features of expressions. We also find among some Modists the idea that attaching (imposing) an expression to what it stands for requires a concept of the expression imposed, as well as a concept of the kind of thing, since it is the expression-type 'cat' that will now be used, e.g., to produce the mental expression that, once imposed and learned by me, pops

---

44) Read (1999), 15.

45) Read (1999), 5.

46) Spade (1988a), 187. As Spade observes, the heyday of modism is actually substantially earlier.

47) For example, Johannes Dacus, *Summa Gramatica*, ed. Otto (1955), 112: '[...] queritur, utrum vox possit componi ex aliis vocibus,' *ibid.*, 113; Consequenter proceditur ad dubitandum circa diuisionem littere, qua diuiditur littera per litteram in scripto et per litteram in pronuntiatione, et queruntur [...] utrum littera in scripto et littera in pronuntiatione sint idem,' *ibid.*, 118; '[...] primum est vtrum aliquid accidat littere, secundum utrum accidentia littere sint vera accidentia littere vel dicantur accidentia sola similitudine,' 118. Similar discussions, though as a rule less philosophically weighted and sophisticated, are to be found throughout thirteenth- and fourteenth-century commentaries on Priscian. Indeed substantial sections of this part of John's *Summa* are lifted verbatim from the Ps. Kilwardby commentary on Priscian Maior (1975). Similar discussions are to be found in the initial sections of Robert Kilwardby's literal commentary on Priscian Minor.

up in my mind as a precondition for my actually saying, 'Put the cat out,' as well as for your understanding who is to be put outside for the night.[48]

Finally, what of our bootstrapping operation? Here, I think, the results are slim, but not disturbing. Richard's practice with respect to personal supposition is, we have observed, in keeping with single theory descriptions of how the theory works. 'Algorithmic hermeneutics' is in fact an apt description for Richard's exposition of examples like <72> *UTERQUE ISTORUM VEL RELIQUUS ISTORUM QUORUM NEUTER DIFFERT AB HOMINE EST ASINUS*, where it is a calculus of reference and cross-reference that is the main focus of logical interest. And the increasing difficulty of the sophismata within each group builds toward just such propositions. With respect to simple supposition, the fact that there does seem to be a move from solid, standard solutions to sophisms that depend on reference to species as something over and above the individuals that belong to them to additional solutions that substitute the notion of a group or a collection seems perhaps to indicate a period of collection and composition for the *Abstractiones* characterized by a developing sensitivity to alternatives to ho-hum realism. Finally, if, as I have suggested, speculative grammar, and in particular modism, has something to do with a rise in consciousness of problems related to material supposition, then the fact that there is no hint in the *Abstractiones* of the acute consciousness of problems with material supposition that we find in Ockham, let alone of the inventions of succeeding generations, suggests composition prior to the heyday of modism and perhaps with minimal exposure to the development of Modist theories.

Bibliography

Primary Literature

Johannes Dacus, *Summa Gramatica*, ed. A. Otto (*Johannis Daci Opera*, I, 2 parts (Corpus Philosophorum Danicorum Medii Aevi, 1; Hauniae 1955)

Ps.-Kilwardby, 'The Commentary on 'Priscianus Maior' Ascribed to Robert Kilwardby', ed. K.M. Fredborg, N.J. Green-Pedersen, L. Nielsen and J. Pinborg, in: *Cahiers de l'institut du moyen âge grec et latin* 15 (1975)

48) Johannes Dacus, *Summa Gramatica*, ed. Otto (1955), 180: 'Sicut enim anima mediante sensu cognitionem habet de rebus, sic habet cognitionem de vocibus, et hoc verum est tam de proferente quam de audiente [...] Cum igitur apud animam rationalem sint intentiones et ymaginationes, consequenter apud eam fit excogitatio. Cogitat enim apud se, quem intellectum per quam vocem debeat significare, et sic patet quod vox mentaliter ante suam prolationem bene fit significativa'.

Richard, Master of Abstractions, *Abstractiones*. Edition by S. Ebbesen, M. Sirridge, P. Streveler (forthcoming)

Guillelmus de Ockham, *Summa logicae*, ed. P. Boehner, G. Gál and S. Brown (Guillelmi de Ockham Opera Philosophica, I. St. Bonaventure, N.Y., 1974)

Secondary Literature

Dutilh Novaes, C. (2007), *Formalizing Medieval Logical Theories*—Suppositiones, Consequentia *and* Obligationes (Berlin 2007)

——. (2008), 'An Intensional Interpretation of Ockham's Theory of Supposition', in: *Journal of the History of Philosophy* 46 (2008), 365-393

Ebbesen, S., and J. Pinborg (1982), 'Thott 581 4°, or 'De ente rationis', 'De definitione accidentis', 'De probatione terminorum'', in: A. Maierù, ed., *English Logic in Italy in the 14th and 15th Centuries*. Acts of the 5th European Symposium on Medieval Logic and Semantics. Rome, 10-14 November 1980) (Napoli 1982, 111-146)

Ebbesen, S. (1988), 'Concrete Accidental Terms: Late-Thirteenth-Century Debates about Problems Relating to such terms as 'album'', in: N. Kretzmann, ed., *Meaning and Inference in Medieval Philosophy. Studies in Memory of Jan Pinborg* (Dordrecht-Boston-London 1988, 107-161)

Huelsen, R. (1988), 'Concrete Accidental Terms and the Fallacy of Figure of Speech', in: N. Kretzmann, ed., *Meaning and Inference in Medieval Philosophy* (Dordrecht 1988, 175-185)

Kneepkens, C.H. (2003), 'Aristotle's Transposition and the Twelfth-Century Commentaries on *De interpretatione* 20b1-12. An Exploratory Study', in: H.A.G. Braakhuis and C.H. Kneepkens, eds., *Aristotle's* Perihermeneias *in the Latin Middle Ages. Essays on the Commentary Tradition* (Artistarium, Supplementa, X; Groningen-Haren 2003, 363-411)

Kretzmann, N. (1988), *Meaning and Inference in Medieval Philosophy. Studies in Memory of Jan Pinborg* (Synthese Historical Library, 32; Dordrecht 1988)

Matthews, G.B. (1997), 'Two Theories of Supposition?,' in: *Topoi* 16 (1997), 35-40

Normore, C.G. (1997), 'Material Supposition and the Mental Language of Ockham's *Summa logicae*', in: *Topoi* 16 (1997), 27-33

Ors, A. d' (1993). "Hominis asinus/Asinus hominis', in: S. Read, ed., *Sophisms in Medieval Logic and Grammar*. Acts of the Ninth European Symposium for Medieval Logic and Semantics, held at St.Andrews, June 1990 (Dordrecht-Boston-London 1993, 382-397)

Quine, W.V.O. (1966), 'Three Grades of Modal Involvement', orig. in: R. Gibson Jr., *The Ways of Paradox* (New York 1966, 156-174) (repr.: *Quintessence: Basic Readings from the Philosophy of W.V.O. Quine* (Cambridge, Mass., and London 2004, 379-397)

Read, S. (1999), 'How is Material Supposition Possible?', in: *Medieval Philosophy and Theology* 8 (1999), 1-20

—— (1993), *Sophisms in Medieval Logic and Grammar* (Dordrecht-Boston-London 1993)

Read, S., and G. Priest (1981), 'Ockham's Rejection of Ampliation', in: *Mind* 90 (1981), 274-279

Scott, T.K. (1966), *John Buridan: Sophisms on Meaning and Truth*. Translation and Introduction (New York 1966)

Spade, P.V. (1977), 'General Semantic Closure,' in: *Journal of Philosophical Logic* 6 (1977), 209-221.

——. (1988), 'The Logic of the Categorical: The Medieval Theory of Descent and Ascent' in: N. Kretzmann, ed., *Meaning and Inference in Medieval Philosophy* (Synthese Historical Library, 32; Dordrecht-Boston-London 1988, 187-224)

——. (1997), 'Walter Burley on the Simple Supposition of Singular Terms', in: *Topoi* 16 (1997), 7-13.

Streveler, P. (1993), 'A Comparative Analysis of the Treatment of Sophisms in mss Digby 2 and Royal 12 of the Magister Abstractionum', in: S. Read, ed., *Sophisms in Medieval Logic and Grammar.* Acts of the Ninth European Symposium for Medieval Logic and Semantics, held at St.Andrews, June 1990 (Dordrecht-Boston-London 1993, 144-184)

Tabarroni, A. (1994), 'Figure of Speech and Aristotle's Division of Fallacies', in: C. Cellucci, M. C. Di Maio and G. Roncaglia, eds., *Logica e Filosofia della Sciencza: Probleme e Prospettive.* Atti del Congresso triennale della Societa Italiana di Logica e Filosofia delle Scienze (Pisa 1994, 15-24)

The Role of Discrete Terms in the Theory of the Properties of Terms

Julie Brumberg-Chaumont
CNRS, Paris

Abstract

Discrete supposition occurs whenever a discrete term, such as 'Socrates', is the subject of a given proposition. I propose to examine this apparently simple notion. I shall draw attention to the incongruity, within a general theory of the semantic variation of terms in a propositional context, of the notion of discrete supposition, in which a term usually has a single semantic correlate. The incongruity comes to the fore in those treatises that attempt to describe discrete supposition as a sort of personal supposition, although the same term cannot be in simple supposition in another propositional context, because it has no significate distinct from its suppositum. This shows a fundamental link between common signification, simple supposition and predicability, three notions that rely on the existence of a significate distinct and independent from the suppositum of the term. The connection is to be seen especially in William of Sherwood's *Introductiones*, the only author of a terminist *Summa* who recognizes the existence of simple supposition for discrete terms.

Keywords

discrete supposition, predicability, individuals, proper name, individual form, universal

Introduction[1]

What is a Discrete Term?

Discrete supposition is generally described through the occurrence of a discrete term such as 'Socrates' as the subject of a given proposition. This definition

[1] For brevity appellation has been excluded from this presentation. For the same reason, only the treatises belonging to the first period of terminism are studied here.

depends on the way singular propositions are described in the *Peri hermeneias*.[2] A discrete term is mostly identified with a proper noun, a paradigmatic example, but it can also be a pronoun such as *iste* or a complex expression formed by the addition of a deictic pronoun to a common noun, such as *hic homo*. All those expressions are suited to form a singular proposition.[3] Their signification is both singular and substantive because they can be the subject of a proposition and can identify a singular subject to which the predicate is applied.

As proper nouns were the models according to which discrete terms were conceived, grammatical theories played an important part in terminist discussions. The commentators on Priscian's *Institutiones* in the twelfth century taught that the pronoun and the proper noun had a common function[4] (the identification of a determinate individual as the subject of discourse), but also that each of them performed its referential function by distinct semantic means. The pronoun has a deictic force (*deixis*, *demonstratio*) which leads directly to the person and signifies only substance,[5] whereas the proper noun,

---

2) In *Peri hermeneias*, ch. 7, the presence of the same non-equivocal singular name in two opposite propositions (one being the negation of the other) makes them a pair of contradictory propositions because the same predicate is negated and affirmed of the same subject. This is not the case in propositions where the subject term is universal because of the ambiguity of its reference; the pair of opposed indefinite propositions are not contradictory, while the quantified propositions (universal particular) are contradictory only under precise conditions. The theory of supposition is partially intended to account for cases of ambiguity in universal names not dealt with by the Aristotelian theory of contrariety and contradiction.

3) There are as many examples of this definition as there are terminist treatises. See the *Dialectica Monacensis, ca.* 1220: 'Singularis est ista in qua subicitur terminus discretus, ut 'Sor currit', vel terminus communis ad discretionem redactus mediante pronomine demonstrativo, ut 'iste homo currit', ed. De Rijk (1967) (from now on referred to as LM II/1), 469; *Ars Emmerana, ca.* 1200: Singularis est in qua subicitur proprium nomen vel aliquid loco proprii nominis, ut 'Socrates legit', 'hic homo disputat', LM II/2, 154; *Logica 'Ut dicit'*, ca. 1220: Singularis est illa in qua subicitur terminus discretus, hoc est nomen proprium, ut 'Socrates currit', vel in qua subicitur pronomen demonstrativam (*sic*!), ut 'ille homo currit' , LM II/2, 383; Peter of Spain, *Tractatus, ca.* 1230-40: Propositio singularis est illa in qua subicitur terminus singularis vel terminus communis iunctus cum pronomine demonstrativo, ut 'Socrates currit', vel 'iste homo currit'. Terminus singularis est qui est aptus natus de uno predicari , ed. De Rijk (1972), I, 8, pp. 4-5; William of Sherwood, *Introductiones in Logicam*, 1230-1240: 'Singularis est, in qua subicitur terminus discretus, et hoc potest esse proprium nomen vel pronomen demonstrativum, ut 'Socrates currit', vel 'iste currit', ed. Brands and Kann (1995), 14.

4) This theory is generally known to the logicians of the thirteenth century through Peter Helias' authoritative *Summa grammaticae* and through Petrus Hispanus (non Papa)'s commentary on Priscian Minor, see Kneepkens (2000), 373-403.

5) See for instance *Institutiones grammaticae, Grammatici Latini* III, 129, 12-17.

like all nouns, signifies 'substance with a quality':[6] it signifies the individual substance with its proper quality, and in fact identifies this individual through the quality. For the terminist treatises which use this grammatical theory there is in both cases a 'substantive' signification (they both signify substance), so that the expression can function as a subject in a proposition and have the property of supposition. Yet the difference between the two semantic structures raises a problem if proper nouns and pronouns are to belong to the same logical category ('discrete terms'), because some semantic properties, such as personal and simple supposition, imply the existence of a signified form. This signified form cannot exist in the case of pronouns, which signify only substance without a quality (or form). For different reasons, linked to the metaphysical implications of semantic properties, this problem is also encountered in proper nouns: here an independent form might exist no more than in the case of pronouns, even if this form is presupposed by the theory itself, as we shall see in William of Sherwood's *Introductiones*.

Some Remarks on the Historiography of Terminism

Studies in the history of terminism currently suggest that inquiry into semantic variation in propositional context in fact leads to a theory of common nouns that, with varying degrees of success, is then applied to singular terms. When analysing William of Sherwood's *Introductiones*, W. and M. Kneale emphasize this phenomenon:

> It is a little surprising, however, that all terms should be assumed to signify forms, and we may perhaps take this assumption as an indication that the theory was first conceived as a doctrine about general terms such as *homo* and only later extended to singulars terms such as *Socrates* and *ille*. There is in fact no place in the theory for these latter except in connexion with one of the subdivisions of *suppositio*.[7]

L.M. De Rijk's paper on the origins of the theory of properties of terms in the *Cambridge History of Later Medieval Philosophy* might as well have been entitled: 'The Origins of the Theory of the Properties of Common Terms'. His arguments are only to be applied to common terms. When he comes to the influence of grammar on the theory, De Rijk identifies the grammatical couple 'substance/quality' with the couple 'singular thing/universal nature', because

---

[6] This is the formula expressing the signification proper to noun according to Priscian, *Institutiones grammaticae, Grammatici Latini* II, 55, 6.
[7] Kneale (1962), 247.

supposition is the signification of substance. This leaves no place for proper nouns in the theory:

> *Substantia*, according to the medieval interpretation, is nothing but the individual thing, and the *qualitas* meant here is the universal nature in which the particular thing participates.[8]

In the final summary of his *Logica Modernorum* II/1, De Rijk refers to the analysis of the *Dialectica Monacensis* about the lack of interest in any discussion of the supposition of invariant terms (such as discrete terms), and formulates this conclusion:

> As is easily seen, it is the very variability of meaning of the common terms which is considered and evaluated in the theory of supposition.[9]

Discrete Supposition: An Anomaly

The notion of context is crucial to the method of *logica modernorum*, as De Rijk often emphasized. The object of the theory of properties of terms is semantic variation in a propositional context. Yet most authors of treatises agree that singular terms, although they are part of the theory, do not vary semantically. This idea can be understood in a weak sense, when one says that the same individual is always supposited for, or in a strong sense, when one says that the singular term supposes only in one way. This can be expressed in different ways: the singular term can only supposit for its significate; it lacks simple supposition; its suppositum and its significate (even its appellatum) are the same, and so on.

The problem I want to investigate is the way in which our authors try with great difficulty to force an invariant term into a theory of semantic variation in a propositional context. Discrete supposition functions as an incongruous notion within the framework of the theory of properties of terms, especially when it is turned into a general semantics in thirteenth century *summae*, and consequently produces various internal anomalies.

It should be noted that although singular terms are not semantically variant terms, and therefore not the immediate objects of the theory, they certainly are its instruments. Singular terms are necessarily included in the thirteenth century *summae* not only for structural reasons—linked to their analytical

---

8) De Rijk (1982) (pp. 161-173), 163; cf. De Rijk, LM II/1, 521, 522, 523, 525-527 (substantia = individual thing; qualitas = the universal nature (forma)).

9) LM II/1, 582.

and synthetic method which relies on the reconstruction of all components of syllogisms, the main subject of logic—but also for textual reasons, i.e. the need to include all the material inherited from the Aristotelian *Organon* in a new theoretical order. Yet the need to define singular propositions is not only to be explained by this systematic aim of the *summae*: singular propositions play a specific role in the theory of properties of terms. Singular propositions do indeed appear as a fundamental tool in the chapters on supposition for analysing and resolving the ambiguities of universal terms, as may be observed when the 'descent to singulars' is used to test the scope of universal propositions. But when we analyse the properties of singular terms, we find that they are neither restricted nor amplified by the action of the verb or any other addition of a term. The marginal status of discrete terms is revealed when one realizes that the treatises are unable to place discrete supposition in relation to simple, natural, accidental or personal supposition, and, more generally, that they cannot find a place for singular terms within the tree of the divisions of modes of supposition.

Various attitudes toward those difficulties should be distinguished. There are four main ways in which treatises deal with discrete terms:

1) Some treatises do not mention the supposition of discrete terms, thus sticking to the idea, already mentioned, that the theory of the properties of terms is intended only for common terms. This does not mean that discrete terms play no part in these approaches. This position is represented in a number of treatises of the *Logica Modernorum*, particularly in the *Tractatus de univocatione Monacensis* (*ca.* 1200) and in the *Tractatus Anagnini* (*ca.* 1200). For obvious reasons, I will not undertake an analysis of these texts here.
2) Other texts mention the supposition of discrete terms, but ultimately exclude it from the theory. A division between common supposition and discrete supposition appears. At this stage, discrete supposition remains on the margins of the system, since it constitutes the top division of supposition. No mode of supposition is attributed to it: it is neither simple nor personal (neither accidental nor natural, when these modes of supposition are part of the theory). We find this position in Peter of Spain's *Tractatus* and in some treatises of the *Logica Modernorum*.
3) Some authors try to locate the division between common and discrete supposition within the tree. Discrete supposition is not only the supposition of discrete terms but a genuine mode of supposition. This can be done in two different ways, depending on the theoretical assumptions

of each author: discrete supposition is personal, or it is accidental and personal, when the author recognizes the division between natural and accidental suppositions. This approach causes a series of inconsistencies between the divisions of supposition and their respective definitions. The *Logica 'Ut dicit'* and the *Logica 'Cum sit nostra'* are striking examples of this incoherent treatment.[10]

4) Other authors start from different assumptions and argue that the supposition of discrete terms can vary, so that discrete terms can have simple supposition as well as personal supposition. This position is rather rare. Like the first approach, which excludes singular terms from the theory because of their invariance, it is more consistent than the second and third approaches. But this position is not without serious difficulties, to be linked to the metaphysical presuppositions of the theory of predicables with which it is connected. This position is represented by William of Sherwood and by William Arnaud, a commentator on Peter of Spain.[11] His metaphysical presuppositions are completely different from those of William, but not by that token better suited to handling the problem of discrete supposition, as we shall see.

1. The Treatises of the *Logica Modernorum*

A careful reading of the first terminist treatises shows that the set of problems which are at the origins of the development of the theory—as delineated in all its aspects by De Rijk's *Logica Modernorum*—and the scope of the response that this theory seeks to provide as a general logic (in the words of De Libera: in its 'summulist form')[12] should be carefully distinguished. The study of the difficulties encountered in the integration of discrete termes suggests a shift from the original scope and relevance of the theory of properties of terms, which did not originally include discrete terms, to the general scope claimed by the broad logical synthesis offered in the period of the great *summae*. The first terminist treatises consider singular terms at best as instruments in a doctrine

[10] According to these treatises discrete supposition is a sort of personal supposition. The *Summa Lamberti* (*ca.* 1250) is another good example of the difficulties encountered by this approach: it says that discrete supposition is both accidental and personal, whereas it can be neither natural nor simple in various propositional contexts. I have discarded this text for sake of brevity.

[11] Roger Bacon's *Summule dialectice*—which are not studied for the same reason as previously formulated—are to be included in this category.

[12] See De Libera (1982a), 213-234.

whose legitimate subject is clearly common terms. Singular terms are a fixed points of reference from which the phenomenon of semantic variation of common terms is measured. But the treatises do not study the appellation or the supposition of singular terms, nor do they delineate the rules for their restriction or ampliation. The first *summae* retain the operational role of singular terms (especially with the descent to singulars), but also tend to treat them as subjects of the theory. Discrete terms consequently appear as terms, but are thought of as devoid of most of the properties belonging to common terms, that is to say, as anomalies.

1.1. *A Place for Discrete Supposition in the Margins of the Theory of Supposition (Type 2)*

The treatises of type 2, where the terms 'discrete' and 'discrete supposition' are introduced in the margins of the theory of modes of supposition, are represented by the *Dialectica Monacensis*, the *Summe Metenses*, the *Summule antiquorum* and the *Tractatus de proprietatibus sermonum.*[13]

The Tractatus de proprietatibus sermonum *(ca. 1200)*
This treatise is the only one in the period to state explicitly that proper nouns do not have the same meaning as the others, since they signify things as they are received in the imagination and not in the intellect:

> Videtur autem omnia propria nomina significare imaginationi. Nomen enim proprium nominat rem sensibilem sub collectione proprietatum sensibilium. Nomina autem universalia generum et specierum, ut 'homo', 'animal', et consimilia, videntur significare intellectui, hoc est prout apprehensa sunt in virtute intelligibili. Ita enim nomina 'homo', 'animal' non determinant aliqua accidentia sensibilia vi sue prolationis. Et ita quedam nomina significant intellectui, quedam imaginationi.[14]

This passage provides, in a sense, a noetic and semantic foundation for the peculiar logical status of proper nouns, even though the idea is only sketchy and the link between the two aspects is never explained. The author does not draw any consequences from this description for the theory of supposition, although this theory in fact applies only to common terms. This is the case of personal supposition:

---

13) The *Introductiones Parisienses* (*ca.* 1170, but see S. Ebbesen, *Early Supposition Theory II* in this volume, who dates the tract 1190-1210) are also to be situated in this category.
14) De Rijk, LM II/2, 709-710.

> Subpositio termini communis quando non suscipit predicatum pro communi sed pro persona, idest aliquo particulari.[15]

All the subsequent divisions follow the same path. The author explicitly points out the absence of semantic change in discrete terms, so that a theory of properties of terms is unnecessary:

> Et cum suppositio quedam sit discreta quedam communis, et discreta, cum certa sit, non possit variari per aliquod adiunctum, notandum est quod cum suppositio termini communis vaga sit et incerta, quandoque habet variari per aliquod adiunctum, quandoque non habet variari per adiunctionem signi.[16]

The Dialectica Monacensis (*ca. 1220*)

The *Dialectica Monacensis* defines supposition as follows:

> Supponere [...] est substantive <rem> significare et per se et sine dependentia tali que est in principali significatione.[17]

This property is linked to the grammatical description of what is proper to nouns (i.e., as meaning "substance with quality"):

> Dicitur autem terminus ille supponere qui nullam dependentiam habet in principali sua significatione. Ut patet in hoc termino 'homo', qui substantiam cum qualitate finite significat.[18]

The first division is between common and discrete supposition, or rather between the things signified by discrete terms and by common terms, with explicit reference to the way the singular subject and the universal subject of a proposition are distinguished in chapter 7 of the *Peri hermeneias*.

> Suppositio communis est que fit in termino communi, ut 'homo est animal'. Suppositio discreta est que fit in termino discreto, ut 'Sor currit'. Et hec divisio fit penes divisionem rerum significatarum per tales terminos. Que talis est: rerum alie universales, alie singulares. In qua divisione non est medium. Quare nec in hac: suppositionum alia communis, alia discreta, quia omnis terminus aut significat rem universalem, aut singularem [...] Relicto

15) De Rijk, LM II/2, 713.
16) De Rijk, LM II/2, 722.
17) De Rijk, LM II/2, 606.
18) De Rijk, LM II/2, 606.

de suppositione discreta, quia non variabilis est, dicendum est de divisione suppositionis communis, quia illa multis modis variatur.[19]

It is quite remarkable that the first division in fact concerns terms and, because of that, supposition. The notion of discrete supposition is introduced, but it is not placed on the same footing as other properties. It does not properly exist, since it is not a division of supposition, like simple or personal supposition, but a division of terms. This is also partially true for the notion of common supposition. The situation is different, however, because the supposition of common terms is subjected to varying modes of supposition. Behind the homogeneity of the divisions, one should recognize the following implicit thesis: of terms, some are singular, others common; in a sense discrete terms have a supposition that can be called 'discrete supposition' in order to unify the definitions, but in fact what is proposed is a theory of supposition for variant terms, i.e., a theory of common supposition.

The Summe Metenses *(ca. 1220)*

We here follow the edition of De Rijk, who attributes the work to Nicholas of Metz and thinks it was written around 1220. As Braakhuis has shown, the author should be identified with Nicholas of Paris, who also produced a treatise on syncategorematic terms.[20] The editor had stressed the fact that the text is characterized by a grammatical approach. The definition of supposition endangers the unity of the category of discrete terms, since it applies only to nouns (and therefore not to pronouns):

Suppositio est substantiva rei designatio, idest per nomen substantivum.[21]

Yet pronouns do have discrete supposition:

Dividitur ergo suppositio in communem et discretam. Communis suppositio est quam habent nomina appellativa a forma communi a qua imponuntur, ut 'homo', 'animal' etc. Suppositio discreta est quam habent termini discreti, ut propria nomina et pronomina demonstrativa, ut 'Sor', 'Plato', 'iste'. Et discreta dicitur quia ad unum determinata est nec ampliari potest nec coartari.[22]

---

19) De Rijk, LM II/2, 607-608.

20) Braakhuis (1979), 318 and 514.

21) De Rijk, LM II/1, 455.

22) De Rijk, LM II/1, 455.

Here is the division of the modes of supposition proposed:[23]

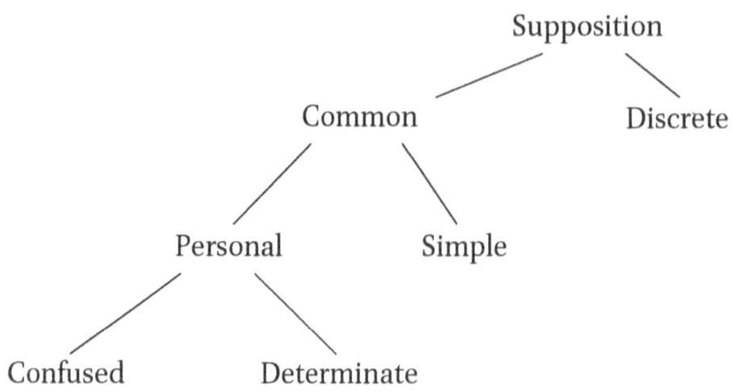

The singular proposition and, therefore, the discrete term are important because the descent to singulars is an essential tool in the description of confused supposition. The author introduces an asymmetry between common and singular terms: common terms are homogeneous from a grammatical point of view (they are all appellative nouns), while singular terms involve the combination of two different parts of speech on grammatical grounds, i.e., proper nouns and pronouns. As a consequence, supposition concerns pronouns, although it is unclear how the author would explain this point. Common supposition is based upon the existence of a common form, while a term having discrete supposition is simply stated, without any 'deep' underlying structure, such as a 'singular form' that would be what is signified. Discrete supposition is simply the supposition of discrete terms, because they are discrete. This redundant definition is a recurrent feature of the treatises of the *Logica modernorum.* Nothing is known of the semantic structure of discrete terms, or about the nature of their imposition. We only know that they are of a different type. Is this because the form according to which they were imposed is itself singular or is it because they are not imposed according to a form, a peculiarity that would explain the grouping of nouns with pronouns in one logical category?

The Summule antiquorum[24]

The *Summule* do not contain an elaborate discussion of the role of discrete supposition, but they offer an interesting reflection, though not always a coherent

23) The diagram appears in De Rijk, LM II/1, 456.

24) The text has been edited by De Rijk according to the manuscript London, *British Museum*, Royal Mss 8 A VI. It probably belongs to Parisian circles and is the main source of Peter of Spain's *Tractatus*. See De Rijk (1968), 8.

one, on the properties of discrete terms. As in previous texts, the division of supposition into discrete and common supposition is the first division and it simply depends on the presence of a common or a discrete term.[25]

The chapter on restriction and ampliation allows a better understanding of discrete terms and their properties. It appears that the recognition of the specific status of the division between discrete and common suppositions is consistent with the idea that discrete terms have a distinct semantic structure. This is stated in the chapter on restriction, where there is clearly a problem in establishing the text. It is perfectly obvious that the edited text prints *ampliatio* for *appellatio* and *ampliare* for *appellare*, whatever the origin of this confusion may be:[26]

> Restrictio est species suppositionis, quia personalis suppositio dividitur per restrictam et ampliatam. Restrictio est coartatio termini communis a maiori suppositione ad minorem. Ut cum dicitur: 'homo albus', hoc adiectivum 'albus' restringit hominem ad supponendum pro albo. Dico autem 'termini communis', quia terminus discretus neque restringitur, neque ampliatur. Differt autem restrictio ab ampliatione [=appellatione] quia restrictio est de re existente et non existente. Ampliatio [=appellatio] autem est termini suppositio pro re existente. Et dividitur per communem et discretam. Ampliatio [=appellatio] discreta est suppositio termini significantis rem existentem, ut 'Sor'. Unde terminus discretus significans rem existentem idem significat et supponit et ampliat [= appellat], ut 'Sor' significat rem et supponit pro Sorte et ampliat [=appellat] Sortem. Si vero significet rem non existentem, tunc nil ampliat [=appellat], sed solum significat et supponit, ut 'Cesar' significat Cesarem et supponit pro Cesare et nil ampliat [=appellat], ut 'Antichristus' et similia. Appellatio vero communis est suppositio termini communis pro re existente. Et sciendum quod terminus communis habet simplicem suppositionem, tunc idem significat quod <quando> supponit et ampliat [=appellat]. Ut cum dicitur: 'homo est species', iste terminus 'homo' supponit pro homine in communi et significat hominem et appellat hominem in communi [...]. Quando autem terminus communis habet suppositionem personalem, tunc terminus non idem significat et appellat.[27]

Once the text restored, we can observe that the explanation for the semantic structure of discrete terms is different from that of common terms, and that this structure changes, depending on whether the noun refer to an existing thing or not. For terms such as 'Socrates', the appellatum, the significate and the suppositum are the same thing, i.e., the existing thing Socrates. For terms such

25) "Suppositio alia communis, alia discreta. Suppositio communis est que fit per terminum communem, ut 'homo'. Suppositio discreta est que fit per terminum discretum, ut 'Sortes' vel 'iste homo'", De Rijk (1968), 9.

26) See Goubier (2003), 55.

27) *Summule Antiquorum*, ed. De Rijk (1968), 13. Our emphasis.

as 'Caesar', there is no appellatum and the significate and the suppositum are identical: the non-existent Caesar.[28]

The *Summule antiquorum* draw a clear relationship between the marginal status of discrete supposition and the existence of a specific semantics for discrete terms. All nouns do not have the same semantic structure. Common terms such as 'man' have a significate (the 'common man'), one or several supposita, individual men, when in personal supposition, and one or several appellata, the existing individual men. Discrete terms such as 'Socrates' have a significate, a suppositum and an appellatum which are identical, i.e., the existent individual. Terms such as 'Caesar' or 'chimera' have a significate and a suppositum which are identical, the individual non-existent, but no appellatum.

The problem raised by this classification is that it distinguishes types of discrete terms: those that appellate something and those that do not. But the principle of this classification is based upon the state of the world. One must therefore account for the existence of terms that will change categories when the thing named, originally existing, comes to be corrupted. What will happen when the existing thing signified, appellated and supposited for by 'Socrates' becomes non-existent? The appellatum, the suppositum, and the significate

---

28) There is another formulation of the same idea in the treatise. The chapter on appellation says the same thing as the chapter on restriction but in a more allusive and confused way: a signification and a supposition are both attributed and refused to terms such as 'Caesar'; appellation is a property of common terms, and yet discrete terms do have an appellation); 'Appellatio est acceptio termini communis pro re existente. Dico autem 'pro re existente' quia terminus significans non ens nichil significat <nec> appellat, ut 'Cesar', 'nil' et 'chimera', et sic de aliis. Differt autem appellatio a suppositione et significatione, quia appellatio est tantum de re existente, suppositio et significatio tam de re existente quam de non existente. Ut 'Antichristus' significat Antichristum et supponit pro Antichristo, sed nil appellat, 'homo' autem significat hominem de natura sua <et> supponit tam pro existentibus quam pro non existentibus et appellat tantum existentes homines. Appellationum autem alia est termini communis, ut 'hominis', alia est termini singularis, ut 'Sortis'. Terminus singularis idem significat et supponit et appellat, quia significat rem existentem, ut 'Petrus' et 'Johannes'. Item. Appellationis termini communis alia est termini communis pro ipsa re in communi, ut quando terminus habet simplicem suppositionem. Ut cum dicitur 'homo est species' vel 'animal est genus'; et tunc terminus communis idem significat et supponit et appellat, ut 'homo' significat hominem in communi et supponit pro homine in communi et appellat hominem in communi. Alia autem est termini communis pro suis inferioribus, ut quando terminus communis habet personalem suppositionem. Ut cum dicitur 'homo currit', tunc 'homo' non idem significat et supponit et appellat, sed significat hominem in communi, sed pro particularibus supponit et appellat particulares homines existentes (*Summule antiquorum*, ed. De Rijk (1968), 18). This text can be seen as a truncated and disordered repetition of the text previously quoted from the chapter on restriction. The parallelism cannot be further studied here, but it is important to underline the fact that Peter of Spain relied on this (bad) text, but removed its obvious inconsistencies.

are initially identical. But what is identical to what? If to the appellatum, this means that the corruption of the thing named leads to the corruption of the meaning of the name. In this case, 'Socrates' would lose its signification upon the death of the individual. Our text says that 'Socrates' signifies, supposes and appellates the existing thing, strictly speaking, the appellatum. Yet 'Caesar' continues to suppose for and to signify Caesar, and can probably be inserted in a true past tensed proposition such as 'Caesar was killed by Brutus'. We must therefore conclude that the assertion of the identity of the suppositum, the appellatum and the significate in the case of discrete terms does not imply that the meaning of singular terms is transient. They mean the existing thing, when there is an existing thing to be named, and they supposit for or appellate nothing other than this existing thing. But the name means the non-existent suppositum if the thing named has ceased to exist. This theory does not seem very economical, since the proper name that loses its bearer undergoes not only a change in its significate (since it no longer means the existing thing) but also in its system of signification, since it no longer enjoys a semantics of direct reference to the thing. One must assume a transfer of imposition and signification when the individual named ceases to exist.

1.2. *Discrete Personal Supposition (Type 3)*

Type 3 is well represented by the *Logica 'Ut dicit'* (*ca.* 1220).[29] Here is the diagram of the modes of supposition offered by the treatise:

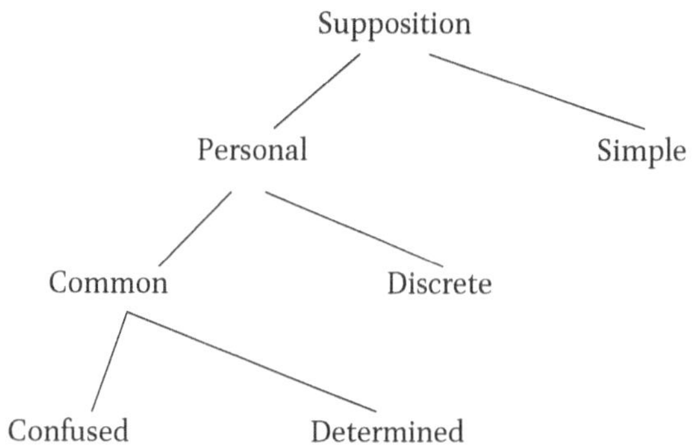

The definitions that correspond to the different types of supposition only speak of the first part of the division, the one that is not further divided:

[29] For a brief presentation of the text, see De Rijk, LM II/1, 446.

> Suppositionum alia simplex, alia personalis. Simplex quando terminus communis subponit pro forma propria a qua inponitur nomen et non pro aliquo inferiori eius, ut 'homo est dignissima creaturarum creatura.' [...] Suppositionum personalium alia communis, alia discreta. Discreta est illa quando supponitur terminus discretus vel pronomen demonstrativum, verbi gratia: 'Socrates est animal' vel 'iste homo est animal'. Suppositionum communium alia determinata, alia confusa.[30]

This device allows the author to avoid defining the type of supposition that is common to discrete and common terms: personal supposition. As we shall see, he has good reason to do so. Simple supposition belongs only to common terms. If one were to reconstruct the definition of personal supposition, one would have to find something that common and discrete supposition have in common, and this can only be the reference to individuals. The obvious problem with this solution is that personal supposition is no longer directly opposed to simple supposition since it is not a property of common terms. The other option would be to take the reverse of the definition of simple supposition. In this case, personal supposition would be defined as supposition not for the form from which the noun was imposed, but for the 'inferiors', i.e., the individuals under this form. But this cannot be applied to discrete terms. The notion is thus problematic.

1.2.1. *The Logica* 'Cum sit nostra'[31] *(ca. 1200)*

The treatise '*Cum sit nostra*' relates the signification of substantive nouns, i.e., substance and quality, to the logical properties of those nouns. Proper substantive nouns supposit but do not appellate:

> Substantivorum quedam supponunt tantum, ut propria nomina, quedam supponunt et appellant, ut nomina communia.[32]
> [Common substantive nouns] supponunt gratia materie sive substantie et appellant gratia qualitatis.[33]

This text is a combination of the innovations of the *Logica modernorum* and of the teaching of the previous century's grammarians about the possibility of applying to all nouns the proper of the noun according to Priscian, "the signification of substance with quality". Since common nouns seem to have an appellation because they signify quality, the logical conclusion of this passage

---

30) De Rijk, LM II/2, 409.

31) For a detailed presentation of this text, see De Rijk, *LM* II/1, 416 f. De Rijk thinks it may be the work of Wilhelmus Montanus (*ibid.*, 442-443).

32) De Rijk, LM II/2, 446.

33) De Rijk, LM II/2, 446.

should be that the reason why proper nouns have no appellation is because they signify only substance, and not quality. If the text goes so far as to say that adjectival nouns signify only quality, it does not state that proper nouns signify only substance, nor does it say that they signify quality.[34] This silence about the signification of quality in the case of a proper noun is consistent with the whole treatise, where neither predicability proper nor simple supposition is attributed to singular terms.

The treatise adopts the same doctrine as the one previously studied, but does not hesitate, in its definitions, to express the inconsistencies connected to the concept of personal supposition. Here is the diagram of the divisions of supposition:

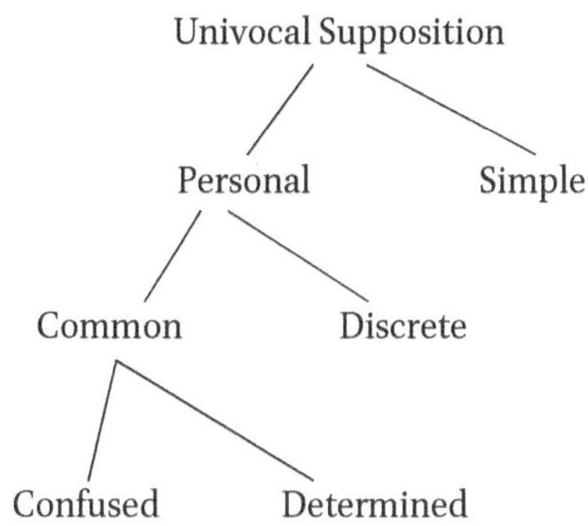

The definitions are completely inconsistent with it:

> Suppositio univoca sic dividitur: alia simplex, alia personalis. Suppositio simplex est quando terminus communis supponit pro forma communi in qua nomen ponitur, ut 'homo est dignissima creatura creaturarum'. Suppositio personalis est quando terminus communis supponit pro aliquo inferiori, ut 'homo currit'; hic li 'homo' supponit pro illo et pro isto. Suppositio personalis sic dividitur: alia communis, alia discreta. Suppositio communis est quando subicitur terminus communis secundum quod habet respectum ad ea que sunt communia. Suppositio discreta est quando subicitur terminus discretus, ut 'Socrates currit', vel pronomen demonstrativum, ut 'ille vel iste currit'.[35]

The inconsistencies are obvious: personal supposition is the supposition of common terms, but it is divided into the supposition of common terms and the supposition of discrete terms.

The asymmetry between the definitions in the two cases can easily be seen. The definition of discrete supposition includes only the presence of the discrete

[34] For a study of this problem in twelve century grammar and in the case of proper nouns, see Brumberg-Chaumont (2011) and Brumberg-Chaumont (2007), 137-166

[35] De Rijk, LM II/2, 447.

term in subject position, but not the relationships between the term and what is supposited for, nor the nature of the thing supposited for. The definition of common supposition refers to the fact that the common noun is taken in its relationship to 'common things'. This is probably to be understood as the things that have the noun in common. If one compares the definition of common supposition and that of personal supposition, the dividing line is very difficult to trace. It is rather a question of emphasis, since he speaks of the same things as 'inferiors' in the definition of personal supposition and as 'common' in the definition of common supposition. Personal supposition is common personal supposition and discrete supposition is not personal, because it cannot be a division of personal supposition, which is a supposition of common terms.

The two treatises describe supposition as a general property of terms, and theoretically isolate its operation from the contexts of semantic variability, so that discrete terms and discrete supposition are artificially introduced as problematic objects for the theory of supposition. This phenomenon is to be systematically encountered in the great syntheses of the thirteenth century where the question of properties of terms is understood within the context of a general logical theory. Unlike what we have observed in the two treatises mentioned above, the difficulties raised by the role of singular terms are often directly and explicitly dealt with, but this does not mean that the solutions adopted are satisfactory.

The problem of influences and currents is worth mentioning before we leave the earliest treatises of the *Logica Modernorum*. Many of the treatises on the properties of terms edited by De Rijk belong to the Oxonian tradition,[36] especially those which influenced Roger Bacon and William of Sherwood: the *Logica 'Ut dicit'*, the *Logica 'Cum sit nostra'* and the *Introductiones Parisienses*. In Parisian logic, influences are to be found in the *Summule antiquorum*, the *Introductiones Antique*,[37] in some aspects of the *Dialectica Monacensis*, in the *Appellationes*[38] of Johannes Pagus, and in the treatises of Nicholas of Paris. These considerations on the difference between Oxford and Paris are not always very useful for the precise question of the properties of discrete terms. We can, however, notice a constant position in the Oxonian tradition, according to which discrete supposition is seen as a type of personal supposition, with a greater or lesser awareness of the challenges that this position can generate. Seen from this point of view, the case of William of Sherwood is remarkable,

---

36) See Pinborg (1979), 26, who insists on the great influence of the *Logica 'Cum sit nostra'*. See also De Libera (1982b), 175-177.

37) Edited with the *Summule antiquorum*.

38) See De Libera (1985), 193-255.

because he is the only logician to take on all the implications of this position and to attribute to discrete terms a real capacity for semantic variation. Discrete supposition can be not only personal, but simple as well. On the Parisian side, there is relative consistency in excluding discrete supposition from all the modes of supposition.

2. The *Tractatus* (ca. 1230-1240) of Peter of Spain (Type 2)

Peter of Spain's position belongs to type 2, as does its source, the *Summule antiquorum*. Supposition is a property of 'substantive terms',[39] which would suggest that pronouns are not included in the theory of supposition. In fact, they are absent from the examples of singular terms. The division of supposition puts the distinction between common and discrete supposition at the top of the tree. Here are the corresponding definitions:

> Suppositionis alia communis, alia discreta. Suppositio communis est que fit per terminum communem, ut 'homo'. Suppositio discreta est que fit per terminum discretum, ut 'Sortes' vel 'iste homo'.[40]

Common supposition is then divided into natural and accidental, then accidental supposition into personal and simple, according to the following diagram:

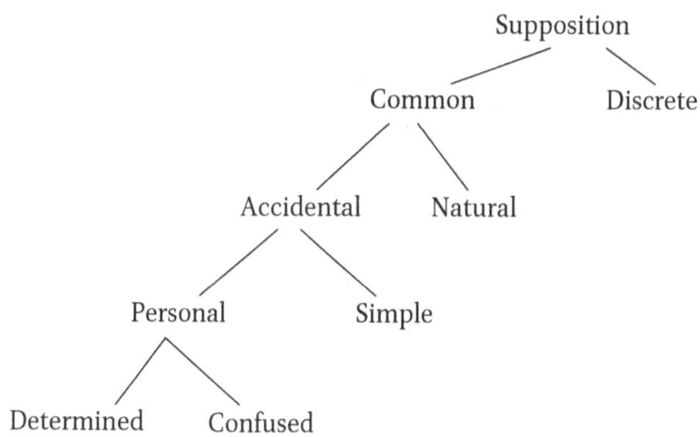

The definitions are consistent with this pattern, since discrete terms are not mentioned and the definitions are intended for common terms alone:

---

[39] Peter of Spain, *Tractatus*, ed. De Rijk (1972), 80: 'Acceptio termini substantivi pro aliquo'.

[40] Peter of Spain, *Tractatus*, ed. De Rijk (1972), 80.

> Suppositio naturalis est acceptio termini communis pro omnibus a quibus aptus natus est participari, ut 'homo' per se sumptus de natura sua supponit pro omnibus hominibus qui fuerunt et qui sunt et qui erunt. Accidentalis autem suppositio est acceptio termini communis pro eis pro quibus exigit adiunctum. Ut 'homo est', iste terminus 'homo' supponit pro presentibus; cum autem dicitur 'homo fuit', supponit pro preteritis [...] Et ita habet diversas suppositiones secundum diversitatem eorum que ei adiunguntur.[41]
> Simplex suppositio est acceptio termini communis pro re universali significata per ipsum. [...] Personalis suppositio est acceptio termini communis pro suis inferioribus.[42]
> Dico autem 'termini communis', quia terminus discretus, ut 'Sortes' non restringitur neque ampliatur.[43]

The concepts of predicability, of simple supposition and of significate all fit together—this will be discussed further in the case of William of Sherwood. Seen from this perspective, the identity of the suppositum and the significate explains both the absence of simple supposition and the impredicability of discrete terms. The first point has already been established. The second point can be demonstrated since discrete terms are not predicable in a strict sense. The theory of predication involves a close relationship between the predicate term and the form signified. The absence of a predicable form in the case of singular terms is to be understood on the basis of the distinction between predicability in a wider sense, which applies also to discrete terms,[44] and predicability proper, which applies only to universal terms, a distinction observed in most treatises. Several elements show that predicability proper is not just a property of the term, but a property the term gets from the universal signified, while predicability in a wider sense is a property of the term itself, because there is no significate distinct from the suppositum. The link between the predicate and the form signified can be seen when Peter of Spain divides simple supposition into supposition of the subject and supposition of the predicate. If the predicate is in simple supposition,[45] it supposits for its significate, that is to say, for the universal form signified. The fact that a discrete term has a particular semantic structure, in which the significate and the appellatum are identical, explains both why it cannot be in simple supposition and why a discrete term cannot be truly predicated.

The identity of the significate, the suppositum and the appellatum is explicitly stated by Peter of Spain:

---

41) Peter of Spain, *Tractatus*, ed. De Rijk (1972), 81.

42) Peter of Spain, *Tractatus*, ed. De Rijk (1972), 81-82.

43) Peter of Spain, *Tractatus*, ed. De Rijk (1972), 194.

44) Peter of Spain, *Tractatus*, ed. De Rijk (1972), 19: 'Individuum est quod de uno solo predicatur'.

45) Peter of Spain, *Tractatus*, ed. De Rijk (1972), 81.

> Appellationum autem alia est termini communis, ut 'hominis', alia termini singularis, ut 'Sortis'. Terminus singularis idem significat et supponit et appellat, quia significat rem existentem, ut 'Petrus' vel 'Johannes'.[46]

The specificity of singular terms and their inability to find a place within a theory of properties of terms cannot be more strongly emphasized. Discrete terms have neither ampliation, nor restriction. The supposition of discrete terms is neither natural nor accidental, neither simple nor even personal. They have only one semantic relation, that to the thing named. They have no distinct significate that could be predicated in attribute position and for which they could supposit in the subject position in simple supposition.[47] On Peter of Spain's view, two separate semantic patterns are to be constructed for singular and universal terms.[48]

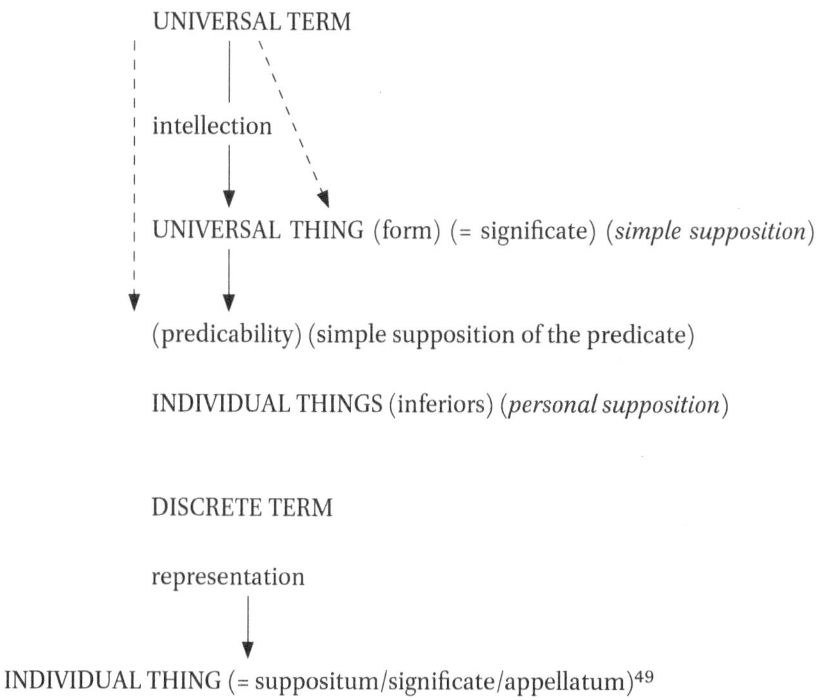

[46] Peter of Spain, *Tractatus*, ed. De Rijk (1972), 197-198.

[47] The major problem this theory faces is the problem of empty names, a problem superficially adressed previously about the *Summule antiquorum*, and dealt in more details in Brumberg-Chaumont (2005) and Brumberg-Chaumont (forthcoming).

[48] The full arrows represent signification in a broad sense, and the dotted arrows represent supposition.

[49] If there is an *appellatum*.

3. William of Sherwood, *Introductiones in Logicam* (Type 4)

3.1. *The Theory of Supposition: Internal Difficulties*

According to William of Sherwood, supposition is present in words that 'signify the thing as subsistent and that can be placed under another',[50] a definition which includes pronouns. The appellation does not concern pronouns, 'because they do not signify a form, but only the substance' .[51] Here is how formal supposition is divided into simple and personal:

> Et est simplex, quando dictio supponit suum significatum pro significato, ut 'homo est species'. Personalis autem, quando supponit suum significatum, sed pro re, quae subest, ut 'homo currit'. Cursus enim inest homini gratia alicuius singularis.[52]

But formal assumption is divided again:

> Est alia divisio suppositionis formalis, scilicet quod quaedam est communis et quaedam est discreta. Communis, quae fit per terminum communem, ut 'homo currit' ; discreta, quae fit per terminum discretum, ut 'Socrates currit', vel 'iste'.[53]

This gives the following diagram (a bush rather than a tree):

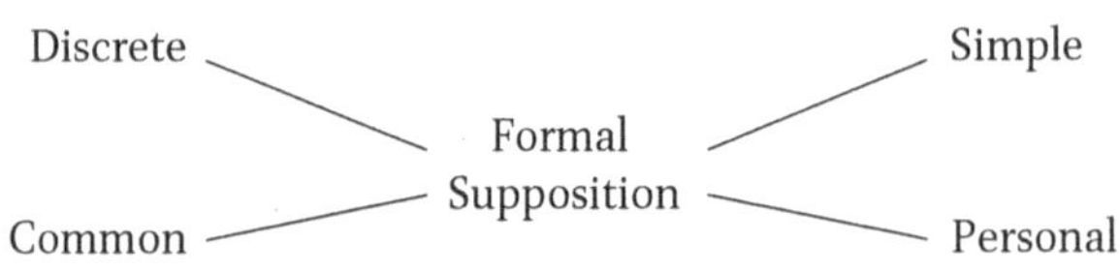

Each member of these two divisions includes all cases of formal assumption:

> Omnis enim dictio supponens aut est communis aut discreta. Item. Aut accipitur pro forma significata, et tunc est simplex suppositio, aut pro re deferente formam, et tunc personalis.[54]

---

50) William of Sherwood, *Introductiones in Logicam*, ed. Brands and Kann (1995), 134: 'Significant rem ut subsistentem et ordinabilem sub alio'.

51) William of Sherwood, *Introductiones in Logicam*, ed. Brands and Kann (1995), 134: 'Quia non significant formam aliquam, sed solam substantiam'.

52) William of Sherwood, *Introductiones in Logicam*, ed. Brands and Kann (1995), 136.

53) William of Sherwood, *Introductiones in Logicam*, ed. Brands and Kann (1995), 136.

54) William of Sherwood, *Introductiones in Logicam*, ed. Brands and Kann (1995), 136.

The distinction between personal supposition and simple supposition is reformulated in different terms, focusing on the relationship between the form signified and the thing: simple supposition is the supposition for the signified form, while personal supposition is the supposition for the thing under this form. William is well aware of his originality, since he mentions the opinion of his opponents:

> Volunt enim quidam, quod haec divisio, alia simplex—alia personalis, sit divisio communis suppositionis, quia non cadit haec diversitas in discreta suppositione. Non enim est ibi nisi personalis. Semper enim supponitur individuum in tali, scilicet in discreta suppositione.[55]

William's answer evidently concerns the status of discrete supposition, but also the nature of personal supposition in general. If it were defined solely by reference to the individual(s), any discrete supposition would be personal, since the singular term always supposits for an individual. To oppose this view, the *Introductiones* connects the distinction between personal supposition and simple supposition with the grammatical description of the proper signification of the noun, "the signification of substance with quality":

> Dicendum, quod hoc non facit personalem suppositionem, scilicet quod supponitur individuum, sed quod supponitur res deferens formam significatam per nomen. Et hoc potest accidere in nomino proprio, cum significat substantiam cum qualitate, ut cum dico 'Socrates currit', respicitur pro sua re. Cum dico 'Socrates est praedicabile de uno solo', respicitur pro forma significata per nomen.[56]

From a semantic point of view, William offers a unified theory of signification for all subject terms, which takes into account both their meaning and reference, as well as the link between these two semantic properties and the metaphysical structure of the things signified and referred to. It implies a relationship between the individual and the form that enables the term to be predicated (in predicate position) and to supposit for its significate (in simple supposition). We can give the following diagram of this semantics:

---

55) William of Sherwood, *Introductiones in Logicam*, ed. Brands and Kann (1995), 140.

56) William of Sherwood, *Introductiones in Logicam*, ed. Brands and Kann (1995), 140.

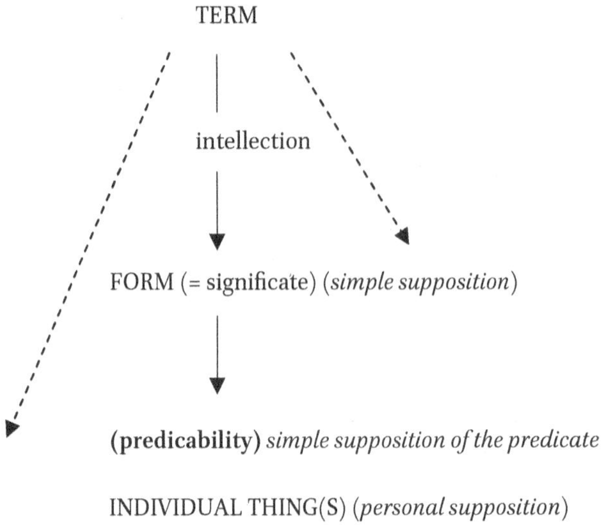

William's argument raises several problems. First of all the reconstruction of his opponents' thesis is not an easy matter. If we compare the position criticized by William of Sherwood with the theories previously studied, we can see that the *Logica 'Ut dicit'* and the *Logica 'Cum sit nostra'* do deny simple supposition for discrete terms, so that discrete supposition is always a personal supposition, which can be defined only by reference to individuals in order to suit discrete as well as common terms (if the theory were to be consistently reconstructed). This is precisely the definition William puts into the mouth of his opponents. The problem is that, in both texts, the distinction between personal and simple supposition is not a division of common supposition (the position William attributes to his opponents), but rather the division between common and discrete supposition is a division of personal supposition. We also have other treatises for which the distinction between simple and personal supposition is a division of common supposition, such as all texts of the second group of the *Logica Modernorum* and Peter of Spain, but they are careful enough not to state in addition that the discrete supposition is personal. The opinion that William attributes to his opponents is not consistent: they want the distinction between personal and simple supposition to be a division of common supposition, but they also want discrete supposition to be a personal supposition. On the other hand, this inconsistency is quite similar to what we observed in *Logica 'Cum sit nostra'*, whatever the explicit intentions of the author of the treatise may be: in fact, discrete supposition is a sort of personal supposition even though personal supposition is defined only for common terms. The existence of a sort of personal supposition that is not on a par with simple supposition raises questions about the very nature of these properties: how can a term have personal supposition if it cannot have simple supposition in other propositional contexts? Are we talking about types of terms or about modes of supposition?

The position of William's opponents comes down to the distinction between two types of personal supposition: a personal supposition such that a term may only have personal supposition if it can otherwise have simple supposition, and a personal supposition that is simply the fact that the term stands for an individual thing (or a person), as can be seen in the following diagram:

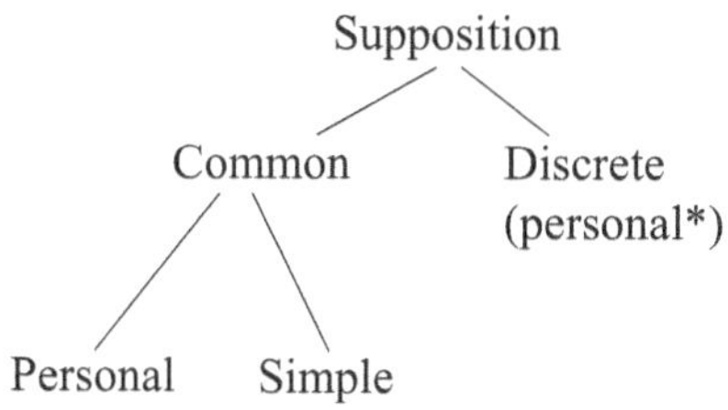

On the contrary, William of Sherwood preserves consistency by allowing discrete terms both a personal and a simple supposition, thus avoiding a duplication of the concept of personal supposition. But he must also face important challenges, some that are internal to the theory, and others that are related to the interactions between semantic theory and metaphysical issues—a topic dealt with in detail later on.

The existence of discrete supposition for pronouns raises problems internal to the theory. Discrete simple supposition cannot be found in pronouns such as 'iste', yet they are always mentioned as examples of discrete terms. As previously observed, William adopts the thesis of traditional grammar according to which pronouns signify substance but not quality. That is why they cannot have an appellation.[57] We should therefore build a new tree in which the supposition of discrete pronominal terms comes before the division of discrete supposition into personal and simple, unless we exclude pronouns from the list of discrete terms, which would contradict William's explicit assertions on that subject. The impossibility of constituting a homogeneous class of discrete terms and discrete supposition appears as the inevitable counterpart of the integration of singular terms within the theory of the properties of terms.

[57] See quotation above, note 51.

3.2. *Predicability and Simple Supposition*

Predicability in a Wider Sense and Predicability Proper

The question of individual predicability implies contexts in which discrete terms are in simple supposition, that is to say the existence of propositions where discrete terms actually have simple supposition. One must find an example in which a proper name appears in subject position and supposits not for an individual, but for the form belonging to that individual. William of Sherwood gives as an example of discrete simple supposition: 'Socrates is predicable of only one'. This property is linked to the grammatical description of the noun as 'signifying substance with quality'. William manages to combine the grammatical inheritance—where a proper name designates a single thing because the quality signified is proper (*propria*) to one individual—with the legacy of Porphyry's *Isagoge*—where the individual is what is predicable of only one—as reformulated by Boethius: individual predicability notably implies the fact that the proper name refers to a single individual, but it also relies on the existence of an individual quality signified by the name, which is predicated of the individual named and can belong to no other individual.[58] But is there such a thing as an individual form signified by the name and predicated of the individual?

William of Sherwood is uneasy about the metaphysical counterpart of predicability. He takes the predicability of the individual seriously, since for him it is the property of the signified form in his example of simple discrete supposition, but he refuses such a structure in his chapter on predicables. Compare 1) the text on simple discrete supposition previously quoted and 2) the text on the individual predicable in the chapter on the *Categories*:

1) Cum dico 'Socrates currit', respicitur pro sua re. Cum dico 'Socrates est praedicabile de uno solo' respicitur pro forma significata per nomen.[59]
2) Praedicabile autem dicitur communiter et proprie. Communiter dicitur praedicabile omne, quod mediante hoc verbo 'est' potest alii adiungi, sive sit commune sive individuum; proprie praedicabile solum est commune. Et est individuum, quod de uno solo est praedicabile, ut est nomen proprium et pronomen et dictio communis cum pronomine. Dicitur enim individuum eo quod non dividitur in partes subiectivas. Commune autem

[58] Boethius, *In Porphyrium, editio prima*, 81-82: 'Natura autem indiuiduorum haec est, quod proprietates indiuiduorum in solis singulis indiuiduis constant et in nullis aliis transferuntur atque ideo de nullis aliis praedicantur'.

[59] See above note 56.

> et universale idem sunt. Dicitur enim commune, quia unit multa simul, id est in unam naturam; universale autem quod plura vertit in unum, quale est hoc nomen 'homo'.[60]

If William were asked to show how the individual form signified is what is attributed when the individual is predicated, he would have to attribute predicability in a narrow sense to proper names. But William refuses to do so, in accordance with the *Categories*, since nothing can stand in the relation of subject to an individual, and individuals cannot be predicated. Indeed William does not assign 'individual predicability' to signified forms, but to linguistic expressions, although this is in contradiction with the description of simple discrete supposition he gives in the chapter on supposition. A proper name or a pronoun, the individual predicables according to William, are not what is signified by the linguistic expression: they are themselves linguistic expressions. When one says 'Socrates is predicable of only one [thing]', 'Socrates' should be taken in material supposition, according to the chapter on predicables: it supposits for the name 'Socrates' to which predicability is attributed, and not for the individual form of Socrates. We must understand that it is the proper name 'Socrates' that is predicated of only one thing—i.e. it signifies only one thing.[61] The fact that there is no form predicated is confirmed by the fact that the set of what is predicable of only one thing, according to the chapter on the *Categories*, and the set of what can be in simple discrete supposition, according to the chapter on supposition, do not coincide. As seen, a pronoun cannot supposit for a form, since it does not signify a form, but it may be an individual predicable according to William (and Porphyry).

It appears that predicability is a logical and a metaphysical property of the significate in the case of common terms, but it is a linguistic property of the term itself in the case of discrete terms. The latter simply means that singular terms can occupy the place of the predicate in a linguistic structure of the form 'S is P', without being literally predicable. An example of this would be identity statements such as 'Tully is Cicero'.

At the cost of this inconsistency, William escapes the need to explain the metaphysical nature of the individual form signified by proper names. The individual form perfectly fulfils its semantic role: it can maintain proper names in the grammatical and logical category of nouns, which implies the signification

60) William of Sherwood, *Introductiones in Logicam*, ed. Brands and Kann (1995), 48.

61) On the equivalence between *significatio* and *praedicatio* in some texts of the Middle Ages and late Antiquity (*katègorein, sèmainein, legein*), see De Rijk, 'Categorization as a Key Notion in Ancient and Medieval Semantics', in: *Vivarium* 26 (1988), 1-18.

of a form, precisely because it is devoid of any metaphysical content. We must conclude that for a term to be predicable of only one thing simply means that it refers to a single individual. A semantic interpretation of this notion must therefore prevail in the case of singular terms. The proper name can be P in 'S is P', but only insofar as the pronoun 'iste' may also be P in 'S is P' (e.g., 'Socrates est iste'): that is to say, it can be predicated but without attributing a form. These terms can be predicates, but they do not attribute anything.

Form Signified and Predicate

This coordination of predication and signification of a form is also found in what we can reconstruct of the theory of predication in William's *Introductiones*. The logician teaches that the term in predicate position 'says its form', as the form of the substance and not as the form taken in itself:

> Omne enim nomen significat solam formam, et non absolute, sed in quantum informat substantiam deferentem ipsam. Et sic aliquo modo dat intelligere substantiam. Nomen ergo in praedicato dat intelligere formam, dico, ut est forma substantiae subiecti. Et ideo cum illa substantia intelligitur in subiecto, non intelligitur iterum in praedicato. Unde praedicatum solam formam dicit. Nec tamen vere dicitur 'species est homo', quia haec dictio 'homo' significat humanitatem ut est forma individuorum. Et ideo non praedicatur de specie, quia non est forma substantiae speciei. Et notandum quod quia praedicatum dicitur inesse subiecto, semper praedicatur forma, ut est inhaerens et informans. Subiectum autem quandoque supponit formam absolute, quandoque autem non, et hoc est secundum exigentiam praedicati.[62]

The interpretation of predication as an affirmation of the inherent form signified confirms the inconsistency of William's position about the supposition of discrete terms. In order to be in simple supposition, a term must be able to

---

62) William of Sherwood, *Introductiones in Logicam*, ed. Brands and Kann (1995), 144. Here William is talking about the three types of simple supposition he has previously distinguished: in the first type ('homo est species') the significate is taken without any relation to things or as an abstract species; in the second case ('homo est dignissima creaturarum') predication concerns the species as 'it is in the things' or 'the significate in relation to things' and this can happen in two ways: either the significate is really 'saved' in a thing of which it is predicated of as in 'homo est dignissima creaturarum', or because 'it is referred to any one in a vague and general manner'; in a third way, as in 'piper venditur hic and Romae', the term 'supposits for the species as it can be instantiated (*signabilis*) in its individuals but not really instantiated (*signata*)'. It is taken for pepper *simpliciter*. The text quoted is intended to prove that the predicate term can predicate the form *simpliciter* without making it necessary to admit a conversion such as 'species est homo', see *Introductiones, op. cit.*, 143-144.

suppose for the signified form. This is something discrete nominal terms can in principle do, since they signify substance with quality (a form). But according to the example given in the chapter on supposition, this form must be what ensures the predicability of the term, something William is not willing to accept for discrete terms in the chapter on the predicable, as already seen. The description of predication confirms this point by attaching to the predicate the attribution of the signified form. Everything said about the signification of the form by the term in predicate position obviously concerns only the form signified by universal terms.

It seems impossible for William to demonstrate how a discrete term can actually have simple supposition in a proposition. The only example given ('Socrates is predicable of only one') is false, since the supposition is then material, the linguistic expression being what is predicable. An anonymous commentary on the *Introductiones* (*ca.* 1270) noticed these difficulties and is obviously not at all convinced by William's doctrine:

> Contra, in singularibus idem significatum et suppositum. Etiam singularia non predicantur proprie, cum non significent substantiam potentem qualificare illud de quo dicitur. [...] Sed singulare predicat formam particulatam que idem est cum supposito secundum omnimodam idemptitatem.[63]

For William of Sherwood the supposition of discrete names is to be situated within the divisions of supposition, not on its margins as is the case with most other authors. But the integration of singular terms has an important theoretical cost: it throws open the class of discrete terms, and it leads to an incoherent treatment of simple discrete supposition, no actual example of which can be found in a proposition. The singular form signified is an empty shell, since none of the traditional instruments for determining this concept is mobilized by William: neither the proper quality of the grammarians, nor Porphyry's unique collection of properties (or accidents as explicitly found in Boethius), nor Boethius' platonity (corresponding to a singular name such as 'Plato' in the same manner as humanity is signified by the universal name 'man').

---

63) Pinborg and Ebbesen, 'Thirteenth Century Notes on William of Sherwood's Treatise on Properties of Terms. An Edition of *Anonymi Dubitationes et Notabilia circa Guilelmi de Shyreswode Introductionum Logicalium Tractatum V* from ms Worcester, Cath. Q. 13', in: *Cahiers de l'Institut du moyen-âge grec et latin* 47 (1984), (104-141), 126. For a detailed study on this topic and the question of empty names (linked to the identity of the *significate* and the *suppositum*), see Brumberg-Chaumont, La nomination du singulier dans les *Quaestiones* sur la *Métaphysique* de Geoffroy d'Aspall, *Archives d' histoire doctrinale et littéraire du moyen-âge* 72, (2005), 47-103.

The relationship between a singular name and the individual is explained neither by direct reference nor by a mediated signified form, because the relationship between this form and the individual referred to is not and cannot be subjected to a metaphysical explanation.

4. William Arnaud, Commentator on Peter of Spain's Tractatus (Type 4)

William Arnaud's[64] commentary on the *Tractatus* was written in the last quarter of the thirteenth century.[65] It is worth studying here for several reasons. The commentator is in favour of the existence of simple discrete supposition. He understands it as supposition for the intention signified in the soul, thereby ensuring proper names a signification even when no individual is named, so that a general description of the semantics of proper names, whether they are empty or not, can be provided. For the same reason he rejects the idea that the predicate has simple supposition because 'omnis homo est animal' would always be false if the proposition were to predicate a concept of all men. It would lead to a fallacy of accident: the predicate term must be in personal supposition since the subject term is in personal supposition.[66] On all these points, William opposes Peter of Spain, for whom discrete supposition is neither personal nor simple, simple supposition is supposition for the (metaphysical) form signified, the predicate has simple supposition, and non-empty proper names signify the suppositum-appellatum-significate (so that we do not know how they keep their signification when they lose their bearer). At first glance, William adopts the same position as William of Sherwood, since both authors defend the existence of simple discrete supposition, but they do so for very different reasons. The relationships between form, signification

[64] In his introduction to Thomas Aquinas's *Expositio Libri Peryermeneias* (Thomas Aquinas, *Opera Omnia* I /1 (1989), 52* and 69*) Gauthier has challenged de Rijk's assertion that William was a master at the University of Toulouse in the 1240s (de Rijk (1969), p. 120-162, 126). His activity can only be situated at the end of the thirteenth century, since his commentary on Peter of Spain makes explicit reference to Thomas Aquinas's commentary on the *Peri hermeneias*, composed between 1270 and 1271. His lectures on *Logica vetus* were edited in an old edition, falsely attributed to Giles of Rome, see [Ps-]Giles of Rome, *Expositio dominii Egidii in Artem Veterem*, ed. Venetiis (1507; reprint Frankfurt 1968), Venetiis (1582), Bergomi (1592). Tabarroni has shown to what extent William is influenced by Thomas Aquinas in his commentary on the *Peri hermeneias* (Tabarroni (1988), 371-427). For a more detailed presentation of the author and his work, see Brumberg-Chaumont (2011).

[65] That is, after Thomas Aquinas finished his commentary on the *Peri hermeneias*.

[66] *Lectura Tractatuum*, ed. de Rijk (1969), 147.

and predicability are deeply altered by the fact that these issues are understood by William Arnaud in terms of the Avicennian concept of intention, as mediated by the use Thomas Aquinas makes of it in his analysis of essential predication.[67]

The commentator rejects the view he finds in Peter of Spain according to which discrete terms are invariant. He points to what he considers as an incorrect ordering of discrete supposition in the tree of the divisions: it should rather be divided into simple and personal, as is the accidental supposition of common terms. But the distinction between discrete and common supposition is still necessary, since personal supposition of common terms is a division of accidental supposition, which is not the case for discrete personal supposition.

This is why the following diagram must be reconstructed:

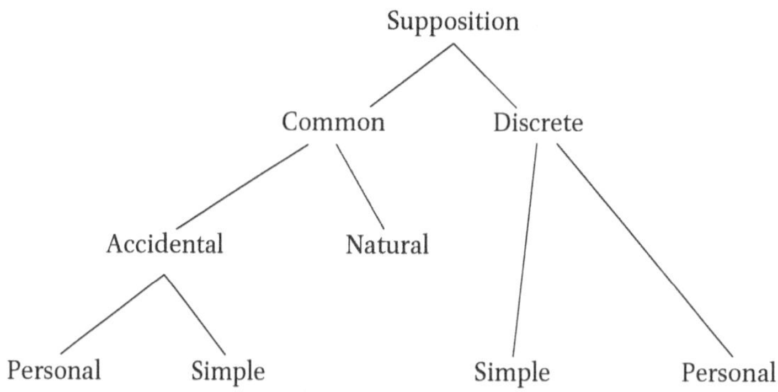

Here is the way William defends the idea of a discrete simple supposition:

[67] I cannot demonstrate this point in detail here. The strong influence Thomas had on William as a commentator of Aristotle's logic has already be mentioned. This also applies to his commentary on Peter of Spain, so that the semantic and metaphysical theories of Thomas are reformulated in the vocabulary of supposition. As we will try to establish in detail in a subsequent study, the way William denies the predicate being in simple supposition is perfectly consistent with the way Thomas Aquinas in *De ente et essentia* and also in the commentary on *Peri hermeneias* clearly distinguishes the meaning of 'man' when it is predicated of many individuals in an essential predication ('Socrates is a man', 'Plato is a man' etc.), from that of 'man' when universal predicability is attributed to it ('Man is a species'), *i.e.*, a logical property extrinsic to the essence (universal predicability). For William, the term in the first case is in personal supposition, while in the second it is in simple supposition: it is the thing as it has its being in the intellect that receives the logical property of being universally predicable.

> Et dicit quod simplex subpositio est acceptio termini communis pro re universali significata per ipsum [. . .] Unde <notandum>, ut meliores dicunt, quod terminus habet simplicem subpositionem quando stat pro intentione que est in anima. Nam res non habet nisi duplex esse, scilicet in anima et in re extra. Et quando stat pro esse quod habet in anima, tunc est subpositio simplex, ut 'homo est species'. Similiter dicitur personalis quando atst terminus pro re extra, ut 'homo currit'. Unde si dicatur: 'Sor est individuum' potest dici quod ibi habeat simplicem subpositionem, cum stet pro intentione in anima.[68]

This formulation removes the reference to 'common terms' and to 'inferiors' in the definitions of simple and personal supposition William has found in Peter of Spain. The replacement of the universal form by the intention in the soul allows a generalization of the existence of a distinct significate, which is what all terms, including discrete terms, supposit for in simple supposition. This implies the existence of an intention in the soul that constitutes a singular intellection of the individual. It explains the fact that the signification of a singular noun remains stable, regardless of the state of the world: an identical semantic structure is maintained despite the disappearance of the thing named. To someone who asks whether the vocal sound retains its meaning when the thing named is destroyed, the commentator answers:

> Significare est intellectum constituere. Sed destructa re ipsa vox de eodem constituit intellectum. Ergo destructa re potest remanere significativa. Unde ista vox 'Petrus' que significat aliquem hominem, tunc corrupto illo quod significat, adhuc de eodem constituit intellectum et sic idem significat.[69]

Without analysing in detail the general semantics proposed by William,[70] we can propose a simplified diagram of the relationships between signification, supposition, and intellection. One can see how William Arnaud proposes a uniform semantics for all terms, common or discrete, as William of Sherwood does, and in contrast to what Peter of Spain proposes, but he considers simple supposition as supposition for the intention, not for the metaphysical form:

---

68) *Lectura Tractatuum*, ed. De Rijk (1969), 146.

69) *Lectura Tractatuum*, ed. De Rijk (1969), 143.

70) He develops a theory of the signification of the *res ut intellecta* with implications in terms of theories of predication and reference that are too complex to be explained here.

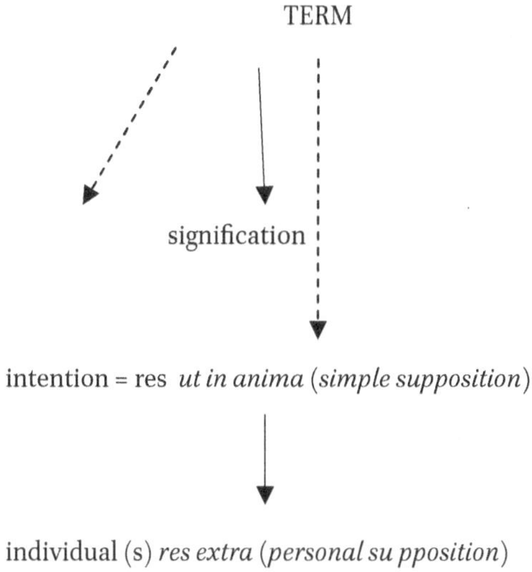

Our commentator has realized that the division between discrete and common suppositions is not to be placed on the same level as the other divisions. This allows him to consider simple and personal suppositions as internal divisions of two opposites (discrete and common). He justifies this change in classification as follows:

> Et tu arguis quod quando aliqua duo ex opposito distinguuntur, inpossibile est quod illud quod continetur sub uno contineatur sub alio. Et subpositio dividitur per communem et discretam, et sic illud quod continetur sub communi non poterit contineri sub discreta. Sed simplex continetur sub communi quia sub accidentali. Ergo non continetur sub discreta. Dicendum sustinendo quod dictum est quod si dicatur 'Sor est individuum', ibi est simplex subpositio. Nam sicut accidentalis dividitur per simplicem et personalem, ita subpositio discreta potest dividi in simplicem, ut 'Sor est individuum' et personalem ut 'Sor currit'. Et ad argumenta, dicendum quod prima divisio per communem et discreta non est data ex parte rei, sed potius ex parte vocis. Et ideo quantum ad illud [...] quod dividitur illud quod continetur sub uno, non continetur sub alio. Et ita quia [...] res non dividuntur, potest ista divisio per simplicem et personalem utrique competere tam communi quam discrete.[71]

When the text says that it is not the things, but the words that are divided, one should probably understand that it is not supposition, nor modes of supposi-

[71] *Lectura Tractatuum*, ed. De Rijk (1969), 146.

tion, but terms that are divided into discrete and common. We have already seen this in the texts of the *Logica Modernorum*: the assertion was made in order to marginalize discrete terms in the theory. Here the argument has utterly a different goal, since justifies a twofold division of personal and simple supposition. William must indeed explain the existence of two kinds of personal supposition and two kinds of simple supposition: the first ones are divisions of common accidental supposition, the others are divisions of discrete supposition, which is neither accidental nor natural. This preserves the unity of the definition of these doubled suppositions. But the fact that simple and personal discrete supposition are not accidental remains enigmatic, since it seems that they are semantic variations linked to different propositional contexts, and this corresponds to what one may think of as accidental supposition.

It is therefore legitimate to ask why simple and personal supposition of discrete terms are not divisions of accidental supposition. If this were the case, it would presumably compel the logician to attribute natural supposition to discrete terms as well. But this is impossible because the commentator emphasizes the metaphysical definition of natural supposition he finds in Peter of Spain, since it adds a reference to a 'common form' preserved in the individuals supposited for, whether they exist or not, an expression not found in the *Tractatus*. Our author shows some familiarity with the problem of the compound and of the quiddity as the subject of the definition found in *Metaphysics* Z/ /7. Following Thomas Aquinas's teaching, he develops a clear position on the use he intends to make of these notions from a semantic point of view: the name of a compound such as 'man' does not signify the quiddity alone, but also what possesses the quiddity. On the other hand, he gives a clear noetic interpretation of what the term supposits for when it is in simple supposition: it is the intention in the soul. We can therefore easily see why he tends to isolate his idea that there is simple discrete supposition from such a metaphysical context: otherwise he would be compelled to admit the existence of an essential form of the individual compound, an individual form the name 'Plato' would signify, as well as signifying Plato himself.

The distinctions between natural and accidental supposition, on the one hand, and between simple and personal supposition, on the other, do not therefore rely on the same conception of the thing signified: in the one case it is a metaphysical form, in the other an intention in the soul. This is not completely coherent because the simple supposition of a common term, which is a supposition for the form as an intention, is an accidental supposition. For Peter of Spain, the thing signified is always a metaphysical form.

Conclusion

We are thus faced with two forms of consistency. In the first one, discrete terms are considered fully as terms so that they can be placed within the tree of the divisions of supposition. This is the position of William of Sherwood and William Arnaud. In the second one, discrete terms are denied simple supposition, so that there is no distinction between what they signify and what they suppose for. They have a special status and the division between discrete and common supposition is primitive. This is the position of Peter of Spain. Each position has a theoretical cost.

On the one hand, Peter of Spain gives up the attempt to integrate discrete terms in his theory of the properties of terms. This solution is probably consistent and economical, but less ambitious in terms of general logic, since it simply excludes the troublemakers. Although Peter of Spain does not acknowledge it, in the chapters of the *Tractatus* where the definitions of fundamental notions (signification, supposition, appellation) are given, these notions cannot be applied to singular nouns in the same sense as they are applied to to universal nouns.

On the other hand, William of Sherwood wants to apply the general definitions he gives (for noun, term, signification, and supposition) to all terms, including discrete terms. These terms show no other peculiarity than that of signifying only one thing and not several, as opposed to universal nouns. But this position is completely inconsistent with William's conception of predicability. William's semantics implicitly relies on the existence of an individual form that he avoids on metaphysical grounds, so that he is unable to find a single proposition in which the discrete term is truly in simple supposition. If supposition is to be defined through the occurrence of the word in a propositional context, one must wonder if there is such thing as a simple discrete supposition.

William Arnaud is able to assign real modes of supposition (simple, personal) to discrete terms, avoiding the difficulties of the notion of an individual metaphysical form, but he does so in such a way that the notion of propositional context is again undermined. Normally, the change from one mode of supposition to another in two different propositions depends on what is required by the predicate (or a term added to the subject term); this is what the theory labels as 'accidental supposition'. But William defines accidental supposition more precisely as a restricted supposition within the scope of a supposition already circumscribed by natural supposition: this includes all the *supposita* in which the form signified can be 'saved', whether they exist or not. William explicitly refuses to consider that personal discrete supposition is accidental, and the same applies to simple discrete supposition. But he should also deny it of common simple supposition if he is to be coherent, because what the term supposits for in simple supposition, the intention, is by no means included in the set of *supposita* the term has in natural supposition, the set from which the term's supposition is supposed to be selected by the predicate (or another term that is added).[72] What the term supposits for in simple supposition is not included in this set; the form itself from which it is circumscribed, the metaphysical form, is not the form for which the terms supposes in simple supposition, the intention. Common simple supposition is simply no more 'accidental', i.e. the result of a restriction from all the natural *supposita*, than simple discrete supposition is, yet the latter is not a division of accidental supposition, whereas the former is.

Should we then consider that the position of Peter of Spain is the only one to be consistent, and offers a serious answer to the problem of discrete supposition? His position should probably be radicalized, so that notion of discrete supposition and the division between discrete and common supposition would in fact disappear. The distinction between what is common and what is discrete concerns types of terms, not modes of supposition. In other words, it is not enough to put the division between common and discrete suppositions at the top of the tree: it has to be utterly removed because it is the consequence of a category mistake. One moves surreptitiously after the first branch of the tree from a division of types of terms to a division of modes of supposition. The integration of discrete terms into the theory of supposition presupposes a notion of discrete supposition that was absent from the first treatises on supposition. It creates an inconsistency that threatens the whole structure.[73] The concept of discrete supposition appears just as a thread artificially connecting discrete terms to the theory of supposition. What Peter of Spain in fact means by the concept of 'discrete supposition' is a referential capacity, a semantic property of the word, and not a logical property of the term, since it is independent of

72) This is a recurring problem in the division between natural and accidental supposition, because it is considered as being upstream from the division between simple and personal supposition when it in fact describes an internal division within personal supposition (natural *supposita* are in fact personal *supposita*): one can find it also in Lambert of Auxerre, for example.

73) Goubier emphasizes the purely lexical nature of discrete supposition as opposed to common supposition, see Goubier (2003), 84.

the propositional context, an incongruity also encountered in the appellation of discrete terms.

Bibliography

Primary Literature

Anonymus, *Lectura tractatuum*, ed. L.M. de Rijk, in: id., 'On the Genuine Text of Peter of Spain's *Summule*. IV. The *lectura tractatuum* by Guillelmus Arnaldi, Master of Arts at Toulouse (1235-1244). With a Note on the Date of Lambertus of Auxerre's *Summule'*, in: *Vivarium* 7 (1969) 120-162

Boethius, *In Isagogen Porphyrii commenta*, ed. S. Brandt (Corpus Scriptorum Ecclesiasticorum Latinorum, 48; Wien-Leipzig 1906, repr. New York-London 1966)

Guillelmus Arnaldi: William Arnauld (Ps-Giles of Rome), *Expositio dominii Egidii in Artem Veterem* (Venetiis 1507, Venetiis 1582, Bergomi 1592; repr. Frankfurt 1968)

Guillelmus de Shireswood: William of Sherwood, *Introductiones in Logicam. Einführung in die Logik*, ed. H. Brands and C. Kann (Hamburg 1995)

Petrus Hispanus: Peter of Spain, *Tractatus, called afterwards Summulae logicales*. First Critical Edition from the Manuscripts with an Introduction, by L.M. de Rijk (Philosophical Texts and Studies, XXII; Assen 1972)

Priscianus, *Institutiones grammaticae*, ed. M. Hertz, in: *Prisciani grammatici Caesariensis Institutionum grammaticorum libri XVIII*, 1-II (Grammatici Latini ex recensione Henrici Keilii II-III; Lipsiae 1855-1859, repr. Hildesheim 1961)

Thomae de Aquino, *Expositio libri Peryermeneias*, ed. altera retractata, cura et studio Fratrum Praedicatorum (Sancti Thomae de Aquino Opera omnia, 1* 1; Roma-Paris 1989)

William Arnaud *vide s.n.* Guillelmus Arnaldi.

Secondary Literature

Braakhuis, H.A.G. (1979), *De 13de eeuwse tractaten over syncategorematische termen: Inleidende studie en uitgave van Nicolaas van Parijs' Sincategoreumata*. 2 vols. (Dissertation Leiden, Nijmegen 1979)

Brumberg-Chaumont, J. (2005), 'La nomination du singulier dans les *Quaestiones* sur la *Métaphysique* de Geoffroy d'Aspall', in: *Archives d'histoire doctrinale et littéraire du moyen âge* 72 (2005), 47-103

——. (2007), 'Sémantiques du nom propre: sources anciennes et discussions médiévales à l'époque d'Abélard', in: *Histoire épistémologie langage* XXXIX, fascicule 1 (2007), 137-166

——. (2011), 'Grammaire et logique du nom dans les Gloses sur Priscien de Guillaume de Conches', in: I. Caiazzo ans B. Obrist eds, *Guillaume de Conches*, Firenze, 2011, p. 377-465.

——. Brumberg-Chaumont. (forthcoming), 'Consequences from universal to singular propositions and existential import in the sophisma "omnis homo de necessitate est animal"', in F. Goubier, A. de Libera, and L. Gazziero eds, Sophismata : histoire d'une pratique philosophique, Paris, Vrin.

——. (2011), 'William Arnaud', in: *Encyclopedia of medieval philosophy, Philosophy Between 500 and 1500*, ed. H. Lagerlund (Dordrecht-Heidelberg-London-New York 2011: 1393-1395)

Goubier, F. (2003), *Une sémantique fragmentée. Référence et détermination au XIIIe siècle* (Paris VII université 2003) (PhD Dissertation)

Kneale, W. and M. (1962), *The Development of Logic* (Oxford 1962).

Kneepkens, C.H. (2000), 'The *Absoluta Cuiuslibet* Attributed to P.H. Some Notes on its transmission and the use made of it by Robert Kilwardby and Roger Bacon', in: I. Angelelli et P. Pérez-Ilzarbe, eds., *Medieval and Renaissance Logic in Spain*. Acts of the 12th European Symposium on Medieval Logic and Semantics (Philosophische Texte und Studien, 54; Hildesheim/New-York 2000, 373-403)

Libera, A. de (1982a), 'Textualité logique et forme summuliste', in: L. Brind'amour et E. Vance, eds., *Archéologie du Signe* (Toronto 1982, 213-234)

—— (1982b), 'The Oxford and Paris Traditions in Logic', in: N. Kretzmann, A. Kenny and J. Pinborg, eds., E. Stump, ass. ed., *The Cambridge History of Later Medieval Philosophy. From the Rediscovery of Aristotle to the Desintegration of Scholasticism 1100-1600* (Cambridge-London-New York-New Rochelle-Melbourne-Sydney 1982, 174-187)

—— (1984), 'Les *Appellationes* de Jean le Page', in: *Archives d'histoire doctrinale et littéraire du moyen âge* 51 (1984), 193-255

Pinborg, J. (1979), 'English Logic before Occam, in: *Synthese*, 40 (1979), 19-42

Pinborg, J., and S. Ebbesen (1984), 'Thirteenth Century Notes on William of Sherwood's Treatise on Properties of Terms. An Edition of *Anonymi Dubitationes et Notabilia circa Guilelmi de Shyreswode Introductionum Logicalium Tractatum V* from ms Worcester, *Cath*. Q. 13', in: *Cahiers de l'institut du moyen âge grec et latin* 47 (1984), 104-141

Rijk, L.M. de (1962-1967), *Logica Modernorum. A Contribution to the History of Early Terminist Logic*. I. *On the Twelfth Century Theories of Fallacy*; II/1: *The Origin and the Early Development of the Theory of Supposition* II/2: *Texts and Indices* (Assen 1962-1967)

—— (1968), 'On the Genuine Text of Peter of Spain's *Summule Logicales*. Part I. General Problems Concerning Possible Interpolations in the Manuscripts', in: *Vivarium* 6 (1968), 1-34

—— (1969), 'On the Genuine Text of Peter of Spain's *Summule Logicales*. Part IV. The *Lectura tractatuum* by Guillelmus Arnaldi, Master of Arts at Toulouse (1235-1244). With a Note on the Date of Lambertus of Auxerre's *Summule*', in: *Vivarium* 7 (1969) 120-162

—— (1982), 'The Origins of the Theory of the Properties of Terms', in: A. Kenny, N. Kretzmann, I. Pinborg, eds., E. Stump, ass. ed., *The Cambridge History of Later Medieval Philosophy. From the Rediscovery of Aristotle to the Desintegration of Scholasticism 1100-1600* (Cambridge-London-New York-New Rochelle-Melbourne-Sydney 1982, 161-173)

—— (1988), 'Categorization as a Key Notion in Ancient and Medieval Semantics', in: *Vivarium* 26 (1988), 1-18

Tabarroni, A. (1988), 'Lo Pseudo Egidio (Guglielmo Arnaldi). Un'inedita continuazione del commento di Tommaso al 'Peryermeneias'', in: *Medievo* 14 (1988), 371-427

Suppositum between Logic and Metaphysics Simon of Faversham and his Contemporaries (1270-1290)

Dafne Murè
University of Rome, 'La Sapienza'

Abstract

This article is the result of research on the occurrences of the terms *suppositio*, *supponere* and their linguistic derivations in the literature on fallacies (comments on the *Sophistical Refutations*) of the second half of the thirteenth century. The authors analysed are Albert the Great, Giles of Rome, Simon of Faversham, the so-called Incerti Auctores (Anonymous C and SF), the Anonymous of Prague (P) and John Duns Scotus. The central elements that emerge are the role played by the notion of *suppositum* and by the linguistic context (*adiuncta, determinatio*) to determine the denotation of an expression, and the importance of the metaphysical problem of the unity and identity of *suppositum* in both the theory of predication and the theory of inference. Both subjects, obviously, are closely connected.

Keywords

suppositum, adiunctum, determinatio

Introduction

In the literature on fallacies of the second half of the thirteenth century, the term *suppositio* and the active forms of the verb *supponere* are not used frequently. Their occurrences only rarely refer to the functions or properties of the *terminus* as such: the 'theory of *suppositio*' is rarely used to solve linguistic problems and fallacies. The active forms of the verb *supponere* tend to be replaced by other expressions (*stare pro, teneri pro, esse pro, accipere, sumere*). Instead, *suppositum* is the key term. This term is not used in the context of a theory of *suppositio*, but in a great variety of philosophical, logical and linguistic contexts: semantic theories in the strict sense, categories, the 'tree of

Porphyry', and metaphysics (*quidditas, essentia, natura* and their relation with individuals).

This research amply confirms the change in doctrines and logical terminology that took place in the second half of the thirteenth century, at least as regards the 'Paris tradition': the decline of a terministic paradigm and the new emphasis on epistemology and metaphysics in logical and linguistic thought. Gauthier, for example, has shown that most linguistic problems, including those involving fallacies, were dealt with by Thomas Aquinas without recourse to the theory of supposition. As the texts show, however, this does not mean that the theory had been forgotten or was unknown.[1]

In this perspective, a precise understanding of the notion and the term *suppositum* is important, especially with regard to its logico-linguistic function. The essential elements of Pinborg's analysis of the category of Modism still hold good: we can follow him in taking the term *suppositum* as 'what is denoted' by a signifying term. In contrast with the early theory of *suppositio*, the main function of a signifying term was no longer 'denotation', but *significatio*; the *significatio* of a term establishes the boundaries of the possible *supposita* to which it can refer within and outside the proposition: *terminus supponit illud quod significat.*[2] Radulphus Brito claims:

> [From the point of view of linguistic properties], no determination varies the meaning of the term. Thus the term always supposes the same thing, whatever determination is added to it, and this as regards what the term supposes.[3]

1) Thomas Aquinas, *Expositio libri Peryermeneias*, ed. Roma-Paris (1989), 51*-56*. See also, for example, Schoot (1993), Park (1999).

2) On modism and its evolution, see, e.g., Pinborg (1967), (1972), (1982), (1984); De Rijk (1968), (1971); Ebbesen (1977), (2000); De Libera (1981), (1982); Rosier (1983), (1994); Marmo (1994), (2006) and the related bibliography. About 'denotation' as the main function of a signifying term, I follow in particular Pinborg's and De Rijk's interpretations. I mean that, with regards to a terministic doctrine of *suppositio* as opposed to a modistic approach, the main function of a signifying term is to point at a referent inside a proposition (what I call 'denotation'), and not to have a meaning (*significatio*) apart from its fuction in a propositional context. The point will be developed in the article (cf. section 1, n. 27 and the whole section 2).

3) Radulphus Brito, *Quaestiones super Analytica Priora*, I, q. 46, ed. Pinborg (1984), III, 73: '[...] <de virtute sermonis> modo nulla determinatio variat significatum termini. Ergo terminus semper idem supponit, quaecumque determinatio sibi adiungitur, et hoc quantum ad illud quod supponit'. See, e.g., Incertorum Auctorum *Quaestiones super Sophisticos Elenchos*, Anonymus SF, ed. Ebbesen (1977), 102.

Radulphus' description can certainly be interpreted as being opposed to a terministic doctrine of *suppositio*. Nevertheless, the texts show that the signifying term can denote various types of entity: for example, the *suppositum* (both the so-called *suppositum per se* and the *suppositum per accidens*) is not always identical to an individual; it is not always clear what type of unity and reality the *supposita* possess; nor is their ontological structure always clear.[4]

The literature on fallacies clearly brings out all the terminological and doctrinal problems raised: in particular, the theory of fallacies *extra dictionem*[5] and the distinction between different types of syllogism.[6] The term *suppositum*, however, always refers to the problem of the unity and identity of logical terms, and their metaphysical principles: in fact, the question of the unity and multiplicity of terms is a constant of Aristotle's exegesis in the *Sophistical Refutations* and involves every type of taxonomy and definition; besides, the commentators connect the question of the unity and / or multiplicity of terms to the transcendental *ens* and *unum*.[7]

The 'modistic' and realist texts that have been analysed also reveal something interesting. As we have said, the 'modistic' logico-linguistic doctrine is in many respects opposed to a terministic theory of *suppositio*. Nevertheless—precisely in relation to the idea of *suppositum*—the linguistic context continues to play a role in determining the referent. The recognition of the function of the context and the strategies for identifying the referent are nevertheless closely linked with ontological and metaphysical problems. In the following

4) See, for example, Marmo (2006), 252-253 and 265-271; Marmo (1999).

5) The *fallaciae extra dictionem* are the flaws in the argument which do not depend on the properties of language, while the *fallaciae in dictione* depend on them. In section 2 we analyse this distinction.

6) De Rijk (1962), Ebbesen (1981b, vol. 1), (1987), (1993).

7) Simon of Faversham seems to be the most explicit author on the subject, setting the problem in relation to the transcendental *ens* and *unum*: 'Intelligendum quod secundum opinionem Alexandri multitudo et multiplicitas differunt, quia multitudo est diversitas aliquorum prout non conveniunt in aliquo uno, multiplicitas autem est diversitas aliquorum ut conveniunt in aliquo uno. [...] unum enim eodem modo dividitur sicut ens; ens autem dividitur in substantiam et qualitatem et quantitatem; ideo similiter unum. [...] Unum in substantia dicitur identitas. [...] Nam ubicumque est latentia plurium significatorum sub aliquo uno, [...] ibi est vera multiplicitas.' (Simon of Faversham, *Quaestiones super libro Elencorum. Quaestiones novae*, ed. Ebbesen *et al.* (1984), 145). See also Albertus Magnus, *Liber Elenchorum*, ed. Borgnet (1890), I, t. 3, c. 20, 596; Incertorum Auctorum *Quaestiones super Sophisticos elenchos*, Anonymus C, ed. Ebbesen (1977), q. 835, 350-351; Johannes Duns Scotus, *Quaestiones super librum Elenchorum Aristotelis*, ed. R. Andrews *et al.* (2004), qu. 15, 8-10, pp. 333-334

sections we shall try to show this link (sect. 1) and discuss some of the possible logical consequences (sect. 2) by analysing some texts.

1. Terminological and Metaphysical Aspects: Simon of Faversham

Like his contemporaries, Simon of Faversham[8] rarely uses the terms *suppositio* and *supponere*. Albert the Great, for example, still uses the device of *multitudo suppositionum* (multiplicity of suppositions) to solve some fallacies, both those *in dictione* and those *extra dictionem*,[9] and sometimes uses the theory of *descensus* by opposing *suppositio confusa* (confused supposition) to *suppositio determinata* (determinate supposition).[10] The opposition occurs just once in the Anonymous of Prague, and nothing of particular importance can be found in the comments of Giles of Rome and Duns Scotus on the *Sophistical Refutations*.[11] In general, as we have said, the frequent expressions '*stare pro, esse pro*' replace the previous terminology: the signifying term stands for (*stat pro* or sometimes *supponit pro*) its *significatum* (the meaning) or its *suppositum*, without any distinction between types of supposition or between the transitive and intransitive uses of the verb *supponere*. In his published texts on logic, Simon, unlike Scotus, never introduces the question of a possible connection between different *acceptiones* or *rationes intelligendi* (ways of conceiving) and the transitive or intransitive use of the verb *supponere*, on the one hand, and different ways of suppositions (*modi supponendi*), on the other.[12] What emerges first of all is rather the connection with categories and metaphysical questions.

8) For the date of Simon of Faversham's works, see Simon de Faversham, *Quaestiones super libro Elenchorum*, eds. S. Ebbesen *et al.*, (1984), 5-6; Donati (1990) and related bibliography.

9) See, e.g., Albertus Magnus, *Liber Elenchorum*, ed. Borgnet (1890), I, t. 4, c. 1, 603; c. 2, 604; c. 7, 612; t. 5, c. 1, 616.

10) Albertus Magnus, *Liber Elenchorum*, ed. Borgnet (1890), II, t. 2, c. 7, 681-683. On this topic and the technical terminology, see Maierù (1972).

11) Ms Praha, *Knihovna Metropolitní Kapitoly*, L. 6, f. 85rb. Simon of Faversham, *Quaestiones super libro Elencorum. Quaestiones novae*, ed. Ebbesen *et al.* (1984) 169: '[...] genus contains the different natures of all its species, but according to one *ratio* which is confused and determinate' ('[...] genus importat naturas diversas omnium suarum specierum, tamen sub ratione una confusa et determinata').

12) See, in this volume, Marmo, 'Scotus on Supposition', section 1. 4. Simon makes only a few general remarks, perhaps, on the possibility of these correspondences: see, e.g., Simon of Faversham, *Quaestiones super libro Perihermeneias*, ed. Mazzarella (1957), 163, ll. 30-35.

The term *suppositum* is certainly linked up with the metaphysical theory of substance, as we can see from some texts by Aquinas.[13] As we know, logically and semantically, the *vox significativa* (the sound which has a meaning, *significatio*) stands for (*supponit pro, stat pro*) its *significatum* (what is meant), or for the common nature (*natura communis*) or essence (*essentia*) expressed by the definition.[14] This assumption certainly involves every aspect of the logical functions, including predication and the term predicated. For Simon of Faversham, what is predicated is specifically the *res sola*, which is a nature:

> The term predicates what it means and nothing else; on the other hand, the term means what is expressed by means of its definition. [...] Thus only the nature of the animal is predicated here, saying: 'man is an animal'. But neither the aggregate of the thing and the intention is predicated, nor the intention, but the thing alone [...][15]

The reasoning that underlies a claim of this kind, and that in some way forms the basis for the author seeing the context as possibly having a role, is the following. The categories (*praedicamenta*) are distinguished on the basis of the different ways of predication (*modi praedicandi*), which are assumed (*sumuntur, accipiuntur*) starting from the different ways of being (*modi essendi*).[16] However, both the *esse* and the *essentia*[17] of the genus and the species may

13) See, e.g., De Rijk (1970), (1981). Cf. Thomas Aquinas, *Summa theologiae* III, q. 2, a. 3. *Ibid.*, III, q. 2, a. 2. See also *ibid.*, I, q. 29, a. 2. For the distinction between *homo* and *humanitas* in semantic debates (XIIIth century), see Ebbesen (1988).

14) On this typically modistic point, see also Thomas Aquinas, *Expositio libri Posteriorum Analyticorum*, I. 4, l. 9, n. 3 *and* nn. 3-4. Giles of Rome often uses the term *suppositum* as a synonym for *subiectum*, indicating in particular the subject of the definition (see Aegidius Romanus, *Quaestio quid sit medium in demonstratione*, in Pinborg (1984), III, 255-268). For modistic semantic doctrine, cf. n. 2.

15) Simon of Faversham, *Quaestiones super libro Porphyrii*, ed. Mazzarella (1957), 33, ll. 27-35: 'Terminus predicat quod significat, et nihil aliud; terminus autem significat id quod exprimitur per suam diffinicionem. [...] Ergo sola natura animalis hic predicatur dicendo 'Homo est animal'. Non autem predicatur aggregatum ex re et intentione, nec intencio sed res sola [...]'. See also *ibid.*, 42, ll. 20 ff.; id., *Quaestiones super libro Perihermeneias*, ed. Mazzarella (1957), 155, ll. 9-29; 153, ll. 27-30.

16) Simon of Faversham, *Quaestiones super libro Praedicamentorum*, ed. Mazzarella (1957), 142, ll. 13-15; see also *ibid.*, 132, ll. 15-22.

17) A distinction between *esse essentiae* and the *esse existentiae* is set out by Simon in Simon of Faversham, *Quaestiones super libro Porphyrii*, ed. Mazzarella (1957), 64-65, at 65, ll. 24-25: the definition of 'accident' is possible from the point of view of the *esse essentiae*. The reference to a composition of *esse* and *essentia* is there in Simon of Faversham, *Quaestiones super libro Praedicamentorum*, ed. Mazzarella (1957), 93, ll. 27 ff.

be divided into different types of multiplicity (*multitudo*). The final level of multiplicity is due to the 'division of matter' (*divisio materiae*), or the 'division of quantity' (*divisio quantitatis*). However, the genus, the species and the *essentiae* do not lose their unity and real nature. In fact, being one means being *indivisum aliqua natura*.[18] Thus, if we ask if Socrates and Plato are a single man (*unus homo*), the term *unus* can express the *indivisio* either within human nature, or absolutely (*absolute*), or inasmuch as human nature is determined (*determinatur*) to its *suppositum*.[19] The reason that '*homo est animalitas*' is false is not because '*animalitas*' is only a part of the definition of man, but because it means something that is not determined to its *supposita*.[20] On the other hand, several things or entities, as in the case of the multiplicity of *supposita* (*multitudo suppositorum*), can constitute a unity on the basis of a 'real agreement' (*convenientia*) that they possess in a single real *intentio*:

> By virtue of what is the number essentially one? [...] for in the third book of his *Metaphysics* Avicenna says this: that things that are many in number constitute a unity (*unum*) because of some real agreement that they possess in some real intention. [...] That is why the great philosopher rightly says that six is six once and not three twice. Three consequences follow from this: [first] that the matter of a number is its units, which are parts of the number and have being in a single form. [...][21]

On this basis, *determinatio ad suppositum* comes about through the context (*adiuncta*):

> But 'man', from the point of view of linguistic expression (*de vi vocis*), does not include nature and *supposita*, because it does not include *supposita*. [...] I do not say that, with the name 'man', both the *quiditas* and what the *quiditas* possesses are not signified: however,

---

18) Simon of Faversham, *Quaestiones super libro Porphyrii*, ed. Mazzarella (1957), q. 26, 46 ff. Here the *multitudo suppositorum* is required (*requiritur*) *secundum aptitudinem* (*ibid.*, 48, l. 4).

19) Simon of Faversham, *Quaestiones super libro Porphyrii*, ed. Mazzarella (1957), 52-53, ll. 32-18.

20) In marked contrast with Thomas Aquinas, see Simon of Faversham, *Quaestiones super libro Porphyrii*, ed. Mazzarella (1957), 55, ll. 23 ff.; 33, ll. 31-38.

21) Simon of Faversham, *Quaestiones super libro Praedicamentorum*, ed. Mazzarella (1957), 123-125, ll. 28-10: 'Quid ergo est id per quod numerus est essentialiter unus? [...] Propter hoc dicit Avicenna tertio Metaphysice sue quod multa numero constituunt unum propter aliquam convenientiam realem quam habent in aliqua una intentione reali. [...] Propter hoc dicit bene egregius Philosophus, quod sex sunt semel sex et non bis tria. Et ex hoc sequuntur tria, scilicet quod materia numeri sunt ipse unitates, que sunt partes numeri et habent esse in forma una. [...]' The *quantitas* is able to produce *aliquid essentiale*, i.e., some parts which before its *adventus* were only potentially in *quiditas*: it is a specific property of quantity the other categories have not (*ibid.*, 117, ll. 13 ff.).

> what the *quiditas* possesses is not signified there in that it is determined to something that is one in number; nor is it determined to more things; and neither to all those things as *supposita*. However, it is determined to that by means of additions: it is determined to one man in number, saying 'this man'; it is determined to several men saying 'several men'; it is determined to all men saying 'every man'. Thus, the name of 'man' includes what possesses the *quiditas*, not as *suppositum*, but only as informed by the *quiditas*.[22]

This passage comes from the discussion of the truth value of the proposition '*Coriscus est alter ab homine*', which was considered as an example of fallacy of accident (*fallacia accidentis*); we shall analyse it below. In the *Sophistical Refutations*, however, the problem emerges in different contexts, for example, as part of the classical discussion of analogical terms or equivocal terms of the first type. It is formulated through the justification of how different 'natures' are divided in their *supposita*. The solution, as we know, is based on the idea of a 'universal whole' (*totum universale*) and the existence of a *ratio universalis* shared by the same *supposita*.[23] But there are at least three other contexts in the

---

[22] Simon of Faversham, *Quaestiones super libro Elencorum. Quaestiones novae*, ed. Ebbesen *et al.* (1984), 154: 'Sed 'homo' de vi vocis non importat naturam et supposita, quia supposita non. [...] Non dico quod nomine hominis non significetur quiditas et habens quiditatem: sed tamen habens quiditatem non significatur ibi ut est determinatum ad unum numero; nec ad plura; nec ad omnia sub ratione qua supposita sunt, sed ad hoc determinatur per adiuncta: determinatur autem ad unum numero dicendo 'hic homo', ad plures dicendo 'plures homines', ad omnes dicendo 'omnis homo'. Unde nomine hominis importatur habens quiditatem, non tamen ut suppositum, sed solum ut informatum quiditate.' See also Simon of Faversham, *Quaestiones super libro Porphyrii*, ed. Mazzarella (1957), 53, ll. 10-11. Simon uses the distinction between *suppositum* and 'what possesses *quiditas*' (*habens quiditatem*), which is also used in metaphysical contexts by medieval scholars (see for example the quoted references to Aquinas' *Summa theologiae*, note 13). It is a well-known distinction, though of delicate interpretation. The term *suppositum* seems to refer only to individuals, while 'what possesses *quiditas*' is something which cannot be, without determinations, reduced to a specific numerical quantity. From a metaphysical point of view, we cannot here resolve what it is exactly meant (for example, whether the compound of matter and form, or its universality). In the *Quaestiones* on Porphyry's *Isagoge*, Simon applies the term *suppositum* explicitly to individuals and the primary sense of *unum* is numerical (see Simon of Faversham, *Quaestiones super libro Porphyrii*, ed. Mazzarella (1957), 32, l. 18; 47, l. 9; 53, ll. 1-3). Sometimes, however, a kind of *numerositas essentiae* is admitted: in this case too there are references to the determination of the referent through the context (see, e.g., *ibid.*, q. 20, 41 ff., at 44, ll. 35-37). In general, he gives more weight to the central books of *Metaphysics* than to the *Categories*; the species or second substances, Simon claims, *se habent sicut numeri* (cf. Aristotle, *Metaphysica* VIII, chs. iii and vi); the unity and identity between first and second substances, between *quidditas* and *essentia*, are also underlined more strongly (see Simon of Faversham, *Quaestiones super libro Praedicamentorum*, ed. Mazzarella (1957), 107, ll.1-2; 109, ll. 11-20; 95, ll. 24-40; 97, l. 20).

[23] Simon of Faversham, *Quaestiones super libro Elencorum. Quaestiones novae*, ed. Ebbesen *et al.* (1984), 124, ll. 62-78. On the equivocal terms of the so-called first mode, see, e.g., Simon of

Sophistical Refutations where the ontological strategies are relevant for resolving the associated paralogisms: a) '*sedentem ambulare est possibile*'; b) '*quinque sunt duo et tria*'; c) the third mode of the figure of speech (*figura dictionis*).[24] In each of these three cases the solution to the paralogisms rests on two conditions: 1) the term has a referent; 2) the criteria for identifying the referent must be consistent with the metaphysical structure of the substance. The two conditions are respected even at the expense of the rigid non-contextual rules imposed and determined by the *significatio*.

The discussion of the meaning and semantic content of the expressions *quale quid* and *hoc aliquid* (case c) is well known.[25] For Simon, the same *natura* can be expressed by *diversae rationes*; in particular, the nature signified by the *terminus communis* is related to the *suppositum* in the same way that the *forma* is related to the *materia*.[26] In '*sedentem ambulare est possibile*', the term 'sedentem' that functions as logical subject can be understood (*accipitur*) both as *subiectum* and as the *forma* of the *subiectum*, depending on whether we understand the phrase as compound or divided.[27]

However, in the case of the proposition '*quinque sunt duo et tria*', which is also a traditional example of fallacy of composition and division (*fallacia com-*

Faversham, *Quaestiones super libro Elencorum. Quaestiones veteres*, ed. Ebbesen *et al.* (1984), 69 and 76.

24) *Figura dictionis* is the last of *fallaciae in dictione* in Aristotle's *Sophistical Refutations* (see *infra*, sect. 2).

25) On this terminology from a semantic point of view, see Ebbesen (1975).

26) Simon of Faversham, *Quaestiones super libro Elencorum. Quaestiones novae*, ed. Ebbesen *et al.* (1984), 148-149. The argument relies on Aristotle's *Metaphysics*, VII, 10 and on Avicenna. The notion of 'universal compound' (*aggregatum sumptum universaliter*) is once more employed (see *infra*).

27) Cf. Simon of Faversham, *Quaestiones super libro Elencorum. Quaestiones novae*, ed. Ebbesen *et al.* (1984), 134-135: 'Et ideo alii dicunt quod ista oratio 'sedentem ambulare est possibile' distinguenda est de virtute sermonis, ex quo li 'sedentem' potest accipi pro subiecto vel pro forma [...] Licet autem ego ponam ispam esse veram in sensu divisionis, tamen ponam esse falsam per se, quia ratione formae accidentalis importatae per hunc terminum accidentalem qui est 'sedentem'; est tamen vera per accidens, quia ratione subiecti'. See also Simon of Faversham, *Quaestiones super libro Porphyrii*, ed. Mazzarella (1957), 63; Simon of Faversham, *Quaestiones super libro Praedicamentorum*, ed. Mazzarella (1957), q. 32, 112. In contrast, Johannes Duns Scotus, *Quaestiones super librum Elenchorum Aristotelis*, ed. R. Andrews *et al.* (2004), q. 27-28, pp. 406-7. The sentence is obviously false if it means 'it is possible that who is sitting is walking' (compound sense, by which the *forma* of the subiect is meant), but it can be true if it means 'who sits can walk' (divided sense, by which only the *subiectum* is meant). Therefore, this kind of explanation does not consider the possible different uses of the modal term *possibile*, but only metaphysical notions.

positionis et divisionis), Simon offers a different solution. The unity and essence of the number five is threatened if we give a true value to the compound sense (*sensus compositionis*): according to Simon the proposition is false in both senses, formally speaking (*formaliter loquendo per se*).[28] There must then be another sense in which the proposition can be in some way true, even if not *formaliter loquendo per se*: a sort of potential unity of the proposition. Following a line of reasoning similar to that of the *Categories*, Simon argues:

> The compound, from a universal point of view, is different from the parts that compose it, just as the Philosopher states in the eighth book of the *Metaphysics*. [...] To that I answer that two and three potentially are five in number, and two and three are potentially one in number. [...] However, two and three are not a single number actually (*in actu*), but are actually different numbers [...I state that they are a certain number potentially in that they constitute a certain number, in the same way that flesh and bone are a man, because they constitute a man, and that is so only from a material point of view.[29]

In the same context of the discussion of the *fallacia compositionis et divisionis* the contemporary Anonymous of Prague (1270-ca. 1290)[30] instead justified absolutely and explicitly the extension or restriction of the range of possible referents, using the notions of *actus* and *potentia*:

> However, some are predicates for which the term stands only for the *supposita per se*, while others are predicates for which the term stands for *supposita* of both types [*i.e.*, both

28) Simon of Faversham, *Quaestiones super libro Elencorum. Quaestiones novae*, ed. Ebbesen *et al.* (1984), 140-141. See also, e.g., Incertorum Auctorum *Quaestiones super Sophisticos Elenchos. Anonymus SF*, q. 58, ed. Ebbesen (1977), 158-162.

29) Simon of Faversham, *Quaestiones super libro Elencorum. Quaestiones novae*, ed. Ebbesen *et al.* (1984), 134-135; Simon of Faversham, *Quaestiones super libro Porphyrii*, ed. Mazzarella (1957), 63; Simon of Faversham, *Quaestiones super libro Praedicamentorum*, ed. Mazzarella (1957), q. 32, 112: 'Compositum enim universaliter est aliud a partibus componentibus ipsum, sicut Philosophus vult VIII° *Metaphysicae*. [...] Dico ad hoc quod duo et tria in potentia sunt quinque in potentia et duo et tria in potentia sunt unus numerus. [...] Tamen duo et tria non sunt unus numerus in actu, sed sunt diversi numeri in actu. [...] Dico quod sunt aliquis numerus in potentia in quantum veniunt in constitutionem alicuius numeri, sicut caro et os sunt homo, quia veniunt in constitutionem hominis, et hoc est tantum materialiter.'

30) The manuscript Praha, *Knihovna Metropolitní Kapitoly*, L. 6, ff. 81ra-91vb contains a set of 42 *quaestiones* on Aristotles's *Sophistical Refutations*. The work is now supposed in all probability to have been composed between 1270 and 1290 and it seems to be linked to a 'modistic' trend, because of its key notion of *virtus sermonis*. See Lohr (1968), 219; Ebbesen (1993), 138 (n. 24), 170; Ebbesen (2000).

> *supposita per se* and *supposita per accidens*]. Therefore these added predicates do not make the term equivocal, but bring into effect what is only potentially in it.[31]

1.1. *A Short Conclusion*

However problematically, it should be clear that the *determinatio ad suppositum* in both 'modistic' writers, Simon of Faversham and the Anonymous of Prague, can be determined by the context (*adiuncta*), even in the case of *fallaciae in dictione*, that is to say, of flaws in the argument depending on the properties of natural language (*virtus sermonis*) and which concern the meaning of the terms. Secondly, this possibility, when it occurs, seems to be based on precise metaphysical and category requirements (sometimes—as we shall see later—different from each other), while the preceding theory of *suppositio* seems to play no part. These statements are sufficient in order to move on to the next section, in which we shall see more clearly the semantic and the logical aspects of this kind of approach.

2. The *variatio medii*: the Anonymous of Prague and his Contemporaries

From the second half of the thirteenth century, both the *fallaciae in dictione* and the *fallaciae extra dictionem* were regarded as a false variation of the middle term (*variatio medii*), which produces the so-called *quaternio terminorum* (the presence of four logical terms in a syllogism). In this way both types of fallacy produce syllogisms that are formally faulty (*syllogismi peccantes in forma*): in fact, to obtain a valid syllogism, identity of expression (of the *voces*) is not enough, but identity of the things (of the *res*) is also required.[32] Giles of Rome claimed:

---

31) Ms Praha, *Knihovna Metropolitní Kapitoly*, L. 6, q. 16, f. 84vb: '[...] quedam autem sunt predicata respectu quorum tenetur pro suppositis per se, quedam uero respectu quorum tenetur pro suppositis utrisque. Vnde talia predicata sibi addita non faciunt ipsum equiuocum, sed hoc quod in ipso est in potentia ducunt in actum.'

32) See, e.g., Simon of Faversham, *Quaestiones super libro Elencorum. Quaestiones novae*, ed. Ebbesen *et al.* (1984), 173: 'Ergo similiter ad bonitatem syllogismi non requiritur identitas vocis sed identitas rerum. Et propter hoc requiritur ad bonum syllogismum quod medius terminus et alii similiter realiter sumantur pro eodem in diversis propositionibus.' See also Albertus Magnus, *Liber Elenchorum*, ed. Borgnet (1890), I, t. 4, c. 1, 602-603.

> [...] in illo autem falso loco non potest reseruari principium dici de omni et dici de nullo [...], fere omnis fallacia aliquo modo variat medium et variatio medii arguit quasi in quattuor terminis.[33]

Fallacies *in dictione* depend on the properties of natural language (*virtus sermonis*). In modistic terminology these properties (*significatio, modi significandi*) are fixed once and for all by the original *impositio*, and a linguistic sign (*dictio*) cannot change its meaning (*significatio*) by virtue of context (*de virtute adiuncti*): that is possible only in the case of a specific prescription of the *impositor* or from the point of view of the interpreter (*de bonitate intelligentis*). In this way different meanings (*plures significationes*) produce different linguistic signs (*plures dictiones*), which are also different logical terms (*plures termini*).[34] Thus the *fallaciae in dictione* lead to a *quaternio terminorum*. In the case of the fallacies *extra dictionem*, however, the *causa apparentiae* directly concerns the identity of *res* (*identitas rei*) irrespective of the properties of language (*significatio* and *modi significandi*): therefore, the *quaternio terminorum* and the false variation of the middle term concern the different *res* indicated by the same linguistic sign (*dictio*) and the same meaning (*significatio*). In Pinborg's terminology, then, it seems to be a case of a variation of referent, which—as we shall see—may be determined by the (predicative or inferential) context.

The term and concept of *suppositum* is essential for identifying the *quaternio terminorum*. As a variation of the *suppositum* could determine a *quaternio terminorum*, it has significant logico-formal consequences. To bring out some aspects of these consequences, as well as the importance of their metaphysical implications, a short account follows of the treatment of fallacy of accident, which is a fallacy *extra dictionem* in Aristotle's text, and which by the Latin tradition is regarded as the 'fallacy of the middle term' (*fallacia medii*) par excellence.

---

33) Aegidius Romanus, *Expositio supra libros Elenchorum*, ed. Venetiis (1496), f. 9vb. Cf. Simon of Faversham, *Quaestiones super libro Elencorum. Quaestiones veteres*, ed. Ebbesen *et al.* (1984), 51: 'et ideo in paralogismo secundum quamcumque fallaciam est variatio medii termini [...] et similiter sunt quattuor termini. [...]'. Johannes Duns Scotus, *Quaestiones super librum Elenchorum Aristotelis*, ed. R. Andrews, q. 49, 4, p. 494. There is not such a clear link between *quaternio terminorum* and *syllogismus simpliciter* either in Albert the Great or in *Incerti Auctores*: see, e.g., Albertus Magnus, *Liber Elenchorum*, ed. Borgnet (1890) I, t. 3, c. 6, 571, Ebbesen (1977), 114.

34) For the treatment of *fallaciae in dictione* in Modistic thinking, and its logico-semantic consequences, see in particular Marmo (1994).

2.1. *The Definition of fallacia accidentis*

As we know, the fallacy of accident (*fallacia accidentis*) is the first of the seven types of *fallaciae extra dictionem*. Exegetical tradition concentrates on three main passages of Aristotle's text. The first is his definition of fallacy:

> (A) Ergo secundum accidens [*parà tò symbebēkòs*] quidem paralogismi sunt quando similiter [*homoíōs*] assignatum fuerit [*hotioûn axiōthêi*] rei subiectae [*tôi prágmati*] et accidenti inesse. Nam quoniam eidem multa accidunt, non necesse est omnibus praedicatis, et de quo praedicantur, haec [*taûta* codd. *Λ*; see *SE* ed. Ross: *tautá*] omnia inesse. Nam omnia sic erunt eadem, ut si Coriscus alterum est ab homine, ipse a se alter; est enim homo. Aut si a Socrate est alter, Socrates autem homo, alterum ab homine aiunt concessum esse, eo quod accidit, a quo dixit alterum esse, hunc esse hominem.[35]

The second is Aristotle's claim that sophistical syllogisms *secundum accidens* are missing a 'part' of the very definition of a syllogism, and not a part of the definition of a contradiction (*contradictio*):

> (B) Paralogismi secundum accidens, *diffinito syllogismo*, manifesti sunt.[36]

The last is the claim that all those same determinations seem to belong only to those things which do not differ according to their substance and which are one:

> (C) Solis enim his quae secundum substantiam sunt indifferentia et quae unum sunt, omnia videntur eadem inesse.[37]

Each of the three texts presents problems of interpretation and is connected with other well-known and complex passages in Aristotle.[38] In addition, Aristotle's examples in the *Refutations* also include arguments that do not seem to correspond easily to Boethius' distinction between predication *in subiecto* and predication *secundum accidens*.[39] Nevertheless, we can say that, taken

---

35) Aristoteles Latinus, *De sophisticis elenchis*, transl. Boethii, ed. Minio-Paluello (1975), 11, ll. 8-16 (Aristotle, *Sophistici Elenchi* V, 166 b 29-37).

36) *Ibid.*, 15, ll. 20-21 (Aristotle, *Sophistici Elenchi* V, 168 a 35).

37) *Ibid.*, 47, ll. 9-13 (Aristotle, *Sophistici Elenchi* XXIV, 179 b 38-39: '*mónois gàr toîs katà tēn ousían adiafórois kaì hèn oûsin hápanta dokeî tautà hypárkhein*', ed. Ross (1958)).

38) See, e.g., Aristotle, *Praedicamenta* III, 1 b 10-11; V, 3 b 13-14; Aristotle, *Topica* I 7, 103 a 9-10, V 4, 133 b 31-36; Aristotle, *Posterior Analytics* I 4; Aristotle, *Metaphysica* VII 6, 1031 a 24-29.

39) Boethius, *In Categorias Aristotelis libri quattuor*, ed. Migne (1860), 175D-17A; *Summa Sophisticorum Elenchorum*, ed. De Rijk, in: *Logica Modernorum* (1962-1967), I, 356; *Glose in Aristotelis*

together, the passages bring out two main themes: I) the fallacy of accident is presented as a 'syllogistic' fallacy; II) the relations between first and second substances, and the problem of the so-called 'accidental compounds', is one of the central problems.[40] For the Latin medieval exegetical tradition the problem is essentially one of explaining and justifying how the false identity of terms is contained in the premises, particularly within the different predicative relations between *minor extremitas* and middle term. The general idea that the fallacy of accident involves a variation of the middle term goes back to the twelfth century: indeed Albert the Great tells us that even in his day it was still commonly referred to as the 'fallacy of the middle term' (*fallacia medii*).[41] However, since the twelfth century, the authors had disagreed on various points: a) whether the middle term is the *res subiecta* or the *accidens* of Aristotle's text; b) how the term *accidens* is to be understood, particularly if as a *praedicatum* (what is predicated), or as an *extraneum* ('extraneous'), or as a *non necessarium* (not necessary), or as opposed to the '*per se*'; c) if the variation (*variatio*) of the terms concerns only the middle term, and/or how this variation is to be interpreted; d) lastly, how fallacy should and can be reduced and understood in each of the syllogistic figures.[42] Starting from the *Dialectica Monacensis* the connection between the fallacy of figure of speech (*figurae dictionis*, which is the last of the *fallaciae in dictione* in Aristotle's text) and the fallacy of accident also became common: during the thirteenth century the

Sophisticos Elenchos, ed. De Rijk, in: *Logica Modernorum* (1962-1967), I, 215, *Logica 'Ut dicit'*, ed. De Rijk, in: *Logica Modernorum*, (1962-1967), II/2, 388 *et* n. 1. Cf. Maierù (1972), 55 and n. 23. For the *Refutations*, see, for example: 'Neque si triangulus duobus rectis aequales habet, accidit autem ei figuram esse vel primum vel principium, quoniam figura vel principium vel primum hoc. Non enim in eo quod triangulus demonstratio. Similiter et in aliis.' (Aristoteles Latinus, *De sophisticis elenchis*, transl. Boethii, ed. Minio-Paluello (1975), 15, ll. 25-16, ll. 2 (168 a 37-168 b 4); Aristotle, *Sophistici Elenchi* VI, 168 b 1-4).

40) See, e.g., Lewis (1982), 17; Matthews (1990), 259; Williams (1985), 72. The connections between *fallacia accidentis* and metaphysical subjects were highlighted only by the Aristotelian literature. Some contemporary logical inquiries into Aristotle's text make the fallacy a particular kind of 'false generalization', rather than a syllogistic defect against syllogistic rules (see, e.g., Bueno (1988); Copi and Cohen (1990) 381). With regard to the Latin tradition, Gelber (1987) claims particularly on the basis of the analysis of the *De fallaciis* by the Ps. Aquinas, that the 'pre-Ockhamist' interpretation regards fallacy as a false extension of the *dictum de omni* to the case of accidental predication, thus as a special case of syllogistic fallacy.

41) Albertus Magnus, *Liber Elenchorum*, ed. Borgnet (1890), II, t. 3, c. 3, 563; see, e.g., *Dialectica Monacensis*, ed. De Rijk, in: *Logica Modernorum* (1962-1967), II/2, 585, 23-28.

42) *Ibid.*, 585, 591. *Fallaciae Londinenses*, ed. De Rijk, in: *Logica Modernorum* (1962-1967), II/ 2, 669. Peter of Spain, *Tractatus*, ed. De Rijk (1972), VII, 146-147. See Albertus Magnus, *Liber Elenchorum*, ed. Borgnet (1890) II, t. 3, c. 3, 561. Ps. Thomas Aquinas, *De fallaciis* (1954), 233-234, n. 677.

syllogisms '*homo est species, Sor est homo* etc.', and/or '*Coriscus est alter ab homine, Coriscus est homo* etc.', were often treated as cases of both fallacies, along with the so-called argument of the third man, an Aristotelian example of figure of speech.[43] In general, the type of identity (*ydemptitas*), i.e., only accidental (*accidentalis*) and/or partial (*partialis*), of the terms of the syllogism was gradually considered the *causa apparentiae* of the fallacy of accident.[44]

Trying to summarize the various positions, the Anonymous of Prague defines *fallacia* in these terms:

> The third sense of *variatio medii* is diversity by incompatibility. And this type of diversity causes the fallacy of accident [...] The reason why diversity by incompatibility causes fallacy of accident is the following: fallacy of accident is caused by the fact that the major extreme is attributed to the middle term under a mode of being in which it (the major extreme) does not apply to the minor extreme and in which the first (the major extreme) is different from the latter (the minor extreme), e.g., as if we say '*man is a species*': the *intencio* 'species' is here attributed to the term '*man*' under a mode of being in which it is absolutely different from Socrates. Therefore there is incompatibility and diversity; therefore it is caused not by any variation of the middle term, but only by a variation which is incompatibility; it is clear in

---

43) *Dialectica Monacensis*, ed. De Rijk, in: *Logica Modernorum* (1962-1967), II/2, 591. See Peter of Spain, *Tractatus*, ed. De Rijk (1972), VII, 135; *Fallaciae Vindobonenses*, ed. De Rijk, in: id., *Logica Modernorum* (1962-1967), I, 527, ll. 25-27; *Tractatus de dissimilitudine argumentationum*, ed. De Rijk, in: *Logica Modernorum* (1962-1967), II/2, 489, l. 34; Albertus Magnus, *Liber Elenchorum*, ed. Borgnet (1890) I, t. 3, c. 3, 560; II, t. I, c. 7, 681; Aegidius Romanus, *Expositio supra libros Elenchorum*, ed. Venetiis (1496), f. 17rb, 15ra; Simon of Faversham, *Quaestiones super libro Elencorum. Quaestiones novae*, ed. Ebbesen *et al.* (1984), 149; Radolfus Brito, *Quaestiones super Sophisticos elenchos*, q. 35; Johannes Duns Scotus, *Quaestiones super librum Elenchorum Aristotelis*, ed. R. Andrews (2004), qq. 41 and 42, pp. 459-467; Incertorum Auctorum *Quaestiones super Sophisticos elenchos*, Anonymus C, ed. Ebbesen (1977), q. 80, 179, 183; About Kilwardby, see Tabarroni (1993), 201 n. 28. On the Greek tradition, see Ebbesen (1981b), I, 224-234; II (Ps. Alexander-2), 168, ll. 56-59 (*ad Sophistici Elenchi* VI, 168 b 28-30); III, 201-205, 211, 218; Lloyd (1971); De Rijk (1981).

44) See, e.g., *Dialectica Monacensis*, ed. De Rijk (1962-1967), II/2, 585; *Fallaciae Londinenses*, ed. De Rijk (1962-1967), II/2, 669; *Fallaciae Magistri Willelmi*, ed. De Rijk (1962-1967), II/2, 691; Anonymus Parisiensis, *Compendium Sophisticorum Elenchorum*, ed. Ebbesen (1990), 288; Peter of Spain, *Tractatus*, ed. De Rijk (1972), VII, 101 (p. 145), 106 (p 148), 107 (p. 149); Lambertus Autissiodorensis, *Summa*, ed. Alessio (1971), VII, 175, ll. 7-13; Guillelmus de Sherwood, *Introductiones in Logicam*, ed. Brands and Kann (1995), *Tractatus*, VI, 196, ll. 402-410; Ps.Thomas de Aquino, *De fallaciis*, ed. Spiazzi (1954), n. 680, 234b; Albertus Magnus, *Liber Elenchorum*, ed. Borgnet (1890), I, t. III, cc. 1-3, 557-60 (see Ebbesen (1981), 96-97); Aegidius Romanus, *Expositio super libros Elenchorum*, ed. Venetiis (1496), 16rb; Simon of Faversham, *Quaestiones super libro Elencorum. Quaestiones novae*, ed. Ebbesen *et al.* (1984), q. 20, 151, ll. 26-27; *Incertorum auctorum Quaestiones super Sophisticos elenchos, Anonymus SF*, ed. Ebbesen (1977), q. 82, 189, ll. 34s; *Incertorum auctorum Quaestiones super Sophisticos Elenchos, Anonymus C*, ed. Ebbesen (1977), q. 839, 368.

> each figure that anywhere and in each figure the fallacy of accident is always a variation of the middle term, and by incompatibility.[45]

Referring to the classification of the figure of speech proposed by Giles of Rome, the anonymous claims that the transition from one category to another and from the *quale quid* to the *hoc aliquid* is faulty only when it is not allowed by the properties of natural language (*de virtute sermonis*).[46] Secondly, the text points out, fallacy of accident arises only when there is a *variatio medii cum repugnacione*, that is, a variation which contains an incompatibility: in this case the variation and the incompatibility (*repugnacio*) do not, however, have a specifically semantic character, in the sense that they do not depend on the incompatibility of the *significationes* and/or the proper *modi significandi* of the terms.[47] According to the anonymous, then, the inference

> *omnis homo est animal*
> *quoddam homo est album*
> *ergo quoddam album est animal*

cannot be a case of fallacy of accident, as there is no *variatio medii cum repugnacione*. It can only be considered a case of figure of speech. In particular, there will be a case of figure of speech whenever *omnis* and *quoddam* are understood as distributed only in relation to the *supposita per se*, an impossible distribution *de virtute sermonis*.[48] The syllogism

> '*homo est species,*
> *Sor est homo,*
> *ergo Sor est species*',

[45] Ms Praha, *Knihovna Metropolitní Kapitoly*, L. 6, q. 31, f. 88va: 'Tertia est uariatio medii diuersitas cum repugnacione. Et hec diuersitas causat fallaciam accidentis. [...] Quod autem diuersitas que est repugnantia causet fallaciam accidentis, huius ratio est, quia ex eo causatur fallacia accidentis, quod maior extremitas attribuitur termino medio sub hoc esse sub quo non competit minori extremitati et distinguitur ab illa, ut si dicatur 'homo est species': hec intencio 'species' attribuitur isti termino 'homo' sub hoc esse sub quo distinguitur omnino a Socrate. Ergo est repugnantia et diuersitas; ergo causatur non ex qualibet uariatione medii, sed tali uariatione que est repugnantia; et hoc patet in omni figura, quod vbicumque et in quacumque figura fallacia accidentis semper est uariatio medii, et hoc cum repugnantia'.

[46] Aegidius Romanus, *Expositio super libros Elenchorum*, ed. Venetiis (1496), q. 28, f. 87vb.

[47] The consequences about the rules of *conversio* are openly underlined by Siger de Courtraco, *Ars priorum*, ed. Wallerand (1913), 12 *et passim*.

[48] Ms Praha, *Knihovna Metropolitní Kapitoly*, L. 6, q. 31, f. 88vb.

by contrast, is fallacious because of an incompatible variation of the middle that does not depend on the linguistico-semantic characteristics of the terms: here, the transition from the *quale quid* to the *hoc aliquid* is allowed, for the anonymous and his contemporaries, *de virtute sermonis*.

Apart from the more technical aspects of the arguments used by the anonymous and his contemporaries, their general strategy is to theoretically establish and justify that the middle term—like any other term that undergoes a (faulty) variation—is not the same within the premises; even though there is identity of expression (*vox*) and identity of meaning (*significatio*) on the level of natural language (*de virtute sermonis*), a meaning that, strictly speaking, remains unvaried by the context, there can nevertheless be a plurality—numerical and otherwise—of logical terms.[49] More simply, even in the presence of the same words and even if these same words are not equivocal, there can still be a plurality of logical terms, which causes the error in the reasoning. This thesis, of course, admits a certain role for the context in determining both the referent in the broad sense and the logical functioning of the proposition and the syllogism, although the meaning of the term cannot strictly speaking be modified *de virtute sermonis*.[50] If this is true, then exactly what type of variation is it that we find in fallacy of accident?

2.2. Homo est species *and* Coriscus est alter ab homine

An analysis of the propositions '*Homo est species*' and '*Coriscus est alter ab homine*' may not solve the problem, but it will help us to examine it by showing how

[49] Discussion on this crucial topic was mixed with other related arguments that we cannot examine here. In particular there were arguments as to whether the conclusion of the syllogism is or is not involved in the *variatio termini*, or if the falsity of an argument is contained wholly, or not, in its premises. It is *opinio communis* that not only the *variatio medii* can cause fallacy in every figure, but also a *variatio* of the *extremitates*, which involves the conclusion as an essential part of the syllogism. Cf. Simon of Faversham, *Quaestiones super libro Elencorum. Quaestiones novae*, ed. Ebbesen *et al.* (1984), q. 20; *Incertorum auctorum Quaestiones super Sophisticos elenchos, Anonymus SF*, ed. Ebbesen (1977), q. 83; Albertus Magnus, *Liber Elenchorum*, ed. Borgnet (1890), II, t. 3, c. 4, 563; Aegidius Romanus, *Expositio super libros Elenchorum*, ed. Venetiis (1496), 17rb; see also Boethius de Dacia, *Quaestiones super libros Topicorum*, ed. Pinborg and Green Pedersen (1976), 32, ll. 34-42.

[50] It could be said that the middle term is the same, keeps the same meaning (*significatio*) *de virtute sermonis*, but a different aspect is highlighted because of the adjunct. The fact that, according to the authors, a *quaternio terminorum* also takes places in the fallacies not depending on the meaning of terms induced me to say that the middle term is not the same and to get the following statement: the meaning does not change, but the middle term does. Of course, a further terminological adjustment would be necessary.

the influence of context is possible, and how, at the same time, the ontologico-metaphysical distinctions play a fundamental role in this direction.

Since the second half of the thirteenth century, as we have seen, the proposition '*Homo est species*' was dealt with in the discussion of *fallacia accidentis*, outside the traditional theoretical tools of *suppositio*. First of all, however, if the proposition is regarded as the premise of a syllogism, identifying the variation of the term (*variatio termini*) produces complicated formal consequences: if it were always the middle that varied in the fallacy of accident, for example, there could never be a fallacy of accident in the second figure, in which the middle term is always the subject. The solutions given by different logicians vary.

On the basis of a definition by Giles of Rome,[51] the Anonymous of Prague, diverging from *communis opinio*, uses the verb *attribuere* in a broad sense, both for the subject term and for the predicate term: in this way the verb *attribuere* indicates, simply, the relation between the terms within the proposition. The consequence of this interpretation is that in a second-figure syllogism and the corresponding first-figure syllogism it is not the same term that varies, as the middle term in the two syllogisms is simply not the same word:

> (First figure)
> *homo est species*
> *Sor est homo*
> *Sor est species*
>
> (Second figure)
> *homo est species*
> *Sor non est species*
> *Sor non est homo*

It is clear that the same proposition (*homo est species*) in this interpretation has a different logical function depending on the syllogism of which it is a part.[52]

The possibility that the way in which the middle term is taken is different in the two premises, with the consequences underlined, is theoretically justified by the Anonymous of Prague, from both a logico-linguistic and an

[51] Aegidius Romanus, *Expositio super libros Elenchorum*, ed. Venetiis (1496), f. 17rb: 'Dicendum quod duplex est subiectum videlicet propositionis et intentionis siue attributionis: quotiescumque enim trahitur aliquid ad standum pro aliqua intentione vel sub aliquo attributo potest dici quod subicitur illi intentioni'.

[52] See also Anonymus Tabbaroneus, in: *Sophismata and Physics Commentaries*, ed. S. Ebbesen, in: *Cahiers de l'institut du moyen âge grec et latin* 65 (1995), 315-318.

ontologico-metaphysical point of view. In particular, according to the anonymous, the same (*secundum se et absolute*) semantic and conceptual content can be grasped by the intellect and referred to different ontological unities, depending on the different relations between subject term and predicate term. When these different ontological unities are absolutely (*omnino*) different, the assumption of the middle term proves to be false, or the middle term is no longer the same, 'at least as regards its *esse*' (*variatur secundum esse saltem*):

> I assert that animal is one in respect of its species because of its essence, but on the other hand it is different in its species as regards its being: in fact animal, in respect of this being in which it is within man, is absolutely different from that being it has in a goat or in an ass; so it is clear that the middle term varies at least as regards its being. So, at last, we must face the following example:
> Man is a species
> Sor is not a species
> therefore Sortes is not a man.
> [...] Yet we have to understand that here there is a middle term variation: in fact, when we say within the major premise 'man is a species', there the intention is affirmed about the human form; on the other hand, when we say 'Sor is not a species' by the minor premise, there the intention is denied of a principle which individuates Sortes absolutely. But that is not to assume 'species' in the same way. So it must be understood that 'species' is intended in two ways: in the first as in itself and absolutely, in the second as compared to the extremes to which it actually links up; therefore I assert that 'species' does not vary according to itself, but it varies as long as it is compared to the major and the minor extremes, as we saw; therefore a fallacy of accident is at work here.[53]

Compared to his contemporaries, the anonymous' solution—sharply criticized by Duns Scotus[54]—is certainly more radical and explicit, but it clearly

53) Ms Praha, *Knihovna Metropolitní Kapitoly*, L. 6, q. 31, f. 88vb: '[...] dico quod quamuis animal per essentiam suam sit vnum respectu suarum specierum, tamen secundum esse aliud et aliud est in speciebus: nam animal sub hoc esse sub quo est in homine est omnino diuersum ab illo esse quod habet in capra uel asino; et ideo patet quod uariatur medium secundum esse saltem. Tunc ultimo dicendum ad istud exemplum: homo est species; Sor non est species; ergo Sor non est homo. [...] Sed intelligendum quod ibi fit uariatio medii termini: cum enim in maiori propositione dicitur 'homo est species', ibi intencio affirmatur pro humana forma; sed cum dicitur 'Sor non est homo' in minori, ibi intencio negatur a principio simpliciter indiuiduante. Sed hoc non est eodem modo accipi 'speciem'. Vnde intelligendum est quod 'species' consideratur dupliciter: vno modo in se et absolute, alio modo in respectu ad extremitates quibus actu coniungitur; ex hoc dico quod, quamuis 'species' secundum se non variatur, tamen, prout comparatur ad extremitatem maiorem et minorem, sic uariatur, ut uisum est; ergo est ibi fallacia accidentis.'

54) The solution is strongly criticized by Scotus for breaking the unity of the proposition (cf. Johannes Duns Scotus, *In libros Elenchorum quaestiones*, ed. R. Andrews *et al.* (2004), q. 45, p. 481; see also Ps.-Scotus, *Quaestiones super libros Posteriorum*, ed. Wadding-Vivès (1891), I, q. 23, p. 257.

documents the debate on the problem of the influence of context and its theoretical justification in the case of logical errors not dependent on natural language (*fallaciae extra dictionem*). From this point of view, the treatment of the proposition '*Coriscus est alter ab homine*' is a further interesting field to consider, as the ontologico-metaphysical theme takes on even more importance for the argument.

In the second half of the thirteenth century it was a common view that the proposition '*Coriscus est alter ab homine*' is false, while there were various positions on the truth value of '*Coriscus est alter a Socrate*'. In the first case the proposition was considered false in a primary sense (*principaliter*), but there was a tendency to distinguish distinct ways in which a term could be taken (*acceptiones*) by which the same proposition might be considered true. Simon of Faversham, for example, suggested an analogy between the relation between '*Coriscus*' and '*homo*' and that between '*album*' and '*albedo*'. On this basis and in line with the ontological-category arguments expounded in the *Categories*, Simon claimed in the *Quaestiones novae*:

> Yet if Coriscus is grasped as regards his accidents that individualize the human nature, so Coriscus is different from man; [...] thus I will say that Coriscus and Coriscus' essence are not completely the same, because Coriscus' essence is man, and Coriscus adds something to man.[55]

It was precisely on the basis of distinctions of this kind that the second proposition (*Coriscus est alter a Socrate*) was generally considered true. By contrast, the Anonymous of Prague claimed it was absolutely false, as part of a radically realist and essentialist position.

In the first place, the Anonymous of Prague rejected the possibility of attributing different truth values to the proposition according to a distinction between different possible referents. Unlike Simon of Faversham and the anonymous C and SF, the Anonymous of Prague claimed that no referent, strictly speaking, verifies the proposition: the proposition, in whatever way the proposition is understood (whether *pro significato*, or *pro supposito*, or *pro forma*), is

[55] Simon of Faversham, *Quaestiones super libro Elencorum. Quaestiones novae*, ed. Ebbesen *et al.* (1984), 155: 'Si autem consideretur Coriscus quantum ad accidentia individuantia naturam humanam, hoc modo Coriscus est alter ab homine, [...] sic dicam quod Coriscus et quod quid est Corisci non sunt totaliter idem, quia quod ‹quid› est Corisci, est homo, et Coriscus addit aliquid supra hominem.' See also Aegidius Romanus, *Expositio super libros Elenchorum*, ed. Venetiis (1496), f. 55rb.

false.[56] On the basis of the authority of Porphyry and Boethius, the Anonymous claimed that Coriscus is essentially identical to and accidentally different from not only Socrates, but also essentially identical to and accidentally different from his human essence, that is to say, the 'universal' compound that constitutes the essence of the individual man.[57] In this context, once again, the term and notion of *suppositum* become central, both logically and metaphysically; according to the Anonymous of Prague's argument, it is precisely because the *suppositum* represents a *hoc aliquid signatum*[58] that it is identified essentially with the quiddity: *suppositum* and *quidditas* are essentially the same (*essentialiter idem*). The fact that there is in any case an accidental difference between *suppositum* and *quidditas* is irrelevant:

> The Philosopher does not understand that statement in the sense that *quidditas* and *suppositum* are different because of their essence, but they are different only accidentally, i.e., the *suppositum* includes individuating principles, yet the quiddity of the thing is assumed under being which is not demarcated; so they are different as *demarcated* and *not-demarcated*: yet this is to be different only *per accidens*. However, that proposition expresses a diversity in substance and an essential one, but not an accidental one; therefore it does not enunciate the same way it is in reality. Therefore, it should be judged as false.[59]

The application of the Anonymous of Prague's solution to the argument of the third man has important consequences, which his contemporaries probably

---

56) The distinction, also rejected by Incerti Auctores and Simon, between acceptance *pro forma* and acceptance *pro supposito*, goes back to Lambert of Lagny (cf. Lambertus Autissiodorensis, *Summa Lamberti*, ed. Alessio (1972), VII, 181). Cf. *Incertorum auctorum Quaestiones super Sophisticos elenchos*, Anonymus SF, ed. Ebbesen (1977), q. 85, 196; *Incertorum auctorum Quaestiones super Sophisticos elenchos*, Anonymus C, ed. Ebbesen (1977), q. 840, 371; Simon of Faversham, *Quaestiones super libro Elencorum. Quaestiones novae*, ed. Ebbesen *et al.* (1984), q. 20, 151-153.

57) Ms Praha, *Knihovna Metropolitní Kapitoly*, L. 6, q. 32, f. 89ra. See Simon of Faversham, *Quaestiones super libro Elencorum. Quaestiones novae*, ed. Ebbesen *et al.* (1984), 148.

58) The distinction between *hoc aliquid signatum* and *hoc aliquid vagum* is used by many authors to solve the paralogisms of the 'third man', often with aims diametrically opposed to that indicated by Anonymous of Prague. Cf. Aegidius Romanus, *Expositio super libros Elenchorum*, ed. Venetiis (1496), ff. 54rb-55va; Albertus Magnus, *Liber Elenchorum*, ed. Borgnet (1890), II, t. 1, c. 7, 684; see also Radulphus Brito, *Quaestiones super Sophisticos Elenchos*, ed. Ebbesen, q. 35. See also Robertus Kilwardby, *De ortu scientiarum*, ed. Judy (1973), 151, n. 438.

59) Ms Praha, *Knihovna Metropolitní Kapitoly*, L. 6, q. 32, f. 89ra: 'Philosophus illam auctoritatem non intelligit sic quod quidditas et suppositum differant per essentiam suam, sed tantum differunt per accidens, scilicet quod suppositum includit principia indiuiduantia, et quidditas rei accipitur sub esse non signato; et sic differunt sicut signatum et non signatum: sed hoc est differre per accidens tantum. Ista autem propositio dicit diuersitatem in substantia et essentialem, sed non accidentalem; ergo non enuntiat hoc modo sicut est in re. Et ideo est falsa iudicanda'.

knew and regarded as evident, and which provided more fuel for the lively debate. Applying the Anonymous of Prague's interpretation to the argument of the third man produces the following argumentative chain:

Coriscus est alter a Socrate [F]
Coriscus est homo
Coriscus est alter ab homine [F]

Coriscus est alter ab homine [F]
Coriscus est homo
Coriscus est alter a se [F]

The truth of the two conclusions is undermined in the first place by the falsity of one of the premises, as Albert the Great said earlier.[60] The Anonymous of Prague makes clear that it may be a case of a fallacy of accident despite the falsity of the premises. However, the syllogism must be false from a formal point of view: its falsity, in the case of the fallacy of accident, lies, as we have seen, in a variation of the middle term which implies an incompatibility (*variatio medii cum repugnacione*), so that the middle term is not the same in both premises. In particular, the term '*homo*'—which is traditionally understood here as a middle term (first figure)—must refer to entities that are, ontologically, absolutely (*omnino*) different, and so incompatible. These absolutely different entities or unities are, as is evident from the syllogism, the following: 1) on the one hand, the essential unity between *quidditas* and *suppositum*, i.e., Coriscus as a man (*Coriscus in quantum homo*); 2) on the other, the purely accidental unity between the *quidditas* and the main accidental individuals (*principia individuantia accidentalia*, i.e., *forma, locus* etc.), included in the *suppositum* itself. Coriscus as a man and Coriscus as Coriscus, in the anonymous' interpretation, thus seem to remain absolutely different entities, just as, for the anonymous, the *esse animalis* of one species compared with another are absolutely different.

Conclusion

Despite the significant differences of position, the texts discussed seem to confirm the doctrinal importance of the notion of *suppositum* and the problematic role of context in determining the 'referent', as well as the variety of consequences that different metaphysical conceptions can produce in form and

[60] Albertus Magnus, *Liber Elenchorum*, ed. Borgnet (1890), I, t. 3, c. 4, 565.

logic (and/or *e converso*). The reasons behind the different positions are not only logico-semantic in nature, but also strictly ontological, and the two planes are hard to distinguish: logico-linguistic doctrine and ontologico-metaphysical doctrine seem to operate explicitly and systematically within the same theoretical framework. Apart from the overall coherence of the doctrines and the possibility of reconstructing them systematically, which is not the purpose of this article, it is nonetheless within this 'logico-metaphysical' framework, and not through the tools of the theory of *suppositio*, that these texts sought to provide a justification of the different use of terms within a proposition.

One of the points worth noting, however, is that the semiotic aspects of language—according to which the definition of the linguistic sign (*dictio*) and its strictly semantic properties (*significatio, modus significandi*) are prominent—are not the only properties of the language studied: apart from the meaning (*significatio*), or when the meaning (*significatio*) is no longer relevant for the analysis, as in the case of the *fallaciae extra dictionem*, a purely 'logical' plan emerges that can justify the formal correctness and falsity of the reasoning, identify the terms of the logical reasoning and explain the different uses of the same tense within different propositional or inferential contexts. Only a detailed and systematic reconstruction can determine the exact historical and theoretical significance of the positions and texts we have discussed. Nevertheless, in a study of that kind the term and notion of *suppositum*, including its metaphysical implications, is a fundamental field of investigation.

Bibliography

Primary Literature

Aegidius Romanus, *Expositio supra libros Elenchorum* (Venetiis 1496, http://capricorn.bc.edu/siepm/books.html#14; reprint 1516)

Ps. Alexander, *Commentum deperditum. Fragmenta*, ed. S. Ebbesen, *Commentators and Commentaries on Aristotle's Sophistici Elenchi. A Study of Post-Aristotelian Ancient and Medieval Writings on Fallacies.* I: The Greek Tradition; II: Greek Texts and Fragments of the Latin Translation of 'Alexander's' Commentary; III: Appendices, Dannish Summary, Indices (Corpus Latinorum Commentariorum in Aristotelem Graecorum. De Wulf-Mansion Centre, VII, 1,2,3; Leiden 1981)

Albertus Magnus, *Liber Elenchorum*, ed. A. Borgnet (Alberti Magni Opera Omnia, II; Parisiis 1890, 525-713)

Anonymus Parisiensis, *Compendium Sophisticorum Elenchorum. The Uppsala Version*, ed. S. Ebbesen, in: *Cahiers de l'institut du moyen-âge grec et latin* 66 (1996), 253-312

Anonymus Tabbaroneus, in: *Sophismata and Physics Commenaries*, ed. S. Ebbesen, in: *Cahiers de l'institut du moyen âge grec et latin* 65 (1995), 303-318

Aristoteles Latinus, *De Sophisticis Elenchis*. Translatio Boethii. Fragmenta translationis Iacobi et recensio Guillelmi de Moerbeka, VI/1-3, ed. G. Dod (Leiden-Bruxelles-Paris 1975)

Aristoteles, *Topica et Sophistici Elenchi*, ed. W.D. Ross (Oxford 1958, new edition 1986)

Boethius, *Commentarius in Aristotelis Categorias*, ed. J.P. Migne (Boethius, Opera Omnia, 64; Paris 1860, coll. 159 A- 294 C)

Guillelmus de Shireswood: William of Sherwood, *Introductiones in Logicam. Einführung in die Logik*, ed. H. Brands and C. Kann (Hamburg 1995)

Incertorum auctorum *Quaestiones super Sophisticos Elenchos*, ed. S. Ebbesen (Copenhagen 1977)

Johannes Duns Scotus, *Quaestiones super librum Elenchorum Aristotelis*, ed. R. Andrews, O. Bychkov, S. Ebbesen, G. Etzkorn, G. Gál, R. Green, T. Noone, R. Plevano and A. Trever (Opera philosophica, II; St. Bonaventure, N.Y. and Washington DC 2004)

Lambertus Autissiodorensis, *Logica* (*Summa Lamberti*), ed. F. Alessio (Firenze 1971)

Petrus Hispanus: Peter of Spain, *Tractatus, called afterwards Summulae logicales*. First Critical Edition from the Manuscripts with an Introduction, by L.M. de Rijk (Philosophical Texts and Studies, XXII; Assen 1972)

Radulphus Brito, *Quaestiones super Sophisticos Elenchos. Quaestio 35*, ed. S. Ebbesen, in: id., '*Hoc aliquid-Quale quid* and the Signification of Appellatives', in: *Philosophia* 5-6 (1975-1976), 389-391

Robertus Kilwardby, *De ortu scientiarum*, ed. A. G. Judy (London-Toronto 1973)

Siger de Courtraco: Siger of Courtrai, *Ars Priorum*, ed. G. Wallerand, in: id., *Les oeuvres de Siger de Courtrai* (Les Philosophes Belges. Textes et Études; Louvain 1913, 1-77)

Simon de Faverisham: Simon of Faversham, ed. P. Mazzarella in: Opera Omnia I (Padova 1957)

Simon de Faverisham: Simon of Faversham, *Quaestiones super libro Elenchorum*, eds. S. Ebbesen, Th. Izbicki, J. Longeway, F. del Punta, E. Serene and E. Stump (Toronto 1984)

Thomas Aquinas, *Summa theologiae* ed. Leonina (Thomae Aquinatis Opera omnia iussu impensaque Leonis XIII P. M. edita; Romae 1888-1906, t. 4-12)

Thomas Aquinas, *Expositio libri Peryermeneias*, ed. altera retractata, cura et studio Fratrum Praedicatorum (Sancti Thomae de Aquino Opera omnia, 1* 1; Roma-Paris 1989)

Ps. Thomas Aquinas, *De fallaciis ad quosdam nobiles artistas* ed. R.M. Spiazzi (Thomae Aquinatis Opuscula Philosophica; Taurini 1954)

Secondary Literature

Bueno, A. (1988), 'Aristotle, the Fallacy of Accident, and the Nature of Predication: A Historical Inquiry', in: *Journal of the History of Philosophy* 26 (1988), 5-24

Copi, I. and C. Cohen (1990), *Introduction to Logic* (New York 1990)

Donati, S. (1990), 'Studi per una cronologia delle opere di Egidio Romano. I. Le opere prima del 1285: I commenti aristotelici. Parte I', in: *Documenti e studi* 1 (1990), 1-111

Ebbesen, S. (1977), 'Can Equivocation be Eliminated?' in: *Studia Mediewistyczne* 18 (1977), 103-124

——. (1981a), 'Albert (the Great?), Companion, to the Organon', in: A. Zimmerman, ed., *Albert der Grosse, sein Werk, seine Wirkung* (Miscellanea Mediaevalia, 14; Berlin-New York 1981, 89-103)

——. (1981b), *Commentators and Commentaries in Aristotle's 'Sophistici Elenchi'. A Study of Post-Aristotelian Ancient and Medieval; Writings on Fallacies*. I: *The Greek Tradition*; II *Greek Texts and Fragments of the Latin Translation of 'Alexander's' Commentary*; III: *Appendices, Danish Summary, Indices* (Corpus latinum commentariorum in Aristotelem Graecorum, VIII; Leiden 1981)

——. (1987), 'The Way the Fallacies were Treated in Scholastic Logic', in *Cahiers de l'institut du moyen âge grec et latin* 55 (1987), 107-134

——. (1988), 'Concrete Accidental Terms: Late-Thirteenth-Century Debates about Problems Relating to such terms as 'album'', in: N. Kretzmann, ed., *Meaning and Inference in Medieval Philosophy. Studies in Memory of Jan Pinborg* (Dordrecht-Boston-London 1988, 107-161)

——. (1993), 'Medieval Latin Glosses and Commentaries on Aristotelian Logical Texts of the Twelfth and Thirteenth Centuries', in: *Glosses and Commentaries on Aristoteian Logical Texts. The Syriac, Arabic and medieval Latin Traditions*, edited by Ch. Burnett (The Warburg Institute, University of London, London, 1993, pp. 129-177)

——. (2000), 'Anthony, Albert, Anonymus Mazarineus and Anonymus Pragensis on the *Elenchi*', in: *Documenti e studi* 11 (2000), 271-295

Gelber, H. (1987), 'The Fallacy of Accident and the *dictum de omni*: Late Medieval Controversy over a Reciprocal Pair', in: *Vivarium* 25 (1987), 110-145

Lewis, F.A., 'Accidental Sameness in Aristotle', in: *Philosophical Studies* 42 (1982), 1-36.

Libera, A. de (1981), 'Supposition naturelle et appellation: aspects de la sémantique parisienne au XIIIe siècle', in: *Histoire Epistémologie Langage* 3/1 (1981), 63-77

—— (1982), 'The Oxford and Paris Traditions in Logic', in: N. Kretzmann, A. Kenny and J. Pinborg, eds., E. Stump, ass. ed., *The Cambridge History of Later Medieval Philosophy.* From the Rediscovery of Aristotle to the Desintegration of Scholasticism 1100-1600 (Cambridge-London-New York-New Rochelle-Melbourne-Sydney 1982, 174-187)

Lloyd, A.C. (1971), 'Neoplatonists' Account of Predication and Medieval Logic', in: *Le néoplatonisme. Actes du Colloque de Royaumont*, 9-13 juin 1969 (Paris 1971, 357-364)

Lohr, C. (1968), 'Medieval Latin Aristotle Commentaries G-I', in: *Traditio* 24 (1968), 149-245

Maierù, A. (1972), *Terminologia logica della tarda scolastica* (Rome 1972)

Marmo, C. (1994), *Semiotica e linguaggio nella scolastica: Parigi, Bologna, Erfurt 1270-1330. La semiotica dei Modisti* (Nuovi Studi Storici, 26; Roma 1994)

——. (1999), 'The Semantics of the Modistae', in: S. Ebbesen and R. Friedman, eds., *Medieval Analyses in Language and Cognition.* Acts of the Symposium *The Copenhagen School of Medieval Philosophy*, January 10-13, 1996 (Copenhagen 1999, 83-104)

——. (2006), 'La funzione del contesto: teorie 'continentali' e 'inglesi' a confronto sull'eliminazione dell'equivocità tra fine XIIIe inizio XIV secolo', in: S. Caroti, R. Imbach, Z. Kaluza, G. Stabile and L. Sturlese, eds., *'Ad Ingenii Acuitionem'. Studies in Honor of Alfonso Maierù* (Louvain-La-Neuve 2006, 249-280)

Matthews, G.B. (1990), 'Aristotelian Essentialism', in: *Philosophy and Phenomenological Research* 50 Supp. (1990), 251-262

Park, S.C. (1999), *Die Rezeption der mittelalterlichen Sprachphilosophie in der Theologie des Thomas von Aquin mit besonderes Beruchtsichtigung der Analogie* (Leiden 1999)

Pinborg, J. (1967), *Die Entwicklung der Sprachtheorie im Mittelalter* (Münster-Kopenhagen 1967)

——. (1972), *Logik und Semantik im Mittelalter. Ein Überblick*, mit einem Nachwort von H. Kohlenberger (Stuttgart-Bad Cannstatt 1972)

——. (1982), 'Speculative Grammar', in: N. Kretzmann, A. Kenny, J. Pinborg, eds., E. Stump, ass. ed., *The Cambridge History of Later Medieval Philosophy . From the Rediscovery of Aristotle to the Disintegration of Scholasticism 1100-1600* (Cambridge-London-New York-New Rochelle-Melbourne-Sydney 1982, 253-269)

——. (1984), *Medieval Semantics. Selected Studies on Medieval Logic and Grammar.* ed. S. Ebbesen (London 1984)

Rijk, L.M. de (1962-1967), *Logica Modernorum. A Contribution to the History of Early Terminist Logic.* I. *On the Twelfth Century Theories of Fallacy*; II/1: *The Origin and the Early Development of the Theory of Supposition* II/2: *Texts and Indices* (Philosophical Texts and Studies, VI et XVI; Assen 1962-1967)

—— (1968), 'On the Genuine Text of Peter of Spain's *Summule Logicales*. Part I. General Problems Concerning Possible Interpolations in the Manuscripts', in: *Vivarium* 6 (1968), 1-34

—— (1970), 'Die Bedeutungslehre der Logik im 13. Jahrhundert und ihr Gegenstück in der metaphysischen Spekulation', in: A. Zimmermann, ed., *Methoden in Wissenschaft und Kunst des Mittelalters* (Miscellanea Medievalia, 7; Berlin 1970, 1-22)

—— (1971), 'The Development of *suppositio naturalis* in Medieval Logic. Part I. Natural Supposition as Non-contextual Supposition', in: *Vivarium* 9 (1971), 71-107

—— (1981), 'Die Wirkung der Neuplatonische Semantik auf das Mitellalterliche Denken über das Sein', in: A. Zimmermann, ed., *Sprache und Erkenntnis im Mittelalter* (Miscellanea Mediaevalia, 13/1; Berlin 1981, 19-36)

Rosier-Catach, I. (1983), *La grammaire spéculative des modistes* (Paris 1983)

——. (1994), *La parole comme acte. Sur la grammaire et la sémantique au XIII*[e] *siècle* (Paris 1994)

Schoot, H.J.M. (1993), 'Aquinas and Supposition: the Possibilities and Limitation of Logic *in divinis*', in: *Vivarium* 31 (1993), 193-225

Tabarroni, A. (1993), '*Omnis phoenix est*: Quantification and Existence in a New sophisma Collection (MS Clm 14522)', in *Sophisms in Medieval Logc and Grammar*. ed. S. Read, Acts of the Ninth European Symposium for Medieval Logic and Semantics, St. Andrews, June 1990. Dordrecht-Boston-London (1993, 185-201)

Williams, C.J.F. (1985), 'Aristotle's Theory of Descriptions', in: *The Philosophical Review* 94 (1985), 63-80

XIVTH CENTURY

Scotus on Supposition

Costantino Marmo
University of Bologna

Abstract

In his commentaries on Porphyry and Aristotle's *Organon* (*Categories*, *Peri hermeneias*, *Sophistici elenchi*, and *Topics*) and in his other works, John Duns Scotus shows his knowledge of both the modistic theory of language and the theory of supposition. My contribution sheds some light on the relationship between Scotus' philosophy of language and the theory of supposition, collecting and commenting on all the passages in which he makes use of it or discusses some theoretical points. I take into special account the almost unknown commentary on the *Topics*, which is preserved in a Vatican manuscript.

Keywords

history of semantics, *significatio*, terminism, *suppositio*, *appellatio*, *copulatio*, modism, *modi significandi*, *modi intelligendi*, *modi supponendi*, John Duns Scotus

Introduction

While studies of Scotus' philosophy of language published in the last 30 years are not rare,[1] it is hard to find in them any observation about the presence and actual use of supposition theory in Scotus' writings. Furthermore, the editions of Scotus' works are not always helpful on this point: while the (old) Commissio Scotistica makes reference in the *apparatus fontium* to Peter of

[1] See, for instance, Dahlstrom (1980); Vos (1985); Bos (1987); Marmo (1989); Perler (1993), (2003); Pini (2001), (2002) and (2004). The authenticity of the logical works usually ascribed to Scotus (i.e., *Quaestiones* on Porphyry's *Isagoge*, on Aristotle's *Praedicamenta*, on *Perihermeneias* and on *Sophistici elenchi*) is definitively assessed by the editors (cf. Andrews *et al.* (2004), 31-33). I will also use Scotus' litteral commentary on Aristotle's *Topics*, leaving aside the discussion about its authenticity (see Marmo (2010)). I am deeply indebted to Chris Martin for his thorough revision of my English version.

Spain's *Tractatus*[2] every time Scotus uses the term *suppositio* in his theological works, the St. Bonaventure University team, working on the edition of Scotus' philosophical works, more appropriately (as I will show below), points to William of Sherwood's or Roger Bacon's treatises as Scotus' sources.[3]

In what follows, I shall review all the passages in Scotus' philosophical and theological works that are relevant for drawing any conclusion about Scotus' logico-semantic sources and about his actual use of supposition theory. Special attention will be devoted to the relationship between terminist theory and the Modists' semantic views in Scotus' philosophy of language.

1. Scotus on the Properties of Terms

Although Scotus made use of the specific terminology of the theory of the properties of terms, as far as I know he neither wrote a treatise on them nor referred to any particular text belonging to the terminist tradition. In the first section, I will illustrate his knowledge of the definitions of the different properties and, in particular, of the various types of supposition, so that it will be possible to link with certainty his knowledge to a peculiar branch of the terminist tradition. I will also try to show how Scotus' writings might call into question the traditional opposition between the modistic approach to language and the theory of the properties of terms.

1.1. Significatio

First of all, Scotus distinguishes between *significatio* and *suppositio*, and defines *significatio* as 'repraesentatio alicuius ex impositione'[4] in his commentary on the *Categories*, where he opposes it to *dare intelligere* (i.e., to signify secondarily or to allude) and not to *supponere*. This definition actually shows nothing new: the whole medieval grammatical and logical tradition always underlined the dependence of signification on the primitive act of imposition. In his commentary on the *Elenchi*, however, Scotus adds that 'significare est aliquid

---

2) See, for instance, John Duns Scotus, *Ordinatio* I, d. 21, q. unica, n. 30, ed. Commissio Scotistica (1959), 337, n. 2, where the Commissio Scotistica refers to Peter of Spain, who held a completely different position on the supposition ascribed to the predicate term of a universal affirmative proposition (see below).

3) Scotus, *Quaestiones in libros Perihermenias*, I, q. 13, ed. R. Andrews *et al.* (2004), 131, *apparatus*.

4) John Duns Scotus, *Quaestiones super Praedicamenta Aristotelis*, ed. Andrews *et al.* (1999), q. 8, n. 19, 319.

intellectui repraesentare',[5] indicating the other term of the double relation of signification, that is the addressee (or his/her intellect) to whom something is signified by a speaker/writer.[6] Scotus adds some explanations of the role of the intellect, insofar as either the *impositor nominum*, or the speaker (*proferens*) or the hearer (*audiens*) is concerned. From the point of view of the *impositor*, the kind of knowledge required—according to Scotus—coincides more or less with etymology.[7] From the point of view of the speaker, Scotus gives a peculiar reading of Bacon's 'direct signification model',[8] holding that a word signifies directly the *res*, not as far as it exists (*ut existit*), that is as an individual, but rather as it is understood (*ut concipitur*), i.e., as a common nature or essence, which is the first object of the intellect.[9] From the point of view of the hearer, Scotus somehow recovers the 'indirect signification model', especially in his *Reportata Parisiensia*, where he defines the *vox significativa* as 'signum ad placitum rememorativum' and adds that it produces the actual knowledge (or the concept) of the signified things that are known *habitualiter*: in this way the phonic expression also signifies the concept, as a sign serving to remind.[10] In the last text, Scotus also rejects the view that ascribes to the *vox significativa* an inner quality that would act 'naturally' on the hearer and describes how every phonic expression modifies the sense of hearing, inducing the intellect to consider again the thing for which that vocal sound was imposed, and makes it therefore pass from a habitual knowledge of that thing to actual knowledge.

Scotus' discussions of signification do not allow us to place his thought in any specific semantic tradition, since both the English and the Continental traditions make use of the notion of *repraesentatio*:[11] on the one hand, William of Sherwood defines signification as 'praesentatio formae ad intellectum',[12] and,

---

5) John Duns Scotus, *Quaestiones super librum Elenchorum Aristotelis*, ed. Andrews *et al.* (2004), q. 15, n. 17, 336.

6) On the theory of the signification as a double relation, see Rosier (1994), 112-122.

7) See Ashworth (1980); Boulnois (1995).

8) See Perler (2003), 164.

9) On the debate and its interpretations, see Marmo (1989), Pini (2001), (2002) and (2004); Perler (2003).

10) John Duns Scotus, *Reportata Parisiensia*, II, d. 42, q. 4, n. 20, XI. 1, ed. Wadding, Parisiis (1891), 414b; cf. Marmo (1989), 168-170. This is the way William of Ockham describes how every sign, in a broad sense, works: it transforms habitual knowledge into actual knowledge (see *Summa logicae*, I.1, ed. Boehner *et al.* (1974), 15-16).

11) On the two traditions of terminist semantics, see in particular De Rijk (1962), (1967), (1968), (1971), (1972), (1982); Braakhuis (1977), (1979); Ebbesen (1970), (1981), (1985); Ebbesen and Rosier (2000); Spade (1982); De Libera (1980), (1981), (1982), (1986), (1997); Goubier (2000).

12) William of Sherwood, *Introductiones in Logicam*, V, ed. Brands and Kann (1995), 132.

on the other, Peter of Spain 'rei per vocem secundum placitum repraesentatio'.[13] Furthermore, his polemical stance against a 'naturalistic' conception of meaning may be linked both to his (implicit) refusal of the modistic view of signification as substantial form of the *dictio*, displayed in his commentary on the *Elenchi*,[14] and to his position against the Thomistic doctrine of the sacrament's effectiveness due to an inner *virtus* brought about by consecration.[15] Scotus' relation to the Modists, however, does not generally appear polemical: in the same context in the commentary on the *Elenchi* Scotus specifies the relationship between *significare* and *intelligere* and adopts the famous modistic scheme of the derivation of the *modi significandi* from the *modi intelligendi*, which in turn derive from the *modi essendi*: 'intelligere sequitur esse, et significare intelligere', ergo modus intelligendi modum essendi, et modus significandi sequitur modum intelligendi'.[16] The connections to both the modistic and the terminist traditions that will emerge again in what follows appear to be characteristic of Scotus' approach to language and need further investigation.

1.2. Suppositio

Suppositio, as a property of terms, appears to be closely related to signification. In Scotus' texts, however, I could not find any explicit definition of supposition. The property of supposition is rather described, in some of his texts, in terms of *acceptio* and of *ratio intelligendi*. Here is a series of passages from Scotus' commentary on the *Categories*:

> 'homo' de se est indifferens ad multas *acceptiones*, scilicet pro voce, intentione, et pro suppositis. Signa ergo, ut 'omnis', 'aliquis' sibi addita, indifferentiam ad intentiones et ad vocem tollunt, et determinant ipsum ad acceptionem tantum pro suppositis.[17]
> Terminus in una propositione non habet nisi unam suppositionem, quia una propositio est signum unius intellectus compositi. Et diversae suppositiones termini sunt diversae *rationes intelligendi* terminum respectu tertii; et omnes diversae suppositiones in suo genere sunt

---

13) Peter of Spain, *Tractatus*, VI, ed. De Rijk (1972), 79.

14) See Marmo (1994), 121 and (2006), 279-280.

15) See Rosier-Catach (2004), ch. 2, for a thorough presentation of the different models used in explaining the causality of the sacraments: Scotus' explanations have these discussions as their framework.

16) John Duns Scotus, *Quaestiones super librum Elenchorum Aristotelis*, ed. Andrews *et al.* (2004), q. 15, n. 15, 335-336.

17) John Duns Scotus, *Quaestiones super Praedicamenta Aristotelis*, ed. Andrews *et al.* (2004), q. 12, n. 13, 358; cf. *Lectura in librum primum Sententiarum*, I, d. 4, q. unica, n. 4, ed. Commissio Scotistica (1960), 408 (italics are mine).

> oppositae . . . impossibile est idem in una propositione habere diversas suppositiones. Indifferens igitur est terminus ad diversas suppositiones, licet non in una propositione.[18]

In this passage there is a mysterious reference to a 'third (element)' ('diversae suppositiones termini sunt diversae rationes intelligendi terminum respectu tertii'), and one wonders what it might be. Another text (from his commentary on the *Elenchi*) casts some light on it:

> terminus, quod primo significat, supponit respectu praedicati.[19]

The third element is then the predicate. Every term, on the one hand, signifies something and, on the other, stands for it in relation to the predicate term. This implies that the term in question occurs as the subject of a proposition. Here it is only implicit, but explicitly in a passage of his second commentary on the *De interpretatione* one finds some typical elements of the English terminist tradition: firstly, *suppositio* is a property of a term only when it occurs in a proposition (it also bears traces of the grammatical use of the term, where it was mainly conceived of as the property of being the subject of a verb); secondly, *suppositio* and *significatio* are closely related, so that the term, in a proposition, supposits for what it first signifies. Translating the transitive use of *supponere* that is found in the text quoted above is very difficult: this acceptation of the verb, however, reflects exactly the phrase used by William of Sherwood when, in his *Introductiones*, he defines the different types of supposition. Showing which types of supposition Scotus knows and makes use of will be the object of the next section.[20]

Before dealing with Scotus' knowledge of the types of supposition, I would like to highlight another point of his theory. As seen above, he describes *suppositio* in terms of *acceptio*. Again, this term does not connect Scotus' texts to any particular semantic tradition: the term *acceptio* is used both by Peter of Spain in order to define *suppositio*,[21] and by some Modists as equivalent to *causa*

18) John Duns Scotus, *Quaestiones super Praedicamenta Aristotelis*, ed. Andrews *et al.* (2004), q. 13, n. 33, 373; cf. *Lectura in librum primum Sententiarum*, I, d. 21, q. unica, n. 13, ed. Commissio Scotistica (1960), 299 (italics are mine).

19) John Duns Scotus, *Quaestiones super librum Elenchorum Aristotelis*, ed. Andrews *et al.* (2004), q. 15, n. 7, 333; cf. *Quaestiones in libros Perihermeneias*, ed. Andrews *et al.* (2004), I, q. 13, n. 32, 131.

20) By the way, the intransitive usage of the verb, stemming from grammar, is also found in Scotus' texts; i.e., the phrase '*supponere* verbo' in the sense of 'playing the role of subject in relation to a verb' (cf. *Quaestiones in libros Perihermenias*, q. 9, ed. Andrews *et al.* (2004), 97 ff.).

21) Peter of Spain, *Tractatus*, ed. De Rijk (1972), VI, 80: 'acceptio termini substantivi pro aliquo'.

veritatis;[22] in the same sense, but also as equivalent to *suppositio*, the term *acceptio* is used by Scotus in his qq. 12 and 13 on the *Categories*.[23] Furthermore, the use of the term *ratio intelligendi*, which is synonymous with *modus intelligendi*, suggests a different picture of the relation between the two traditions in Scotus' writings. Other passages confirms that, according to the young Scotus, *suppositio* is not simply a property of a term, but rather a way of understanding it (*modus intelligendi terminum*) that is determined by some other syntactic component (such as syncategoremes, quantifiers and so on). Thus in a proposition such as *aliquis homo est animal*,

> 'aliquis' dicit tantummodo modum intelligendi hominem pro suppositis.[24]

Again, in his *Lectura*, Scotus repeats the point that a term in a proposition can have but one kind of supposition, as well as one meaning:

> sicut terminus in propositione habet unum significatum et unum intellectum essentialem, ita etiam et unum modum intelligendi et supponendi.[25]

Conceiving the supposition of a term as a way of understanding it opens up the possibility of discovering an unsuspected link between the traditions of terminism and of modism. The theory of supposition, far from being incompatible with or alternative to the Modists' views on semantics, in Scotus' works appears to be rather a part of it, as a peculiar *modus intelligendi*. The fact that it has no corresponding *modus significandi* should be interpreted as the simple acknowledgement that *suppositio* does not concern second order (or grammatical) signification but only first order signification (i.e., lexical semantics), specifying the rules of reference to individuals. I would suggest that we see it as a counterpart of the *modi praedicandi*, which are another kind of *modi intelligendi* (obviously linked to the predicate), aimed—according to some

---

22) Cf., e.g., Anonymus-SF, *Quaestiones super Sophisticos Elenchos*, q. 48, in *Incertorum auctorum Quaestiones super Sophisticos Elenchos*, ed. Ebbesen (1977), 106-107; Radulphus Brito, *Quaestiones super Sophisticos Elenchos*, q. 23, ms Bruxelles, *Bibliothèque Royale* 3540-47, 504va-b; Radulphus Brito, *Quaestiones super Posteriorum Analyticorum libros*, I, q. 46, in: Pinborg (1976), 272-275.

23) John Duns Scotus, *Quaestiones super Praedicamenta Aristotelis*, ed. Andrews *et al.* (2004), q. 12, n. 13, 358; n. 16, 359; n. 23, 360; q. 13, n. 17, 369; n. 31, 372.

24) John Duns Scotus, *Quaestiones super Praedicamenta Aristotelis*, ed. Andrews *et al.* (2004), q. 12, n. 23, 360; cf. n. 28, 362.

25) John Duns Scotus, *Lectura in librum primum Sententiarum*, ed. Commissio Scotistica (1960), I, d. 21, q. unica, n. 13, 299.

Modists—at defining the ten Aristotelian categories in a modistic framework.[26] *Suppositio*, in this respect, would therefore be a way of understanding the subject term—and by extension the predicate term—determined by its occurrence in a proposition. Scotus makes clear, however, that even if the *modus supponendi* of the subject term is produced by the propositional context, it does not strictly depend on the relation between the subject and the predicate. Again he is in perfect agreement with the Modists, who held that equivocation cannot be eliminated by the predicate (or by the verb), because it has no immediate syntactic link to the subject term.[27] In his first commentary on the *Peri hermeneias* Scotus says that 'praedicatum nullam rationem supponendi tribuit subiecto',[28] and in his commentary on Aristotle's *Topics* he specifies that

> predicatum nullam talem acceptionem pro suppositis tribuit subiecto, quia non cedit in eundem extremum (*ms.* intellectum) cum ipso, sicut oportet determinationem cedere cum determinabili.[29]

Suppositio would thus be a way of understanding a term determined by the quantifier, by the negation or by a syncategoreme that is directly linked to it. This characterization of supposition does not explain, however, the distinction between simple and personal supposition. I think there is an explanation and I will try to give it later (§1.4).

1.3. Copulatio, Appellatio *and* Restrictio

Question 13 of Scotus' first commentary on the *De interpretatione* deals with a problem also discussed in some modistic *sophismata* and commentaries on the *Prior Analytics*, i.e., the problem of the *restrictio* of the scope of supposition induced by the predicate on the subject term. His solution, however, appears to be incompatible with what he says in some previous questions (qq. 9-11) about a predicate not determining the type of supposition of the subject and a common term standing in every proposition for all its *supposita per se* (i.e., individuals existing either in the present, in the past or in the future), unless

---

[26] See Marmo (1999), 85-89.

[27] See Marmo (2006), 259-264.

[28] John Duns Scotus, *Quaestiones in libros Perihermenias*, ed. Andrews *et al.* (2004), I, qq. 9-11, n. 25, 104.

[29] John Duns Scotus, *Notabilia in libros Topicorum*, II, in: ms *Bibliotheca Apostolica Vaticana*, Ottob. Lat. 318, f. 258rb. I will not discuss here, however, the reasons why I hold that this work too belongs to the Aristotelian commentaries written by the *Doctor subtilis*. On this question, see Marmo (2010).

it is 'contracted' by an immediate determination that specifies in which time those individuals exist. In q. 13 Scotus holds the opposite position: the predicate (or the verb) produces a restriction of the scope of the subject term, with no further specifications:

> Ideo ad quaestionem dicendum quod terminus supponens verbo de praesenti restringitur. Et causa est actualis inhaerentia praedicati ad subiectum.[30]

The discussion that comes before the author's solution concerns only the most plausible explanation of the main conclusion, as if the opposite conclusion would not even be taken into consideration. Furthermore, at its beginning this question does not display the usual series of arguments pro and contra. To sum up, in terms of both content and form, the authorship of this question seems questionable. I will not push the argument further, however, because the solution may not be as straitforward as it appears at first glance.[31] As a matter of fact, here Scotus never mentions the reference to presently existing things as the result of *restrictio*; he just specifies what should be understood as the actual inherence of the predicate in the subject. Furthermore, he denies the reference to presently existing things in propositions such as *Homerus est poeta* or *homo est cursurus*, even if the verb is in the present tense. The conclusion, where Scotus distinguishes between actual inherence and actual composition of the terms of a proposition, is a doubt:

> Hic potest dubitari per quid significatur inhaerentia, quae est causa restrictionis.[32]

Immediately after this last *dubitatio*, Scotus adds some remarks about supposition theory,[33] noting

(1) that a common term might stand either for an absolute common nature (without taking into account its relation to its *supposita*), or for the same nature but as it exists in its supposita (for example in propositions such as *piper venditur hic et Rome* and *homo est dignissima creaturarum* are true), or for its supposita (like in *homo currit*);

---

30) John Duns Scotus, *Quaestiones in libros Perihermenias*, ed. Andrews *et al.*, I, q. 13, n. 14, 126.

31) As far as I know, the editors did not point out any problem of attribution.

32) John Duns Scotus, *Quaestiones in libros Perihermenias*, ed. Andrews *et al.* (2004), I, q. 13, n. 28, 130.

33) John Duns Scotus, *Quaestiones in libros Perihermenias*, ed. Andrews *et al.* (2004), I, q. 13, nn. 29-33, 130-132.

(2) that in *piper venditur hic et Rome* the subject has personal supposition and, coherently with his position as expressed in other works (see below), that if one takes the conjunction between *hic* and *Rome* as forming one single predicate (*de praedicato copulato*) then the proposition is false; if one interprets the proposition as a copulative proposition equivalent to *piper venditur hic et piper venditur Rome*, then it is true;

(3) *suppositio* in a proper sense concerns the subject term: only in a wider sense can one ascribe supposition to the predicate, which properly has *copulatio*;

(4) a common term has a triple relation (a) to the predicate, (b) to its supposita (and Scotus specifies that this reference to *supposita* is called *appellatio*; he does not, however, say 'reference to presently existing *supposita*'), and (c) to the *supposita* of the predicate term.

This text, at least on matters of terminology, is in agreement with other texts from Scotus dealing with the supposition of terms, and adds further evidence of Scotus' knowledge of the whole theory of the properties of terms.

1.4. *Types of* suppositio

Various texts by Scotus attest to his knowledge and use of a certain classification of the types of supposition. As seen above, a term can be used in at least three different ways: 'pro voce, (pro) intentione, et pro suppositis'.[34] These three ways correspond to the three basic types of supposition: *materialis*, *simplex* and *personalis*. I could not find, however, any other passage where Scotus alludes to, defines or makes use of *suppositio materialis*. In his commentary on the *Categories*, on the contrary, one can read passages about the relationship between *suppositio simplex* and *personalis*. While discussing a question about the identity in meaning of concrete and abstract terms, and offering arguments for the position which maintains that a concrete term signifies the subject of inherence of the form signified by the abstract term, Scotus says that a term signifies what it supposits (*supponit*, in the transitive form) and that a variation in supposition does not change its meaning.[35] As is well known, Scotus sides with the Modists on this problem, holding that concrete and abstract terms have

34) John Duns Scotus, *Quaestiones super Praedicamenta Aristotelis*, ed. Andrews *et al.* (1999), q. 12, n. 13, 358.

35) John Duns Scotus, *Quaestiones super Praedicamenta Aristotelis*, ed. Andrews *et al.* (1999), q. 8, n. 2, 314: 'terminus supponens, ex hoc quod supponit, non habet novum significatum'. This position was held by the Modists, but not by Roger Bacon (see Ebbesen (forthcoming)).

the same meaning but different *modi significandi*.[36] Furthermore, he replies to the arguments supporting the subject-signification position in this way:

> Concedo quod terminus illud supponit quod significat. Sed non semper pro eo supponit quod significat, sed tantum supponitur significatum pro significato quando est suppositio simplex; sicut commune supponit pro suppositis et non suppositum, quia non significat suppositum.[37]

Here Scotus makes clear that *supponere* in its transitive use and *supponere pro* are not equivalent, so that *supponere significatum* and *supponere pro significato* are the defining characterics of a particular type of supposition such as *suppositio simplex*; while a common term that supposits for (*supponit pro*) its individuals, does not supposit them, i.e., it does not signify them and does not posit them in every proposition in which it occurs. From these premises he concludes that a concrete term 'supponit pro subiecto, sed non subiectum', i.e., it stands for the subject (of inherence), but does not signify it. The other general kind of supposition, i.e., personal supposition, is alluded to in the quoted text, without naming it. From other texts, however, it comes out very clearly that Scotus would define it as the property of a term of standing for its individuals (*supposita*).[38] Implicitly Scotus suggests that a term, even if taken in *suppositio personalis*, 'supponit significatum' (transitively). That is why, on the one hand, there is no equivocation between a term taken in simple supposition and the same term taken in personal supposition and, on the other, the distinction between them is not to be explained by the immediate addition of a syntactic element. They are just different ways of understanding and using a common term occurring in a proposition, where it stands for the *natura communis*, in the first case, and for its individuals, in the case of personal supposition.

In the history of supposition theory there is at least one author who makes use of this kind of phrasing, i.e., the transitive use of *supponere*, and indicates this feature as characteristic of a type of supposition that is common to simple

36) See Ebbesen (1988) and (forthcoming).

37) John Duns Scotus, *Quaestiones super Praedicamenta Aristotelis*, ed. Andrews *et al.* (1999), q. 8, n. 24, 321 (the first sentence was probably a commonly held position in Paris: see Anonymus-SF, *Quaestiones super Sophisticos Elenchis*, ed. Ebbesen (1977), q. 48, 102: 'terminus idem supponit quod significat'; note that the anonymous commentator on the *Elenchi* does not refer to supposition as the property of standing for something, but talks about an *acceptio* of the term; see below); cf. q. 13, n. 16, 369.

38) Cf. John Duns Scotus, *Quaestiones super Praedicamenta Aristotelis*, ed. Andrews *et al.* (1999), q. 13, n. 14, 368; n. 17, 369; n. 37, 374; n. 46, 377.

and personal supposition. It is William of Sherwood, who defines *suppositio formalis* this way:

> Formalis est, quando dictio supponit suum significatum.[39]

In Sherwood's text, the distinction between simple and personal supposition follows the lines already seen in Scotus' texts, grounding it in the transitive vs. intransitive use of the verb *supponere*:

> Et sic dividitur: alia simplex, alia personalis. Et est simplex, quando dictio supponit significatum pro significato, ut 'homo est species'. Personalis autem, quando supponit significatum, sed pro re que subest, ut 'homo currit'.[40]

I have found similar definitions and a comparable use of the verb *supponere* not in the terminist tradition (Roger Bacon, Peter of Spain or Lambert of Lagny), but rather in the modistic milieu. The so-called Anonymus-SF (edited by Ebbesen), in his commentary on the *Elenchi*, discusses whether the inference *canis currit, ergo latrabile currit* is correct, and quotes as an old saying that 'terminus idem supponit quod significat'.[41] In the solution to the following question, where he discusses whether different uses of a term produce equivocation or not, he adds that

> diversae enim acceptiones non ponunt aequivocationem in hoc termino 'homo', dicendo 'homo est species' et 'homo currit'; istae duae acceptiones, quarum una est pro forma, alia pro suppositis, non ponunt aequivocationem. [...][42]

Other subdivisions of personal supposition are found in Scotus' texts. Even if there is no mention of the distinction between *suppositio discreta* and *communis*, Scotus discusses at length how proper names, such as Caesar, refer to individuals. The various types of supposition he refers to are usually ascribed to common terms.

In both logical and theological texts, one can read passages dealing with or just mentioning *suppositio determinata* as a kind of personal supposition. In his commentary on the *Categories*, for instance, Scotus says that a common term having this kind of supposition 'stat pro multis disiunctive' and can be

---

39) William of Sherwood, *Introductiones in Logicam*, ed. Brands and Kann (1995), V, 136.

40) *Ibidem*.

41) Anonymus-SF, *Quaestiones super Sophisticos Elenchos*, ed. Ebbesen (1977), q. 47, 102.

42) Anonymus-SF, *Quaestiones super Sophisticos Elenchos*, ed. Ebbesen (1977), q. 48, 106-107.

exemplified by the subject term of a particular or indefinite proposition, such as *homo est animal* or *aliquis homo est animal*.[43] In other passages he quite obviously calls *descensus* the inference from such a proposition to the disjunction of singular propositions or, more generally, from a general proposition to a singular one under the subject or under the predicate.[44]

In Scotus' text one can also find references to *suppositio confusa*, but more frequently to the kind known as *confusa tantum* than to the kind known as *confusa et distributiva*. Concerning the latter, I will just mention a couple of passages from his commentary on the *Categories*, where he says, first, that the quantifier *omnis* makes the common term stand for its individuals (*pro suppositis*),[45] and, second, that in the proposition *aliquis homo non est aliquis homo*, which contradicts *omnis homo est aliquis homo*, the predicate term is 'confusum confuse et distributive'.[46] Finally, in Scotus' theological works one can read passages that appeal to *suppositio confusa tantum*, where he concludes that this kind of supposition does not allow any inference to the *supposita*:

> Ideo dico aliter quod subiectum exclusivae affirmativae supponit confuse tantum, sicut predicatum universalis affirmativae [...] et sub termino sic stante—scilicet confuse tantum—non licet descendere.[47]

While completely absent from the Parisian tradition, the position which holds that the predicate of a universal affirmative proposition has *suppositio confusa tantum* is typical of the English development of supposition theory,[48] and Scotus' use of it confirms his close connection with this tradition.

Scotus' involvement with the English tradition of supposition theory finds further corroboration in his commentary on Aristotle's *Topics* (furthermore, the discussion of *suppositio vaga* there is perfectly consistent with the attribution

43) John Duns Scotus, *Quaestiones super Praedicamenta Aristotelis*, ed. Andrews *et al.* (1999), q. 13, n. 37, 374.

44) John Duns Scotus, *Quaestiones super Praedicamenta Aristotelis*, ed. Andrews *et al.* (1999), *passim*.

45) John Duns Scotus, *Quaestiones super Praedicamenta Aristotelis*, ed. Andrews *et al.* (1999), q. 12, n. 13, 358.

46) John Duns Scotus, *Quaestiones super Praedicamenta Aristotelis*, ed. Andrews *et al.* (1999), q. 12, n. 30, 362.

47) John Duns Scotus, *Ordinatio*, ed. commissio Scotistica, I, d. 21, q. unica, n. 30, 337. Cf. *Quaestiones super Praedicamenta Aristotelis*, ed. Andrews *et al.*, q. 12, n. 3, 355: 'praedicatum universalis affirmativae confunditur tantum confuse'; cf. Walter Burley, *De puritate artis logicae tractatus longior*, I.1.4, ed. Boehner (1955), 21.

48) See for instance both William of Sherwood, *Introductiones in Logicam*, ed. Brands and Kann (1995), V, 138, and Roger Bacon, *Summulae dialectices*, ed. De Libera (1986), 266.

of the commentary to Scotus). The passage in question, drawn from his commentary on the second book of Aristotle's *Topics*, deals with the kind of supposition ascribed to the subject term of a proposition such as *piper venditur hic et Romae*, and parallels similar discussions that can be found in both William of Sherwood's and Roger Bacon's treatises.

William holds that *piper* has simple supposition, the third kind to be precise; the first kind is exemplified according to William by the proposition *Homo est species*, where the meaning is signified in its abstraction, the second kind by *Homo est dignissima creaturarum*, where the meaning is signified insofar as it is compared to the things and it is actually preserved in every subordinate species or individual. In the third kind—William adds—the meaning is signified in comparison to the things, but as far as it refers to them *communiter et vage*: 'unde solet dici quod hec est vaga suppositio'. The same kind of simple supposition can be found in the answer to a question about the animal that is useful for ploughing ('quid animal est utile aratro?'): when we reply *bos*, we do not refer to any particular individual, but simply to one of the members of that species, taken at random.[49]

Roger Bacon, when listing in his *Summulae dialectices* the examples of simple supposition, does not mention the *piper* proposition. This example actually appears in a kind of addition to the discussion about the traditional types of *suppositio* and *appellatio*: besides the above-mentioned types—he says—one also has to mention the double (*gemina*) supposition, the antonomastic and the methonimic (*methonomastica*) ones. The example for the *gemina suppositio* is exactly that of *piper venditur hic et Romae*. While the *antonomastica suppositio* (ex.: *Plato deus Philosophorum*) is always personal and determinate, methonymic supposition (ex.: *bibe ciphum*) is usually personal, but might be simple by accident. There is some discussion (*dubitatio*) about the *gemina* type: on the one hand, while the term *piper* refers to an individual, it cannot be here and in Rome, so it has no personal supposition; on the other, because the verb *vendere* refers to an operation that can be performed only by individuals and not by species, it cannot let the subject term stand for a species (and therefore have simple supposition). Bacon's solution is double: in the first case, if one takes the conjunction (*et*) as connecting propositions, then the term *piper* must occur twice, once for each proposition, and in each occurrence it stands personally (in this case, as Roger Bacon adds, it makes no sense wondering about the supposition of the term taken only once, *semel positum*). In the second case, if one interprets the conjunction (*copulatio*) as holding between

---

49) *William of Sherwood, Introductiones in Logicam*, ed. Brands and Kann (1995), V, 142.

terms, thus producing a proposition with a conjunctive predicate (*de copulato praedicato*), then if the term *piper* has personal supposition, the proposition is false; if it has simple supposition, then it is true.[50] Bacon does not mention here, however, any *vaga suppositio*.

I have been dwelling on Bacon's positions only because Scotus, in his commentary on the second book of Aristotle's *Topics*, appears to side with Bacon in his solution to the *dubitatio*, even if he makes use of different terminology. According to Scotus, in the case of *piper venditur hic et Rome* or of *hec herba crescit hic et in orto meo*, there is no need to appeal to any *vaga suppositio* that 'solet ab aliquibus poni', because the proposition can be taken in the composite or divided sense: in the composite sense, it is equivalent to a proposition with a conjunctive predicate (*hic et Rome*), and, since the subject stands for the same thing, the proposition is always false (Scotus implicitly holds that the subject term has personal supposition);[51] in the divided sense, it is equivalent to a conjunction of propositions in which each subject stands for different individuals, so that the conjunction may be true:

> Ista etiam ratio impedit suppositionem vagam que solet ab aliquibus poni in talibus propositionibus 'piper venditur hic et Rome' [...]. Non oportet autem proprietatem talium propositionum 'piper venditur hic et Rome' et 'hec herba crescit hic et in orto meo' ponere suppositionem vagam. Sunt enim huiusmodi distinguende secundum compositionem et diuisionem: in sensu compositionis possunt poni false sine inconuenienti et tunc sunt de predicato copulato et subiectum sumitur pro aliquo eodem; in sensu diuisionis sunt vere et tunc sunt copulatiue uel subiectum iteratum in duabus cathegoricis copulatis ad invicem potest supponere pro diuersis, licet ipsum non distributum non possit in eadem propositione pro diuersis supponere.[52]

To sum up, for Scotus this kind of *suppositio* is useless. As a matter of fact, one does not find any other mention of it in his other works.

2. Actual Use of the Theory in Scotus' Logical Works

In his commentaries on the *ars vetus* and ars *nova*, the young John Duns Scotus not only mentions the different kinds of *suppositio*, but also appeals to them in

50) Roger Bacon, *Summulae dialectices*, 2.2, in: De Libera (1986), 288.

51) As seen above, he explicitly says that in his first commentary on the *De interpretatione*, q. 13.

52) John Duns Scotus, *Notabilia in libros Topicorum*, II, in: ms *Biblioteca Apostolica Vaticana, Ottob*. 318, f. 258ra-b. Cf. Walter Burley, *De puritate artis logicae tractatus logicae*, ed. Boehner, I.1.4, p. 21, who perfectly agrees with Scotus on this point.

order to argue in favour of or against some positions, building on them various *rationes* or *determinationes*. In what follows I will examine how he actually uses supposition theory in arguments, starting with his questions on Porphyry.

2.1. Quaestiones in Porphyrium

In his commentary on Porphyry, Scotus makes use of the terminology of supposition in at least two questions: q. 10 (where he discusses the truth of the proposition *homo est universale*) and q. 24 (where he examines the truth of the proposition *Socrates senex differt a seipso puero*). In the first question, among the arguments against the truth of the proposition, he lists the following:

> Subiectum significat veram naturam, ergo istam supponit. Praedicatum non praedicat veram naturam, quia nec significat. Ergo, praedicatur non vera natura de vera natura; igitur oppositum de opposito.[53]

In his solution to the previous question, Scotus had distinguished three ways of considering the meaning of a common term signifying a real essence (*vera natura*): in the first way, this nature is considered insofar as it exists in its individuals, or in its *esse materiale*; in a second way, it is taken insofar as it has an *esse quidditativum*, that is, without considering its actual existence in its individuals (*absolute*); in a third way, it is considered insofar as it is understood by the intellect, and only in this sense is the proposition *homo est universale* true, because the intellect is induced by some property of that nature (such as *esse unam in multis et de multis*) to produce a second intention that can be truly predicated of the term signifying that very nature. Finally, in his reply to the argument quoted above, Scotus makes clear that the three ways of considering an essence correspond to three distinct types of supposition.[54] As a matter of fact the first type corresponds to personal supposition, the second and third with the first two kinds of *suppositio simplex* listed by William of Sherwood in his *Introductiones*.[55] He adds, however, that these differences in supposition do not

---

53) John Duns Scotus, *Quaestiones in librum Porphyrii Isagoge*, ed. Andrews *et al.* (1999), qq. 9-11, n. 7, (1999), 45.

54) John Duns Scotus, *Quaestiones in librum Porphyrii Isagoge*, ed. Andrews *et al.* (1999), q. 9-11, n. 24, 49: 'vera natura potest supponi tripliciter.' (I would correct the edited text because here the expression *vera natura* is not just mentioned but used, or in other terms, it has personal supposition and not material supposition); see also Duns' *Quaestiones in libros Perihermeneias Aristotelis*, ed. Andrews *et al.* (2004), I, q. 13, n. 29, 130.

55) William of Sherwood, *Introductiones in Logicam*, ed. Brands and Kann (1995), V, 140-142. The last one, being illustrated by the subject term of the proposition *piper venditur hic et Romae*, is not

yield equivocation or ambiguity,[56] but rather a *fallacia figurae dictionis*, and what is *multiplex* according to *figura dictionis* is not *realiter multiplex* but only *phantastice* (according to the famous *dictum* ascribed to Alexander of Aphrodisia). The first observation does not bring anything new: one can read similar assertions in both summulist[57] and modist works, such as the Anonymous-SF's commentary on the *Elenchi* and those of Radulphus Brito:[58] all of them agree on the fact that variation in supposition does not produce equivocation. The difference between Scotus and Radulphus, who devotes a question to that problem, lies in the type of fallacy they indicate as yielded by this kind of multiplicity: according to Brito, the difference in supposition produces a fallacy of accident.[59]

In question 24, discussing the truth of *Socrates senex differt a seipso puero*, the reference to *suppositio* is very quick and is used by Scotus to specify what can be inferred from this proposition in virtue of the negation that is implicit in the verb *differre*: while it is possible to infer *Socrates senex et Socrates puer differunt*, and then furthermore *Socrates senex et Socrates puer sunt multa*, it is not possible to further infer correctly *Socrates et Socrates sunt multa*, because there is a fallacious inference from a term taken *secundum quid* (*Socrates senex* or *Socrates puer*) to the same term taken *simpliciter*. The first inference (that is *Socrates senex differt a seipso puero, ergo Socrates senex et Socrates puer differunt*) is correct, on the contrary, because the subject term keeps its supposition and the negation included in the verb has no influence whatsoever on it.[60]

2.2. Quaestiones super Praedicamenta

In Scotus' commentary on the *Categories*, the references to the theory of supposition are to be found in questions 8, 12, and 13. I have already mentioned above the role played by the first reference in Scotus' discussion about the problem of the different meanings of concrete and abstract terms linked by

mentioned by Scotus, and its irrelevance might be better understood taking into consideration his discussion about *suppositio vaga* in his commentary on the *Topics* (see above).

56) John Duns Scotus, *Quaestiones in librum Porphyrii Isagoge*, ed. Andrews *et al.* (1999), q. 9-11, n. 24, 49: 'nec [. . .] propositio est distinguenda.'

57) Cf. William of Sherwood, *Introductiones in Logicam*, ed. Brands and Kann (1995), V, 138.

58) See for instance Anonymus-SF, *Quaestiones super Elenchos*, ed. Ebbesen (1977), q. 48, 106-107; Johannes Dacus, *Summa grammatica*, ed. Otto (1955), 371; Radulphus Brito, *Quaestiones super Elenchos*, q. 23, ms Bruxelles, *Bibiothèque Royale* 3540-47, ff. 504va-505ra.

59) Radulphus Brito, *Quaestiones super Elenchos*, q. 23, in: ms Bruxelles, *Bibliothèque Royale* 3540-47, f. 505ra.

60) John Duns Scotus, *Quaestiones in librum Porphyrii Isagoge*, ed. Andrews *et al.* (1999), q. 24, n. 12, 153.

a paronymy relation: the concrete term stands for the subject of inherence of the form signified by the abstract term, but signifies that very form, i.e., the concrete term 'supponit pro subiecto, sed non subiectum'.[61]

In question 12, where he discusses whether *aliquis homo* is a first substance (because of the example given by Aristotle in *Categories* V and translated by Boethius as *aliquis homo*), Scotus' answer is in the negative: *aliquis homo* is not a first substance, because it is not equivalent to *iste homo*, the phrase that best expresses the Aristotelian notion of first substance. In order to give a sound interpretation of the words used by Aristotle, however, Scotus adds first that a common term such as *homo* is indifferent to its various uses, that is, *pro voce* (*suppositio materialis*), *pro intentione* (which corresponds, in the English tradition, to one kind of *suppositio simplex*), or *pro suppositis* (*suppositio personalis*), and second that the quantifiers (*omnis* or *aliquis*) exclude the material and simple type of supposition, determining that the supposition of the common term is personal (he actually uses the phrase *personalis acceptio*). Even in this case, though, the term is indifferent to the reference to any individual ('est indifferens ad quodcumque suppositum').[62] Aristotle's text, therefore, ought to be interpreted in this sense: since he wanted to distinguish between first and second substance, while keeping the property of being the subject of the intention 'second substance', he decided to use the phrase *aliquis homo*, 'id est quidlibet quod est iste homo'.[63] He also gives an alternative solution that allows us to qualify *aliquis homo* as first substance: in this sense, the phrase indicates a first substance, but in an indeterminate way ('indeterminate tamen'), or, better, it does not signify any first substance, either in a determinate or in an indeterminate way, even though it stands for it when it is the subject of an indefinite proposition. The latter is the solution Scotus prefers, because it is consistent with what he says in the following question about the fact that a common term, when it is the subject of an indefinite proposition, stands for individuals, but does not signify them.[64]

In question 13, Scotus faces another interpretative problem: he wonders whether the inference made by Aristotle in chap. V, that is 'animal praedicatur de homine, ergo de aliquo homine', holds *formaliter* or not. His conclusion is

---

61) John Duns Scotus, *Quaestiones super Praedicamenta Aristotelis*, ed. Andrews *et al.* (1999), q. 8, n. 24, 321.

62) John Duns Scotus, *Quaestiones super Praedicamenta Aristotelis*, ed. Andrews *et al.* (1999), q. 12, n. 13, 358.

63) *Ibidem*.

64) John Duns Scotus, *Quaestiones super Praedicamenta Aristotelis*, ed. Andrews *et al.* (1999), q. 12, n. 15, 359.

that this is a sound inference, but that the predicate term must be more general (*superior*) than the subject and have personal supposition.[65] One of the arguments against this conclusion says that a common term is indifferent to the various types of supposition, therefore the inference in question, going from what is absolute or indifferent (*homo*) to what has a kind of supposition or to what is determinate (*aliquis homo*), is fallacious.[66] In his reply, Scotus makes clear that a common term, taken by itself, is indifferent to the various types of supposition, but it is no more indifferent when it occurs in a proposition: it is impossible for a term in a proposition to have different kinds of supposition, because they are opposite ways of understanding a thing.[67] In the same question, again in the replies to arguments against the author's position, we read also that the subject of an indefinite or particular proposition has *suppositio determinata*, and that it is not possible to 'descend' under the predicate of a universal affirmative proposition (because it has *suppositio confusa tantum*).[68]

2.3. Quaestiones super Peri Hermeneias

In both of Scotus' series of questions on the *De interpretatione*, there are many references to the theory of supposition. In question 6 of his first commentary (where he discusses whether there are simple individuals falling under a universal term that signifies a real nature, other than the individuals actually existing), Scotus takes for granted the link between supposition and the verification of propositions, and offers his definition of the individual or the *suppositum per se* of a common term. It should be noted that the same problem was discussed from the 1270s on in modistic milieux, adopting the same terminology:[69] the *suppositum per se* is the real nature conceived as non-predicable of many (that is, as an individual instantiation of the common nature), apart from its existence or non-existence; this real nature so conceived is signified by phrases

65) John Duns Scotus, *Quaestiones super Praedicamenta Aristotelis*, ed. Andrews *et al.* (1999), q. 13, n. 15, 368-369.

66) John Duns Scotus, *Quaestiones super Praedicamenta Aristotelis*, ed. Andrews *et al.* (1999), q. 13, n. 3, 365: 'fit fallacia consequentis'.

67) John Duns Scotus, *Quaestiones super Praedicamenta Aristotelis*, ed. Andrews *et al.* (1999), q. 13, n. 33, 373.

68) John Duns Scotus, *Quaestiones super Praedicamenta Aristotelis*, ed. Andrews *et al* (1999), q. 13, nn. 34 and 37, 374.

69) Cf. Ps. Boethius of Dacia, *Quaestiones in libros Analyticorum Posteriorum*, I, q. 26, in: Pinborg (1971), 53; Peter of Auvergne, *Sophisma* V, quoted in Marmo (1999), 95, n. 37; Simon of Faversham, *Quaestiones in libros Analyticorum Posteriorum*, I, q. 60, quoted in Marmo (1999), 97, n. 40; Radulphus Brito, *Quaestiones in libros Analyticorum Posteriorum*, I, q. 46, in: Pinborg (1976), 272-273.

composed by a demonstrative adjective and a common term, such as *iste homo*, or by proper names (*Caesar* or *Antichristus*) insofar as they are imposed to signify men or 'natura humana concepta ut haec'.[70] Scotus' solution to some other questions (qq. 9-11) dealing with the set of individuals referred to by a common term which is the subject of a present-, past- or future-tensed verb (*supponit verbo de praesenti*, etc.) is consistent with this position and adds some other features that it shares with modistic semantic syntax:

> Terminus communis in quacumque propositione, sive de praesenti sive de praeterito sive de futuro, supponit pro quibuscumque suppositis, sive exsistentibus sive non-exsistentibus, quando sibi non additur immediate aliqua determinatio contrahens ad supposita unius differentiae temporis.[71]

As hinted above, the syntactic views of the *modistae* (based on binary relationships between syntactic components) had some direct consequences in semantics, when applied to the possible role of context in the elimination of equivocity. In their opinion (at least *de virtute sermonis*), equivocity could not be eliminated except by a *determinatio* immediately linked to the term (either *a parte subjecti* or *a parte praedicati*), like in *canis latrabilis currit*. A mediately connected *determinatio* such as a predicate, according to them, had no effect on the equivocal term, like in *canis est latrabilis*: here the term *canis* keeps the totality of its meanings (*de virtute sermonis*). In the text quoted from Scotus the distinction between *determinatio* immediately or mediately added to the common term plays a major role in allowing him to conclude that a common term, taken with no temporal *determinatio* immediately added to it, refers to all its *supposita per se*, apart from their existence in the present, in the past, or in the future. Consequently, Scotus' solution to the problem is that a common term with no immediately added temporal determination is distributed over all its *supposita per se*.[72]

As we saw above, in the first series of questions on the *De interpretatione* (q. 13), other properties of terms are mentioned: *appellatio*, *copulatio* and *restrictio*. In the second series of questions on the *De interpretatione*, *copulatio*

70) John Duns Scotus, *Quaestiones in libros Perihermenias*, ed. Andrews *et al.* (2004), I, qq. 5-8, n. 44, 83.

71) John Duns Scotus, *Quaestiones in libros Perihermenias*, ed. Andrews *et al.* (2004), I, qq. 9-11, n. 24, 103.

72) John Duns Scotus, *Quaestiones in libros Perihermenias*, ed. Andrews *et al.* (2004), I, qq. 9-11, n. 44, 111.

as the property of verbs and predicate terms is again referred to, without, however, any particular development of the theory.[73]

2.4. *Commentaries on the* Topics *and the* Sophistici Elenchi

As we saw above, Scotus in his commentary on Aristotle's *Topics* takes a stand against the notion of *suppositio vaga* as applied to the subject term of the proposition *piper venditur hic et Romae*. His discussion is part of the analysis of the following *locus* (or *consideratio*, according to the terminology used in Aristotle's *Topics*): 'Si de aliquo subiecto predicetur genus [...] de eodem predicatur species'.[74] Against this version of the *locus a genere*, two objections are raised and rejected: 1) if one can truly say that *Socrates et asinus sunt animalia*, no species can be predicated of the subject; 2) if a shield is half white and half black, one can say that it is coloured, but no specific colour can be predicated of it. In particular, the discussion of the second objection leads Scotus to his rebuttal of *suppositio vaga*. He wonders what kind of supposition might have the term *coloratum* in the proposition: *scutum est coloratum*. In his answer, Scotus says that it cannot have simple supposition, because otherwise it would refer to the form, verifying therefore the proposition *scutum est color*; if it stands personally, then it stands for one individual *sub disiunctione ad alia*, but not *pro multis actu*, unless a distributive sign is added to it. The reference to the example of *piper* should make clear that the term *coloratum* in that proposition (referring to a shield half black and half white) cannot be substituted for by any species of coloured things (*album* or *nigrum*), but rather by a complex predicate made up of a disjunction or conjunction of terms: 'et sic conceditur—thus Scotus concludes—quod 'scutum est album et nigrum'' (f. 258 rb) in the composite sense is true, while in the divided sense it is false. Since the two propositions *scutum est album* and *scutum est nigrum* are both false. Aristotle's observation is therefore to be understood and rewritten this way (f. 258 rb):

> Consideratio est sic intelligenda quod illud quod denominatur a genere denominatur ab aliqua specie uel absolute sumpta uel sub disiunctione seu copulatione ad aliam prout propositio est de copulato predicato et non copulativa.

In Scotus' questions on the *Elenchi*, very briefly, I will only mention the fact that he specifies what kind of fallacies are determined when one infers from

---

73) John Duns Scotus, *Quaestiones in libros Perihermenias*, ed. Andrews *et al.* (2004), I, I, q. 6, 167 ff.

74) John Duns Scotus, *Notabilia in libros Topicorum*, in: ms *Bibliotheca Apostolica Vaticana*, Ottob. 318, II, f. 257va.

a proposition where a term has a certain kind of supposition to a proposition where it has another kind. Consistently with what he had already stated in his questions on the *Categories*, Scotus remarks that when one term has personal supposition in the antecedent and simple supposition in the consequent, or vice versa, the inference is fallacious *secundum figuram dictionis*.[75] Discussing whether changing the construal of a term from *quale quid* to *hoc aliquid* produces a *fallacia figurae dictionis*, Scotus specifies that this fallacy can be exemplified by the inference 'Homo est species, Socrates est homo; ergo etc. (*scil.* Socrates est species)',[76] where *species* is predicated of a common term insofar as it is common, and then it is predicated of a singular term as such. Scotus specifies that *figura dictionis*, in this case, is produced when the confusion is based on the similarity between the words and not on the unity of the thing signified (which causes a *fallacia accidentis*, as Radulphus Brito also held).[77] As far as I can see, there is no mention of the properties of terms in Scotus' other philosophical commentaries, such as his questions on the *De anima* and the *Metaphysics*.

3. On the Actual Use of the Theory in Scotus' Theological Works

In his theological works, a rather sparse use of the terminology of the theory of *suppositio* can be observed, and it can be noted that it is always aimed at the rejection of contrary arguments, rather than at building up any *determinatio quaestionis*. In general, Scotus' remarks are consistent with the doctrine he followed in his Aristotelian commentaries. He repeats, for instance, that the predicate of a universal affirmative proposition stands *confuse tantum* for its individuals and that it is not possible to 'descend' under a term having such a supposition.[78] I will not review all of them here, but just present an interesting case: the comparison between the version of distinction 21, *quaestio unica* in his *Lectura* on the first book of the *Sentences* and in his later *Ordinatio*. The question at issue here is whether the proposition *solus pater est deus* is true. One of the arguments in support of its truth runs as follows: *solus deus est deus,*

75) John Duns Scotus, *Quaestiones super librum Elenchorum Aristotelis*, ed. Andrews *et al.* (2004), q. 15, n. 11, 334-335; q. 41, n. 15, 462.

76) John Duns Scotus, *Quaestiones super librum Elenchorum Aristotelis*, ed. Andrews *et al.* (2004), q. 41, n. 11, 461.

77) Radulphus Brito, *Quaestiones in Elenchos*, q. 23, in: ms Bruxelles, *Bibliothèque Royale* 3540-3547, f. 505ra.

78) John Duns Scotus, *Ordinatio* I, ed. Commissio Scotistica. (1959), d. xxi, q. unica, n. 30, 337.

ergo solus pater est deus. The inference may be demonstrated by appealing to the fact that the first proposition is indefinite and therefore it is possible to infer a singular proposition concerning its *supposita*, in disjunction from other similar singular propositions, such as *solus Pater est deus* or *solus Filius est deus* and so on. In his answer, Scotus discusses an argument allegedly advanced by Bonaventure against this inference, in his commentary on the same distinction, saying that in the antecedent the subject term has two relations: one to the *dictio exclusiva* that makes it stand in simple supposition, and one to the predicate that makes it stand in personal supposition.[79] According to Bonaventure—who does not actually discuss this inference—only the first relation offers a solution, because (following Peter of Spain)[80] he holds that a common term that follows a *dictio exceptiva* has simple supposition and that under a term having such a supposition it is not possible to descend to the individuals (such as in the consequent of the inference in question). Scotus, however, interprets Bonaventure as saying that the same term in such a proposition might have two different kinds of supposition, a possibility that he rejects here as he does also in his commentary on the *Categories*.[81] Every term occurring in a proposition has just one kind of supposition (*modum supponendi*) as well as only one meaning and one way of being understood (*modum intelligendi*), and in this particular case the inference is blocked by the *suppositio* determined by the *nota exclusionis*:

> terminus cui additur nota exclusionis, confunditur confuse tantum ab exclusione, sicut in propositione universali in quam convertitur stat confuse tantum; et ideo non convenit descendere ad aliquod suppositum nec actualiter nec sub disiunctione.[82]

This text confirms again that Scotus' frame of reference is not Peter of Spain's theory of supposition, but rather William of Sherwood's, and shows a deep connection with his logical texts. The discussion that Scotus offers in his *Ordinatio*, even if it repeats the same pattern (rejection of the double supposition solution and support of the *suppositio confusa tantum* option), does not insist on the

79) Cf. Bonaventure, *In primum librum Sententiarum*, I, d. 21, a. 1, q. 1, ad 2um, ed. Collegium S. Bonaventure (1882), 380b.

80) Peter of Spain, *Tractatus*, ed. De Rijk (1972), VI, 81 (even if here Peter deals with *praeter* and not with *solus*).

81) John Duns Scotus, *Quaestiones super Praedicamenta Aristotelis*, ed. Andrews *et al.* (1999), q. 13, n. 33, 373.

82) John Duns Scotus, *Ordinatio* I, dist. xi-xxv, ed. Commissio Scotistica (1959), d. xxi, q. unica, n. 13, 299.

link between unity of the meaning, on the one hand, and unity of the *modus intelligendi* and *modus supponendi*, on the other. The link to Scotus' philosophical works seems to be weaker in this context. Furthermore, he adds a different solution, suggesting that *in divinis* things go differently than *in creaturis*, that is, without worrying too much about how human language works: the term *deus* determined by the syncategoreme *solus* might stand for *hoc deus* which is common to the three persons of the Trinity, just like the subject of *deus est pater et filius et spiritus sanctus* (previously discussed in distinction 4).[83]

4. Conclusions

Here I'll try to draw some conclusions from this review of the passages where Scotus makes use of the theory of supposition.

1) Scotus' theory of supposition is clearly linked to the English tradition: *suppositio* is therefore a property that terms have only when they occur in propositions; furthermore, Scotus (just like Roger Bacon) tries to lessen the role of simple supposition in favour of the *suppositio personalis confusa tantum*, in order to keep the referential commitment of nominal phrases placed in syntactical positions that block the *descensus ad supposita*;
2) the theory of *suppositio terminorum* does not appear to be a key tool in Scotus' theology, and even less in his general philosophical project, but quite obviously it plays a relevant role only in his philosophy of language;
3) other theoretical tools drawn from the modistic framework, however, play a role in Scotus' philosophy of language; in particular, he shares with the Modists the idea that the *modi significandi* derive from the *modi intelligendi* and from the *modi essendi*; supposition theory might find a place in the modistic framework as a particular kind of *modi intelligendi*, together with the *modi praedicandi*, that define the ten categories;[84]
4) in agreement with the majority of the Modists (and, at least on this topic, in contrast with the English line of thought, represented by William of Sherwood and Roger Bacon), Scotus maintains that a common term,

83) John Duns Scotus, *Ordinatio* I, dist. xi-xxv, ed. Commissio Scotistica (1959), d. xxi, q. unica, n. 31, 337; cf. John Duns Scotus, *Ordinatio* I, dist. iv-x, ed. Commissio Scotistica (1956), d. iv, q. unica, n. 11, 5.

84) Cf. Marmo (1999).

followed by a past-, present- or future-tensed verb, stands for all its individuals, whether they exist or not at the time of its utterance;[85]

5) finally, I would suggest that the last position is a consequence of the conception of signification that Scotus shares with the Modists: the meaning of a universal term is a common nature, indifferent to existence and non-existence, and it is 'proposed' (to the hearer's intellect) by this term wherever it occurs; the reference to the individuals and its extension (or *volume*, to use the nice French expression proposed by Goubier 2000, 44) are therefore just a matter of accidental conditions.

Bibliography

Primary Literature

Anonymus-SF, *Quaestiones super Sophisticos Elenchos*, ed. Ebbesen, in: *Incertorum Auctorum Quaestiones super Sophisticos Elenchos* (Corpus Philosophorum Danicorum Medii Aevi 7; Hauniae 1977)

Boethius of Dacia, *Quaestiones in libros Analyticorum Posteriorum*, ed. J. Pinborg (1971), in: id., 'Bezeichnung in der Logik des XIII Jahrhunderts', in: A. Zimmermann, ed., *Der Begriff der Repraesentatio im Mittelalter. Stellvertretung—Symbol—Zeichen—Bild* (Miscellanea Mediaevalia, 8; Berlin-New York 1971, 237-281)

Bonaventura, *Commentaria in quatuor libros Sententiarum*, ed. Collegium S. Bonaventurae (Bonaventurae Opera omnia, I; Quaracchi 1882-1902)

Gualterus Burlaeus, *De Puritate Artis Logicae Tractatus Longior, with a Revised Edition of the Tractatus Brevior*, ed. Ph. Boehner O.F.M. (Louvain-Paderborn 1955)

Guillelmus de Ockham, *Summa logicae*, ed. Ph. Boehner, G. Gál and S. Brown (Guillelmi de Ockham Opera Philosophica, I, St. Bonaventure, N.Y. 1974)

Guillelmus de Shireswood: William of Sherwood, *Introductiones in Logicam. Einführung in die Logik*, ed. H. Brands and C. Kann (Hamburg 1995)

Johannes Dacus, *Summa Gramatica*, ed. A. Otto (Johannis Daci Opera, I, 2 parts (Corpus Philosophorum Danicorum Medii Aevi 1; Hauniae 1955)

Johannes Duns Scotus: John Duns Scotus, *Ordinatio* I, dist. i et ii, ed. Commissio Scotistica, praeside C. Balić (Ioannis Duns Scoti Opera Omnia, XVI, Civitas Vaticana 1950)

Johannes Duns Scotus, *Ordinatio* I, dist. iv-x, ed. Commissio Scotistica, praeside C. Balić (Ioannis Duns Scoti Opera Omnia, XVI, Civitas Vaticana 1956)

——, *Ordinatio* I, dist. xi-xxv, q. unica, ed. Commissio Scotistica, praeside C. Balić (Ioannis Duns Scoti Opera Omnia VICivitas Vaticana 1959)

——, *Quaestiones in libros Perihermenias Aristotelis*, eds. R. Andrews, G. Etzkorn, G. Gál, R. Green, T. Noone, R. Plevano, A. Traver and R. Wood; *Quaestiones super librum Elenchorum Aristotelis*,

[85] It should be noted, however, that q. 13 of his first commentary on the *De interpretatione* appears to go in the opposite direction.

eds. R. Andrews, O. Bychkov, S. Ebbesen, G. Etzkorn, G. Gál, R. Green, T. Noone, R. Plevano and A. Traver; *Theoremata*, eds. M.Dreyer and H. Möhle, in coorperation with G. Krieger (Ioannis Duns Scoti Opera Philosophica II, St. Bonaventure, N.Y.—Washington DC 2004)

——, *Reportata Parisiensia* II-IV, ed. L. Wadding (Ioannis Duns Scoti Opera Omnia, editio nova iuxta editionem Waddingi [...] a patribus franciscanis de observantia accurate recognita I, 224-272; Parisiis 1891, Lugduni 1639, XXII 513-XXIV)

——, *Lectura in librum primum Sententiarum*. Prologus et distinctiones a prima ad septimam, ed. Commissio Scotistica, praeside C. Balić (Ioannis Duns Scoti Opera Omnia, XVI, Civitas Vaticana 1960)

——, *Notabilia in libros Topicorum* II, in: ms *Bibliotheca Apostolica Vaticana*, Ottob. Lat. 318

——, *Quaestiones in librum Porphyrii Isagoge* et *Quaestiones super Praedicamenta Aristotelis*, eds. R. Andrews, G. Etzkorn, G. Gál, R. Green, T. Noone and R. Wood (Ioannis Duns Scoti Opera Philosophica I, St. Bonaventure, N.Y. 1999)

Peter of Auvergne, *Sophisma* V, ed. C. Marmo, in: id., 'The Semantics of the Modistae', in: S. Ebbesen and R. Friedman, eds., *Medieval Analyses in Language and Cognition*. Acts of the Symposium *The Copenhagen School of Medieval Philosophy*, January 10-13, 1996 (Copenhagen 1999, 83-104)

Petrus Hispanus: Peter of Spain, *Tractatus, called afterwards Summulae logicales*. First Critical Edition from the Manuscripts with an Introduction, by L.M. de Rijk (Philosophical Texts and Studies, XXII; Assen 1972)

Radulphus Brito, *Quaestiones in libros Analyticorum Posteriorum* I, q. 46, ed. J. Pinborg, in: id., 'Some Problems of Semantic Representation in Medieval Logic', in: *History of Linguistic Thought and Contemporary Linguistics*, ed. H. Parret (Berlin-New York 1976, 254-278)

——, *Quaestiones super Sophisticos Elenchos. Quaestio 35*, ed. S. Ebbesen, in: id. (1975-6), '*Hoc aliquid-Quale quid* and the Signification of Appelatives', in: *Philosophia* 5-6 (1975-6), 389-391

Simon de Faverisham Simon de Faverisham, *Quaestiones in libros Analyticorum Posteriorum*, I, q. 60, quoted in: C. Marmo, 'The Semantics of the Modistae', in: S. Ebbesen and R. Friedman, eds., *Medieval Analyses in Language and Cognition*. Acts of the Symposium *The Copenhagen School of Medieval Philosophy*, January 10-13, 1996 (Copenhagen. 1999, 83-104)

Walter Burley, *vide s.n.* Gualterus Burlaeus

William of Ockham, *vide s.n.* Guillelmus de Ockham

William of Sherwooid, *vide s.n.* Guillelmus de Shireswood

Secondary Literature

Ashworth, E.J. (1980), "Can I Speak More Clearly than I Understand?' A Problem of Religious Language in Henry of Ghent, Duns Scotus and Ockham', in: K. Koerner, H.-J. Niederehe and R.H. Robins, eds., *Studies in Medieval Linguistic Thought*, dedicated to G.L. Bursill-Hall (Amsterdam 1980, 29-38) (also in *Historiographia Linguistica* 7/1-2 (1980), 29-38)

Bos, E.P. (1987), 'The Theory of the Proposition According to John Duns Scotus' Two Commentaries on Aristotle's *Perihermeneias*', in: L.M. de Rijk and H.A.G. Braakhuis, eds., *Logos and Pragma: Essays on the Philosophy of Language in Honour of Prof. Gabriel Nuchelmans* (Nijmegen 1987, 121-139)

Boulnois, O. (1995), 'Représentation et noms divins selon Duns Scot', in: *Documenti e studi sulla tradizione filosofica medievale* 6 (1995), 255-280

Braakhuis, H.A.G. (1977), 'The Views of William of Sherwood on Some Semantical Topics and Their Relation to Those of Roger Bacon', in: *Vivarium* 15 (1977), 111-142

——. (1979), *De 13de eeuwse tractaten over syncategorematische termen: Inleidende studie en uitgave van Nicolaas van Parijs' Sincategoreumata*. 2 vols. (Dissertation Leiden, Nijmegen 1979)

Dahlstrom, D.O. (1980), 'Signification and Logic: Scotus on Universals from a Logical Point of View', in: *Vivarium* 18 (1980), 81-111

Ebbesen, S. (1970), 'Roger Bacon and the Fools of his Times', in: *Cahiers de l'institut du moyen âge grec et latin* 3 (1970), 40-44

——. (1981), 'Early Supposition Theory (12th-13th Cent.)', in: *Histoire Epistémologie Langage* 3/1 (1981), 35-48

——. (1985), 'OXYNAT: A Theory about the Origins of British Logic', in: P.O. Lewry, *The Rise of British Logic*. Acts of the Sixth European Symposium on Medieval Logic and Semantics, Balliol College, Oxford, 19-24 June 1983 (Toronto 1985, 1-17)

——. (1988), 'Concrete Accidental Terms: Late-Thirteenth-Century Debates about Problems Relating to such terms as 'album'', in: N. Kretzmann, ed., *Meaning and Inference in Medieval Philosophy. Studies in Memory of Jan Pinborg* (Dordrecht-Boston-London 1988, 107-161)

——. (forthcoming), '*Modi significandi* in Logic and Grammar', in: B. Löwe, C. Kann, C. Rode and S. Uckelman, eds., *Modern Views of Medieval Logic* (Recherches de théologie et philosophie médiévales—Bibliotheca) (forthcoming)

Ebbesen, S. and I. Rosier-Catach (2000), 'Robertus Anglicus on Peter of Spain', in: I. Angelelli and P. Pérez-Ilzarbe, eds., *Medieval and Renaissance Logic in Spain*. Acts of the 12th European Symposium on Medieval Logic and Semantics, held at the University of Navarre, Pamplona, 26-30 May 1997 (Philosophische Texte und Studien 54; Hildesheim-Zürich-New York 2000, 61-95)

Goubier, F. (2000), 'Influences prédicatives et conséquences référentielles: un aspect de l'approche terministe de la première moitié du XIIIe siècle', in: *Cahiers de l'institut du moyen âge grec et latin* 71 (2000), 37-70

Libera, A. de (1980), 'On Some 12th and 13th Century Doctrines of Restriction', in: K. Koerner, H.-J. Niederehe and R.H. Robins, eds., *Studies in Medieval Linguistic Thought, Dedicated to G.L. Bursill-Hall* (Amsterdam 1980, 131-143) (also in *Historiographia Linguistica* 7/1-2 (1980), 131-143)

——. (1981), 'Supposition naturelle et appellation: aspects de la sémantique parisienne au XIIIe siècle', in: *Histoire Epistémologie Langage* 3/1 (1981), 63-77

——. (1982), 'The Oxford and Paris Traditions in Logic', in: N. Kretzmann, A. Kenny and J. Pinborg, eds., E. Stump, ass. ed., *The Cambridge History of Later Medieval Philosophy*. From the Rediscovery of Aristotle to the Desintegration of Scholasticism 1100-1600 (Cambridge-London-New York-New Rochelle-Melbourne-Sydney 1982, 174-187)

——. (1986), 'Les *Summulae dialectices* de Roger Bacon. I-II. De termino. De enuntiatione', in: *Archives d' histoire doctrinale et littéraire du moyen-âge* 53 (1986), 139-289

——. (1997), 'Roger Bacon et la logique', in: J. Hackett, ed., *Roger Bacon and the Sciences*. Commemorative Essays (Leiden 1997, 103-132)

Marmo, C. (1989), 'Ontology and Semantics in the Logic of Duns Scotus', in: U. Eco and C. Marmo, eds., *On the Medieval Theory of Signs* (Foundations of Semiotics, 21; Amsterdam-Philadelphia 1989, 143-193)

——. (1994), *Semiotica e linguaggio nella scolastica. Pariggi, Bologna, Erfurt 1270-133): la semiotica dei Modisti* (Nuovi Studi Storici, 26; Roma 1994)

——. (1999), 'The Semantics of the Modistae', in: S. Ebbesen and R. Friedman, eds., *Medieval Analyses in Language and Cognition*. Acts of the Symposium *The Copenhagen School of Medieval Philosophy*, January 10-13, 1996 (Copenhagen 1999, 83-104)

——. (2006), 'La funzione del contesto: teorie 'continentali' e 'inglesi' a confronto sull'eliminazione dell'equivocità tra fine XIII e inizio XIV secolo', in: S. Caroti, R. Imbach, Z. Kaluza, G. Stabile and L. Sturlese, eds., *'Ad Ingenii Acuitionem'. Studies in Honor of Alfonso Maierù* (Louvain-La-Neuve 2006, 249-280)

——. (2010), 'Scotus' Commentary on Aristotle's *Topics*', in: *Lo scotismo nel Mezzogiorno d'Italia. Atti del Congresso Internazionale (Bitonto 25-28 marzo2008), in occasione del VII Centenario della morte di Giovanni Duns Scoto*, edited by Francesco Fiorentino (Turnhout 2010, 153-170)

Perler, D. (1993), 'Duns Scotus on Signification', in: *Medieval Philosophy and Theology* 3 (1993), 97-120

——. (2003), 'Philosophy of Language', in: Th. Williams, ed., *The Cambridge Companion to Duns Scotus* (Cambridge-New York 2003, 161-192)

Pinborg, J. (1971), 'Bezeichnung in der Logik des XIII Jahrhunderts', in: A. Zimmermann, ed., *Der Begriff der repraesentatio im Mittelalter, Stellvertretung—Symbol—Zeichen—Bild* (Miscellanea Mediaevalia, 8; Berlin-New York 1971, 237-281)

——. (1976), 'Some Problems of Semantic Representation in Medieval Logic', in: H. Parret, ed., *History of Linguistic Thought and Contemporary Linguistics* (Berlin-New York 1976, 254-278)

Pini, G. (2001), 'Signification of Names in Duns Scotus and Some of His Contemporaries', in: *Vivarium* 39 (2001), 20-51

——. (2002), *Categories and Logic in Duns Scotus. An Interpretation of Aristotle's Categories in the Late Thirteenth Century* (Leiden 2002).

——. (2004), 'Il dibattito sulle specie intelligibili alla fine del tredicesmo secolo', in: *Medioevo* XXIX (2004), 267-306

Rijk, L.M. de (1962-1967), *Logica Modernorum. A Contribution to the History of Early Terminist Logic*. I. *On the Twelfth Century Theories of Fallacy*; II/1: *The Origin and the Early Development of the Theory of Supposition* II/2: *Texts and Indices* (Assen 1962-1967)

——. (1968), 'On the Genuine Text of Peter of Spain's *Summule Logicales*. I. General Problems Concerning Possible Interpolations in the Manuscripts', in: *Vivarium* 6 (1968), 1-34

——. (1971), 'The Development of *suppositio naturalis* in Medieval Logic. I. Natural Supposition as Non-contextual Supposition', in: *Vivarium* 9 (1971), 71-107

——. (1972), *vide Peter of Spain. Tractatus*, ed. L.M. de Rijk (Assen 1972)

——. (1982), 'The Origins of the Theory of the Properties of Terms', in: A. Kenny, N. Kretzmann, I. Pinborg, eds., E. Stump, ass. ed., *The Cambridge History of Later Medieval Philosophy. From the Rediscovery of Aristotle to the Desintegration of Scholasticism 1100-1600* (Cambridge-London-New York-New Rochelle-Melbourne-Sydney 1982, 161-173)

Rosier-Catach, I. (1994), *La parole comme acte. Sur la grammaire et la sémantique au XIIIe siècle* (Paris 1994)

——. (2004), *La parole efficace. Signe, rituel, sacré* (Paris 2004).

Spade, P.V. (1982), 'The Semantics of Terms', in: N. Kretzmann, A. Kenny and J. Pinborg, eds., E. Stump, ass. ed., *The Cambridge History of Later Medieval Philosophy. From the Rediscovery of Aristotle to the Desintegration of Scholasticism 1100-1600* (Cambridge-London-New York-New Rochelle-Melbourne-Sydney 1982, 188-196)

Vos, A. (1985), 'On the Philosophy of the Young Duns Scotus: Some Semantical and Logical Aspects', in: E.P. Bos, ed., *Medieval Semantics and Metaphysics* (Artistarium, Supplementa, II; Nijmegen 1985, 195-220)

Supposition and Predication in Medieval Trinitarian Logic

Simo Knuuttila
University of Helsinki

Abstract

Many fourteenth-century logicians took affirmative propositions to maintain that the subject term and the predicate term stand or supposit for the same. This is called the identity theory of predication by historians and *praedicatio identica* (or one form of *praedicatio identica*) by Paul of Venice and others. The identity theory of predication was an important part of early fourteenth-century Trinitarian discussions as well, but what was called *praedicatio identica* by Duns Scotus and his followers in this context was something different. After some remarks on Scotus's view and its background, I shall analyse Adam Wodeham's explanation of Scotus's *praedicatio identica* and how he understood the assumptions pertaining to supposition in the Scotist approach. I also describe Wodeham's own solution to Trinitarian sophisms, which did not deviate from the identity theory of predication.

Keywords

dici de omni et nullo, identity, identity theory of predication, *praedicatio identica*, Trinity, supposition

Many fourteenth-century philosophers and theologians discussed the question of the universal validity of logic. The practical background of this interest was the concern over whether various theological doctrines might be incompatible with the principles of logic, as is shown by numerous treatises on the logical problems of Trinitarian formulations. There were often-repeated Trinitarian arguments called paralogisms or sophisms in which doctrinally true syllogistic premises seemingly implied a refutable conclusion. The most-discussed examples were formulated by Roger Roseth as follows:

> Et primo videtur quod nullus syllogismus expositorius tenet in divinis. Nam iste syllogismus non tenet: 'Haec essentia divina est Pater, haec essentia divina est Filius; ergo Filius est

> Pater', quia conclusio est falsa et praemissae sunt verae. [...] Praeterea, in primo modo non valet syllogismus, nam iste syllogismus non valet: 'Omnis essentia divina est Pater, omnis Filius est essentia divina, ergo omnis Filius est Pater.'[1]

These syllogisms were found particularly interesting because they seemingly denied the validity of the expository syllogism and the *dici de omni et nullo*, which many logicians regarded as the constitutive principles of syllogistic theory. In dealing with these sophisms, medieval thinkers developed some new ideas on how singular or universal premises should be read.[2]

These reformulations became very influential, with even Leibniz employing them in his logic.[3] My aim is to make some remarks on supposition and predication in this context and to shed light on the somewhat complicated history of the notion of *praedicatio identica*, which was also employed in these discussions.

1. Identity and Difference among the Oxford Realists

In his treatise *On universals* Paul of Venice puts forward his version of the realist interpretation of universals developed by John Wyclif and his followers in late fourteenth-century Oxford.[4] In arguing for the realist position, which implies, for example, that the common human being is really the same as individual human beings, Paul first explains the distinction between personal or suppositional and essential or simple supposition, the former being associated with terms standing for non-universal singular beings and the latter with terms standing for universal or common things. While some terms, such as proper names, only have personal supposition and some others essential supposition, the abstract terms 'humanitas' or 'animalitas', for example, there are terms which may have personal supposition or essential supposition; for example, 'homo' and 'animal' have a personal supposition in some propositions, essential supposition in others, and indifferently one or the other in some propositional contexts. If the terms of two syllogistic premises do not have the same sort of supposition, the conclusion should not be assumed to have only one or

1) Roger Roseth, *Lectura super Sententias*, 3-5, ed. Hallamaa (2005), 3.1, 67-68.

2) See, e.g., Maierù (1981); id. (1984); id. (1985); id. (1986); id. (1988). See also Gelber (1974); Hallamaa (2003), 84-119; Knuuttila (2007), 69-87.

3) See Knuuttila (2007), 71-72.

4) The *Quaestio de universalibus* is partially edited by Conti as an appendix to Johannes Sharpe, *Quaestio super universalia* (1990), 199-207.

the other. For example, the premises 'Human nature is Socrates' and 'Human nature is Plato', having terms which refer to common and singular things, do not imply that Plato is Socrates; the correct conclusion is: 'Plato is something that is Socrates'. The added term something (*aliquid*) stands indifferently for common nature and individual singular beings. The argument then goes on:

> Et per hoc potest solvi omnis paralogismus factus in divinis, dicendo quod non sequitur: 'Omnis divinitas est Pater, sed Filius est divinitas; ergo Filius est Pater', sed solum sequitur quod Filius est aliquid quod est Pater. Etiam non sequitur: 'Omnis Pater generat; sed divina essentia est Pater; ergo divina essentia generat', sed solum sequitur quod aliquid quod est divina essentia generat, ubi ly 'aliquid' indifferens est ad utramque suppositionem.[5]

I shall not comment on this summary of fourteenth-century solutions of Trinitarian paralogisms; it is sufficient to say that there were many who, like Paul of Venice, thought that they had a general logical theory which solved these problems. I shall return to this view when discussing Adam Wodeham, who was one of the proponents of the blessing of logic in theology.

After his remarks on supposition, Paul explains the distinction between identical predication (*praedicatio identica*) and formal predication (*praedicatio formalis*):

> Quarto notandum est istud: quod praedicationum ad hanc materiam pertinentium quaedam dicitur identica et quaedam formalis. Praedicatio identica est illa per quam denotatur idem esse realiter significatum subiecti et significatum praedicati, ut 'Homo est animal'; et communiter talem faciunt praedicationem termini primae intentionis de se invicem absolute praedicantes. [...] Praedicatio autem formalis est illa per quam denotatur subiectum et praedicatum invicem convenire secundum eandem rationem formalem; et talem praedicationem faciunt termini primae intentionis cum addito [...] qui sunt 'formaliter', 'per se', 'inquantum', 'ut sic' et ita de aliis, ut 'Homo per se est animal'.[6]

Following the identity theory of predication, which was formulated by William Ockham, John Buridan and others, Paul regards the basic affirmative proposition as an assertion in which things signified by the subject term are maintained to be the same as those (or some of those) signified by the predicate term. The sameness of identical predication is the sameness of supposita as such, not sameness in the sense of sharing something in common as in formal predication:

---

5) Paul of Venice, *De universalibus*, ed. Conti (1990), 201.

6) Paul of Venice, *De universalibus*, ed. Conti (1990), 201.

> Sed hic est advertendum quod, licet illa sit praedicatio identica 'Homo est asinus', non tamen aequivalet isti 'Idem est homo et asinus', quia ista est vera et altera falsa. Unde in prima subiectum non supponit pro aliquo communi sibi et praedicato, nec etiam praedicatum pro aliquo communi sibi et subiecto, [...] sed in secunda ly 'idem' ratione sue communitatis supponit pro aliquo communi homini et asino, [...] numquam tamen concedo quod Socrates est identice asinus, nec homo est realiter asinus, quia illi termini 'identice' et 'realiter' non supponunt nec faciunt subiectum et praedicatum supponere pro aliquo eis communi.[7]

This terminology was summarized by Augustine of Ferrara, an early fifteenth-century Italian Franciscan, in his questions commentary on Aristotle's *Categories*.[8] According to Augustine, identical predication in a broad sense is a predication in which the subject and the predicate 'have some identity between them, formal or real and so on'. For example, there can be an identical predication between a human being and an animal, because these are formally the same. The second use is more limited, referring to a predication in which the subject and the predicate are in all respects the same, such as 'This human being is a human being'. In the third use, which is said to be the most proper, the subject and the predicate 'are not absolutely the same in reality but are somehow the same through a relative identity, namely with respect to a third'. An example of this is 'A white thing is sweet' because of the identity these have with respect to the subject, namely milk.

According to Augustine, any predication can be called an identical predication in the broad sense of the term, i.e., all predications involve some sort of identity. Insofar as the most proper use of the term means that the subject and the predicate are said of the same suppositum, this third use could be taken to have the same extension as the broad use, as in Paul of Venice, but since it is regarded as a special case, the proper sense of the *praedicatio identica* pertains to the predications which do not represent formal or simple identity. It is

[7] Paul of Venice, *De universalibus*, ed. Conti (1990), 203. Paraphrasis: 'A human being is a donkey' is not equivalent to 'The same (*idem*) is a human being and a donkey' nor is 'Socrates is Plato' equivalent to 'The same is Socrates and Plato'. The former propositions, which are false, involve an identical predication, and the latter ones, which are true, involve a formal predication, for 'idem' in these stands for something which is common to many things. The sameness of identical predication can be explicated using the adverbial terms '*identice*' and '*realiter*', which do not refer to anything shared by the subjects in the way 'idem' could be taken to do. These signs of 'real identity' can be thought to be added to propositions involving identical predication.

[8] Augustine of Ferrara, *Quaestiones super librum Praedicamentorum Aristotelis*, ed. Andrews (2000), 44-45.

merely the case of identical predication; in fact it is called the *praedicatio identitate identica*.[9]

I mentioned the fourteenth-century logical theory that an affirmative proposition maintains that the subject term and the predicate term stand or supposit for the same subject. This is called the identity theory of predication by historians, as is the similar theory which Peter Abelard employed in discussing Trinitarian formulations in his logical treatises—in fact there are also some indirect historical links here.[10] It is of some interest that what Paul of Venice calls *praedicatio identica* is just the fourteenth-century identity view of predication. This theory of predication was an important part of early fourteenth-century Trinitarian discussions as well, but what was called *praedicatio identica* by Scotus and his followers was something different. After Paul of Venice and Augustine of Ferrara, this term continued to have various uses.

2. From Abelard to Scotus

Abelard calls the extensional numerical identity expressed by 'An A is B' an *idem quod* identity or *identitas essentiae*, as distinct from the intensional identity expressed by synonymous terms, which he calls *identitas proprietatis*. In the first case, the terms name things which are numerically identical particular beings—this is the kernel of the Abelardian identity theory of predication. Taking a waxen image as an example, Abelard says that the lump of wax and the waxen image are essentially the same, even though they have different properties, as is clear from the fact that the wax, as distinct from the image, is not created from wax.[11]

Abelard's most detailed discussions of the identity view of predication are found in his theological treatises. According to him, many Trinitarian formulations are meant to express the identity of essence:

> Ad identitatem quidem essentiae tantum, non etiam proprietatis, hae spectant: Pater est Deus, Filius est Deus, Spiritus est Deus. [...] Hae uero enuntiationes: Pater est Filius uel Spiritus Sanctus, uel Pater est genitus siue procedens, ad identitatem proprietatis aspiciunt, ideoque omnino falsae sunt. Ibi quippe solummodo ostendebatur idem esse rem praedicati et subiecti termini, ueluti idem esse Deum quod est Pater, etc. Hic vero rem subiecit termini

9) For the notion of identical identity in Petrus Thomae, see Maierù (1988), 253.

10) For identity in predication, see Peter Abelard, *Logica 'Ingredientibus'*, ed. Geyer (1919-27), 60.13; William of Ockham, *Summa logicae*, ed. Boehner *et al.* (1974), II, chs 2-4, pp. 249-266; John Buridan, *Tractatus de consequentiis*, ed. Hubien (1976), I. 5, pp. 25-26.

11) Knuuttila (2007), 74-79.

> proprietatem quoque praedicati termini sic habere, ut hoc sit huius personae proprium quod illius, vel eadem insit proprietas huic personae quae illi [...].[12]

This analysis is taken to resolve most logical queries associated with the Trinity, one of these being the question of whether the propositions 'The Father is God' and 'God is the Son' syllogistically imply that 'The Father is the Son'. Abelard remarks that if the premises are read according to essential identity, the conclusion should be understood in the same way. This is not problematic, since all these propositions are true in this sense. If the conclusion is wrongly understood in the sense of intensional identity, it does not follow from the premises, which are true in the sense of essential identity.

After Abelard, many theologians distinguished between various kinds of identity in the doctrines of the Trinity and the Hypostatic union. A distinction often quoted in this context was Bonaventure's division between predication *per inhaerentiam* and *per identitatem*. According to him, if the terms of a true Trinitarian proposition are nouns, they form a predication *per identitatem*. The meaning of these terms includes a reference to the subject to which the signified form belongs, and the identity expressed by the predication pertains to the sameness of these subjects. Adjectives and verbs said of a subject constitute a predication *per inhaerentiam*, which asserts the inherence of what is predicated.[13]

John Duns Scotus commented on a syllogistic argument similar to that in Abelard in his longer explanation of why Trinitarian propositions do not violate the principle of logic that 'things that are the same as one and the same thing are also the same as each other.' Like Abelard, he also argues that the argument is valid or not, depending on how identity is understood in the premises and the conclusion. Scotus distinguishes between two kinds of identity as follows:

> Quaecumque uni et eidem sunt simpliciter eadem, inter se sunt simpliciter eadem, [...] sed quaecumque sunt in natura divina, sunt simpliciter eadem eidem simpliciter, quia naturae divinae. [...] Dico quod si accipitur medium secundum identitatem essentialem extremis, sequitur quod extrema habeant unam essentiam eis communicatam, et ideo sequitur quod Pater sit idem quod Filius, non tamen sequitur quod Pater sit Filius, quia tunc concluderetur identitas formalis vel suppositiva. [...] Ex hic patet ad illud sophisma 'hic Deus est Pater,

---

12) Peter Abelard, *Theologia christiana*, ed. Buytaert (1969), IV, 52-53, p. 288.

13) Bonaventura, *Commentaria in quatuor libros Sententiarum*, ed. Collegium S. Bonaventurae (1882), I, dist. 5, a.1, q.1, ad 2-3.

> hic Deus est Filius, igitur Filius est Pater': medio existente 'hoc aliquid', necesse est extrema coniungi.[14]

When two divine persons, the Father and the Son, are said to be the same, this can be understood in the sense of essential identity (*identitas essentialis*)—the medium is one essence which, as numerically one thing, is communicated to two subjects, and consequently the correct conclusion is 'The Father is the same as the Son', namely, the single essence with which both are numerically the same. The premises and the conclusion are true when understood in this way. The false conclusion, 'The Father is the Son', which is said to express formal identity (*identitas formalis*) or suppositional identity (*identitas suppositiva*), does not follow from the premises expressing essential identity; it would follow if the premises expressed formal identity, but they would then be false. 'The essence is the Father' and its converse are true, but there is a formal non-identity between the Father and the essence.

These remarks are in agreement with Scotus' general division between two types of predication in Trinitarian theology: identical predication (*praedicatio identica*), which is based on essential identity, and formal predication, which is based on formal identity. Even though he often refers to this distinction, there is no detailed explanation of it. In some places formal predication is characterized as a *per se* predication, all other predications being apparently identical.[15]

Scotus's use of the notion of essential identity in the quotation above is not the same as that of Abelard, who refers to the sameness of the subject of which the terms are expressed, the ultimate subject being one and the same in all Trinitarian propositions of the *idem quod* type. While this was regarded as true in medieval theology, it was an abstract idea and not helpful in analysing the inner structure of the Trinitarian entity which Scotus tries to explain, which is why he suggests that the divine essence as the basis of the unity is treated like a common predicate rather than as a common subject. Scotus explains this by referring to a distinction between 'hoc aliquid' and 'quale quid':

> Sicut in creaturis commune se habet ut 'quale quid', singulare ut 'hoc aliquid', ita hic, essentia communis personis habet rationem 'qualis quid', et persona habet rationem 'huius

---

[14] John Duns Scotus, *Lectura* I, ed. Commissio Scotistica, vol. xvi (1960), I, 2.2.1-4, nn. 136, pp. 280 and 281.

[15] John Duns Scotus, *Ordinatio* I, ed. Commissio Scotistica, vol. iv (1956), I.5.1, q. unica, nn. 32-33; and I.8.1, q. 4, nn. 218-222, pp. 282-289 and pp. 274-277 respectively.

> alicuius'. Medium igitur hic est 'quale quid' et non 'hic aliquis'. Concluditur autem identitas extremorum in conclusione ac si medium esset 'hoc aliquid'.[16]

According to Scotus, the essence is a communicable individual with three incommunicable *supposita*. Insofar as the notion of essence behaves like the terms which refer to common natures or qualities, the persons can be said to have the same nature without a numerical identity between them. While this is one part of essential identity, the divine essence differs from finite common natures because it is an individual entity which is not numerically different from the persons. Thus while the persons are numerically identical with one divine essence—otherwise there would be a quaternity instead of a trinity—the persons themselves are numerically different. Scotus thinks that the mistaken syllogism is formed without realizing that the essence is a metaphysically simple individual and also communicable like created common natures, albeit different from these by existing in itself as an individual.[17] One might ask whether these distinctions of metaphysical theology qualified the logical form of predication which Scotus understood in accordance with the identity theory.[18] There is no such proposal; Scotus regarded it sufficient to state that theological *praedicatio identica* propositions, while in agreement with the principle of non-contradiction, were not syllogistic.

3. Identity and Predication in Wodeham

While Scotus's considerations did not directly contribute to the development of the logical analysis of predication which Abelard had introduced, it reinforced questions about the relationship between theological and logical analysis. This discussion was carried on by fourteenth-century logicians who combined the identity theory of predication with a suppositional view of distribution. In this connection the logical problems associated with Trinitarian syllogisms encouraged certain new formulations of the structure of the proposition. According to William Ockham, the Trinitarian Barbara paralogism is solved by the correct reading of the universal proposition which runs as follows: 'Of whatever this subject A is said of the same B is said'. Referring to theological examples, Buridan also states that syllogistic premises should be formulated with the circumlocution *qui est* or *quod est* of the subject terms, because this is how they

---

16) John Duns Scotus, *Ordinatio* I, ed. Commissio Scotistica, vol. ii (1950), I, 2.2.1-4, n. 415, p. 363.

17) See also Cross (2003).

18) For Scotus's view of predication, see Pini (2004).

are in accordance with the *dictum de omni et nullo*.[19] Adam Wodeham argued that the circumlocutionary reading was what Aristotle in fact meant in order to avoid problems associated with Plato's common natures.

> Propter quod dico tertio [...] quod Aristoteles non semper fecit talia argumenta in terminis generalibus. [...] In multis passibus philosophiae habet aliam modum loquendi, puta istum vel alium similem vel equivalentem: omne in quo b ei in quo a idem, omne in quo c ei in quo b idem, igitur omne in quo c ei in quo a idem, quod est dictu: omne idem ei quod est b est idem ei quod est a, sed omne idem ei quod est c est idem ei quod est b, igitur omne idem ei quod est c est idem ei quod est a. [...] Et ulterius secundum istum modum respondendi esset dicendum quod ubicumque termini simplices equivalent terminis talis circumlocutionis valeret illatio ex forma in terminis simplicibus sine circumlocutione et ubi non non, hoc semper memoriter habito pro regula quod si per maiorem denotetur quod cuicumque conveniat subiectum et predicatum et bene sumatur sub erit syllogismus bonus et regulatus per dici de omni.[20]

Wodeham thought that the 'Aristotelian' circumlocution was meant to keep logic universally valid with respect to the Platonic ontology of universals, according to which Socrates and Plato were numerically the same as the common human nature but otherwise separate individuals. This was also the case with the divine essence and the persons—the persons were numerically identical with the essence but really distinct from each other. The circumlocutional explication demonstrated that the propositions whose subject terms referred to common natures were often false if read as universal circumlocuted propositions and could not be used in constructive syllogistic arguments.[21]

When Wodeham explains Scotus's formulation quoted above, he equates Scotus's distinction between identical and formal predication with Bonaventure's distinction between predication through identity and inherence.[22]

---

19) See Ockham, *Summa logicae*, ed. Boehner *et al.* (1974), III/1, 4, pp. 370-371; John Buridan, *Tractatus de consequentiis*, ed. Hubien (1976), III.1.4.1, p. 86.

20) Adam Wodeham, *Lectura super Sententias* (Oxford), I.33.3.2, ms Vaticano, *Biblioteca Apostolica Vaticana*, Vat. lat. 955, ff. 1r-208v (book I); see also Maierù (1981), 487-489; most of the Wodeham quotations are also found in the not too reliable *Super quattuor libros Sententiarum*, abbreviated by Henry Totting of Oyta (1512), I, 33.

21) For the Platonic example in Adam and his followers, see Maierù (1981), 486-487; Shank (1988), 80-81, 90-94; for the non-transitivity of numerical identity in medieval Oxford realism, see Spade (2005).

22) Adam Wodeham, *Lectura super Sententias* I.33.1.2, in: ms. Vaticano, *Biblioteca Apostolica Vaticana*, Vat. lat. 955, f. 177v: 'Aliter respondet Scotus, d. 2 primi, q. 4, in solutione primi argumenti principalis, quod maior est vera, sed conclusio quae concludit Patrem esse Filium concludit aliud genus ydemptitatis quam sit illud genus ydemptitatis quo Pater et Filius sunt idem in essentia,

He then adds a separate paragraph for explaining the notion of identical predication, employing the suppositional analysis of logic rather than the metaphysical considerations of Scotus:

> Predicatio ydemptica vocatur illa ubi illud pro quo subiectum supponit est illud pro quo predicatum supponit et tamen predicatum non supponit pro eo pro quo subiectum nec denotatur pro eo supponere, sicut hic: Essentia est generans, sumpto predicato substantive eo modo quo propositio est vera. Predicatum enim ibi precise supponit pro Patre, et quia essentia est Pater, pro quo predicatum supponit, ideo est propositio ista vera. Predicatio formaliter sive per inherentiam vocatur quando predicatum et subiectum denotantur supponere pro eodem ut hic: Pater est generans. Si dicas: tanta potest secundum istud inferri ydemptitas extremorum vel talis inter se quanta vel qualis in medio, sed quelibet personarum est eadem realiter deitati, ergo una est realiter eadem alteri, concedit iste doctor et bene quod potest inferri ydemptitas essentialis, non formalis, id est denominativa sive suppositiva et ideo non debet inferri quod Filius est Pater, quia ibi denotatur ex vi sermonis ydemptitas hypostatica, sed sic debet inferri: Filius est idem Patri vel cum eo quod est Pater vel Filius est illud quod est Pater.[23]

Wodeham argues that, in identical predication, that for which the subject supposits is that for which the predicate supposits, but the predicate does not supposit for that for which the subject supposits and is not denoted to supposit for this, as in 'The essence is generating', taking the predicate substantively in the way in which the proposition is true. In a formal predication the supposita of the subject and the predicate are not merely the same, but the predicate supposits for and is also denoted as suppositing for the suppositum of the subject.

---

et ideo conclusio non sequitur ex premissis. Pro quo notandum quod secundum eum et doctores antiquos ut Bonaventura, d. 5 primi, q. 1, in divinis duplex est modus predicandi, scilicet per ydemptitatem et per inherentiam, per ydemptitatem quidem ut cum dicitur: Essentia est Pater, per inherentiam sive denominationem sicut faciunt adiectiva et verba. Et hanc distinctionem sepe exprimit Scotus, dicens [...] quod in divinis est duplex predicatio, scilicet ydemptica et formalis, id est denominativa, et ista distinctio necessaria est in multis propositionibus circa divina. Nam ista: Essentia divina est generans, vera est si sit predicatio ydemptica ad hunc sensum: essentia divina est ista res que generat, scilicet Pater; sed si sit predicatio per inherentiam, scilicet denominativa seu formalis vel adiectiva, falsa est quia tunc denotatur quod essentia generat. Et tunc ultra secundum eum premisse verificantur secundum ydemptitatem essentialem quae exprimitur per praedicationem ydempticam, sed conclusio notat predicationem formalem, id est denominativam et ydemptitatem personalem. [...] Ideo conclusio falsa est nec sequitur ex premissis quia significat maiorem ydemptitatem extremorum inter se quam significabatur in premissis. Et ista responsio sic exposita est rationalis apud me.'

23) Adam Wodeham, *Lectura super Sententias* (Oxford) I.33.1.2, in: ms Vaticano, *Biblioteca Apostolica Vaticana*, Vat. lat. 955, ff. 177v-178r.

Wodeham assumes that a singular affirmative proposition maintains that the supposita of the terms are numerically one, which is the truth-condition of such a proposition. This was the idea of the identity theory of predication in general, which Wodeham takes as his starting-point. In a true singular formal predication, the proposition is meant to be understood so that the terms stand for the same, as is clear when the predicate is a verb or an adjective. A true singular identical predication is not meant to be understood so that the predicate stands for what the subject stands for, even though these references are not numerically different. Through this somewhat complicated formulation Wodeham specifies the Trinitarian *praedicatio identica* as a predication which is exemplified by all those true propositions which are not Trinitarian formal predications.

This account seems to be the same as Augustine of Ferrara's third use of the *praedicatio identica,* according to which the references of the terms which have independent meanings are the same. However, Wodeham restricts his explication to those cases in which the true singular affirmation is associated with a division between the supposition of the subject and the predicate. This has the consequence that the numerically single reference can be treated as two and the numerical sameness does not have the property of transitivity. From the point of view of the identity theory of predication, it was natural to think that it was precisely the transitivity of identity which was formulated by the constitutive principles of syllogistic logic, the *dici de omni* and the expository syllogism. Now, because of the peculiarities of numeral identity in the Trinity, many theologically true singular propositions ceased to be true if they were circumlocuted into the *dici de omni* forms like 'Everything that is the Essence is the Father', instead of 'The Essence is the Father' or 'Every Essence is the Father'. Formulating propositions in this way could be used in refuting paralogisms in which heretical consequences were derived from Catholic doctrines.

This idea, found in many early fourteenth-century authors, was particularly developed by Wodeham. In discussing the Trinitarian Barbara paralogism, he argues that only those readings of the first premise are syllogistic which are regulated by the *dici de omni et nullo.* These have the circumlocutional logical structure which, as stated in the text quoted above, is spelled out as *Omne in quo B ei in quo A idem* or *Omne idem ei quod est B est idem ei quod est A*. These explications are characterized as the forms which Aristotle had in mind, even though he did not explicitly use them; it is also considered possible that they are not found in the Latin text because of the poor quality of the translations.[24] Wodeham thought that the allegedly problematic Trinitarian syllogisms were mostly solved by this treatment of a universal proposition, with some necessary further qualifications discussed below. It was also applied to expository syllogisms which were reduced to the *dici de omni et nullo* by transposing the singular premises to universal propositions.[25] In this approach, the major premise of seemingly problematic Trinitarian syllogisms was false. If theological propositions were treated as true, they were not 'sufficiently universal' from the syllogistic point of view and the syllogism was not sufficiently regulated by the *dici de omni et nullo*.[26]

While this was the approach Wodeham dealt with most extensively in his discussion of Trinitarian syllogisms, he also sketched another analysis, 'more subtle and perhaps more true'. The central idea was to treat the universalisation of the first Barbara premise as restricted to the supposita of the subject term and to check whether the subject of the minor premise was correctly subsumed.

> Secundus modus subtilior et forte verior solvendi est concedere omnes istas maiores que videntur esse universales esse sufficienter universales et de omni ad propositum et tamen illi paralogismi non regulantur per dici de omni quia non fiunt sub in minore pro quo fiebat distributio in maiore nec pro quo supponebat subiectum maioris. Si ergo dictus modus non placeat, subtilius et forte verius diceretur ad paralogismos ex universalibus quod non regulantur per dici de omni vel de nullo, quia illa universalis est de omni in qua subiectum pro nullo supponit vel denotatur ex vi sermonis supponere quin sibi conveniat predicatum, sed non oportet quod conveniat omni illi quod est res significata per subiectum, si universalis sit, nec oportet quod hoc denotet. Et tunc secundum hoc ista 'Omnis deitas est Pater' est universalis et de omni nec habet nisi unam singularem et veram, hanc scilicet, 'Hec deitas communis etc.', et tunc minor que dicit quod omnis Filius in divinis est deitas non sumitur sub quia licet Filius sit deitas ista, tamen pro Filio non supponebat subiectum maioris nec pro Patre nec pro alia persona [...] subiectum istius maioris 'Omne illud quod est Pater [est Pater]', licet Filius sit illud quod est Pater, tamen pro Filio non supponit. Maior enim vera est et per consequens non denotatur ibi predicatum convenire alicui nisi cui vere convenit, cuius<modi> non est Filius, nec supponit subiectum istius pro Filio licet vere conveniat Filio, quia nec pro Filio distribuitur, et tamen ibi distribuitur pro quolibet pro quo supponit,

---

24) Adam Wodeham, *Lectura super Sententias* (Oxford) I.33.3.2, in: ms Vaticano, *Biblioteca Apostolica Vaticana*, lat. 955, f. 188r.

25) Adam Wodeham, *Lectura super Sententias* (Oxford) I.33.3.1, in: ms Vaticano, *Biblioteca Apostolica Vaticana*, lat. 955, f. 186r.

26) Adam Wodeham, *Lectura super Sententias* (Oxford) I.33.3.2, in: ms Vaticano, *Biblioteca Apostolica Vaticana*, lat. 955, f. 187v.

> cuius<modi> sunt deitas et spiratio activa et omne in divinis commune tribus personis, quia omne tale in divinis est Pater.[27]

The universality of the first premise is now interpreted to mean that the subject does not supposit for anything of which the predicate is not said, but it is not required that the predicate is said of everything which is signified by the subject. Wodeham assumes that since the subject term may be truly said of things for which it does not supposit, these being numerically the same but supposited for by another term, as in identical predication, it may be reasonable to think that the predicate is meant to be said merely of things for which the subject supposits. For example, while the term 'that which is the Father' supposits for something which is the same as that for which 'the Essence' supposits, it does not supposit for the Son which is the Essence. Why would this be more subtle and true? Perhaps the reason is that it is not necessary to solve the paralogisms by repeating that revelatory propositions are false if read in accordance with the *dici de omni*. Instead of this, one can take the first premise in the form in which it is true and then argue that the minor term is mistakenly subsumed because its subject is not supposited for by the major term.

Even though Wodeham discusses some further aspects of this approach, he concentrates on improving the first one. In distinguishing between these, he remarks that the circumlocution of the universal premise 'Every A is B' as 'Everything which is A' is ambiguous because it can be understood as being restricted to things supposited for by A or as including all things which are the same as those which are A.[28] In order to make this explicit, the latter alternative should be formulated as follows: 'Everything which is that which is A' or 'Everything which is the same as that which is A'. He regards this reading as the basic one in the sense that it explicates the universality which regulates all genuine syllogisms, particularly if the more extensive phrase is associated with the subject term as well as the predicate term. All syllogistic Trinitarian problems are solved when the premises are analysed in this way, i.e., reading 'A is B' as follows: 'Anything which is that which is A is the same as that which is that

---

27) Adam Wodeham, *Lectura super Sententias* (Oxford) I.33.3.2, in: ms Vaticano, *Biblioteca Apostolica Vaticana*, Vat. lat. 955, f. 187v.

28) Adam Wodeham, *Lectura super Sententias* (Oxford) I.33.3.2, in: ms Vaticano, *Biblioteca Apostolica Vaticana*, Vat. lat. 955, f. 188r-v: 'Ex hoc sequitur correlarie quod iste discursus non valet licet habeat subiectum circumlocutum: Omne illud quod generat generat; sed omnis essentia divina est illud quod generat; igitur omnis essentia divina generat, quia maior non est sufficienter universalis de omni cum ipsa sit vera.'

which is B'.[29] Wodeham thinks that the *dici de omni* principle requires that the predicate of the maior premise is said of all those things of which the subject is said. When these are referred to by demonstrative pronouns, it may happen that not all of them are among those which the subject term supposits for. While the scope of simple circumlocutional phrases may remain ambiguous in this respect, the more extensive formulations avoid the problem, the references in Trinitarian identical predications being the supposita of the terms and the things which are numerically identical with these.

Bibliography

Primary Literature

Adam Wodeham, *Lectura super Sententias* (Oxford), in: ms Vatican, *Biblioteca Apostolica Vaticana*, Vat. lat. 955, ff. 1r-208v (book I)

——, *Quaestiones in libros Sententiarum*. Editio media, ed. Henry Totting of Oyta, published by John Major (Paris 1512)

Augustine of Ferrara, *Quaestiones super librum Praedicamentorum Aristotelis*, ed. with an Introduction by R. Andrews. Acta Universitatis Stockholmiensis (Studia Latina Stockholmiensia, 45; Stockholm 2000)

Bonaventura, *Commentaria in quatuor libros Sententiarum*, ed. Collegium S. Bonaventurae (Bonaventurae Opera omnia, I; Quaracchi 1882-1902)

Guillelmus de Ockham, *Summa logicae*, ed. P. Boehner, G. Gál and S. Brown (Guillelmi de Ockham Opera Philosophica, I, St. Bonaventure, N.Y. 1974)

Johannes Buridanus, *Tractatus de consequentiis*, ed. H. Hubien (Philosophes médiévaux, 16; Louvain-Paris 1976)

Johannes Duns Scotus, *Ordinatio* I, dist. iv–x, ed. Commissio Scotistica, praeside C. Balić (Ioannis Duns Scoti Opera Omnia, IV; Civitas Vaticana 1956)

——, *Ordinatio* I, dist. xi-xxv, ed. Commissio Scotistica, praeside C. Balić (Ioannis Duns Scoti Opera Omnia, V; Civitas Vaticana 1959)

——, *Lectura* I, prologus et dist. i-vii, ed. Commissio Scotistica, praeside C. Balić (Ioannis Duns Scoti Opera Omnia, XVI; Civitas Vaticana 1960)

Paulus Venetus, *Quaestio de universalibus*, partially edited by A.D. Conti as an appendix to Johannes Sharpe, *Quaestio super universalia* (Corpus philosophorum medii aevi, testi e studi, 9; Florence 1990, 199-207)

Petrus Abaelardus, *'Logica Ingredientibus'*, part 1: Glossae super Porphyrium, in: *Peter Abaelards Philosophische Schriften. I. Die Logica 'Ingredientibus. 2: Die Glossen zu Porphyrius*, ed. B. Geyer (Beiträge zur Geschichte der Philosophie des Mittelalters, 21.1; Münster 1919)

[29] 'Sexta conclusio est quod circumloquendo tam subiectum maioris quam predicatum per hoc quod dico 'quod est illud quod est' numquam capit instantiam sed circumlocutio est nimis longa.' (Adam Wodeham, *Lectura super Sententias* (Oxford) I.33.1.3.3, in: ms Vaticano, *Biblioteca Apostolica Vaticana*, lat. 955, f. 192r).

——, *Theologia christiana*, ed. E.M. Buytaert (Corpus Christianorum, Continuatio Mediaevalis, 12; Turnhout 1969)

Roger Roseth, *Lectura super Sententias*, 3-5, ed. O. Hallamaa (Helsinki 2005)

Secondary Literature

Cross, R. (2003), 'Duns Scotus on Divine Substance and the Trinity', in: *Medieval Philosophy and Theology* 11 (2003), 181-201

Gelber, H. (1974), *Logic and the Trinity: A Clash of Values in Scholastic Thought 1300-1333* (Wisconsin, Madison 1974) (PhD Dissertation)

Hallamaa, O. (2003), 'Defending Common Rationality: Roger Roseth on Trinitarian Paralogisms', in: *Vivarium* 41 (2003), 84-119

Knuuttila, S. (2007), 'Trinitarian Fallacies, Identity and Predication', in: P. Kärkkäinen ed., *Trinitarian Theology in the Medieval West* (Schriften der Luther-Agricola-Gesellschaft 61; Helsinki 2007, 69-87)

Maierù, A. (1981), 'Logica aristotelica e teologia trinitaria: Enrico Totting da Oyta', in: A. Maierù and A. Paravicini Bagliani, eds., *Studi sul XIV secolo in memoria di Anneliese Maier* (Storia e letteratura 151; Rome 1981, 481-512)

——, (1984), 'Logique et théologie trinitaire: Pierre d'Ailly', in: Z. Kaluza and P. Vignaux, eds., *Preuve et raisons à l'Université de Paris. Logique, ontologie et théologie au XIVe siècle* (Paris 1984, 253-268)

——, (1985), 'A propos de la doctrine de la supposition en théologie trinitaire au XIVe siècle', in: E.P. Bos, ed., *Medieval Semantics and Metaphysics. Studies Dedicated to L.M. de Rijk on the Occasion of his 60th Birthday* (Nijmegen 1985, 221-238)

——, (1986), 'Logique et théologie trinitaire dans le moyen âge tardif: deux solutions en présence' in: M. Asztalos, ed., *The Editing of Theological and Philosophical Texts from the Middle Ages* (Acta Universitatis Stockholmiensis. Studia Latina Stockholmiensia 30; Stockholm 1986, 185-212)

——, (1988), 'Logic and Trinitarian Theology: *De Modo Predicandi ac Sylogizandi in Divinis*', in: N. Kretzmann, ed., *Meaning and Inference in Medieval Philosophy* (Synthese Historical Library 32; Dordrecht 1988, 247-295)

Pini, G. (2004), 'Scotus on Assertion and the Copula: A Comparison with Aquinas', in: A. Maierù and L. Valente, eds., *Medieval Theories on Assertive and Non-Assertive Language*. Acts of the 14th European Symposium on Medieval Logic and Semantics, Rome, June 11-15, 2002 (Florence 2004, 307-331)

Shank, M.H. (1988), *'Unless you believe, you shall not understand'. Logic, University and Society in Late Medieval Vienna* (Princeton 1988)

Spade, P.V. (2005), 'The Problem of Universals and Wyclif's Alleged 'Ultrarealism'', in: *Vivarium* 43 (2005), 111-132

Richard Brinkley on Supposition

Laurent Cesalli
CNRS (UMR 8163) / University of Lille 3

Abstract

This study comments on six *notabilia* found in the general observations (*praemittenda*) with which Brinkley begins his treatise on supposition in his *Summa logicae: i)* the logico-metaphysical explanation of the distinction between *significatio* and *suppositio, ii)* the ontic division principle of supposition, *iii)* the relationship between *supposita* and truth-makers, *iv)* what seems to be a late (and English) resurgence of natural supposition, *v)* a pragmatic suspension of the *regula appellationum* and *vi)* Brinkley's apparently incompatible claims that there are communicable things *and* that there are only singular things, a position that is a medieval form of immanent realism. Based on the two manuscripts that contain the treatise on supposition, an appendix offers a provisional edition of part of Brinkley's *Summa*, a collaboration between the author and Joël Lonfat.

Keywords

significatio, suppositio, suppositio naturalis, appellatio, truth-makers

We know little of the life of Richard Brinkley beyond that he was a Franciscan theologian and philosopher active at the University of Oxford in the middle of the fourteenth century.[1] Although only fragments and abbreviations of his theological works survive, his pertinent writings had a significant impact on Parisian theology throughout the 1360s and 1370s. Working backwards, and given its content and the normal academic cursus of the time, we might guess that Brinkey's *Summa logicae* was written in the late 1340s, although some date it as late at the 1360s. The *Summa logicae* consists of seven parts: 'De terminis in genere', 'De universalibus', 'De predicamentis', 'De suppositionibus',

[1] Cf. Emden (1957), 267 ff., Gál-Wood (1980), 73-77, Cesalli (2004), 205-207, and Cesalli (2011). Many thanks to Joël Lonfat, Frédéric Goubier, and Alain de Libera for their helpful comments and suggestions. I would also like to express my gratitude to the editors of *Vivarium* (and their associated referees) for commenting on earlier versions of this paper.

'De propositionibus', 'De insolubilibus' and 'De obligationibus'.[2] This contribution focuses mainly on the fourth part and present a detailed study on Brinkley's supposition theory.

1. Structure and Content of the *De suppositionibus*

Brinkley's treatise on supposition, containing fifteen chapters, can be divided into two main and uneven parts, the first presenting some preliminary considerations—i.e., the definition of supposition and the exposition of three general semantic rules, themselves called *suppositiones*—and the second addressing the eponymous topic—i.e., the typology of supposition and related issues.[3] The distinction between proper and figurative (or improper) supposition, as well as the separate treatment of the different kinds of suppositions for absolute and relative terms are not original.[4] The ontic foundation of the division principle of supposition found in the second chapter is more interesting; this will be discussed below in section 3. Here is an overview of the structure and content of the *De suppositionibus*:

Praemittenda (c.1)
1. Diffinitio suppotitionis et quibus competit
2. Tres suppositiones

De suppositionibus
1. Divisio suppositionis (c.2)
2. De suppositione propria
    - 2.1. De suppositione terminorum absolutorum
    - 2.11. De suppositione materiali (c.3)
    - 2.12. De suppositione simplici (c.4)
        - 2.121. De convenientia suppositionis materialis cum simplici et personali (c.5)
        - 2.122. Dubia circa suppositionem simplicem (c.6)
    - 2.13. De suppositione personali in genere (c.7)
        - 2.131. De suppositione discreta (c.8)

---

[2] For a more elaborate presentation of the *Summa*, cf. Gál-Wood (1980), 65-72.

[3] In this respect, the treatise on supposition is similar to the one on universals (*Summa logicae* II), where the discussion of Porphyry's five *praedicabilia* is preceded by a discussion about the relation between logical and metaphysical universals, and about the number of universals (*de sufficientia universalium*). Cf. Brinkley (2008), 278, 299-313.

[4] Both distinctions are for example present in Walter Burley's *De puritate artis logicae tractatus longior*, ed. Boehner (1955), I. 4, pp. 28-33, and I. 5 and 6, pp. 45-47, as well as in William of Ockham's *Summa logicae*, I, 76 and 77, ed. Boehner *et al.* (1974), 233-238.

2.132. De suppositione determinata (c.9)
2.133. De suppositione confusa et distributiva (c.10)
2.134. De suppositione confusa tantum (c.11)
2.2. De suppositione terminorum relativorum
2.21. De relativis ydemptitatis (c.12)
2.211. Dubia circa relativa ydemptitatis (c.13)
2.22. De relativis diversitatis (c.14)
3. De suppositione figurativa (c.15)

Concerning the typology of supposition, the following tree can be reconstructed from Brinkley's treatise:

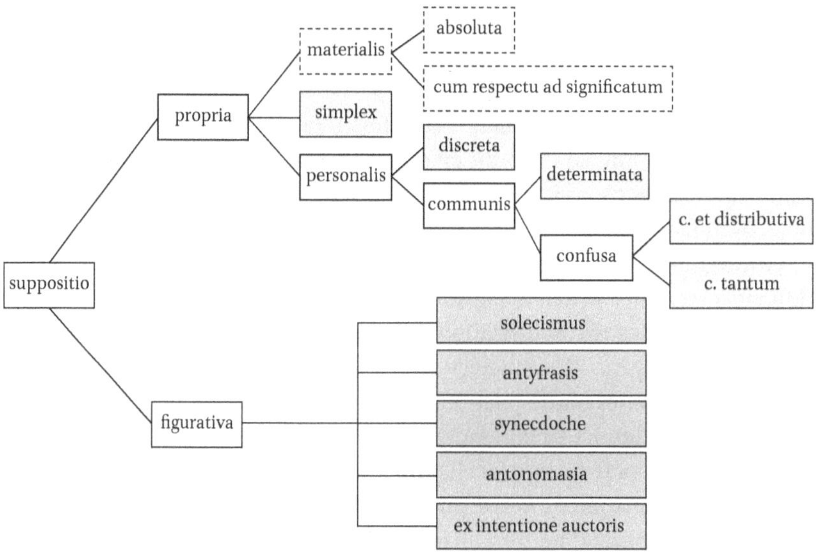

This tree has two interesting but unrelated: first, the doubt expressed by Brinkley about the legitimacy of material supposition as a genuine type of supposition (see below, section 3); secondly, the presence within figurative supposition of a pragmatic category, namely the intention of the author (*ex intentione vel modo loquendi auctoris*, see below, section 6).[5]

[5] Whereas both Ockham and Burley mention the intention of the *usus loquendi* as a cause for figurative or improper supposition, neither of them questions the legitimacy of material supposition.

2. The Distinction between *significatio* and *supposito*

Before giving a formal definition of supposition, Brinkley begins by distinguishing between supposition and signification. He provides two arguments for that distinction: the first concerns the objective character of supposition, as contrasted to the conventionality of signification; the second is based on an analysis of signification and supposition with respect to the number of sub-relations they respectively involve.

Whereas signification, says Brinkley, is a conventional semantic relation, supposition is generated by the nature of the thing: '*Suppositio* [...] *non oritur ex principio voluntario, sed ex natura rei*' (*TDS*, 1.1).[6] Were this not the case, some basic inference rules would not be valid—for example the following 'descent rule': *a termino supponente determinate ad disiunctivam de omnibus suis singularibus est bona consequentia* (*TDS*, 1.1-2). Indeed, the signification of the vocal sound '*homo*' is determined by social agreement, and it is obviously true that the inference 'if a man is running, then either Socrates, or Plato, or [...] (and so on and so forth for every member of the species) is running' is valid independently of any convention. Therefore supposition possesses an objective character that signification lacks.

Furthermore, the two semantic relations of signification and supposition differ insofar as the former is a dyadic, while the latter is a triadic notion. Signification comprises two sub-relations: one to the intellect, and another to the thing that motivates the intellection (*TDS*, 1.3). The signification of '*homo*' involves, on the one hand, a relation from the term to the intellect actually having the concept 'human being' and, on the other hand, a relation from the term to the thing of which the intellect has a concept—in the present case: a human being. The supposition of '*homo*' presupposes its signification and adds something to it: a third relation from the term to the other extreme of the proposition in which it occurs as subject or predicate. The supposition of '*homo*' in '*homo currit*' involves—besides its signification—a relation from the term to the predicate, in the present case: '*currit*'. Brinkley then gives the following formal definition of supposition:

> [...] suppositio est ordinatio alicuius termini de quo alius dicitur vel qui dicitur de alio.[7]

---

6) '*TDS*' stands for '*Tractatus de suppositionibus*', and the numbers refer to the sections of the Latin text given in the appendix to the present article.

7) *TDS*, 1.4.

In other words, Brinkley conceives supposition as a cumulative semantic notion: supposition is syntactically specified signification.

At this stage, two questions can be raised: first, how exactly is the clause '*ex natura rei*' to be understood? Second, is Brinkley's purely syntactical definition of supposition compatible with the claim that supposition has its origin in the nature of things?

The answer to the first question, given Brinkley's realist metaphysics, lies in the ontologically based distinction between personal and simple supposition, which will be discussed below (section 3). The answer to the second question is slightly more complex. There is no incompatibility between the ontic origin and the syntactical determination of supposition. Each one of these two claims is made from a different perspective: the first points to the very 'raison d'être' of supposition—it proposes an ontological explanation for the fact that human language involves an extensional semantic function (supposition) besides an intensional one (signification); the second claim gives the formal characteristics of every occurrence of supposition—it describes what it means for a term in a sentence to have supposition. The fact that sentences are freely formed and pronounced according to what a speaker wants to say does not affect at all the situation that, *once a sentence is formed*, objectively founded relations hold between its terms. The objectivity of these relations is derived both from the essential link between terms and things and from the metaphysical relations between different types of things. This also might explain why Brinkley links the claim of the ontic origin of supposition with the validity of certain inference rules (*TDS*, 1.1-2): since there are indeterminate (universal) and particular things,[8] the validity of the 'descent' from a term suppositing indeterminately to a disjunctive proposition involving all its 'inferiors' is ontologically founded and thus objectively justified.

3. The Ontic Division Principle of Supposition

According to Brinkley, the distinction of the basic types of suppositions is based on an ontological distinction: a term can stand either for a universal thing or for a singular thing. In the first case, the term has simple supposition, in the second, personal supposition. A further distinction leads to the differentiation of personal and material supposition: when a term supposits for a singular

[8] For a detailed discussion of the ontological presuppositions of Brinkley's theory of supposition, see below, section 7.

thing distinct from itself, it supposits personally; when it supposits for a singular thing which is not distinct from itself, it supposits materially (*TDS*, 2.2).

However, Brinkley criticizes this terminology. Properly speaking, simple supposition should rather be called '*communis*' or '*specifica*' because that for which a term supposits in simple supposition is a common nature *ut non contracta*. Along the same lines, personal supposition would be better called '*individua*' or '*materialis*' because its distinctive feature is not to supposit for a person, but for a singular thing. To have '*materialis*' instead of '*personalis*' should not be a problem, Brinkley argues, because what is traditionally called material supposition is not a genuine type of supposition. Two arguments support this claim: first, since supposition (like signification) is relational, if a term really stands for itself, we are in a situation of numerical identity; therefore material supposition would have to be monadic, which is excluded by definition. Furthermore, since words are singular things, the ontic basis for distinguishing between personal and material supposition is missing (*TDS*, 2.2).[9]

4. Supposition and Verification

Having stressed the objective character of supposition compared to signification, Brinkley addresses the topic of the relation between supposition and proposition. In this context, he stipulates what we could call 'the equivalence principle' (*Ep*): in any proposition, both extremes have supposition (*TDS*, 1.5). Brinkley argues as follows: every extreme of every proposition is involved in the three relations needed for supposition (to the intellect, to the thing, to the other extreme—see above, section 2); therefore every extreme in every proposition has supposition. As a consequence of his acceptance of *Ep*, Brinkley criticizes those who think that a term never supposits except for that of which it is verified:

[9] The notion of material supposition Brinkley is criticizing here is the genuine medieval one, which has to be distinguished from the modern (post-tarskian) one, traditionally expressed by the opposition between use and mention. The notion of mention implies that a term has been turned into a name for other occurrences of the same type, whereas medieval material supposition does not imply any nominalization. On that question, see Brands (1990) and Panaccio (2004).

> Igitur, non stat propositionem esse et suppositionem non esse. Falsum est igitur quod aliqui pro regula ponunt, dicentes quod terminus nunquam supponit pro aliquo nisi pro eo de quo verificatur.[10]

The rule quoted by Brinkley—let us call it *R*—corresponds verbatim to a principle formulated and defended by Ockham in chapters 63 and 72 of the first part of his *Summa logicae*,[11] a principle that we could call the 'term-verification principle' (*Tvp*). *Tvp* can be seen as a kind of testing procedure to check whether a term is taken significatively (*significative sumptus*) or not—which, in Ockham's view, amounts to determining whether a term is taken in personal supposition or not: given a sentence 'A is B', A only supposits for a thing *x* such that A can be truly predicated of *x*, that is, such that the sentence '*x* is A' is true. In other words, in '*Socrates currit*', '*Socrates*' only supposits for Socrates, namely for the thing such that '*x est Socrates*' is true. *Tvp* can be formally expressed as follows:

'A is B': Sup. (A, x) ↔ True [A (x)]

Now, in spite of the literal correspondence between *Tvp* and *R*, there are good reasons to doubt that the real target of Brinkley's criticism is *Tvp*. I will first consider and comment on this alternative reading of *R* and then examine the reasons found in Brinkley justifying it.

As a matter of fact, *R* can be understood as expressing another principle, which we could call the 'proposition-verification principle' (*Pvp*). *Pvp* states that in the proposition '*A* is *B*', the term *A* only supposits for a thing *x* if the proposition '*A* is *B*' is true; in other words, *Pvp* states that if '*Socrates currit*' is false, then '*Socrates*' does not supposit for anything. *Pvp* can be formally expressed as follows:

'A is B': Sup. (A, x) ↔ True [B (A)]

Tvp and *Pvp* are fundamentally different principles. First, *Pvp* identifies *supposita* with truth-makers—something that *Tvp* does not—: the testing procedure of *Tvp* does not fail if the proposition is false; on the contrary, it explains why the proposition is false: if '*Socrates est nomen*' is false, it is because

10) *TDS*, 1.5-6.

11) Cf. also Ockham (1974), II.22.

'*Socrates*' supposits for Socrates and not, as *Pvp* says, because '*Socrates*' does not supposit at all. Second, *Tvp* and *Pvp* have different scopes: *Tvp* is not linked to the truth of the primary proposition 'A is B', but only to the truth of the secondary proposition '*x* is A', whereas *Pvp* is only concerned with the truth of the primary proposition and does not say anything about the secondary one. Third, *Tvp* aims at identifying a precise *suppositum*, whereas *Pvp* is concerned with the more general question of the presence or absence of supposition in a given sentence, independently of the identity of possible *supposita.*[12]

The main reason to read *R* as expressing *Pvp* rather than *Tvp* is that *Tvp* is not incompatible with *Ep*—and Brinkley precisely says that if *Ep* is accepted

[12] One of the reasons for reading Ockham as stating *Pvp* (and not *Tvp*) might be found in the very same chapter 63 of the first part of his *Summa logicae*: 'Et sic universaliter terminus supponit pro illo de quo [...] per propositionem denotatur praedicatum praedicari, si terminus sit subiectum; [...] sicut per istam 'homo est animal' denotatur quod Sortes vere est animal, ita quod haec sit vera si formetur 'hoc est animal', demonstrando Sortem.' Indeed, given the primary proposition '*homo est animal*', the secondary proposition '*hoc est animal*' appears—according to the formulas given above for *Tvp* and *Pvp*—not to correspond to [A (x)], but rather to something like [B (x)]. Thus, by construing a secondary proposition in which the predicate B is predicated of the *suppositum* x of the subject term A, Ockham produces something which can easily be read as being *Pvp*. Let us call this principle **Pvp*. But why could **Pvp* not be identical with *Pvp*? Because *Tvp* is incompatible with *Pvp*—and Ockham explicitly accepts *Tvp* (formulated before *and* after **Pvp* in chapter 63)—: even if, according to Ockham, an empty term (that is, a term without suppositum) is a sufficient condition for the falsity of the proposition in which it occurs, there are false propositions in which the terms *do* supposit, but precisely not for the same thing. Furthermore, a secondary proposition of the form [B (x)] for the primary proposition '*homo est animal*' can be interpreted as '*hoc [e.g., Brunellus] est animal*', a proposition which is true, but where x is obviously not a suppositum of *homo*. So, how is **Pvp* to be understood? Two hypotheses can be formulated. Here is the first one, suggested by Frédéric Goubier: **Pvp* must be compatible with *Tvp* and can be interpreted as indicating the *type* of things which are susceptible to make the primary proposition true (**Pvp* would allow us to know if a term in a proposition stays in personal, simple, or material supposition, that is, if it supposits for things, concepts or words), without saying anything about which *tokens* would make it actually true. This hypothesis is supported by the fact that chapter 63 is thematically linked with chapter 65, where Ockham explains precisely under which conditions a term can be taken in personal, simple or material supposition in a given sentence. The second hypothesis (suggested by the author) takes **Pvp* to be a mere specification of *Tvp* for true propositions, where, according to Ockham's definition of truth, subject and predicate supposit for the same thing, so that the two derived propositions [A (x)] and [B (x)] must both be true. This way of reading **Pvp* also explains why Ockham simply repeats *Tvp* after having stated **Pvp*, but also why he goes on explaining that it is false to claim that in a proposition like '*homo est albus*', the predicate supposits for a form and the subject for a substance. Were this true, the specificity of *Tvp* for true propositions would have to be abandoned, since [A (x)] would be true, and [B (x)] false.

then *R* has to be rejected.[13] As a matter of fact, *Tvp* does not entail that there can be propositions without supposition, which is precisely the point stressed by *Pvp*: according to it, false propositions are propositions without supposition. Besides this extrinsic argument, there are textual and intrinsic reasons supporting the *Pvp*-reading of *R*. Brinkley gives three counter-examples against *R*, two of which (see below objections *ii.* and *iii.*) clearly turn to be harmless against *Tvp*, but efficient against *Pvp*. And while objection *i.*, considered for itself, may look like an argument against *Tvp*, both its dialectical context and the general conclusion Brinkley draws from his three counter-examples support its reading as being an objection against *Pvp* and not against *Tvp*. Let us go through the counter-examples one by one.

i. Consider the two propositions '*Antichristus erit*', '*Antichristus non erit*'. In each of them, says Brinkley, the subject term supposits for the Antichrist, although in neither case can it be verified of the Antichrist. Therefore, Brinkley concludes, a term sometimes supposits for something of which it cannot be verified. And it does not make sense, Brinkley adds, to distinguish between a verification now, in the past or in the future, because there are cases—as in Brinkley's second counter-example—where such a distinction does not help.

ii. The proposition '*Deus non est*' is such, says Brinkley, that there cannot be, at any moment of time, anything verifying it. Contrary to '*Antichristus (non) est*', which is only lacking an actual truth-maker, '*Deus non est*' is an impossible proposition: it cannot be made true by anything in virtue of a metaphysical impossibility.

iii. Brinkley's third counter-example considers again an impossible proposition, but this time the impossibility is a logical one. Let us imagine that Socrates forms the following mental proposition: '*Sor non est*'. There is no way for this proposition to be verified by Socrates. Indeed, if Socrates exists, then the proposition can be formulated, but must be false; and if Socrates does not exist, then the proposition should be true, but obviously cannot be formulated.

From this, Brinkley draws the following general conclusion:

> Unde terminus frequenter supponit pro aliquo pro quo respectu eiusdem predicati non verificatur.[14]

Obviously, the verification Brinkley is talking about here concerns the primary proposition 'A is B' and not the secondary one '*x* is A'. The clause '*respectu*

[13] *TDS*, 1.5-6.

[14] *TDS*, 1.7.

eiusdem predicati' compels us to understand Brinkley's conclusion as follows: 'Therefore, a term often supposits for something which cannot verify the proposition in which this term occurs as a subject'. Thus, what Brinkley wants to show with those three counter-examples is that in propositions without actual truth-makers (as in objection *i.*), or in propositions lacking any possible truth-makers (as in objections *ii.* and *iii.*), the subject terms nevertheless supposit. Therefore, it is false to identify *supposita* with truth-makers, and *Pvp* has to be rejected.[15]

Consequently, and in virtue of *Ep*, Brinkley's position is that terms in false and impossible propositions have supposition, just as they have in true and necessary ones. But what do the subject terms in the three counter-instances supposit for? The answer to that question lies in Brinkley's treatment of the *regula appellationum.*

5. A Late (and English) Resurgence of Natural Supposition?[16]

The second part of the *praemittenda*—presented by Brinkley before addressing the more specific question of the typology of supposition—is dedicated to

---

15) The consequence to be drawn is that Brinkley's target is someone who understood *R* as *Pvp*, or that Brinkley himself misunderstood *R* in Ockham and thought of arguing against the *Venerabilis inceptor* while criticizing *Pvp* (see *TDS*, 1.5-6). But since Brinkley basically agrees with *Tvp*, this seems highly improbable. However, some passages in John Duns Scotus and William of Heytesbury are reminiscent of *Pvp*: 'Omne illud pro quo simpliciter terminus potest supponere et verificare propositionem, est simpliciter eius suppositum' (Scotus (2004), q. 6, n. 1); 'Cum homo currit, tunc si haec propositio sit sic significando 'homo currit', iste terminus 'homo' supponit pro aliquo, et si non curreret, tunc iste terminus 'homo' pro nullo supponeret, sicut nec iste terminus 'homo currens'' (Heytesbury (1994), *sophisma* 25, 21 of the electronic edition). On closer inspection, it turns out that what Scotus wants to say is not that the subject term of a false proposition does not have supposition, but rather that a proposition can be made true by something which does not exist now: 'huiusmodi sunt alia ab exsistentibus, ut patet in his propositionibus 'homo fuit', 'homo erit', sumendo 'fuisse' prout dicit actum inchoatum et terminatum. Sic enim non verificatur illa pro aliquo exsistente, et specialiter in hac propositione 'contingit hominem currere', in qua dicit Aristoteles subiectum stare pro eo quod contingit' (*ibid.*) On the contrary, the passage from Heytesbury seems to state *Pvp*: if no man is running, the proposition '*homo currit*' is false and, in that case, '*homo*' does not supposit for anything. Consequently, one of the candidates for the supporters of *Pvp* might be William of Heytesbury (who writes in the 1330s); but this attribution to Heytesbury needs only to be considered as a preliminary hypothesis, until more evidence *pro* or *contra* is found. I am grateful to Alain de Libera for having drawn my attention to the passages in Scotus and Heytesbury.

16) I am grateful to Frédéric Goubier, who first identified the Parisian flavour of Brinkley's treatise and suggested that Brinkley might in fact defend a form of natural supposition.

the discussion of three general semantic rules, the first of which is the so called *regula appellationum*:

> Suppono enim primo quod quando terminus communis supponit respectu verbi de presenti non ampliativi, solum supponit pro hiis que sunt, sicut in ista propositione: 'omnis homo currit'.[17]

The *regula appellationum* is a linguistic restriction rule: it is linguistic because it concerns words (the presence or absence of a certain verb in a certain tense), and it is restrictive because it focuses the supposition on presently existing *supposita*. Now, the very fact that this rule is formulated entails that a term can supposit for more *supposita* than the presently existing ones. Theoretically, we have two possibilities: a word can supposit for more than present existing things *before* becoming a term in a proposition—and Brinkley seems to accept such a kind of supposition when speaking of the ability of a term to supposit *de se* for present, past and future things.[18] Or a term can be 'ampliated' by another in a proposition, so that it supposits for more *supposita* than the presently existing ones (this is suggested by the clause '*verbi [...] non ampliativi*' in the rule)—and in that case, we would have something very close to what Parisian authors of the first half of the thirteenth century such as John Page or Lambert of Lagny called *suppositio naturalis*.[19] Thus, Brinkley seems to work with a kind of supposition which, on the one hand, corresponds to Sherwood's

---

17) *TDS*, 1.8. The expression '*regula appellationum*' is found in John Page's *Appellationes* (cf. De Libera (1981a), 64). Concerning Brinkley's second and third rules, they are the following: 'Alia suppositio sit ista: quando terminus communis supponit respectu verbi de preterito, potest supponere pro hiis que sunt vel pro hiis que fuerunt, sicut in talibus: 'omnis homo fuit', 'omnis asinus fuit Brunellus' [...]. Simile est de termino communi supponente respectu verbi de futuro quod potest supponere pro hiis que sunt [fuerunt, ms.] vel pro hiis que erunt.' (*TDS*, 1.12); 'Tertia suppositio: quando terminus compositus ex determinabili et determinati determinativo supponit respectu verbi de preterito, predicatum potest attribui subiecto pro eo quod supponitur per partem subiecti ut per determinabile, vel pro eo quod supponitur per utrumque, sicut patet in talibus: 'Sor albus fuit' [...].' (*TDS*, 1.13). Similar rules are found in John Page's *Appellationes* as well as in Lambert of Lagny's *Logica*. Cf. De Libera (1981) and (1982).

18) See *TDS*, 1.8. In that sense, he could accept something like William of Sherwood's 'habitual supposition', which is a kind of pre-contextual and thus *potential* supposition (habitual supposition is 'the capacity of a term to supposit for concrete individuals', cf. Jakobi (1980), 158-159). Thus, Brinkley's equivalence principle *Ep* is valid only for actual supposition.

19) On that topic, cf. De Libera (1981, especially, for the definition of natural supposition, p. 63, n. 2). However, the idea of a connection between Brinkley and the French tradition may be less implausible than it appears at first sight. Kaluza has shown, for example, that Brinkley's theological work was influential in Paris—Etienne Gaudet's abbreviation of Brinkley's (lost)

precontextual *suppositio secundum habitum* and, on the other hand, resembles John Page's or Lambert's *suppositio naturalis* as contextually 'ampliated' supposition. At any rate, when the *regula appellationum* does apply—that is: when the verb of the considered proposition is in the present and lacks any ampliative power—the verb exerts an analogue effect on the subject term as the exceptive '*solum*', with the difference that '*solum*' works on quantity (it singles out one individual), whereas '*est*' or another verb in the present works on time (it singles out the present).[20]

6. A Pragmatic Suspension of the *regula appellationum*

Brinkley's discussion of the *regula appellationum* has a positive and a negative part. In the first one, he gives two arguments for the general validity of the rule and specifies the exact conditions under which it is valid; in the second one, he identifies what we could call a pragmatic suspension of the rule, namely when the subject term is a deictic expression such as the demonstrative pronoun '*hoc*'.

i. Two arguments support the validity of the *regula appellationum*. The first argument—called *a priori* because it considers the cognitive and semantic process of producing a proposition—states that the intellect naturally harmonizes the referential scope of the terms (i.e., the supposition of the subject and predicate terms), so that even if a term is able by itself to supposit for more than present things, its composition with a verb in the present naturally restricts its supposition to the corresponding range of *supposita*, namely to the present ones.[21] The second argument—*a posteriori* because it considers prop-

---

commentary on the *Sentences* is one of the very rare traces of Brinkley's theological work. Cf. Kaluza (1989-1990).

20) *TDS*, 1.8: 'Hic iste terminus 'homo' quamvis de se sit indifferens ad supponendum pro hiis qui fuerunt et pro hiis qui erunt disiunctive vel coniunctive, tamen per modum illum quo ad istum contrahitur per 'solum' contrahitur iste terminus 'homo' per hoc predicatum cum verbo de presenti ad supponendum tantum pro hiis que sunt.'

21) *TDS*, 1.9: 'Et ratio huius suppositionis partim sumitur a priori et partim a posteriori. A priori enim eo quod intellectus naturaliter sequendo dictamen rationis componit predicatum pro significato predicati cum subiecto, <vel> quod est idem, cum significato subiecti. Si autem subiectum habeat plura supposita quam predicatum—vel e converso—, sic intellectus sequendo dictamen rationis non componit unum cum alio pro omni suo significato, sed tantum pro significato illo in quo conveniunt respectu verbi de presenti, cum copula in comparatione ad subiectum sit reductibiliter pars predicati, significatum predicati tantum est illud vel illa que sunt presentia. Et sic suadetur suppositio a priori.'

ositions once they are formed—points to the fact that if the rule were not valid, then a proposition like '*omnis homo est*' would be false, because it is obviously false to say of a past or future human being that he or she exists.[22]

The validity of the *regula appellationum*—the restriction of the referential scope to presently existents—depends on two linguistic conditions on the side of the predicate (*i.e.*, the verb alone, or the verb plus its complement): first, the verb has to be in the present tense; second, it cannot be ampliative, which means here that it cannot be a verb expressing a possibility (*posse, contingere*). Two analogous conditions must be fulfilled on the side of the subject: it cannot contain a verb of a non-present tense, nor an ampliative verb. Thus in propositions like '*omne quod fuit est*' or '*omne quod potest esse est*', the *regula appellationum* is not valid anymore.

ii. The non-fulfilment of the conditions just mentioned leads to exceptions to the *regula appellationum*. Since those exceptions are linguistically describable, linguistic rules can be formulated to identify such exceptions. Let us call such rules linguistic second order rules (the *regula appellationum* itself is a linguistic first order rule), for example, the rule: 'if the subject term contains an ampliative verb, then the *regula appellationum* is not valid'. But there are other cases in which the *regula appellationum* is not valid and no linguistic second order rule can be formulated. The reason for this is that these exceptional cases are not identifiable on the base of linguistic features (type of words, tenses of verbs, etc.). In those cases, all the linguistically describable conditions for the validity of the *regula appellationum* are fulfilled, and still the *regula* is not valid. Such cases do not depend on the proposition itself or on its parts, but on the context of its enunciation, and more precisely, on the intention of the speaker.

Brinkley's counter-example to the *regula appellationum* is based on the notion of *demonstratio* in the sense of a gesture pointing towards something. On the linguistic level, a *demonstratio* is achieved by deictic expressions such as demonstrative pronouns or proper names.[23] There are two types of *demonstratio*, says Brinkley, one *ad sensum*, pointing towards existing things, and one *ad intellectum*, pointing towards non-existing things:[24]

---

22) *TDS*, 1.9: 'A posteriori etiam, quia si respectu cuiuscumque verbi cuiuscumque temporis terminus communis supponeret pro omni suo supposito, quelibet propositio universalis affirmativa esset falsa in qua subiceretur terminus communis habens supposita corrupta.'

23) Cf. Bérubé (1964), Ashworth (2004) and Brumberg-Chaumont (2007).

24) On that topic, cf. Marmo (1994) 174-175 and Rosier-Catach (2004), 410-417.

> Nam cum demonstratio sit duplex quedam ad sensum et quedam ad intellectum, ideo per propositionem demonstrativam indifferenter potest demonstrari illud quod non est et illud quod est: illud quod non est intellectui et illud quod est sensui. Et loquor hic de termino demonstrativo qui facit propositionem in qua subicitur, esse singularem.[25]

According to what a speaker wants to point at by using a demonstrative expression, he will perform a *demonstratio ad sensum* or *ad intellectum*. Now, writes Brinkley, suppose that I say '*hoc est*', pointing (now) at the future Antichrist by a *demonstratio ad intellectum*. I would then produce a counter-example to the *regula appellationum*, a counter example in which none of the linguistic first or second-order rules is violated and still the *regula appellationum* does not apply. And indeed, under the considered *casus*, '*hoc*' in '*hoc est*' does not supposit for a presently existing thing. Here is Brinkley's argumentation:

> Facio igitur hanc consequentiam: 'hoc est, igitur aliquid est' et demonstro per subiectum antecedentis antichristum futurum. Constat enim quod consequentia est bona, nam ex opposito formaliter sequitur oppositum; tamen consequentia non valeret si subiectum consequentis tantum supponeret pro hiis que modo sunt, quia stat absque formali et immediata contradictione quod nichil quod modo est esset, et tamen quod antichristus esset.[26]

The general argument can be reconstructed as follows. We start with two premises: the first states that, in virtue of the *demonstratio ad intellectum*, a demonstrative pronoun can point towards a non-existing thing (for example, towards the future Antichrist). The second says that the inference '*hoc est, ergo aliquid est*' is valid in virtue of the *ex opposito*-rule, a rule which states that the opposite of the antecedent may legitimately be inferred from the opposite of the consequent. Thus, since the consequence '*nichil est, ergo hoc non est*' is valid—its antecedent cannot possibly be true and its consequent false—the consequence '*hoc est, ergo aliquid est*' has to be valid as well. If follows that one cannot exclude that '*aliquid*' in '*aliquid est*' supposits for a presently non-existent thing, such as the future Antichrist. Therefore there are exceptions to the *regula appellationum* which cannot be captured by linguistic first or second order rules.

However, there is a third order, non-linguistic rule applying to this complex situation. Let us call such a rule a pragmatic rule:

---

[25] *TDS*, 1.11.

[26] *TDS*, 1.11. A very close argument is used by John Page in his *Appellationes* to show the limits of the validity of the *regula appellationum*. Cf. De Libera (1981a), 69.

> Pro termino igitur communi supponente respectu verbi de presenti et pro aliis terminis de quibus non potest dari regula generalis notanda est regula quam ponit Linconiensis in tractatu suo de individuatione que talis est: in hiis que accidunt vel dicuntur proprie vel minus proprie, semper naturaliter recurrit audientis intelligentia ad sensum magis proprie significatum, nisi ex circumstantia sermonis determinetur quod sermo communiter et non proprie accipiatur. Et ideo ubi terminus communis supponens respectu verbi de presenti non determinatur per aliquod precedens vel subsequens vel etiam ampliatur, sustineri debet quod tantum supponit pro hiis que sunt, cum pro illis maxime proprie supponere videatur.[27]

The distinction between two levels of rules appears clearly in the phrase '[...] and for terms for which no general rule can be given, one has to note the rule [...]'. A general rule for a term would be a linguistic rule of the form 'if a term of this or that type appears in the proposition, then...'. The pragmatic rule does not concern types of words, but the psychological context of enunciation. It has the form 'if a term is used under such and such conditions, then...'. The pragmatic rule quoted by Brinkley says that the intelligence of a hearer has the natural tendency to understand a word according to its standard meaning, unless the context of the word requires an ampliation (*communiter* in the sense of 'common to the present and the non-present') of the standard meaning. In the case of the *regula appellationum*, the standard meaning (*proprie* in the sense of 'according to the tense of the verb') is that which is stated by the linguistic first-order rule: the subject of a proposition in the present supposits for presently existing things. Now, the context of the word (*circumstantia sermonis*) can be understood in two ways: either as its linguistic context, and then the linguistic second-order rules apply—for example, when the verb possesses an ampliating force (*determinatur per aliquod precedens vel subsequens*); or its extra-linguistic context, and then no linguistic rule can apply—for example, when the speaker intends to demonstrate something to the intellect and not to the sense (*vel etiam ampliatur*).

I ended my discussion of the relation between supposition and verification (see above, section 4) by saying that Brinkley's discussion of the *regula appellationum* would provide an answer to the question of the nature of the *supposita* of the subject terms in the three counter-examples *i.* '*antichristus (non) erit*', *ii.* '*Deus non est*' and *iii.* the mental proposition in Socrates' mind '*Sor non est*'. Let us first recall that, in virtue of *Ep*, the terms of any well-formed proposition necessarily have supposition. Now, as I just observed in my examination

---

27) *TDS*, 1.11. No treatise on individuation by Robert Grosseteste is known and I was not able to locate Brinkley's source for that statement.

of the *regula appellationum*, 'having supposition actually' is not equivalent to 'having actual *supposita*'. Indeed, the pragmatic device of the *demonstratio ad intellectum* allows a term to supposit now for something that does not exist now; and, if—as Brinkley showed—this is possible when the verb of the proposition is in the present tense (see '*hoc est*', pointing at the future Antichrist), then it is possible as well (and *a fortiori*) when the verb of the proposition is in the future tense, as in the counter-example *i.* Concerning the counter-examples *ii.* and *iii.*, they are (necessarily) false propositions, but since Brinkley does not accept *Pvp*, nothing prevents him from having '*Deus*' supposit for God in *ii.* and '*Socrates*' for Socrates in *iii.*

7. Singular, Communicable Things: Brinkley's Immanent Realism

Our last *notabile* concerns a metaphysical issue. Two apparently incompatible claims are found in Brinkley's treatise on supposition: according to the first, there are universal or 'communicable' things,[28] while, according to the second, there are only singular things.[29] The tension between these two positions appears clearly in the definitions of simple and personal supposition: a term is taken in simple supposition, says Brinkley, when it supposits for a thing by itself communicable to many insofar as it is communicable (*pro re de se communicabili ut communicabilis est*);[30] a term is taken in personal supposition when it is taken significatively and supposits for a singular thing insofar as it is singular (*sub propria ratione singulari*).[31] Now, Brinkley goes on, personal supposition differs from simple supposition because in simple supposition a term supposits for a singular thing, but not insofar as it is singular (*tamen non ut res singularis*);[32] in other words, in simple supposition, a term supposits for a thing by itself communicable to many—that is: a universal thing—*and* for a singular thing as well. The only way to sort this out without having Brinkley contradict himself—and I want to stick to the principle of hermeneutic charity—is to distinguish between two senses of the word '*res*', namely a holistic and an analytical one. Holistically speaking, a thing is a physical individual (e.g., Socrates);

28) *TDS*, 2.2 and 4.1. See also Brinkley's *De universalibus*, §2.

29) *TDS*, 4.1 and 6.14.

30) *TDS*, 4.1.

31) *TDS*, 7.1. The clause 'taken significatively' distinguishes personal from material supposition. In both cases, the term supposits for a singular thing, but in material supposition the thing at stake is not signified by the term (see above, section 3).

32) *TDS*, 7.1.

analytically speaking, a thing is a metaphysical part of an individual (matter, form, an accident).[33] And indeed, the locutions '*de se*' in '*de se communicabili*', '*ut*' in '*ut communicabilis*' and '*sub propria ratione*' in '*sub propria ratione singulari*' seem to point towards such a distinction. In this respect, the discussion of the following objection is of particular interest:

> [...] quando dicitur quod aliqua sit res que non sit res singularis, hoc nego. Dico enim quod omnis res est singularis et cum dicitur contra sic: nulla res singularis est res communicabilis igitur per conversionem nulla res communicabilis est res singularis, concedo consequens terminis supponentibus personaliter, quia sic nulla res est communicabilis et tamen multe sunt quarum quelibet est de se communicabilis. Exempla: ad omnia ista que dicuntur de suppositione simplici satis expresse habetur in philosophia naturali. Probat enim Philosophus primo *Physicorum* quod materia prima est de se in potentia ad omnes formas generales et hoc in potentia illa que est distincta contra actum. Hoc enim probat eo quod nullam formam materia prima de natura sui sibi determinat. Verumtamen materia prima iam non est nec aliquando fuit in tali potentia, quia semper est et fuit sub actu opposito tali potentie. [...] Stant igitur ille simul 'materia prima de se est in potentia ad omnem formam' et 'materia prima non est in potentia ad omnem formam' semper uniformiter loquendo de potentia, scilicet de potentia ante actum et generaliter de potentia distincta contra actum. Sic dico quod illud quod primo exprimitur per terminum communem ut per illum terminum 'homo' est de se commune sive communicabile ad omne individuum in specie hominis ; et tamen de facto est tantum unum individuum ita quod non communicabile, sed realiter ita incommunicabile, sicut individuum, cum fuit idem.[34]

Brinkley concedes the truth of the two apparently incompatible propositions: 'there is no communicable thing' and 'there are many things by themselves communicable', provided that the subject supposits *personaliter* in the former and *simpliciter* in the latter.[35] Let x be the *suppositum* of '*homo*', and consider two propositions p: '*homo currit*' and s: '*homo est species*'. The term '*homo*' has personal supposition in p and simple supposition in s. Brinkley explains

33) By using this terminology, we want to stress the opposition between physical and metaphysical and not, say, between physical and mental. Another way of rendering the holistic concept of individual in the sense used here would be to say that a physical individual is a *unitary* individual (the unitary sum of the substance, essential and accidental forms composing Socrates).

34) *TDS*, 6.14.

35) Interestingly enough, Brinkley distinguishes between a primary and a secondary significate of terms. Furthermore, a term in simple supposition supposits for its primary significate, whereas in personal supposition it supposits for its secondary significates (*TDS*, 1.11, 4.1, 6.14). Analogous ideas are found, again on the English side of the Channel, in Sherwood's and Burley's conceptions of signification as the imposition of a vocal sound to a universal form. Per definition, such a form must be present in many individuals so that a name necessarily but only secondarily refers to singular things, a step which is explicitly taken by Wyclif. Cf. Cesalli (2007), 338 ff.

that x_s is related to x_p just as primary matter *de se* is related to the matter of any hylomorphic composite or, as another example says, just as an accident to its substance. Thus, what Brinkley wants to say is this: just as matter *de se* and an accident *de se* never factually exist on their own, but nonetheless factually are metaphysical parts of individuals, so communicable things *de se* never exist on their own, but only as metaphysical parts of individuals.[36] We can speak about such metaphysical parts of individuals by producing a proposition in which a term supposits *simpliciter*, but this does not entail that such a communicable thing factually exists on its own. This, I believe, allows us to characterise Brinkley's position as a medieval form of immanent realism.

8. Concluding Remarks

This study has focused on six points of interest identified in Brinkley's preliminary remarks on his typology of supposition. I will now summarize my results according to the three fields to which they pertain, namely *semantics, pragmatics* and *ontology*. However, the most interesting conclusion to be drawn from the preceding considerations concerns the interrelation of those three fields, an idea that seems to be in the background of several of the examined *notabilia*.

Semantics—a) Brinkley's cumulative notion of supposition as syntactically specified signification offers an elegant explanation of the formal and objective character of logical relations holding between linguistic, arbitrarily produced signs. Both signification and supposition contribute to it, the former by determining a context-free and thus very general, intensional relation between a vocal sound and a 'thing'; the latter by resulting in a context-bound, specific and extensional relation between a term in a sentence and a precise thing or class of things (see above, section 2). *b)* In my examination of the relation between supposition and verification, I noted that Brinkley accepts *Ep*—a principle stating the equivalence between having supposition and being a term in a sentence (see above, section 4). In this connection, Brinkley insists that *Ep* is valid for any type of proposition (true, false, necessary, impossible, about the past or the future). This has two remarkable consequences: first, it excludes empty references for a term in a proposition; second (and consequently), 'actually having supposition' does not mean 'having actual *supposita*',

[36]) See the following passage of Brinkley's *De universalibus*, §10: 'Sicut linea que est in Sorte potest intelligi absque Sorte sed non potest esse absque Sorte, ita humanitas que est in Sorte potest intelligi sine Sorte, quamvis non posset esse sine Sorte.'

a feature of Brinkley's theory which reminds of the contextual *suppositio naturalis* defended by some French authors of the thirteenth century (see above, section 5).

Pragmatics—Brinkley's discussion of the so-called *regula appellationum* (see above, section 6) involves a reflection on three different types of rules which I labelled as first, second and third order rules. The first two are given in linguistic terms (e.g., 'if the subject or the predicate contains a word of such and such type, then ...'); first order rules contain positive determinations (e.g., 'a common term with a verb in the present'), and second order rules negative ones (e.g., 'without any ampliative terms'), formally defining exceptions. Third order rules, however, cannot be given in linguistic terms. They concern extra-linguistic, contextual elements such as—in the case discussed above—the intention of the speaker. More precisely, when a speaker uses a word in a sentence in such a way that it performs a *demonstratio ad intellectum* targeting a non-actual thing, then, even if all the first and second order conditions for appellation are satisfied, the *regula appellationum* does not apply. This third order rule, I believe, has a genuine pragmatic character.[37]

Ontology—a) The critique of material supposition is partially based on an ontological consideration: there is no real foundation for the distinction between personal and material supposition because any thing for which a term in a (so-called) material supposition can stand is a singular thing (see above, section 3). *b)* Brinkley's explanations of the relation between simple and personal supposition, as well as the analogies he construes in order to have his readers grasp the relation between universal and singular things,[38] lead to a moderate realist ontology according to which universals are metaphysical parts of individuals. In that sense, Brinkley's ontology can be labelled as a medieval form of immanent realism (see above, section 7).

Semantics, pragmatics and ontology—The relation between the three fields appears most clearly in the description of the origin of supposition (*suppositio oritur ex natura rei*) as well as in its definition (*suppositio est ordinatio alicuius termini de quo alius dicitur, vel qui dicitur de alio*). The *ordinatio* of a term is its actual position (in the active sense of the word) as subject or predicate in a sentence. The equivalence principle *Ep* essentially links supposition and use;

---

37) For seminal studies of pragmatic elements and rules in medieval philosophy of language, cf. Marmo (1994), (1995) and (2006), as well as Rosier-Catach (1994) and (2004).

38) The relation universal / singular is analogous, on one hand, to the relation primary matter / hylomorphic composite and, on the other, to the relation accident / substance; see *TDS*, 6.14 and *De universalibus*, §10, quoted in note 36 above.

and use, in turn, is essentially linked with non-linguistic factors such as the context of enunciation or the speaker's intention. Thus, according to what a speaker wants to say, he will 'posit' a word in a sentence in order to make it supposit precisely for the thing or things about which he intends to say something: if the speaker wants to refer to a universal thing, he produces a sentence such that its subject is in simple supposition. If he wants to refer to a singular thing, he produces a sentence such that its subject is in personal supposition. Consequently, the objective character of supposition—a character which is derived from its origin *ex natura rei*, and explains the validity of logical rules such as the rules of descent and ascent—is subordinated to two arbitrary elements: the conventional signification of words and the intentions of a freely speaking subject.[39]

[39] Another expression of the tension between objective and subjective (or arbitrary) factors in language is the medieval discussions about the two ways of interpreting sentences (*de virtute sermonis* vs. *secundum intentionem loquentis*). On that topic, cf. Courtenay (1984), Thijssen (1997), Hoenen (2002) and Bungs and Goubier (2009).

Appendix

Laurent Cesalli and Joël Lonfat

Below we present a provisional edition of part of the *Tractatus de suppositionibus*. The *Summa logicae* of Richard Brinkley is preserved in three known manuscripts:

Prague, *Státní knihovna ČSR*, Ms 396 8 (III.A.11), ff. 31ra-140ra [P], 71va-81rb for our treatise[40]
Leipzig, *Universitätsbibliothek*, Nr. 1360, ff. 1ra-105vb [L], 53vb-64ra for our treatise[41]
London, *British Library, Harley* 3243, ff. 47ra-56rb [H][42]

The manuscript from the British Library is incompete, containing only the part on obligations. The Prague witness is the better manuscript, and most studies until now have claimed that Leipzig stems from Prague.[43] Nevertheless, Leipzig occasionally has the better reading, as Paul Spade has pointed out, and contains material absent in Prague, so the definitive manuscript study will have to wait for the edition of the entire *Summa*. Below we follow P, with rare and tacit corrections from L. Punctuation has been kept to a minimum.

[1.1] [P 71va; L 53vb] Differt autem significatio a suppositione primo quia aliqua significatio est ad placitum eo quod terminus habet illam ex arbitrio voluntatis, sicut iste terminus 'Sor' non ex natura sed ex imperio voluntatis significat talem rem Sor. Suppositio igitur non oritur ex principio voluntario sed ex natura rei; verbi gratia, non est ex arbitrio voluntatis quod talis consequentia sit bona 'homo currit, igitur iste homo currit vel iste homo currit et sic de aliis'. Unde oritur talis regula generalis: a termino supponente determinate ad disiunctivam factam de omnibus suis singularibus est bona consequentia.
[1.2] Si autem suppositio ad [L 54ra] placitum <determinaretur> [distribueret], nulla talis regula esset necessaria quia mutabilis esset talis descensus ad singularia secundum mutationem voluntatis.
[1.3] Similiter terminus qui significat ratione sue significationis tantum includit duos respectus rationis: unum ad intellectum concipientem, alium ad rem ad quam movet intellectum intelligendam. Terminus autem qui supponit includit illos duos respectus et cum hoc tertiam habitudinem habet ad terminum tertium respectu cuius subicitur vel predicatur ut iste terminus 'homo'

---

40) Described in Truhlář (1905), 154.
41) Described in Fitzgerald (1987), 30.
42) Described in British Museum (1808), 11.
43) See remarks on the *Summa* in Gál-Wood (1980), Gaskin (1997), Fitzgerald (1987), Kaluza (1989-1990) and (1993), Spade (1969), (1991) and (1995) and Cesalli (2002), (2003), (2004) and (2008).

quando significat talem rem homo includit unum respectum ad intellectum cui significat et alium ad rem illam quam significat et quando est extremum propositionis ita quod subiectum vel predicatum, fundat tertium respectum ad extremum respectu cuius subicitur vel predicatur.
[1.4] Et ideo dicitur quod suppositio est ordinatio alicuius termini de quo alius dicitur vel qui dicitur de alio. Et ista discriptio includit utrumque extremum propositionis quia suppositio, quamvis proprie sit subiecti et appellatio secundum antiquos sit predicati, tamen quia tam subiecto in propositione quam predicato competit ista discriptio de suppositione ideo utrumque extremum rationabiliter potest dici supponere. Et isto modo utendo isto termino 'suppositio' non potest esse suppositio nisi in propositione; quamvis autem multis aliis modis sumatur in physica iste terminus 'suppositio', tamen isto modo prescripto volo uti eo hic et in sequentibus.
[1.5] Unde primo est sciendum quod in omni propositione kathegorica utrumque extremum supponit. Eque bene enim supponit subiectum in propositione falsa sicut in propositione vera et in propositione impossibili sicut in propositione necessaria quia loyce loquendo non stat propositionem esse et suppositionem non esse, quia non stat propositionem esse nisi sit in anima vel extra animam subordinata propositioni in anima ex prius dictis; sed sive propositio sit in anima sive extra animam subordinata propositioni in anima, utrumque extremum illius propositionis habet illos tres respectus qui inferunt terminum supponere; igitur non stat propositionem esse et suppositionem non esse.
[1.6] Falsum est igitur quod aliqui pro regula ponunt dicentes quod terminus nunquam supponit pro aliquo nisi pro eo de quo verificatur. Accipio enim istas duas propositiones 'antichristus erit', 'antichristus non erit'. Iste terminus 'antichristus' in utraque propositione supponit pro antichristo; et in utraque propositione non verificatur pro antichristo; igitur terminus aliquando pro aliquo supponit pro quo non verificatur. Et quando dicitur ab eis quod non semper verificatur per unum eiusdem temporis sed aliquando per unum presentis temporis aliquando per unum preteriti temporis et aliquando per unum futuri temporis hoc non solvit quod ista propositio 'Deus [L 54rb] non est' non potest verificari per unum alicuius temporis cum sit propositio ista impossibilis quia non potest nec potuit nec poterit esse vera.
[1.7] Similiter in ista propositione in anima Sortis 'Sor non est' subiectum supponit pro ipso Sorte et tamen non verificatur pro Sorte quia nec quando Sor est nec quando Sor non est: non quando Sor est quia tunc continue est falsa nec quando Sor non est [P 71vb] quia ista propositio non est; unde terminus frequenter supponit pro aliquo pro quo respectu eiusdem predicati non verificatur.
[1.8] Isto modo sumendo istum terminum 'suppositio' est ipsa in sua significata ulterius dividenda, sed ante eius divisionem volo supponere aliqua ad suppositionem terminorum mihi necessaria. Suppono enim primo quod quando terminus communis supponit respectu verbi de presenti non ampliativi solum supponit pro hiis que sunt sicut in ista propositione 'omnis homo currit'. Hic iste terminus 'homo' quamvis de se sit indifferens ad supponendum pro hiis qui fuerunt et pro hiis qui erunt, disiunctive vel coniunctive, tamen per modum illum quo ad istum contrahitur per 'solum' contrahitur iste terminus 'homo' per hoc predicatum cum verbo de presenti ad supponendum tantum pro hiis que sunt.
[...]
[1.11] Quamvis autem illa suppositio communiter sit vera et ad arguendum et respondendum multum necessaria quia secundum Aristotelem primo *Posteriorum* capitulo primo oportet prescire et specificare significationes terminorum ut sciamus quid quis significat. Et ideo propter faciliorem et breviorem modum investigandi veritatem, supponimus quod terminus communis respectu

verborum diversorum temporum pro diversis supponit ita quod pro aliis respectu verbi unius temporis et pro aliis respectu verbi alterius temporis. Si enim respectu verbi de presenti equaliter supponeret terminus communis pro hiis que fuerunt sicut pro hiis que sunt tunc proposita tali propositione universali 'omnis homo currit' vel ista 'omnis homo est animal'—subiecto supponente pro illo quod secundario significat universalis—simpliciter <haberemus> [haberet] eas negare eo quod quelibet talis multas singulares haberet falsas; et sic non haberemus aliquam propositionem universalem per quam in terminis rectis—supposito quod tantum essent [P 72ra] tres homines quorum quilibet curreret—possemus veritatem illam exprimere requirenti. Et ideo quamvis universaliter verum non fuerit quod quando terminus communis supponit respectu verbi de presenti quod tantum supponit pro hiis que modo sunt—etiam supposito quod verbum illud respectu cuius supponit non sit ampliativum nec aliquod verbum alterius temporis positum sit ex parte unius extremi vel alterius—nec ad hoc probandum meo iudicio ratio cogens poterit inveniri cum aliter respectu verbi de presenti pro aliis supponere necesse concludatur. Nam cum demonstratio sit duplex quedam ad sensum et quedam ad intellectum ideo per propositionem demonstrativam indifferenter potest demonstrari illud quod non est et illud quod est: illud quod non est intellectui et illud quod est sensui. Et loquor hic de termino demonstrativo qui facit propositionem in qua subicitur esse singularem. Facio igitur hanc consequentiam 'hoc est, igitur aliquid est' et demonstro per subiectum antecedentis antichristum futurum. Constat enim quod consequentia est bona [L 54vb] nam ex opposito formaliter sequitur oppositum; tamen consequentia non valeret si subiectum consequentis tantum supponeret pro hiis que modo sunt quia stat absque formali et immediata contradictione quod nichil quod modo est esset et tamen quod antichristus esset. Magis igitur proprie terminus communis respectu verbi de presenti supponit pro hiis que sunt; aliquando tamen et minus proprie pro hiis que fuerunt vel pro hiis que erunt. Pro termino igitur communi supponente respectu verbi de presenti et pro aliis terminis de quibus non potest dari regula generalis notanda est regula quam ponit Linconiensis in tractatu suo *De individuatione* que talis est: in hiis que accidunt vel dicuntur proprie vel minus proprie semper naturaliter recurrit audientis intelligentia ad sensum magis proprie significatum nisi ex circumstantia sermonis determinetur quod sermo communiter et non proprie accipiatur. Et ideo ubi terminus communis supponens respectu verbi de presenti non determinatur per aliquod precedens vel subsequens vel etiam ampliatur sustineri debet quod tantum supponit pro hiis que sunt cum pro illis maxime proprie supponere videatur.

[...]

[2.2] Dividitur igitur suppositio propria secundum divisionem rerum; res enim primaria divisione dividitur in rem universalem et in rem singularem, primo *Perihermeneias* et in [L 55va] libro *Predicamentorum* secundum explanationem Boethii capitulo de substantia. Ideo omnis terminus qui representat nobis in propositione res, vel representat nobis rem universalem vel rem singularem; si universalem tunc terminus pro illa supponens dicitur habere suppositionem simplicem; si singularem ab illo termino distinctam representat nobis terminus tunc debet dici in propositione ubi sic representat habere suppositionem personalem; et si nullam rem ab eo distinctam nobis representat tunc stat materialiter. Sed isti termini non sunt accommodati rebus universaliter representatis per terminos huiusmodi suppositionem habentes quia prorior et accommodatior modus loquendi esset vocare illam suppositionem quando terminus supponit pro natura de se communicabili [P 72vb] ut non contracta suppositionem communem sive specificam quam suppositionem simplicem, et quando terminus supponit pro re singulari suppositionem individuam sive materialem quam suppositionem personalem—multe enim sunt res singulares et individue quarum nulla est persona—, et quando terminus nichil a se facit intellectum intelligere, negare ab illo suppositionem, cum suppositio de vi vocis sit habitudo inter terminum supponentem et suppositum que habitudo proprie est inter distincta. Ne tamen diversitas tantum sit vel videatur vocalis et non realis inter opiniones, volo loqui ut communiter tractantes hanc materiam loquuntur. Et hoc est quod docet Aristoteles secundo *Topicorum* capitulo tertio ubi dicit quod loquendum est ut multi, sentiendum est ut pauci. Sicut enim de vi vocis propositio est pro alio [passio] vel pro aliis positio tamen ubi extrema non stant pro aliis nec significativa est propositio, sic et suppositio ubi terminus non stat significative sed tantum pro se ipso. Dico igitur quod suppositio materialis est quando terminus subicitur vel predicatur de alio termino et non pro aliquo distincto ab illo termino; vel propter protervientes: suppositio materialis est quando terminus pro se ipso subicitur vel predicatur sicut in talibus '*buba* est vox non significativa' et huiusmodi.

[2.3] Suppositio simplex est quando terminus communis supponit pro re ut non contracta sive pro re pluribus communicabili ex parte rei sicut in talibus 'homo est primo risibilis', 'animal est primo sensibile' et huiusmodi.

[2.4] Suppositio personalis est quando terminus supponit pro re singulari ab illo termino distincta que res est singularis ut representatur per illum terminum sicut in talibus 'homo currit', 'homo disputat', 'asinus movetur' et huiusmodi. Et de istis suppositionibus per ordinem est dicendum. Et primo de suppositione materiali.

[...]

[3.2] Est autem sciendum quod quando aliquid predicatur de termino et pro termino absolute sicut in ista '*buba* est vox non significativa' non esset dicendum illum terminum 'buba' habere suppositionem materialem proprie quia talis terminus sic stans nullum respectum habet ad materiam, vel si habet, equaliter habet respectum ad formam sicut ad materiam cum equaliter sit significans sicut significatum. Verumptamen ubi aliquid predicatur de termino et pro termino non absolute sed ut iste terminus habet respectum ad sua significata, isto modo potest dici illum terminum habere suppositionem materialem cum in termino significante rem sit ispa significatio quasi forma et ipse terminus quasi materia recipiens illam significationem. Sed quia illum respectum non includit terminus significans ex impositione ex natura sui sed tantum per arbitrium voluntatis, proprie loquendo neutra suppositio esset dicenda suppositio materialis. Sed propter defectum nominum utimur illo vocabulo in omni suppositione ubi terminus supponit pro se ipso sive absolute sive cum respectu ad significatum.

[...]

[4.1] Post suppositionem materialem dicendum est de suppositione simplici et est sciendum quod suppositio simplex [L 56rb] est quando terminus communis supponit pro re de se communicabili ut communicabilis est sicut hic 'homo est animal', 'asinus est substantia'. Res enim primo significate per extrema istarum propositionum sunt de se communicabiles pluribus quia quamvis homo de facto sit Sor vel Plato et sic de aliis, et nullus sit homo nisi homo singularis nec est aliquis asinus nisi asinus singularis, tamen hoc quod primo significatur per illum terminum 'homo' est de se eque indifferens et indeterminatum ad Sortem sicut ad Platonem et sic de aliis. Consimiliter est de illo quod primo significatur per illum terminum 'asinus' quod de se est indeterminatum ad unum individuum in specie asinina et ad aliud ita quod ad quodlibet individuum in specie asini illud quod primo significatur per illum terminum 'asinus' est indifferens et indeterminatum. Et intelligo hoc 'determinabile de se' ut est conditio restringens terminum cui additur ita quod facit eum supponere prescise pro illo vel pro illis quod vel que insunt suo significato de natura intrinseca illius significati.

[...]

[6.6] Preterea ex ista positione videtur sequi quod sit aliqua res que non sit res singularis quia res de se pluribus communicabilis aut est res singularis aut non; si non: propositum; si sic, contra: omne singulare est primo incommunicabile; igitur nulla res singularis est communicabilis; igitur per conversionem: nulla res communicabilis est res singularis.

[6.7] Item si talis res que de se est communicabilis est res singularis igitur terminus communis quando supponit pro natura de se communicabili supponit pro re singulari, sed quando terminus communis supponit pro re singulari supponit personaliter; igitur quando terminus communis supponit pro natura de se communicabili supponit personaliter.

[...]

[6.14] Ad sextum quando dicitur quod aliqua sit res que non sit res singularis, hoc nego. Dico enim quod omnis res est singularis et cum dicitur contra sic: nulla res singularis est res communicabilis igitur per conversionem nulla res communicabilis est res singularis, concedo consequens terminis supponentibus personaliter quia sic nulla res est communicabilis, et tamen multe [L 59ra] sunt quarum quelibet est de se communicabilis. Exempla: ad omnia ista que dicuntur [P 75rb] de suppositione simplici satis expresse habetur in philosophia naturali. Probat enim Philosophus primo *Physicorum* quod materia prima est de se in potentia ad omnes formas generales et hoc in potentia illa que est disctincta contra actum. Hoc enim probat eo quod nullam formam materia prima de natura sui sibi determinat. Verumptamen materia prima iam non est nec aliquando fuit in tali potentia quia semper est et fuit sub actu opposito tali potentie. Nulla enim est materia que non est actuata per formam nec aliquando fuit materia quando non fuit actualiter per formam, et tamen materia prima de se nec est nec fuit per aliquam formam actuatam sed est et fuit semper de se in potentia ad omnem formam; stant igitur ille simul 'materia prima de se est in potentia ad omnem formam' et 'materia prima non est in potentia ad omnem formam' semper uniformiter loquendo de potentia, scilicet de potentia ante actum et generaliter de potentia distincta contra actum. Sic dico quod illud quod primo exprimitur per terminum communem ut per illum terminum 'homo' est de se commune sive communicabile ad omne individuum in specie hominis; et tamen de facto est tantum unum individuum ita quod non communicabile <et> [sed] realiter ita incommunicabile sicut individuum, cum fuit idem; de hoc alibi magis dictum est. Aliud exemplum potest esse adhuc de subiecto et accidente et precipue de illo quod accidentaliter inest subiecto et non per se ut de Sorte et albedine. Quamvis enim Sor sit albus tamen Sor de se non est albus quia si de se esset albus non posset esse et non esse albus, cum illud synkathegorema 'de se' facit terminum quem determinat prescise stare pro illo vel pro illis sine quo vel quibus absolute impossibile est illam rem esse que per illum terminum significatur. Constat enim quod sine albedine potest Sor esse et ideo hec est falsa 'Sor de se est albus' quamvis hec est vera 'Sor est albus'. Sic quamvis hec sit vera 'homo est Sor' tamen hec est falsa 'homo de se est Sor'. Et si arguitur contra sic: Sor de se est Sor sed Sor de se est homo igitur homo de se est Sor, patet quod consequentia non valet quia in antecedente iste terminus 'homo' supponit personaliter et in consequente simpliciter et sic medium super quod fundari deberet unio extremorum variatur. Sed contra: aliquid de se est Sor quia Sor per datum; aut ergo de se est Sor homo, aut non homo de se est Sor; si non homo de se est Sor igitur Sor est de se non homo per conversionem—item: si sic, quod non est homo est homo, quod est contradictio. Si homo, habetur propositum. Respondeo quod iste terminus 'aliquid' potest supponere cum isto determinabili 'de se' personaliter vel simpliciter eo quod ex se ratione sue significationis non plus determinat sibi naturam de se universalem quam singularem. Sed sic non est de isto termino communi 'homo' qui ex impositione primo imponebatur [L 59rb] ad significandum naturam de se communem ut patet ex precedentibus. Et ideo si iste terminus 'aliquid' in prima propositione supponit personaliter propositio est vera pro Sorte. Et ideo sicut hec est vera 'Sor de se est homo' sic hec est vera 'aliquid de se est homo' subiecto supponente personaliter. Et si arguitur per conversionem 'igitur homo de se est Sor' dico quod non debet sic converti magis quam hec propositio 'omnis homo est animal' debet sic converti 'omne animal est homo'; quomodo autem debent converti huiusmodi propositiones in quibus ponuntur talia synkathegoremata 'de se', 'per se', 'eatenus', 'in quantum', 'ea ratione' et huiusmodi in sequentibus ostendetur. Verumptamen si subiectum in ista propositione 'aliquid de se est Sor' supponat simpliciter, propositio est falsa. Sed tunc valet consequentia 'aliquid de se est Sor et nichil aliud quam homo igitur homo de se est Sor'. Et sicut antecedens est falsum ita consequens. Si autem subiectum supponat personaliter tunc consequentia non valet eo quod iste terminus 'homo' [P 75va] in antecedente supponit personaliter et in consequente simpliciter. Et ideo est fallacia figure dictionis sicut si sic arguatur 'iste homo de se est Sor demonstrando Sortem igitur homo de se est Sor' in cuius antecedente subiectum supponit personaliter et in consequente simpliciter.

[6.15] Ad ultimum quando dicitur: si res de se communicabilis sit res singularis igitur quando terminus communis supponit pro re de se communicabili supponit pro re singulari, hoc enim concedo; nam nulla res est nisi singularis. Et ideo quandocumque terminus aliquis sive communis sive discretus supponit pro re singulari, et ulterius quando infertur quod terminus communis quando supponit pro re singulari supponat personaliter, hoc enim falsum est nisi illa res quatenus exprimitur per illum terminum sit res singularis, quod numquam est quando terminus supponit simpliciter.

[6.16] Propter predicta est dilligenter advertendum quando iste terminus 'de se' vel aliquis ei equivalens determinat totum subiectum in comparatione ad predicatum, et quando non determinat subiectum in comparatione predicati sed tantum unam partem subiecti respectu alterius. Quando enim primo modo determinat tunc facit terminum stare pro illo quod primo significat; verbi gratia sic dicendo 'homo de se est universalis', 'homo de se est indifferens ad Sortem et Platonem', 'materia prima de se est indifferens ad formam que est et ad formam que erit'. Si autem tantum determinet unam partem subiecti respectu alterius tunc illud extremum potest stare indifferenter pro illo quod primo significat sive pro illo quod significat secundario. Unde utraque istarum est vera subiecto supponente diversimode 'homo de se universalis est singularis' et 'homo de se universalis est universalis' et huiusmodi.

[...]

[7.1] [...] Pro primo dico quod suppositio personalis est quando terminus stat significative et supponit pro re singulari sub propria ratione singulari, et voco hic rem quicquid potest terminare actum intellectus quomodo loquitur Aristoteles de ente sexto *Metaphysice* commento ultimo et ens sic acceptum est convertibile cum re per Avicennam primo *Metaphysice* sue capitulo quinto, eo quod utrumque est de primis impressionibus que non possunt per alia priora cognosci. Per primum distinguitur suppositio personalis a suppositione materiali quia quando terminus supponit materialiter non stat significative; per secundum et tertium suppositio personalis distinguitur a suppositione simplici quia quamvis in suppositione simplici terminus supponat pro re singulari, non tamen ut res singularis.

Bibliography

Primary Literature

Manuscripts

Prague, *Státní knihovna* ČSR, Ms 396 8 (III. A. 11), f° 31ra-140ra

Leipzig, *Universitätsbibliothek*, Nr. 1360, f° 1ra-105vb

London, *British Library, Harley* 3243, f° 47ra-56rb

Catalogues

British Museum, *A Catalogue of the Harleian Manuscripts in the British Museum: with Indices of Persons, Places, and Matters*. London 1808

Truhlář, Joseph, *Catalogus codicum manuscriptorum Latinorum qui in C. R. Bibliotheca publica atque Universitatis Pragensis asservantur*. Prage 1905

Editions of Parts and Isolated Chapters of Brinkley's Summa logicae

Richard Brinkley, *Summa logicae*. Part II (*De universalibus*) in : L. Cesalli, 'Richard Brinkley *contra dialecticae haereticos*: une conception métaphysico-logique de l'universel', in: *Documenti e Studi Sulla Tradizione Filosofica Medievale* 19 (2008), 277-333

——, *Summa logicae*. Part V.1 (*De propositione in genere*) in: L. Cesalli, 'Richard Brinkley O.F.M. De propositione (*Summa logicae* V.1-5)', in: *Archives d'histoire doctrinale et littéraire du moyen-âge* 71 (2004), 203-254

——, *Summa logicae*. Part V.2 (*De significato propositionis*) in: M.J. Fitzgerald, *Richard Brinkley's Theory of Sentential Reference* (Leiden 1987)

——, *Summa logicae*. Part VI (*De insolubilibus*) in: P.V. Spade, *An Anonymous Fourteenth Century Treatise on Insolubles. Text and Study* (Toronto 1969)

——, *Summa logicae*. Part VII (*De obligationibus*) in: P.V. Spade and G.A. Wilson, *Richard Brinkley's Obligationes. A Late Fourteenth Century Treatise on the Logic of Disputation* (Münster 1995)

——, *Summa logicae*. Isolated chapters (I, 1: *De significatione*; II, 1: *De hoc termino 'universale'*; III, 1: *De praedicamentis*; III, 6: *De relatione*; IV, 2: *De divisione suppositionis*) in: G.Gál and R. Wood, 'Richard Brinkley and his *Summa logicae*'. in: *Franciscan Studies* 40 (1980), 79-101

Other Medieval Authors

Gualterus Burlaeus, *De Puritate Artis Logicae Tractatus Longior, with a Revised Edition of the Tractatus Brevior*, ed. by Ph. Boehner O.F.M. (Franciscan Institute Publications; Louvain-Paderborn 1955)

Guillelmus de Heytesbury, *Sophismata asinina*, ed. F. Pironet (Paris 1994; Electronic edition at http://ddata.over-blog.com/xxxyyy/4/07/38/29/Ttextdx-Philomed/Heytesbury/W_Heytesbury_Sophismata_Part1.pdf

Guillelmus de Ockham, *Summa logicae*, ed. P. Boehner, G. Gál and S. Brown (Guillelmi de Ockham Opera Philosophica, I; St. Bonaventure, N.Y. 1974)

Guillelmus de Shireswood: William of Sherwood, *Introductiones in Logicam. Einführung in die Logik*, ed. H. Brands and C. Kann (Hamburg 1995)

Johannes Duns Scotus, *Quaestiones in Libros Perihermenias Aristotelis*, eds R. Andrews, G. Etzkorn, G. Gál, R. Green, T. Noone, R. Plevano, A. Trever and R. Wood (Ioannis Duns Scoti Opera Philosophica, II; St. Bonaventure, N.Y, Washington DC 2004)

——, *Quaestiones super librum Elenchorum Aristotelis*, eds. R. Andrews, O. Bychkov, S. Ebbesen, G. Etzkorn, G. Gál, R. Green, T. Noone, R. Plevano and A. Trever (Ioannis Duns Scoti Opera philosophica, II; St. Bonaventure, N.Y. and Washington DC 2004)

Johannes Pagus, *Appellationes*, in: A. de Libera, 'Les *Appellationes* de Jean le Page', in: *Archives d'histoire doctrinale et littéraire du moyen âge* 51 (1984), 193-255

Lambertus Autissiodorensis: Lambert of Lagny, *Appellationes*, ed. A. de Libera, in: id. 'Le *Tractatus de appellatione* de Lambert de Lagny (*Summa Lamberti* VIII)', in: *Archives d'histoire doctrinale et littéraire du moyen âge* 48 (1982), 227-285

Walter Burley, *vide s.n.* Gualterus Burlaeus

William of Ockham: *vide s.n.* Guillelmus de Ockham

William of Heytesbury, *vide s.n.* Guillelmus de Heytesbury

Secondary Literature

Ashworth, E.J. (2004), 'Singular Terms and Singular Concepts: from Buridan to the Early Sixteenth Century', in: S. Ebbesen and R. Friedman, *John Buridan and Beyond. Topics in the Language Sciences 1300-1700* (Copenhagen 2004, 121-151).

Bérubé, C. (1964), *Connaissance de l'individuel au moyen âge* (Paris 1964)

Brands, H. (1990), 'Die zweifache Einteilung der formalen Supposition bei William of Sherwood', in: S. Knuuttila, R. Työrinoja, S. Ebbesen, eds., *Knowledge and the Sciences in Medieval Philosophy. Proceedings of the Eighth International Congress of Medieval Philosophy (SIEPM)* (Helsinki 1990, 445-454)

Brumberg-Chaumont, J. (2007), 'Sémantiques du nom propre: sources anciennes et discussions médiévales à l'époque d'Abélard', in: *Histoire, épistémologie, langage* 29/1 (2007), 137-166

Brungs, A., and F. Goubier (2009), 'Biblical Logicism. Wyclif, *virtus sermonis* and Equivocation', in: *Recherches de Théologie et de Philosophie Médiévales* (2009), 76/1, 199-244

Cesalli, L. (2002), 'Some 14th-Century Realist Theories of the Proposition', in: H. Singh Gill, ed., *Signification in Language and Culture* (Shimla 2002, 83-118)

—— (2003), 'La sémantique des syncatégorèmes chez Walter Burley (1275-1344) et Richard Brinkley (*fl.* 1365)', in: *Histoire, Epistémologie, Langage* 25/II (2003), 115-144

—— (2004), '*Ubi est propositio*'? Richard Brinkley sur les lieux de la proposition (*Summa logicae I, 1-5*)', in: A. Maierù and L. Valente, eds., *Medieval Theories on Assertive and Non-Assertive Language*. Acts of the 14th European Symposium on Medieval Logic and Semantics, Rome, June 11th-15th, 2002 (Firenze 2004, 445-471)

—— (2007), *Le réalisme propositionnel. Sémantique et ontologie des propositions chez Jean Duns Scot, Gauthier Burley, Richard Brinkley et Jean Wyclif* (Paris 2007)

—— (2011), 'Richard Brinkley', in: H. Lagerlund, ed., *Encyclopedia of Medieval Philosophy. Philosophy Between 500 and 1500*. 2 vols. (Heidelberg 2011), 1120a-1123b.

Courtenay, W.J. (1984), 'Force of Words and Figures of Speech: The Crisis over *virtus sermonis* in the Fourteenth Century', in: *Franciscan Studies* 44 (1984), 107-128

Emden, A.B. (1957), *A Biographical Register of the University of Oxford to A.D. 1500*. 3 vols. (Oxford 1957-1959)

Gál, G., and R. Wood (1980), 'Richard Brinkley and his *Summa logicae*', in: *Franciscan Studies* 40 (1980), 59-78

Gaskin, R. (1997), 'Russell and Richard Brinkley on the Unity of the Proposition', in: *History and Philosophy of Logic* 18 (1997), 139-150

Hoenen, M.J.F.M. (2002), 'Jean Wyclif et les *universalia realia*. Le débat sur la notion de *virtus sermonis* au moyen âge tardif et les rapports entre la théologie et la philosophie', in: J.-L. Solère and Z. Kaluza, eds., *La servante et la consolatrice. La philosophie dans ses rapports avec la théologie au Moyen Âge* (Paris 2002, 173-192)

Jacobi, K. (1980), *Die Modalbegriffe in den Logischen Schriften des Wilhelm von Shyreswood und in anderen Kompendien des 12. und 13. Jahrhunderts. Funktionsbestimmung und Gebrauch in der logischen Analyse* (Leiden 1980)

Kaluza, Z. (1989), 'L'œuvre théologique de Richard Brinkley, O.F.M.', in: *Archives d'histoire doctrinale et littéraire du moyen âge* 64 (1989), 169-273

—— (1992), 'Richard Brinkley', in: A. Jacob, *L'encyclopédie philosophique universelle* (Paris 1992, 804-805)

Libera, A. de (1981), 'Supposition naturelle et appellation: aspects de la sémantique parisienne au 13e siècle', in: *Histoire, épistémologie, langage* 3/1 (1981), 63-77

Marmo, C. (1994), *Semiotica e linguaggio nella scolastica. Pariggi, Bologna, Erfurt 1270-133): la semiotica dei Modisti* (Nuovi Studi Storici, 26; Roma 1994)

—— (1995), 'A Pragmatic Approach to Language in Modism' in: S. Ebbesen, ed., *Sprachtheorien in Spätantike und Mittelalter* (Tübingen 1995, 169-183)

—— (2006), 'La funzione del contesto: teorie 'continentali' e 'inglesi' a confronto sull'eliminazione dell'equivocità tra fine XIII e inizio XIV secolo', in: S. Caroti, R. Imbach, Z. Kaluza, G. Stabile and L. Sturlese, eds., *'Ad Ingenii Acuitionem'. Studies in Honor of Alfonso Maierù* (Louvain-La-Neuve 2006, 249-280)

Panaccio, C. (2004), 'Tarski et la *suppositio materialis*', in: *Philosophiques* 31/2 (2004), 295-309

Rosier-Catach, I. (1994), *La parole comme acte. Sur la grammaire et la sémantique au XIIIe siècle* (Paris 1994)

—— (2004), *La parole efficace. Signe, rituel, sacré* (Paris 2004)

Thijssen, J.H.M.M. (1997), 'The Crisis Over Ockhamist Hermeneutic and its Semantic Background: The Methodological Significance of the Censure of December 29, 1340', in: C. Marmo, *Vestigia, Imagines, Verba. Semiotics and Logic in Medieval Theological Texts (XIIth-XIVth century)* (Turnhout 1997, 371-392)

Semantic and Ontological Aspects of Wyclif's Theory of Supposition

Alessandro D. Conti
University of Aquila

Abstract
The relationship between thought and reality was a focal point of Wyclif's reflection. On the one hand, Wyclif believed that thought was linguistically constrained by its own nature; on the other hand, he considered thought to be related to reality in its elements and constitution. Hence he deemed language, thought, and external reality to be of the same logical coherence. Within this context, the theory of supposition was intended to explain the different roles that terms can have in relation to language and the extra-mental world when they appear as extremes in propositions. Characteristically, his theory of supposition provides an account not only of the truth-values of a sentence, but also of its meaning; it is not therefore simply a theory of reference, but a sort of complex analysis of language viewed as a semiotic system whose unique interpretative model was reality itself. It gives clear evidence of Wyclif's realist stance and of his conviction that any kind of linguistic and semantic features must be grounded on ontological structures.

Keywords
essence, individual(s), meaning, predication, signification, supposition, universal(s)

1. Signification, Supposition and Meaning

The relationship between thought and reality was a focal point of Wyclif's reflection. On the one hand, Wyclif believed that thought was linguistically constrained by its own nature; on the other hand, he considered thought to be related to reality in its elements and constitution. Hence he deemed language, thought, and external reality to be of the same logical coherence.[1] Within this context, the theory of supposition was intended to explain the different roles

[1] Cf. Conti (2006), 114-118, and Spruyt (2008), 24-25.

that words (or phrases) can have in relation to language and the extra-mental world when they appear as extremes (that is, as subject or predicate) in propositions.[2] Characteristically, his theory of supposition provides an account not only of the truth-values of a sentence, but also of its meaning; it is not therefore simply a theory of reference, but a sort of complex analysis of language viewed as a semiotic system whose unique interpretative model was reality itself. It gives clear evidence of Wyclif's realist stance and of his conviction that any kind of linguistic and semantic features must be grounded on ontological structures.

In what follows, I shall consider the most important aspects of Wyclif's theory of supposition, trying to set it in relation to the medieval tradition of treatises on signification and supposition and particularly to its main source, the theory expounded by Walter Burley in his *De puritate artis logicae tractatus longior* (composed between 1325 and 1328), which contains an original and intelligent defence of the older view of *significatio* and *suppositio simplex* against Ockham's attacks.[3] Thus, in the first part of this paper I shall present a short outline of Wyclif's definition and divisions of supposition, as developed in the first chapters, and especially in chapter 12, of his treatise on logic (*De logica*, composed around 1360). In the second part, some applications of the supposition-theory to the discussion of some fallacies and sophisms will be analyzed. In the final section, I shall draw some conclusions about the general significance of Wyclif's doctrine in the light of his philosophical programme.

2. Supposition Defined

Wyclif defines supposition as the signification of one categorematic extreme of a proposition (subject or predicate) in relation to the other extreme:[4]

> Supposicio est significacio termini kategor<emat>ici qui est extremum proposicionis, in comparacione ad aliud extremum. Et est extremum in proposicione subiectum vel predicatum.

---

2) Following the medieval usage, in this paper I shall employ the terms 'proposition' and 'sentence' as if they were synonymous.

3) Cf. Burley, *De puritate artis logicae tractatus longior*, ed. Boehner (1955), I, i, iii, p. 7, 6-10. As is well known, according to Ockham, in *suppositio personalis*, which he takes to be the normal case, a term supposits for what it signifies, that is, one or more individual things. Burley argues that, if so, then a common term, like 'man', should signify Socrates, Plato, and any other individual man, and therefore no one could learn the meaning of the term 'man' without learning that it applied precisely to Socrates, to Plato and so on—which is obviously false.

4) Cf. Wyclif, *Tractatus de logica*, ed. Dziewicki (1893-1899), ch. 12, vol. 1, 39.

This definition, which is drawn from Burley's *De suppositionibus* (composed in 1302), sounds partially different from the standard definition of supposition, as it seems to somehow equate signification and supposition, since supposition is considered as a particular kind of signification. On the contrary, according to the most common view, which went back to Peter of Spain's *Summulae logicales*, the *significatio* and *suppositio* of terms were clearly distinct functions, inasmuch as the latter presupposed the former, but was a *proprietas terminorum* totally different from it.[5] In fact, (1) signification consisted in the relation of a linguistic sign to what it signifies apart from any propositional context; (2) a word capable of standing for something else or for itself in a proposition first had to have signification; (3) a term only had supposition in a propositional context; and (4) the kind of supposition a term had depended on its propositional context. In any case, in a traditional realist perspective, supposition served to tell us which things are involved in the truth-conditions of a given sentence: whether they are expressions, real universals, or individuals.

At the very beginning of the chapter on supposition, like Walter Burley,[6] Wyclif divides supposition into improper (*impropria*), in which a term stands for something different from its primary *significatum* by special custom (*ex usu loquendi*), and proper (*propria*), in which a term stands for something by the virtue of the expression itself. So a term has improper supposition when it is used in a figurative speech, as in the case of the term 'cup' in the sentence 'I have drunk a cup [of wine]' ('*bibi ciphum*'). Wyclif divides proper supposition into material (*materialis*), when the term stands for itself or its sound (as it occurs in "'I' is a pronoun" or "'Johannes' is trisyllabic"), and formal (*formalis*), when the term stands for what it properly signifies. Formal supposition is twofold: simple (*simplex*) and personal (*personalis*). Like William of Sherwood, Peter of Spain, and Burley, and against Ockham and his followers, Wyclif affirms that the supposition is simple if the term stands for an extra-mental universal only (*solum assertive supponit pro re universali ad extra*), as it occurs in 'Man can be predicated of every man' ('*homo predicatur de omni homine*') and 'Man is a species' ('*homo est species*'). According to Wyclif, in both cases the term 'man' supposits for the human nature, which is an extra-mental form common to a multiplicity of singulars. Simple supposition is divided into equal (*equa*) and unequal (*inequa*). A term is in simple equal supposition if it stands for the common nature that it directly signifies, as occurs in 'man is a species' ('*homo est species*'). A term is in simple unequal supposition when it stands for

[5] Cf. Maierù (1972), 92 and 218-219.

[6] Cf. Burley, *De puritate artis logicae tractatus longior*, ed. Boehner (1955), I.1.1, p. 2, ll. 17-18.

(1) a less common nature than what it signifies (*pro specie inferiori*), as occurs in 'substance is a species' ('*substancia est species*'), or (2) a concrete accident or the characterizing property (*pro accidente vel proprio primo*), as occurs in 'this universal-man is capable of laughing' ('*hic homo communis est risibilis*')—where the presence of the demonstrative 'this' modifies the *significatum* of the subject-term 'universal-man', so that in the sentence at issue it supposits for that concrete exemplification (the human nature proper to an individual man) which is identical with the subject of inherence (a given human being) of the accidental form, or characterizing property (in the example, the capacity-of-laughing), signified by the predicate-term. Supposition is personal when the term which plays the role of subject in a sentence stands for one or more individuals (*pro uno singulari vel pro multis*). In the first case, the supposition is personal and singular (*suppositio personalis singularis*), as it occurs in 'this man is' ('*hic homo est*'); in the second one, it is personal and common (*suppositio personalis communis*). Personal and common supposition is twofold. If the term stands for many singulars considered separately or for some (that is, at least one) determinate individual named by the common term itself, the supposition is *personalis distincta* (or *determinata*, as Wyclif calls it in the final section of chapter 12), as occurs in 'these (men) are' ('*isti sunt*'—*suppositio personalis communis distincta*). If the term stands for many singulars considered together, supposition is *personalis universalis*. In turn, *suppositio personalis universalis* is divided into *confusa distributiva* and *confusa tantum*. There is *suppositio personalis communis universalis confusa distributiva* when the (subject-)term stands for everything that has the form signified by the term, as occurs in 'every man is' ('*omnis homo est*'). There is *suppositio personalis communis universalis confusa tantum* when the form (or property) signified by the term at issue is affirmed (or not affirmed) equally well of one of the bearers of that form as it is of another, since it applies (or does not apply) to each for exactly the same reasons, as occurs in 'each of them is one of the two' ('*uterque istorum est alter istorum*'), where the expression 'one of the two' has merely confused supposition, since neither of the two can be both of them (*quia non est dare aliquem istorum qui est uterque istorum*). The *suppositiones confusae* are so called since they involve many different individuals, and this is the case for the subject of a universal affirmative proposition.[7]

Wyclif takes a resolutely realist stance, as his own formulation and division of supposition (where simple supposition is described as that possessed by a term in relation to a universal outside the intellect and personal supposition as

[7]) Cf. Wyclif, *Tractatus de logica*, ed. Dziewicki (1893-1899), ch. 12, 39-40.

that possessed by a term in relation to one or more individual) make evident. In this way, he stresses the ontological implications of Burley's theory. In the *De suppositionibus* and *De puritate*[8] the *Doctor Planus et Perspicuus* had adopted a semantic point of view in describing supposition, since he defined formal supposition as the supposition that a term has when it stands for its own *significatum* or for the (individual) *supposita* which fall under it. In the first case, we properly speak of simple supposition, and in the second, we speak of personal supposition. Wyclif makes clear what Burley had stated only implicitly: the *significatum* of a common term is always a common nature (that is, a universal form) really existing outside the intellect. This fits in with his theory of meaning and his ontology.

In the first chapter of his treatise on logic Wyclif maintains that (1) a categorematic term is a *dictio* to which a mental concept, a sign of a thing (*intentio significans pro re*), corresponds in the soul.[9] (2) Categorematic terms are divided into common (namely, general expressions), like 'man' and 'dog', and discrete (namely, singular referring expressions), such as personal and demonstrative pronouns and proper names.[10] (3) Common terms originally and primarily (*principaliter*) signify common natures—for instance, the term 'man' originally and primarily signifies human nature.[11] (4) Categorematic terms can be divided into substantial terms, such as 'man', and accidental terms, such as 'white'. A substantial term signifies a common nature proper to a set of individuals (of which the term is the name) without connoting any accidental property, while an accidental term signifies (but we would rather say 'refers to') a common essence, proper to a set of individuals, and also (we would add: connotes) an accidental property, that is, a property which is not constitutive of the essence referred to.[12] (5) Categorematic common terms can be divided also into abstract and concrete. According to Wyclif, a concrete term, like 'man', is a term which signifies a thing that can *indifferenter* supposit *simpliciter*

---

8) Cf. Burley, *Tractatus de suppositionibus*, ed. Brown (1972), 35-36; and *De puritate artis logicae tractatus longior*, ed. Boehner (1955), I, i, iii, pp. 6-10.

9) Cf. Wyclif, *Tractatus de logica*, ed. Dziewicki (1893-1899), ch. 1, 2.

10) Cf. Wyclif, *Tractatus de logica*, ed. Dziewicki (1893-1899), ch. 1, 2-3.

11) Cf. Wyclif, *Tractatus de logica*, ed. Dziewicki (1893-1899), ch. 1, 3.

12) Cf. Wyclif, *Tractatus de logica*, ed. Dziewicki (1893-1899), ch. 1, 3: 'Terminus substancialis est terminus qui significat naturam rei sine connotacione accidentalis proprietatis; ut iste terminus, *homo*, significat essenciam humanam sine connotacione extranea. [...] Sed terminus accidentalis est diccio significans essenciam rei, connotando accidentalem proprietatem: sicut iste terminus, *albus*, significat substanciam et similiter albedinem, que est proprietas extranea ab essencia, que est substancia.'

and *personaliter*.[13] On the contrary, an abstract term is a term which signifies only a common nature without connoting anything else, like '*humanitas*' and '*albedo*'. It is worth noting that in defining concrete terms Wyclif plainly attributes the capacity for suppositing to things, does not clarify the metaphysical composition of such things signified by concrete terms, and describes the twofold supposition of concrete terms as a sort of signification.[14] (6) Finally, categorematic terms can be divided into terms of first and second intention. A term of first intention is a sign which signifies what it signifies (*significat suum significatum*) without connoting the properties of being-individual or being-universal (*non connotando rationem singularitatis aut universalitatis*) which characterize categorial items. For example, '*Deus*' and '*homo*' are terms of first intention. On the contrary, a term of second intention is a term which connotes such properties and refers to a common nature without naming it. '*Universale*' and '*substantia prima*' are terms of second intention.[15]

As is evident, the basic ideas of Wyclif's theory of meaning are that (1) every simple expression in our language is like a label naming just one *essence*[16] in the world, and (2) distinctions among terms as well as their linguistic and semantic properties are derived from the ontological features of signified things. He affirms that everything which is, signifies in a complex manner that it is something real.[17] He openly claims that supposition is also a property of signified things, and explains the semantic difference between general terms, such as 'man', which can name a set of individuals, and singular expressions, such as 'Socrates' or 'a certain man' ('*aliquis homo*'), which name just one item, by means of the different modalities of existence of their different *significata*. Singular expressions name and signify individuals; general terms name and

---

13) Cf. Wyclif, *Tractatus de logica*, ed. Dziewicki (1893-1899), ch. 1, 5.

14) Cf. Wyclif, *Tractatus de logica*, ed. Dziewicki (1893-1899), ch. 1, 5-6.

15) Cf. Wyclif, *Tractatus de logica*, ed. Dziewicki (1893-1899), ch. 1, 7.

16) In Wyclif's philosophical terminology the term 'essence' is often used in order to designate the categorial items considered as common natures or instances of a certain nature. An essence therefore is a being which has a well defined nature, even if the name 'essence' does not make this nature known (cf. *De materia et forma*, ch. 4, 185-186). As a result, the term 'essence' is less general than 'being' ('*ens*'), but more general than 'quiddity' ('*quidditas*'), since every essence is a being, and not every being is an essence, and every quiddity is an essence, and not every essence is a quiddity, inasmuch as individual items are essences, but not quiddities (cf. *Tractatus de universalibus*, ed. Mueller (1985), ch. 1, 15-16; ch. 6, 116-124, *passim*, expecially 123).

17) Cf. Wyclif, *Tractatus de logica*, ed. Dziewicki (1893-1899), ch. 5, 14: 'Proposicio large loquendo est *ens complexe significans*; et sic quia omne quod est significat complexe se esse, omne quod est satis bene potest dici proposicio.' On this peculiar doctrine, his pan-propositionalism, see Cesalli 2005.

signify common natures. In Wyclif's view, a common term gives a name to a certain set of individuals only by way of the nature that (1) it originally and directly signifies and (2) is common to a certain group of individuals as their own quiddity.[18]

The first three chapters of his *Tractatus de logica* (on terms, universals, and categories respectively) make clear that Wyclif identifies secondary substances (that is, the universals of the category of substance) with the *significata* of general (concrete) terms of that category (such as 'man' or 'animal') and individual substances with the *significata* of singular expressions of that category (such as 'this man', which refers to a single human individual only). Furthermore, he holds that (1) common terms of the category of substance, when used predicatively, specify which kind of substance a certain individual substance is; (2) individual substances are unique physical entities, located at a particular place in space and time; and (3) universal substances are the specific or generic natures proper to the individual substances, immanent in them, and apt to be common to many individuals at the same time. As a result, like Burley, Wyclif thinks of universals and individuals as linked together by a sort of relation of instantiation. In other words, he conceives of individuals as *the* tokens of universal natures, and universal natures as *the* types of individuals. This consequence is common also to many other realist authors of the thirteenth and fourteenth centuries. But, because of his peculiar reading of the relation between universals and individuals, Wyclif derives from it an original conception of the signification and supposition of concrete accidental terms, such as 'white', by which the new theories and divisions of supposition developed in Oxford in the fourteenth and fifteenth centuries were to be inspired. According to them, any concrete accidental term which occurs as an extreme in a proposition can stand for (1) the substrate of inherence of the accidental form that it connotes (*suppositio personalis*), or (2) the accidental form itself (*suppositio abstractiva*), or (3) the aggregate composed of the individual substance, which plays the role of the substrate of the form, and the singular accidental form at issue (*suppositio concretiva*).[19]

---

18) Cf. Wyclif, *Tractatus de logica*, ed. Dziewicki (1893-1899), ch. 1, 7: 'Terminus significat primarie illud quod principaliter apprehenditur per illum; sicut iste terminus, *homo*, primarie vel principaliter significat hominem, scilicet naturam humanam, et secundarie significat *Johannem* vel *Robertum*.'

19) Cf. for instance William Penbygull, *De universalibus*, ed. Conti (1982), 196-197. On this subject see Conti (2005), 177-184.

In his last commentary on the *Categories* (composed in 1337),[20] the *Doctor Planus et Perspicuus* affirmed that, instead of signifying simple categorial items, concrete terms signify aggregates of individual substances along with the form primarily signified by the term itself. In the case of concrete accidental terms, like 'white' ('*album*'), such aggregates are not properly beings (*entia*), since they are lacking in numerical unity, and hence they do not fall under any of the ten categories. The metaphysical constituents of such aggregates (substance and accidental form) are related to the concrete accidental term in different ways: on the one hand, the form is the primary *significatum*, even if the concrete accidental term is not the name of the form; on the other, the concrete accidental term can only supposit for the substance. So concrete accidental terms name substances, but indirectly, through the accidental forms from which they take their names, insofar as substances are the substrates of existence (*subiecta*) in relation to the accidental forms. This fact accounts for the difference between concrete substantial terms (such as 'man') and concrete accidental terms, since the forms that concrete substantial terms primarily signify are the constitutive elements of the essence of the substances that the concrete substantial terms name. Therefore, in this case, the name of the form is just the same as the name of the substances.[21]

Wyclif seems to believe that the *significatum* of a concrete term is a sort of twofold entity formed (1) by one (or more) individual substance(s) and a common nature in the case of a concrete substantial term, and (2) by one (or more) individual substance(s) and an accidental common nature in the case of a concrete accidental term. In the first case, the common nature at issue is really identical with and formally different from the primary substances that the concrete term refers to. In the second one, the accidental nature at issue is really different from the primary substances in which it inheres as well as from the substantial nature (or natures) proper to them. This is the logical consequence of his metaphysical convictions about universals and accidents. As is well known, according to Wyclif, formal universals are common natures, or *veritates*, in virtue of which the individuals that share them are exactly what they are—just as the human species is the truth, or form, or nature, by which every man formally is a man. They are prior, and so 'indifferent' to any division into universals and individuals. Universality (*universalitas* or *communicabilitas*) is,

---

20) Cf. Burley, *Expositio super Praedicamenta Aristotelis*: *de sufficientia praedicamentorum*, ed. Venetiis (1509), f. 21ra; *de substantia*, f. 24rb; and *de relatione*, f. 34rb.

21) Cf. Burley, *Expositio super Praedicamenta Aristotelis*: *de denominativis*, ed. Venetiis (1509), f. 19va-b.

as it were, their inseparable property (*quasi passio*) and not a constitutive mark of the nature itself. Common natures possess it only potentially; it becomes an actual determination when one or more individuals instantiate them. Universals *qua* natures of a certain kind are really (*realiter*) identical to, but formally (*formaliter*) distinct from their individuals, since common natures and individuals share the same empirical reality (that of individuals) but, conceived of properly as universals and individuals, they have opposite principles: the natural-tendency-to-be-common (*communicabilitas*) for universals and the impossibility-of-being-common (*incommunicabilitas*) for individuals. Hence common natures are formal causes in relation to their own individuals, and individuals material causes in relation to their natures, since individuals are *partes subiectivae* of the common natures.[22] All the genera, species, and individuals belonging to the category of substance are therefore really identical (as the individuals which instantiate a certain specific nature instantiate also all the forms superior to it) and, if considered in themselves, formally distinct from each other as well as concrete accidents among them and in relation to the substances in which they inhere.

In fact, the chief feature of Wyclif's treatment of accidents is his twofold consideration of them as abstract forms (quantity and quality) or *respectus* (the other six accidental categories) and as concrete determinations (or modes) of individual substances. In the *De actibus animae* (composed in around 1365) he seems to conceive of them as modes of substance, without actually distinct realities.[23] By contrast, in his *De ente praedicamentali* (composed in 1369) he clearly states that accidents are *essences* really distinct from substance.[24] Indeed, in Wyclif's opinion, accidents, considered in an absolute way, according to their essential being (*esse essentiae* or *esse in genere*), which causes what they are, are abstract forms, really distinct from substances; but, if they are considered from the point of view of their concrete existence, they are not *really* distinct from the substance in which they are present, but only *formally*, since in the latter case they are mere determinations (or modes) of substances.[25]

---

22) Cf. Wyclif, *Tractatus de universalibus*, ch. 1, 15-16; ch. 2, 64; ch. 4, 86-87 and 90-92; ch. 11, 239-240. On Wyclif's theory of universals and individuals and the connected theory of predication see Spade (1985); Kenny (1986); Conti (1997), 150-158; Spade (2005); Conti (2006), 95-102.

23) Cf. Wyclif, *De actibus animae*, ed. Dziewicki (1902), pars II, ch. 4, 122-123 and 127.

24) Cf. Wyclif, *De ente praedicamentali*, ed. Beer (1891), ch. 7, 61. See also his *Tractatus de logica*, ed. Dziewicki (1893-1899), ch. 3, 11.

25) In the *Tractatus de universalibus*, ed. Mueller (1985), ch. 4, 91-92, in defining the formal distinction and its three different kinds, Wyclif maintains that (1) the formal distinction is the difference by which things differ from each other even though they are constitutive elements of the same

As a consequence, there is a difference between Burley and Wyclif in individuating the *significatum* of a concrete accidental term. According to Burley, no universal form can inhere in another one, as inherence concerns only individual items; therefore the *significatum* of a concrete accidental term is the set of all the individual aggregates of a substance and the accidental form signified by the abstract term corresponding to the concrete accidental term at issue—for instance, in the case of the term '*album*', the set constituted by those aggregates of an individual substance and the abstract form of whiteness (*albedo*) that inheres in it. In Wyclif's view, the *significatum* of a concrete accidental term seems to be an aggregate formed by two (or more) natures, one of which indirectly inheres in the other(s), by means of the individuals which are its subjects (or bearers). In the *Tractatus de universalibus*, he accepts that a sentence such as '*species humana est risibile*' is a well formed and true proposition, even though '*species humana est risibilis*' is not, because the human species is really identical to something (the concrete human beings) which is capable of laughing.[26]

Wyclif ends chapter 12 with three *notanda*, by which he completes his treatment of supposition. In the first, he recalls that categorematic common concrete terms can supposit both *personaliter* and *simpliciter* at once (*mixtim*) when the propositions where they occur as subjects are universal affirmative or indefinite. For instance, the term '*animal*' in (1) '*omne animal fuit in archa Noe*' as well as the term '*homo*' in (2) '*homo moritur*' can supposit *personaliter* for every individual animal and man respectively, and if so, the first sentence is false and the second true, and *simpliciter* for every species of animals and the human nature respectively, and then both sentences are true.[27] In the second *notandum*, Wyclif contends that proper names (like '*Johannes*'), personal and demonstrative pronouns (like '*hic*' and '*istud*'), and those terms of second intention by which we speak of singular items considered as such (namely, expressions like '*persona*' and '*individuum*') cannot supposit distributively,

---

single essence or supposit; and (2) among others, this is the case for the concrete accidents inhering in the same substance, as they coincide in the same particular subject, but differ from each other because of their own natures.

26) Cf. Wyclif, *Tractatus de universalibus*, ed. Mueller (1985), p. 240, ch. 11: 'Non est aliquod genus vel individuum accidentis, quin ipsum sit vere praedicabile tam de universali quam de individuo substantiae, diversimode tamen quia utrobique in concreto: de individuo formaliter et de universali secundum essentiam. Ut species humana, quamvis sit risibile, quantum et quilibet homo qualitercumque accidentatus, non tamen est risibilis, quantitative divisibilis, accidenter qualis vel quomodolibet aliter accidentata.'

27) Cf. Wyclif, *Tractatus de logica*, ed. Dziewicki (1893-1899), ch. 12, 40-41.

since they were devised in order to signify *discrete vel singulariter* only.[28] Finally, in the third *notandum*, he lays down the following rules about the supposition possessed by the subject-term and the predicate-term in the Square of Opposition: (1) in every universal affirmative proposition, the subject supposits *mobiliter*, that is, it has confused and distributive supposition (*in universali affirmativa subiectum supponit mobiliter, id est, confuse distributive*), while the predicate has *suppositio confusa tantum* or simple supposition. Supposition is *confusa tantum* if it does not allow for descent to a certain singular or universal (*quando non contingit descendere ad singulare nec universale*)—in other words, a (predicate-)term has supposition *confusa tantum* when it is used attributively of its extension. The supposition is *simplex* if the predicate-term refers to a common nature, as is the case in '*omnis homo est homo*', where the predicate '*homo*' supposits for human nature. (2) Both the subject and predicate of a universal negative proposition have confused distributive supposition, if they are common terms, as occurs in '*nullus homo est lapis*'. (3) In particular affirmative propositions, such as '*aliquis homo est animal*', both the subject and predicate have determinate supposition. (4) In particular negative propositions, the subject-term has determinate supposition and the predicate-term has distributive confused supposition.[29]

With these explanations in our minds, we may now look at some uses and applications of the supposition theory to fallacies and sophisms.

3. Supposition Applied

In the second chapter of the third treatise of his *Continuatio logicae* (composed between 1360 and 1363 according to Thomson,[30] but between 1371 and 1374

---

28) Cf. Wyclif, *Tractatus de logica*, ed. Dziewicki (1893-1899), ch. 12, 41. Dziewicki's text reads '*simpliciter*' instead of '*singulariter*', but this does not make any sense, as, in Wyclif's view, those terms are *discreti* (we would say 'atomic'), and therefore they cannot stand for a multiplicity of things in a proposition. But if they could supposit *simpliciter* they would signify a *common* nature, and through this common nature the set of individuals which share it. So, ultimately, they could supposit distributively for those singular items which instantiate the nature at issue, just as any other term which has simple supposition. In the case of terms such as '*persona*' and '*individuum*', this would imply the existence of an individual common nature, that is, an (auto-contradictory) entity present in all the individuals as the cause of their being individuals—an entity that Wyclif could not admit within his world, as Alyngton, Whelpdale, Penbygull, and Tarteys after him explicitly argued. On this point see Conti (1999).

29) Cf. Wyclif, *Tractatus de logica*, ed. Dziewicki (1893-1899), ch. 12, 41-42.

30) Cf. Thomson (1983), 5-6.

according to Mueller)[31] Wyclif deals with the problem of the truth-conditions of copulative sentences (that is, propositions of the form: p and q). It is just in order to solve it that he examines some fallacies connected to the question of universals and introduces further interesting remarks about supposition.

The starting point is the analysis of the truth-condition of the following sentence:

(A) *Sortes est animal et illud est asinus.*

Wyclif observes that, according to the general rules concerning the supposition of relative terms (*termini relati*, or *relativa*—that is, not those terms which fall under the category of *ad aliquid*, but such terms as pronouns, which, in a molecular sentence[32]—refer to other terms present in it), we must consider that sentence equivalent to this one:

(A_1) *Sortes est animal et illud animal quod est Sortes est asinus.*[33]

For this case as well the rule holds that any pronoun necessarily supposits for a subset of that set (of *res*) that the common term to which the pronoun refers (according to Wyclif's terminology, the *antecedens*) stands for (*omne tale relativum limitacius se habet quam suum antecedens*). In fact the propositional context within which the *antecedens* is set (and in particular the verb connected with the *antecedens*) narrows its extension (*omne relativum refert suum antecedens sub habitudine alicuius actus verbi sui antecedentis*). In his opinion, those authors who stated that any pronoun is extensionally and intensionally equivalent to its *antecedens* (*omne relativum converti cum suo antecedente*) were wrong.[34] Their theory entails the following two unsuitable consequences: (1) it cannot supply a valid *de re* interpretation of sentences such as: '*ego scio quis fuit ille homo qui commisit illud furtum*', since they are forced to admit that the sentence at issue is true even if the speaker actually does not know who the thief was, but simply knows that the thief is anyone of the human beings that the term '*homo*' could stand for at that moment:[35]

> Querendo ergo a tali sophista quis fuit ille qui fecit talem turpitudinem, diceret quod ego et quilibet homo mundi est ille.

---

31) Cf. Mueller (1985), xxxv and xxxvii-xxxviii.

32) Namely a sentence composed by two or more elementary sentences.

33) Cf. Wyclif, *Logicae continuatio*, ed. Dziewicki (1893-1899), tr. 3, ch. 2, vol. 2, 26.

34) Cf. Wyclif, *Logicae continuatio*, ed. Dziewicki (1893-1899), tr. 3, ch. 2, 26-27.

35) Wyclif, *Logicae continuatio*, ed. Dziewicki (1893-1899), tr. 3, ch. 2, 27.

(2) It cannot produce any solid argument against the opposite thesis, that is, Wyclif's thesis on this subject. The main argument of his opponents—says Wyclif[36]—is that if one admits the existence of extra-mental universal essences really identical with their own individuals, then one must admit that every relative term, in any propositional context, always refers to the same set of things to which its antecendent refers apart from any propositional context (*absolute*). In Wyclif's view, such an *inconveniens* is prevented by his own formulations of the supposition theory and his solution to the problem of the relationship between universals and individuals, according to which universals are really identical with but formally distinct from their individuals. In fact, thanks to the formal distinction[37] it is possible to explain:

(1) how universals can be distinguished from each other and from their individuals;
(2) how the *ens transcendens* is common to God and His creatures;
(3) why universal essences cannot receive the accidental predications proper to their individuals; and
(4) why universal essences cannot be counted with their individuals.

(1) As far as the first problem at issue is concerned, Wyclif claims that a universal essence *qua* nature of a certain kind is identical with its own individuals (for example, *homo* is the same thing as Socrates), but *qua* properly universal (that is a truth or nature that can exist in many things and can be shared by them) it is distinct from its own individuals, considered *qua* individuals, because of the opposite constitutive principles: *communicabilitas* for universals and *incommunicabilitas* for individuals. Because of this *distinctio formalis* between universals and individuals, the rule of the transitivity of predications among identicals does not obtain: one cannot infer from 'Socrates is a man' and 'man is a species' that 'Socrates is a species', notwithstanding the identity between *homo* and Socrates. On the other hand, this formal difference between Socrates and the human nature does not mean that Socrates and his species are *two* different realities; it simply means that because of two opposite properties, the impossibility-of-being-common proper to Socrates and the natural-tendency-to-be-common proper to the human nature, Socrates is distinct from what is its species:

36) Cf. Wyclif, *Logicae continuatio*, ed. Dziewicki (1893-1899), tr. 3, ch. 2, 28.
37) On Wyclif's theory of identity and distinction, see Conti (2006), 72-78.

Sor differt a specie debet exponi resolvendo hoc verbum differt, significans confuse ad differenciam supradictam; ut si Sor aliqua differencia differt a specie, ergo Sor differt a specie. Et antecedens patet ex hoc quod Sor racione incommunicabilitatis differt a specie.[38]

(2) The second problem runs as follows: if the *ens transcendens* is common to God and the creatures, then, since to be common implies being identical, (a) the *maxima differentia*, the difference between God and the creatures, should be compatible with identity; and (b) God would be His own cause of existence, since Being is the first *causatum* and God is (really identical with) it.[39] Wyclif's reply is that Being is common to God and creatures in the same way as, for instance, the universal-man (*homo communis* or *in communi*—that is, the human nature)[40] is common to every man.[41] Therefore, this syllogism:

(S_1) *ens transcendens est Deus et ens transcendens est aliud a Deo, ergo, aliud a Deo est Deus*

is as invalid as:

(S_2) *homo communis est Sor et ille homo communis est Plato, ergo, Sor est Plato.*

In fact, the universal-man, because of the property of being communicable, is formally distinct from Socrates and from any other man, who are neither communicable nor sharable. The first syllogism (that is, S_1)—Wyclif concludes—is not an expository one, as it appears to be, but a paralogism, since the *ens transcendens* is not a singular entity, like Socrates or a stone. As a consequence the copula in the premisses does not mean identity as in the conclusion. In the second syllogism (that is, S_2), the only logically possible conclusion is not that Socrates is Plato, but that Socrates is (identical with) the same thing which is (identical with) Plato (*ergo, Sor est illud quod est Plato*). In a similar way, the only logically possible conclusion of S_1 is that *quodlibet est ens transcendens,*

38) Wyclif, *Logicae continuatio*, ed. Dziewicki (1893-1899), tr. 3, ch. 2, 35-37.

39) Cf. Wyclif, *Logicae continuatio*, ed. Dziewicki (1893-1899), tr. 3, ch. 2, 38.

40) For the identity between the universal-man and the human nature and in general between a universal-something and a common nature see Wyclif, *Tractatus de logica*, ed. Dziewicki (1893-1899), ch. 12, 42: 'In ista propositione: *omnis homo est homo* iste terminus *homo*, qui est predicatum, supponit pro homine communi vel natura humana, quod idem est; et sic significat quod *homo communis vel natura human est omnis homo*.'

41) Cf. Wyclif, *Logicae continuatio*, ed. Dziewicki (1893-1899), tr. 3, ch. 2, 38.

quod est Deus, even though it is not God (that is, it is not identical to God Himself).[42]

(3) The explanation of the third point is crucial, since it represents Wyclif's reply to Ockham's main argument against realism. As is well known, Ockham had maintained that (1) the assumption of a relation of identity between universals and individuals was inconsistent with the standard definition of (real) identity, which implies transitivity in predication; and that (2) from an ontological point of view, the only kind of distinction which could hold between two created beings was the real one, as (in his opinion) any form of distinction between two created beings necessarily implied a real distinction between them. From these two theses and the presupposition of the truth of Aristotle's statement that there cannot exist real universal forms apart from their individuals, Ockham had derived a rejection of any type of extramental reality for universals. His most general argumentation[43] was that, if universals are something existing *in re*, really identical with their individuals, then whatever is predicated of individuals must be predicated of their universals too, and so a unique universal entity (say, the human nature) would possess contrary attributes simultaneously via the attributes of different individuals, a clearly unacceptable conclusion.

Wyclif acknowledged that Ockham's critique showed that the traditional realist description of the relation between universals and individuals and the traditional notions of identity and difference (or distinction) were inconsistent, but he was convinced that realism as a whole was still defensible. So he tried to remove the aporetic points of the traditional realist theory of universals by elaborating new notions of identity and distinction which he then used to interpret the relation between universals and individuals, and thereby the nature of predication. He thought that not all that is predicated of individuals can be directly (*formaliter*) predicated of universals and *vice versa*. In his opinion, a universal of the category of substance could directly receive only the predications of substantial forms, or essences, more common than itself (namely those forms which are put on a higher level in the *linea praedicamentalis*). On the other hand, he believed that the accidental forms inhering in substantial individuals could be predicated of the substantial form itself (which those individuals instantiated) only indirectly (*essentialiter*), through and in

42) Cf. Wyclif, *Logicae continuatio*, ed. Dziewicki (1893-1899), tr. 3, ch. 2, 39.

43) Cf. Ockham, *Expositio in librum Praedicamentorum Aristotelis*, ch. 8. 1, ed. Gál (1978), 164-168; *Summa logicae*, ed. Boehner *et al.* (1974), I, 15, 50-51.

virtue of the individuals falling under that substantial form.[44] As a consequence, Wyclif distinguished four different kinds of predication, which he conceived as a real relation between metaphysical entities: (i) predication by essence (*secundum essentiam*); (ii) formal predication (*per inherenciam forme*); (iii) causal predication (*secundum causam*); and (iv) habitudinal predication (*secundum habitudinem*).[45]

(i) To speak of predication *secundum essentiam* it is sufficient that the same essence is both the real subject and predicate, even though the formal principle connoted by the predicate-term differs from that connoted by the subject-term. '*Deus est homo*' and '*Universale est particulare*' are instances of predication *secundum essentiam.* In fact, the same essence that is a universal is also an individual, but the forms connoted by the subject-term and by the predicate-term are different. (ii) Formal predication is that predication in which the form signified by the predicate-term is directly present in the essence signified by the subject-term. This happens whenever an item in the categorial line is predicated of something inferior, or an accident of its subject of inherence. In fact, in both of them, the subject-term and the predicate-term refer to the same essence in virtue of the form connoted by the predicate-term itself. Universal essences, such as *homo in communi*, and abstract forms, such as *humanitas*, do not support this kind of predication, since this kind of predication necessarily requires that the real subject of the predication is capable of undertaking a change—something impossible for universal essences and abstract forms. 'Man is an animal' ('*homo est animal*') and 'Peter is musical' ('*Petrus est musicus*') are instances of formal predication. (iii) We speak of causal predication when the form designated by the predicate-term is not present in the essence signified by the subject- term, but is something caused

44) See, for instance, the *Tractatus de universalibus*, ed. Mueller (1985), ch. 11, 239-240.

45) Cf. Wyclif, *Logicae continuatio*, ed. Dziewicki (1893-1899), tr. 3, ch. 2, 40-42. In the second and third chapters of the *Purgans errores circa universalia in communi* (composed between 1366 and 1368) Wyclif lists the following three main types of predication: formal predication, predication by essence, and causal predication; on the contrary, in the *Tractatus de universalibus* (ch. 1, 35-36) causal predication is replaced by habitudinal predication—a kind of predication that Wyclif had already recognized in the *Purgans errores circa universalia*, but whose position within the main division of the types of predication was not clear, as it seems to be a sub-type of formal predication, even though it does not satisfy the criterion of the direct inherence of the form signified by the predicate in the essence signified by the subject. Formal predication, predication by essence, and habitudinal predication are defined almost in the same way in the *Purgans errores circa universalia* and in the *Tractatus de universalibus*, but in the *Tractatus de universalibus* formal predication, predication by essence, and habitudinal predication are described as three non-mutually exclusive ways of predicating, each more general than the preceding one (or ones).

by that entity. '*Dies est lacio solis*' is an example of this kind of predication. (iv) Finally, we speak of predication *secundum habitudinem* when the form connoted by the predicate-term is not present in the essence designated by the subject-term, but simply implies a relation to it, so that the same predicate may be at different times truly or falsely spoken of its subject, without there being any change in the subject itself. According to Wyclif, we use such a kind of predication mainly when we want to express theological truths, like these: that God is known and loved by many creatures, and brings about, as efficient, exemplary, and final cause, many good effects. Universal essences too can support this kind of predication. On the basis of such a division of predication and his theory of supposition, Wyclif denies that universal essences can receive the accidental predications of their individuals. He therefore rejects Ockham's argumentations as well as any syllogism of this form:

(S_3) *hoc albatur, et hoc est illa essencia: igitur illa essencia albatur.*[46]

In fact, in his opinion, the middle term (that is, the pronoun '*hoc*') has different suppositions in the two premisses: personal in the major, where it stands for a singular substance, and simple in the minor, where it stands for a common nature or universal essence. Therefore, the only logically possible conclusion is: '*illa essencia est album*', as the substantival adjective in its neuter form, which plays the role of predicate in the sentence, shows that the form signified by the predicate-term is not directly present in the subject, but is indirectly attributed to it through its individuals. As a consequence, the term has personal, and not simple, supposition and the sentence is equivalent to this: '*illa essencia est illud quod est album*'.[47]

(4) Finally, as to the problem of whether universal essences must be counted with their individuals or not, Wyclif answers that this is impossible, since the universal-man is not another man in addition to the concrete existing human beings. He argues that adding the universal-man as a third man to Socrates and Plato, given that there are only these two individual men in the world, commits a fallacy of equivocation. When a number is added to a term of first intention (like 'man'), the presence of this numerical term modifies the kind of supposition from simple to personal; but one can refer to a universal only with a term with simple supposition. As a consequence the universal cannot be counted

46) Cf. Wyclif, *Logicae continuatio*, ed. Dziewicki (1893-1899), tr. 3, ch. 2, 42.

47) Cf. Wyclif, *Logicae continuatio*, ed. Dziewicki (1893-1899), tr. 3, ch. 2, 43.

with its individuals—and in fact any universal is really identical to each one of its individuals, and so it cannot differ in number from each of them.[48]

It is clear that the final outcome of Wyclif's metaphysical choices is therefore a mixed logical system, where the copula of the standard philosophical sentences of the form '(every or some) A is B' which he deals with can have different values. Causal and habitudinal predications do not require the presence of an absolute form in the real subject nor any kind of identity between the *significatum* of the subject-term and that of the predicate-term, whereas predication by essence does require such an identity. Causal and habitudinal predications involve a loose connection between the real predicate and the real subject, since the form or essence signified by the predicate-term is not present in the real subject and simply entails a reference to it. Though predication by essence indicates a partial identity between the real subject and predicate (which share some, but not all, metaphysical component parts), it excludes that the form connoted by the predicate-term is directly present in the essence denoted by the subject-term. Formal predication, on the contrary, requires such a direct presence. It is intended to be a sort of kind of predication over and above to the standard Aristotelian types, namely essential and accidental predication, as defined in the second and fifth chapters of the *Categories*. It means that the subject-thing in virtue of its nature or by means of one of its inhering forms is a member of a certain set of essences that the predicate-term of the proposition names and signifies. In this way Wyclif was trying to give a logically satisfactory solution to the problem of the relationship between common natures and singular items, which had always been the most difficult issue for medieval Realists. His theory of supposition is aimed at this same goal, as his discussion of the sophism *I promise you a coin that I do not promise* in the third chapter of the third treatise of the *Continuatio logicae* plainly proves.[49]

---

48) Cf. Wyclif, *Logicae continuatio*, ed. Dziewicki (1893-1899), tr. 3, ch. 2, 48: 'Tunc dicitur quod terminus numeralis, additus termino prime intencionis, limitatur ad significandum numerum primo modo dictum (scilicet acceptum pro multitudine singularium). Unde, sicut terminus distribuens limitat speciem specialissimam ad supposicionem personalem, ita ille terminus numeralis limitat terminum prime intencionis, et specialiter speciem specialissimam, ad supposicionem personalem. Cum ergo homo communis sit quilibet hominum singularium, non ponit seorsum in numero cum illis; ideo existente *omni homine, Sorte* vel *Platone*, non superest tercius homo communis ab illis, sed est uterque illorum, et non esset tercius, nec esset tercia persona hominis. Et per idem non oportet, ubique ubi est unus homo, esse duos homines.'

49) On Wyclif's discussion of this sophism see Read (1985).

Like Burley[50] before him, Wyclif defends the claim that what is explicitly promised by 'I promise you one of these coins I have in one of my hands' ('*promitto tibi alterum illorum denariorum in altera manuum mearum*'), is the universal-coin, and not a singular one, even if I can fulfil the promise only by giving any singular coin, since a universal cannot be given or possessed except by a singular.[51] Thanks to his distinction between simple and personal supposition, he is able to explain from a semantic point of view the difference between promising a coin in general and promising a particular coin: in the first case the term 'coin' ('*denarius*') has simple supposition, and therefore the proposition is true if and only if what is said is true of the universal-coin; on the contrary, if the term 'coin' has personal supposition (more precisely, personal and singular supposition), the proposition is true if and only if what is said is true of *a* particular coin. According to him, by promising a singular, a universal is promised *secundarie* and *confuse*, and conversely.[52] So, given two coins in my hands, coin *A* and the coin *B*, the proposition 'I promise you one or the other of these coins' is true, even though, when asked whether I promised coin *A*, my answer is 'No', and so too when asked whether I promised coin *B*. In fact, according to Wyclif, what I promised is the universal-coin, since the phrase 'one or the other of these coins' has simple supposition and therefore stands for a universal, however restricted in its instantiations to one or other of the two coins in my hands.[53] This does not mean, however, that the universal-coin is a sort of third coin over and above the two coins in my hands, since Wyclif had already rejected this mistaken conclusion in the previous chapter of the *Logicae continuatio*.

---

50) Cf. Burley, *Expositio in libros octo Physicorum Aristotelis. Prologus*, ed. Venetiis (1501), f. 8vb.

51) Cf. Wyclif, *Logicae continuatio*, ed. Dziewicki (1893-1899), tr. 3, ch. 3, 62: 'Nemo placitans pro communi promissione denarii vendicat illum denarium vel illum, sed vendicat quod debetur sibi denarius: quod fuit promissum. Sed quia tale commune non potest dari vel haberi nisi per singulare, ideo requiritur promittentem dare singulare; et tunc sequitur ipsam, dando universale, impleri promissionem. Non enim potest quandoque dari vel promitti singulare, nisi in sic faciendo involvatur universale; quia omnes tales predicaciones secundum habitudinem suscipiunt universale a suis singularibus. Et sic conceditur quod habeam communem denarium per ante (si habeam aliquem denarium) non tamen ex illa promissione, ideo vendico illud commune michi dari ab illo qui sic promisit; quia, si posset michi dare illud sine denario singulari, placet michi. Sed cum non potest, ex dacione sua multiplicius habeo illud commune. Quotquot enim denarios quis habuerit, tottupliciter habet communem denarium.'

52) Cf. Wyclif, *Logicae continuatio*, ed. Dziewicki (1893-1899), tr. 3, ch. 3, 64.

53) Cf. Wyclif, *Logicae continuatio*, ed. Dziewicki (1893-1899), tr. 3, ch. 3, 67.

4. Concluding Remarks

If the foregoing account is correct, Wyclif's formulation of the supposition-theory and his theory of universals and predication are linked together, and rest upon a sort of componential analysis, where things substitute for lexemes and ontological properties for semantic features—as shown by his convictions that (1) whatever is is a real proposition (*propositio realis*) and (2) supposition is a property of terms and extra-mental things as well. For Wyclif, everything which is is a real proposition, since everything which is signifies in a complex manner that it is something real, and everything save God is compound (at least of potency and act),[54] and therefore can be conceived of and signified both in a complex (*complexe*) and in a non-complex way (*incomplexe*). When we conceive of a thing in a complex manner we think of that thing considered according to its metaphysical organization, and so as a real proposition (in other words, as a sort of state of affairs). Even the abstract forms, because of their own inner structure, are such—for example, humanity is equal to the 'sum' of the form of animality and that of rationality, which combine as potency and act respectively. As a consequence, we can refer to the same entity by means of various types of linguistic expressions: abstract terms, concrete terms, infinitive expressions (like 'being a man'—'*hominem esse*'), and complex nouns (such as 'universal-humanity'—'*humanitas communis*', 'universal-man'—'*homo in communi*', and 'the species of man'—'*species hominis*'), which have to be considered as synonymous.[55] This is the logical result of Wyclif's idea that the world consists of essences (that is, single items classified into ten different types or categories), which are not simple, but composite, because they are reducible to something else, belonging to a different rank of reality and unable to exist by themselves: being and essence (in the sense of quiddity)[56], potency and act, matter and form, abstract genera, species and differences. For that reason, everything one can speak about or think of is both a thing (we could say: a molecular object) and a real proposition (we could say: a sort of atomic state of affairs), while every true sentence expresses either a simple or a complex real proposition, that is, either the union (if the proposition is affirmative) or the separation (if the proposition is negative) of two (or more) things. In particular, according to him, a singular man (*iste homo*) is nothing but a real proposition, where the actual existence in time as an individual (*ista persona*)

54) Cf. Wyclif, *De ente praedicamentali*, ed. Beer (1891), ch. 5, 38-39.

55) Cf. Wyclif, *Tractatus de universalibus*, ed. Mueller (1985), ch. 3, 70 and 71.

56) See above, note 16. On Wyclif's doctrine on being and essence see Conti (1997), 145-150; Conti (2006), 89-95.

plays the role of subject, the common nature (*natura humana*) plays the role of predicate, and the singular essence (*essencia istius hominis*—that is, that by means of which this individual is *this* man) plays the role of the copula.[57] Hence, in Wyclif's view, everything which is and any constitutive item of its metaphysical reality have the property of being either a real predicate or a real copula or a real subject suppositing for some other entity in the world:[58]

> Terminus concretus est terminus significans rem que indifferenter potest contrahi ad supposicionem simplicem vel personalem; sicut iste terminus, homo, significat in proposicione tam personaliter pro persona, quam eciam simpliciter pro natura.

According to him, only on the basis of this close isomorphism between linguistic expressions and the world can the signifying power of language, the possibility of definitions, and finally the validity and universality of our knowledge be explained and ensured. So the principle that inspires Wyclif's thought is that of the analytic correspondence between the logical connections in discourse and the framework of reality, and the core of his philosophy consists in his trust in the scheme *thing(s)-designation(s)* as the only heuristic and interpretative key for the solution to any semantic and epistemological problem. Wyclif firmly believed that language was an ordered collection of signs, each referring to one of the constitutive elements of the world, and that true sentences were like pictures of the inner structures and mutual relationships of such constitutive elements. He thought of logic as turning on structural forms, independent of both their semantic contents and the mental acts by which they are grasped.[59] It is through these forms that the network connecting the basic constituents of the world (individuals and common natures, substances and accidents, concrete properties, like being-white, and abstract forms, like whiteness) is disclosed to us. As we have seen, he conceived of common natures as real essences shared by many individual items which are necessary conditions for our language to have meaning. He thought that by associating general terms with such universal essences the fact could be accounted for that each common term can stand for many things at once and can name all of them in the same way. For this reason, Wyclif represents common natures as the

[57] Cf. Wyclif, *Tractatus de logica*, ed. Dziewicki (1893-1899), ch. 5, 15. In the *Materia et forma* he develops at greatest length the idea that in all created things the essence corresponds to the Godhead, the matter to the Father, the form to the Son, and the compound to the Holy Spirit; and he calls matter, form, and the compound taken together 'the created trinity.'

[58] Cf. Wyclif, *Tractatus de logica*, ed. Dziewicki (1893-1899), ch. 1, 5.

[59] Cf. Wyclif, *Tractatus de universalibus*, ed. Mueller (1985), ch. 2, 56.

significata of general terms that, because of their presence in singular items and their relations of real identity and formal distinction to them, allow us to pick out the members of the class of *res* which form the extension of general terms themselves.

His peculiar version of the theory of *significatum* and supposition, and especially his idea of two possible kinds of supposition proper to concrete accidental terms, which signify an essence able to supposit both for the individual substance in which the accidental form inheres and for the accidental form itself, restate this *hypostasizing* approach to semantics. Paradoxically, this implies that, for him, the world itself is intrinsically linguistic: a sort of semiotic system where everything is at the same time what it is and *the* natural sign of itself (and of anything else real identical to itself), so that reality could be described as a language of things. This is the opposite of Ockham's nominalism (his polemical target) based on a sharp distinction between things as they exist in the extramental world and the various forms by means of which we think of and talk about them, since for Ockham our (mental) language does not reproduce the world, but merely regards it, as they are logically independent systems.

Bibliography

Primary Literature

Gualterus Burlaeus, *Tractatus de Suppositionibus*, in: S.F. Brown (1972), 'Walter Burleigh's Treatise *De Suppositionibus* and its Influence of William of Ockham', in: *Franciscan Studies* 32 (1972), 15-64

——, *Expositio in libros octo Physicorum Aristotelis* (Venetiis 1501)

——, *De Puritate Artis Logicae Tractatus Longior, with a Revised Edition of the Tractatus Brevior*, ed. Ph. Boehner O.F.M. (Franciscan Institute Publications; Louvain-Paderborn 1955)

——, *Expositio super Praedicamenta Aristotelis*, in: *Expositio super Artem Veterem Porphyrii et Aristotelis* (Venetiis 1509, fols. 17va-50va)

Guillelmus de Ockham, *Expositio in librum Praedicamentorum Aristotelis*, ed. G. Gál, (Guillelmi de Ockham Opera Philosophica, II; St. Bonaventure, N.Y. 1978, 133-339)

——, *Summa logicae*, ed. P. Boehner, G. Gál and S. Brown (Guillelmi de Ockham Opera Philosophica, I. St. Bonaventure, N.Y. 1974)

Guillemus Penbygull, *De universalibus*, in: A.D. Conti, 'Teoria degli universali e teoria della predicazione nel trattato *De universalibus* di William Penbygull: discussione e difesa della posizione di Wyclif', in: *Medioevo* 8 (1982), 137-203

Johannes Wyclif, *Tractatus de logica*, ed. M.H. Dziewicki, 3 vols. (Wyclif's Latin Works 14; London 1893-1899)

——, *De actibus animae*, in: *Johannis Wyclif miscellanea philosophica*, 2 vols., ed. M.H. Dziewicki (London 1902, 1, 1-160)

——, *De ente praedicamentali*, ed. R. Beer (London 1891)
——, *Tractatus de universalibus*, ed. I.J. Mueller (Oxford 1985)
——, *De materia et forma*, in *Johannis Wyclif miscellanea philosophica* (London 1902,. 1, 163-242)
Walter Burley, *vide s.n.* Gualterus Burlaeus
William of Ockham, *vide s.n.* Guillelmus de Ockham
William Penbygull, *vide s.n.* Guilelmus Penbygull

Secondary Literature

Cesalli, L. (2005), 'Le 'pan-propositionnalisme' de Jean Wyclif', in: *Vivarium* 43 (2005), 124-155
Conti, A.D. (1997), 'Analogy and Formal Distinction: on the Logical Basis of Wyclif's Metaphysics', in: *Medieval Philosophy and Theology*, 6/2 (1997), 133-165
—— (1999), 'Second Intentions in the Late Middle Ages', in: S. Ebbesen, R.L. Friedman, eds., *Medieval Analyses in Language and Cognition* (Copenhagen 1999, 453-470)
—— (2005), 'Johannes Sharpe's Ontology and Semantics: Oxford Realism Revisited', in: *Vivarium* 43 (2005), 156-186
—— (2006), 'Wyclif's Logic and Metaphysics', in: I.C. Levy, ed., *A Companion to John Wyclif, Late Medieval Theologian* (Leiden 2006, 67-125)
Kenny, A. (1986), 'The Realism of *De Universalibus*', in: A. Kenny, ed., *Wyclif in His Times* (Oxford 1986, 17-29)
Maierù, A. (1972), *Terminologia logica della tarda scolastica* (Rome 1972)
Mueller, I.J. (1985), 'Introduction', in: Johannes Wyclif, *Tractatus de universalibus*, ed. I.J. Mueller (Oxford 1985)
Read, S. (1985), "I Promise a Penny that I do not Promise'. The Realist/Nominalist Debate over Intensional Propositions in Fourteenth-Century British Logic and its Contemporary Relevance', in: P.O. Lewry, ed., *The Rise of British Logic*. Acts of the Sixth European Symposium on Medieval Logic and Semantics, Balliol College, Oxford, 19-24 June 1983 (Toronto 1985, 335-359)
Spade, P.V. (1985), 'Introduction', in: John Wyclif, *On Universals*. Text Translated by A. Kenny, with an Introduction by P.V. Spade (Oxford 1985)
—— (2005), 'The Problem of Universals and Wyclif's Alleged 'Ultrarealism'', in: *Vivarium* 43 (2005), 111-132
Spruyt, J. (2008), 'The Unity of Semantics and Ontology: Wyclif's Treatment of the *fallacia accidentis*,' in: *Vivarium* 46 (2008), 24-58
Thomson, W.R. (1983), *The Latin Writings of John Wyclif: an Annotated Catalogue* (Toronto 1983)

Thomas Aquinas and Some Italian Dominicans (Francis of Prato, Georgius Rovegnatinus and Girolamo Savonarola) on Signification and Supposition

Fabrizio Amerini
University of Parma

Abstract

Supposition is a controversial logical theory. Scholars have investigated many points of this doctrine such as its historical origin, its use in theology, the logical function of the theory, or the relationship between supposition and signification. In the article I focus on this latter aspect by discussing how some Italian, and in particular Florentine, Dominican followers of Aquinas—Francis of Prato (d. 1348), Girolamo Savonarola (d. 1498), and Georgius Rovegnatinus (d. after 1500)—explained the relation between the linguistic terms' properties of signifying and suppositing, and hence the division of supposition. After sketching out Thomas Aquinas, Hervaeus Natalis, and William of Ockham's positions on the relationship between signification and supposition, I closely examine Francis's criticism of Ockham. Francis follows Walter Burley's account of supposition and considers the statement that a term has simple supposition when (i) it is taken not significatively and (ii) stands for an intention of mind as the weak point of Ockham's explanation of supposition. According to Francis, if this were the case, there would be no semantic basis for differentiating simple from material supposition. Francis is however hesitant about the full subordination of supposition to signification, especially with regards to material supposition, when a term, suppositing for itself, is taken to signify itself besides its meaning. More than one hundred years later, Girolamo Savonarola and Georgius Rovegnatinus have no doubt about the fact that terms may supposit only for what they signify.

Keywords

material supposition, simple supposition, signification

Supposition is a controversial logical theory. Scholars have investigated many points of this doctrine such as its historical origin, its use in theology,[1] the logical function of the theory, and the relationship between supposition and signification.[2] In this article I want to reconsider this last aspect by discussing how some Italian, and in particular Florentine, Dominican followers of Aquinas—Francis of Prato († after 1348) and Girolamo Savonarola († 1498)—approached this topic.

Generally speaking, the first Italian Dominican followers of Aquinas did not show particular interest in the logical theory of supposition. For example, the two major handbooks of logic circulating in Italian Dominican convents during the first two generations after Aquinas, namely the pseudo-Aquinas's *Summa totius logicae Aristotelis* and Gratiadeus of Ascoli's *Logica*, do not reserve any part or section for supposition. Furthermore, no independent treatise on supposition has been ascribed to Aquinas or composed by his followers. For the first Dominican masters, it sufficed to refer to Peter of Spain's *Tractatus* and to use it as the basis for their teaching on supposition. As far as I know, the earliest surviving Italian Dominican treatise on supposition is that of Francis of Prato, which dates to around 1340.

The Italian Dominicans take interest in the theory of supposition after Burley and Ockham's treatments of supposition. In particular, the Dominicans regard Ockham's theory as the starting-point for their own investigations, although Ockham's account of supposition seems to have had no doctrinal influence on them. Francis's treatise on supposition, for example, while critically modeled on Ockham's *Summa logicae*, is doctrinally dependent rather on Walter Burley. Burley's widespread reception in Italy is well-known and Italian Dominicans generally followed Burley in logic. To give just one example, in his *Compendium Logicae*, the upshot of his teaching at the Florentine convent of San Marco (1484), Girolamo Savonarola still accounts for supposition in essentially the same way as Burley and Francis of Prato did.

What philosophical reasons did the Italian Dominicans have for following Burley instead of Ockham? To answer this question, I shall proceed as follows. First, I shall consider what two authoritative masters of the Order, namely Thomas Aquinas and Hervaeus Natalis († 1323), had to say about supposition (1). Then, I shall illustrate why, on the basis of Aquinas's and Hervaeus's account

---

1) Cf. Maierù (1985) and Brown (1993).

2) For an introduction to the medieval doctrine of supposition, see De Rijk (1967), vol. II; De Libera (1983); Parsons (2008). For a useful and up-to-date bibliography, see also http://www.formalontology.it/supposition-biblio-one.htm.

of supposition, Francis of Prato thinks that Ockham provided a bad explanation of the division of supposition, while Burley gave a good one. Francis follows Burley and Hervaeus Natalis and adopts a non-parsimonious realist ontology (2). Finally, I shall conclude by saying something about how Savonarola understands supposition in his handbook of logic (3).

1. Thomas Aquinas and Supposition

It is well known that Thomas Aquinas pays scant attention to supposition. Although Aquinas makes no effort to elaborate a full-fledged theory of supposition, nonetheless there are places in his writings where Aquinas shows himself to be acquainted with the logical mechanisms of this theory. In particular, Aquinas refers to supposition in theology, where he accurately distinguishes supposition from the signification of terms. Putting together the statements scattered in his works, we can say that supposition plays two fundamental roles in Aquinas.

First, supposition serves the function of relating linguistic terms to extra-mental things. For this reason, supposition mainly concerns names and whatever can be the subject of a proposition. This function is to be expected if one remembers that, for Aquinas, signification does the semantic job of relating linguistic terms to the conceptions of the mind. Regardless of how conceptions must be understood, Aquinas is unequivocal in granting terms a two-level signification: terms primarily or immediately signify the conceptions of the mind and, by way of such conceptions, they signify the extra-mental things.[3] In the *Disputed Questions on Power*, Aquinas qualifies such a two-fold signification by stating that terms *formally* signify conceptions of the mind or, more precisely, that which terms have been primarily imposed to signify, while they *materially* signify extra-mental things, or more precisely, that which concretely exemplifies the terms' primary meaning.[4] A term such as 'man', for example, formally signifies the kind of thing that is composed of a body and a rational soul, while materially it signifies each concrete body composed of flesh and bones, and similar biological stuff, which is organized in such a way as to be able to fulfill rational acts. Thus, only nouns can genuinely supposit, while adjectives cannot. Adjectives (and in general every concrete accidental term) signify exclusively an accidental form. The bearer of such a form, which is a substance, is

3) See, for example, Thomas Aquinas, *Expositio libri Peryermenias*, I.2, 10-11, 95-112.

4) See Thomas Aquinas, *Quaestiones disputatae de potentia Dei*, 9.4 and 9.6.

not included in the signification of an adjective, although we can infer it from its signification.[5] Furthermore, Aquinas holds that the modes of signification, and in a special way the distinction between abstract and concrete modality of signification, affect the supposition of terms. While a term such as 'man' signifies a substantial form and can stand for any bearer of it, a term such as 'humanity' signifies a form and can stand only for that same form.

Normally, terms are said to signify conceptions of the mind and to supposit for concrete bearers of the forms represented by such conceptions.[6] Thus, a term such as 'man' is said to signify the form of humanity (*humanitas*) and to stand for the concrete bearer of humanity (*habens humanitatem*). This distinction is a useful instrument for interpreting theological, and especially Christological, sentences. Let me give a pair of examples. When we say that 'Christ is a man' and 'Socrates is a man', someone might remark that 'man' is predicated of Christ and Socrates in an equivocal way, since Christ also has a divine nature while Socrates only a human one.[7] Aquinas replies by reminding the opponent that in whatever proposition 'man' occurs, it always signifies the general form of humanity, while it stands for any concrete *suppositum* of such a form. This distinction helps to understand why the remark is pointless, because equivocation follows signification and not supposition.[8] Since in the above proposition 'man' always signifies the human nature, what changes is only the nature of the *suppositum*: in the case of Christ, 'man' refers to an uncreated hypostasis, while in the case of Socrates and any other man, it refers to a created hypostasis, and this distinction of *supposita* entails no equivocation.[9]

Similarly, Aquinas maintains that the term 'God' has been imposed to signify the divine essence and to stand, *per se*, for this common essence, which is one single nature in reality. When 'God' occurs in a proposition, its signification does not change, but its supposition is narrowed down to the divine essence of a particular Person. So, if we say that 'God generates', 'God' continues to signify the divine essence, although it no longer stands for the same common nature, but now only for the Person of Father; likewise, if we say 'God is generated', 'God' continues to signify the divine essence, but it stands for the

5) Cf., e.g., Thomas Aquinas, *Summa theologiae*, 1.39.5 ad arg. 5; *Expositio libri Metaphysicorum*, 5.9 and 7.5.

6) Thomas Aquinas, *Summa theologiae*, 1.39.5. See also *Scriptum super IV libros Sententiarum*, 3.6.1.2 ad arg. 4.

7) Cf., e.g., Thomas Aquinas, *Summa contra Gentiles*, 4.40, 329, no. 3785.

8) Cf., e.g., Thomas Aquinas, *Quaestiones disputatae de potentia Dei*, 9.4 ad arg. 6.

9) Cf., e.g., Thomas Aquinas, *Summa contra Gentiles*, 4. 49, 339, no. 3847; also *Quaestio disputata De unione Verbi*, 2 ad arg. 4.

Person of Son and not for the divine essence. In these propositions, 'God' does not change its signification, but only its supposition, and this again entails no equivocation.[10] In this question of the *Summa theologiae*, Aquinas makes an interesting note about the different semantic function of the terms 'God' and 'man'. Such a distinction sheds further light on the logical function played by supposition in Aquinas's philosophy. The term 'man'—observes Aquinas—*per se* signifies human nature, but it naturally stands for the concrete bearers of such a nature. Accordingly, when it occurs in a proposition, it stands for each particular man (i.e., it has personal supposition), unless its supposition is modified by the verb or by the predicate[11] (this happens, for example, when we say 'Man is a species': here 'man' precisely stands for the human nature, so it has simple supposition). The reason for speaking this way is that no human nature really exists *as such* outside the mind, with real and numerical unity. God instead is a different case, since the divine nature is a true real entity, so 'God' signifies and naturally stands for the divine nature. The semantic difference between 'man' and 'God' shows that a term can naturally stand only for that which its signification exemplifies. As a consequence, there is not always a perfect overlap between signification and supposition. In particular, this explains the asymmetry existing between the signification and the supposition of the term 'man'.

According to this scheme, supposition properly pertains to concrete and common nouns, and it serves the role of associating the formal meaning of a term to its material meaning. Terms play such a—so to speak—'extensional' or referential role especially when standard categorical propositions are concerned. Nonetheless, Aquinas does not entirely elucidate the nature of this association; thus, it remains unclear whether, for Aquinas, the propositional context does the semantic job of restricting and determining the natural supposition of a term (in the case of 'God', for example, by narrowing down the common divine essence to the divine essence of a particular Person), or rather that of connecting the abstract form signified by a term to the concrete bearer

---

10) Cf., e.g., Thomas Aquinas, *Summa theologiae*, 1. 39. 4 ad arg. 3.

11) In his *Ex divo Thoma suppositionum collectula* (I, f. a ii^{v}), from the end of the fifteenth century, the Dominican Georgius Rovegnatinus lists four places where Aquinas holds the view that the supposition is determined not only by the predicate but also by the verb: 'Praecipue tamen a verbo fit haec determinatio. Interdum etiam a praedicato. [...] Et haec quidem infinitis pene locis Divus Thomas insinuat. Sed tria aut quattuor quod proposito facere potest satis nostro annotasse sufficiat.' Rovegnatinus refers to the *Scriptum super IV libros Sententiarum*, 1.7.2.1 and 3.1.2.5, and to the *Summa theologiae*, 1. 36.3 and 1. 39.4. I thank Stephan Meier-Oeser for indicating this text to me.

exemplifying such a form (in the case of 'God', connecting the divine nature to a particular Person).

Aquinas seems to think that there is a second, related, reason for introducing supposition, based on his conviction that supposition plays its primary role when it concerns terms that occur in a proposition. So Aquinas assigns to supposition the supplementary function of disambiguating propositions and fixing their truth-values. Once again, this more 'intensional' function is significant in theology, where a correct understanding of propositions is crucial for avoiding theologically unwelcome consequences. Thus, if the basic reason for introducing supposition is semantic, i.e., to specify the signification of terms (and this primarily holds for personal supposition), its main scope is exegetic, i.e., to interpret propositions and arguments in order for them to be true, consistent, or unambiguous.[12] We have already noted above the connection between supposition and equivocation. Let me illustrate this point better with two further examples.

In the third part of the *Summa theologiae*, when explaining the sense of the sentence 'Man is made God' (*Homo factus est Deus*), Aquinas makes an explicit reference to supposition. He distinguishes between simple and personal supposition, and relates them to the two-fold meaning of the term 'man'. Such a distinction enables Aquinas to say that if 'man' has simple supposition, the proposition is true; if 'man' instead has personal supposition, the proposition is false. The reason is that, in the first case, 'man' stands for a generic man, so the proposition 'Man is made God' is logically equivalent to the proposition 'It has been made that man be God', which is true; in the second case, instead, 'man' stands for Christ, and Christ has not become God at some time, but he has been God from the eternity. Absolutely speaking, the term 'man' can have both kinds of supposition, but when it occurs in a proposition it can have only one kind of supposition, if the proposition in which it occurs is to be true.[13]

---

12) See, for instance, what Aquinas says in the *Scriptum super IV libros Sententiarum*, 1.4.1.3 ad arg. 4.

13) Cf. Thomas Aquinas, *Summa theologiae*, 3.16.7: 'Respondeo dicendum quod propositio ista, homo factus est Deus, tripliciter potest intelligi. Uno modo, ita quod hoc participium factus determinet absolute vel subiectum, vel praedicatum. Et in hoc sensu est falsa, quia neque homo ille de quo praedicatur est factus, neque Deus est factus, ut infra dicetur. Et sub eodem sensu haec est falsa, Deus factus est homo. Sed sub hoc sensu non quaeritur hic de istis propositionibus. Alio modo potest intelligi ut ly factus determinet compositionem, ut sit sensus, homo factus est Deus, idest, factum est ut homo sit Deus. Et sub hoc sensu utraque est vera, et, homo factus est Deus, et, Deus factus est homo. Sed hic non est proprius sensus harum locutionum, nisi forte intelligatur quod ly homo non habeat personalem suppositionem, sed simplicem. Licet enim hic

In the first part of the *Summa theologiae*, moreover, Aquinas explains the meaning of another sentence, i.e., 'The Son and the Father are one principle with the Holy Spirit' (*Filius et Pater sunt unum principium cum Spiritu Sancto*). In this question, Aquinas distinguishes, within personal supposition, determinate and confused supposition, and quite interestingly he relates such a distinction to the fallacy of figure of speech. To assume that the term 'principle' has confused supposition leads Aquinas to say that the proposition 'The Son and the Father are one principle with the Holy Spirit' is logically equivalent to the conjunction of two propositions: 'The Son is one principle with the Holy Spirit' and 'The Father is one principle with the Holy Spirit'. Understood in this way, the proposition 'The Son and the Father are one principle with the Holy Spirit' is true; understood otherwise, it is false.[14]

These examples show the interconnection existing between supposition and interpretation theory, on the one hand, and supposition and truth theory, on the other. But they also show that Aquinas's sketchy account of supposition is quite standard for his time and that it is technically influenced by Peter of

---

homo non sit factus Deus, quia hoc suppositum, persona filii Dei, ab aeterno fuit Deus, tamen homo, communiter loquendo, non semper fuit Deus. Tertio modo, proprie intelligitur, secundum quod hoc participium factus ponit fieri circa hominem in respectu ad Deum sicut ad terminum factionis. Et in hoc sensu, supposito quod in Christo sit eadem persona et hypostasis et suppositum Dei et hominis, ut supra ostensum est, ista propositio falsa est. Quia cum dicitur, homo factus est Deus, ly homo habet personalem suppositionem, non enim esse Deum verificatur de homine ratione humanae naturae, sed ratione sui suppositi. Suppositum autem illud humanae naturae de quo verificatur esse Deum, est idem quod hypostasis seu persona filii Dei, quae semper fuit Deus. Unde non potest dici quod iste homo incoepit esse Deus, vel quod fiat Deus, aut quod factus sit Deus. Si vero esset alia persona vel hypostasis Dei et hominis, ita quod esse Deum praedicaretur de homine, et e converso per quandam coniunctionem suppositorum, vel dignitatis personalis, vel affectionis, vel inhabitationis, ut Nestoriani dixerunt, tunc pari ratione posset dici quod homo factus est Deus, idest coniunctus Deo, sicut et quod Deus factus est homo, idest coniunctus homini.'

14) See Thomas Aquinas, *Summa theologiae*, 1. 36. 4 ad arg. 4-6: 'Ad quartum dicendum quod haec duo, scilicet, pater et filius sunt unum principium quod est pater, aut, unum principium quod non est pater, non sunt contradictorie opposita. Unde non est necesse alterum eorum dare. Cum enim dicimus, pater et filius sunt unum principium, hoc quod dico principium, non habet determinatam suppositionem, imo confusam pro duabus personis simul. Unde in processu est fallacia figurae dictionis, a confusa suppositione ad determinatam. Ad quintum dicendum quod haec etiam est vera, unum principium spiritus sancti est pater et filius. Quia hoc quod dico principium non supponit pro una persona tantum, sed indistincte pro duabus, ut dictum est. Ad sextum dicendum quod convenienter potest dici quod pater et filius sunt idem principium, secundum quod ly principium supponit confuse et indistincte pro duabus personis simul.' On the connection between supposition and the fallacy of figure of speech, see also Thomas Aquinas, *Scriptum super IV Sententiarum*, 3.1.2.4 ad arg. 6; id., *Summa theologiae*, 3.3.6 ad arg. 3.

Spain. Like Peter, Aquinas seems to think that signification alone is unable to serve both the functions indicated above.

The Minister General of the Order, Hervaeus Natalis, does not significantly modify this view of supposition. Like Aquinas, Hervaeus does not deal expressly with supposition in his theological writings, but confines himself to using supposition to disambiguate some puzzling theological sentences. Nonetheless, Hervaeus stabilizes the vocabulary of supposition. In particular, what Aquinas in the *Disputed Questions on Power* called 'to signify formally/materially' becomes in Hervaeus 'to signify a formal/material meaning' and the Dominicans that follow Hervaeus employ this latter way of speaking. Moreover, unlike Aquinas, Hervaeus endows accidental terms as well with such double signification. On Hervaeus's account, every substantial common noun and every concrete accidental term has two meanings: a formal or primary meaning, i.e., a universal form (substantial or accidental), and a material or secondary meaning, i.e., the (substantial) bearer of the form. In the case of substantial terms, the former is the cognized form of an extra-mental thing, while the latter is the extra-mental thing itself.[15] This simple step enables the followers of Hervaeus more easily to reconcile Aquinas's theological treatment of the signification and supposition of terms with Burley's logical assessment.

Despite such terminological precisions, however, both in Aquinas and Hervaeus, the nature of supposition remains obscure, in particular how it relates to signification. There is an unresolved tension in Aquinas. On the one hand, Aquinas seems to assume that supposition has been introduced in order to *enrich* the signification of terms. If supposition cannot be reduced to signification, it might not be particularly surprising to find Aquinas saying that terms can supposit for things they do not signify or even that non-significant words can have some sort of supposition (presumably, material supposition).

On the other hand, like most medieval philosophers, Aquinas holds that a term maintains its signification even when it occurs in a proposition. If supposition serves the function of disambiguating propositions, especially in order to prevent fallacious arguments and false sentences, it is simple to conclude that supposition has been introduced in order *to specify* rather than to extend the signification of terms within propositional contexts. Moreover, as has been noted, it is not clear whether, for Aquinas, such semantic specification con-

15) See Hervaeus Natalis, *Scriptum super IV Sententiarum* prol., q. 7, ed. Paris (1647), f. 21^{b}B; 1.5.1, f. 51^{b}A-C; 1.23.1, ff. 112^{b}B-113^{a}C; *Quodlibeta*, ed. Venice (1613), I q. 2, ff. 4vb-7va. For Aquinas on natural supposition, see below, the next paragraph.

cerns the signification or the natural supposition of terms.[16] In brief, in one respect, supposition cannot be explained only in terms of signification, but in another respect, supposition can be accounted for as a sort of contextual signification.

What is ultimately Aquinas's position about the relationship between supposition and signification? It is not easy to answer this question. For instance, Aquinas never considers material supposition, so we cannot know what precisely is Aquinas's position when a term is mentioned or linguistically referred to. Nonetheless, some observations scattered throughout his works lead us to say that, for Aquinas, a term can have material supposition when it stands for what it does not signify, i.e., when it stands for itself. I mean that Aquinas often says that a term signifies its meaning (whether it is a form or the bearer of the form), but to my knowledge he never says that a term signifies itself besides its meaning.[17] On the other hand, Aquinas seems to think that supposition is closely related to signification. This seems to be the reason why Aquinas relates simple and personal supposition to the two levels of the signification of terms.

Aquinas's final word, if any, about supposition is uncertain and Hervaeus does not appear to clarify in a significant way this aspect of Aquinas's philosophy. In the accounts of Aquinas and Hervaeus, the value of material supposition lurks in the shadows. Yet there is no doubt that Aquinas makes simple and personal supposition rest firmly upon signification. This seems to be confirmed by Aquinas's conviction that a term has supposition not only when it occurs in a proposition. Taken outside a proposition, a term has what is traditionally

---

16) As to the signification of nouns, an interpreter can find some oscillations in Aquinas's texts. Aquinas seems to make three different claims about the signification of a term such as 'man': first, 'man' signifies human nature; second, 'man' signifies human nature as inherent in a concrete subject; third, 'man' signifies the concrete subject of human nature. In the last two cases, there can be an overlap between signification and supposition. Compare, for instance, Thomas Aquinas, *Summa theologiae*, 3.4.3 and ad arg. 2; 3.17.2, and *Quaestio disputata de unione Verbi*, 3 ad arg. 5, with *Quaestiones disputatae de potentia Dei*, 9. 4 and ad arg. 18, and *De ente et essentia* I. For more details on Aquinas's account of signification, I take the liberty to refer to my forthcoming *Mental Representation and Semantics. Two Essays in Medieval Philosophy*.

17) This is not Rovegnatinus's interpretation. According to him, Aquinas also holds that a term can signify itself: 'Horum igitur unusquisque et materialiter seipsum et formaliter aliud ostendere potest.' (III, f. h i^r). Rovegnatinus substantiates his interpretation by refers to the *Scriptum super IV libros Sententiarum*, 1.38.5, ad arg. 4 and to the *Expositio libri Peryermenias*, 1.5. These texts, however, do not discuss cases of self-signification.

called 'natural supposition' and what Aquinas, following Peter of Spain, sometimes calls *per se* supposition.[18]

Summing up, there is some basis for stating that Aquinas regards supposition as a term's property that is actually different from signification, but there is evidence that, for Aquinas, only signifying terms can be said to supposit in a genuine way, since supposition was introduced in order to specify the signification or even the natural supposition of terms.

2. Ockham's Explanation of the Division of Supposition

A certain tension between supposition and signification can be also found in Ockham's account of supposition. As is well known, Ockham challenges the traditional way of explaining the division of supposition. Fundamentally, he approaches supposition by distinguishing between terms that are taken significatively (*significative*) and terms that are not taken significatively. When a term is taken significatively, it has personal supposition,[19] while when it is not taken significatively, it can have either simple or material supposition.[20] More specifically, Ockham explains the division of supposition by combining a *semantic* with an *ontological* condition, which operates only when terms are taken in the second way. In fact, what distinguishes personal from simple/material supposition is the significativity of terms, and this is a semantic condition. But what distinguishes simple from material supposition is the *kind of things* to which the terms of a proposition refer, and this is an ontological condition. Thus, in Ockham's theory of supposition, there is not one single criterion for deriving all the kinds of supposition. Bringing together these two conditions, Ockham singles out three cases: given a term taken as the subject of a proposition, if the predicate also stands for what the subject signifies (whether it is a real, a mental, or a linguistic item), the subject has personal

[18] Cf. Thomas Aquinas, *Summa theologiae*, 1. 39. 4 ad arg. 2; id., *Scriptum super IV Sententiarum*, 3.1.2.4 ad arg. 6. For more details on Aquinas's views on supposition, see Schoot (1993) and Sweeney (1995).

[19] Cf. William of Ockham, *Summa logicae*, ed. Boehner *et al.* (1974), 1, 63, 194, 33-35: 'Est igitur una regula generalis quod numquam terminus in aliqua propositione, saltem quando significative accipitur, supponit pro aliquo nisi de quo vere praedicatur.'; 1. 64, 195, 4-7: 'Suppositio personalis, universaliter, est illa quando terminus supponit pro suo significato.'

[20] See William of Ockham, *Summa logicae*, ed. Boehner *et al.* (1974), 1, 64, 196, 31-32 and 38-39: 'Suppositio simplex est quando terminus supponit pro intentione animae, sed non tenetur significative. (...) Suppositio materialis est quando terminus non supponit significative, sed supponit vel pro voce vel pro scripto.'

supposition; if the predicate stands for an intention of the mind, but has not been imposed to signify an intention of mind, the subject has simple supposition; finally, if the predicate stands for a spoken or written word, but has not been imposed to signify a spoken or written word, the subject has material supposition. Ockham moreover holds that such a three-fold division of supposition can occur in any kind of language (spoken, written, or mental).[21]

As many contemporary scholars pointed out, Ockham's division of supposition raises various problems.[22] For our argument, it suffices to recall only one of the difficulties with Ockham's explanation of supposition. Ockham's account of the division of supposition clearly entails that the propositional context can modify the signification of a term or, at least, the semantic ability of a term to signify something. Ockham's basic suggestion is that we can establish the kind of supposition a subject-term has by taking into account the signification of the predicate-term to which the subject-term is semantically related. Consider the case of the term 'man'. When 'man' is taken significatively, it refers to every singular man. When 'man' occurs in a proposition, it does not change its signification. Nonetheless, 'man' does not always refer to every singular man. This is the case when the term 'man' is the subject of a proposition such as 'Man is an animal', but when it is the subject of a proposition such as 'Man is a species', for example, the term 'man' cannot be taken significatively, for otherwise the proposition turns out to be false. In this case, 'man' cannot refer to every singular man, so it loses its capacity to signify every singular man.

Ockham's opponents see the relationship between supposition and signification as the critical point of his account of supposition. Personal supposition can be easily explained by means of signification, since the *supposita* of terms having personal supposition can be sorted out from the things signified by those terms. But, as has been said, signification seems to play no role in accounting for simple and material supposition. The basic reason is that Ockham is inclined to deny that a term can signify itself or an intention of the mind beyond its meaning.[23] What semantic function must a term serve, therefore, in order to have simple or material supposition? How does one avoid simple and material supposition being regarded as cases of improper supposition? On Ockham's account of supposition, one can have problems in giving a semantic

---

21) See William of Ockham, *Summa logicae*, ed. Boehner *et al.* (1974), 1, 65, 199, 43-57.

22) See, e.g., Spade (1974), Karger (1982), Normore (1997), Read (1999), Panaccio-Perini Santos (2004). For further details on the relationship between signification and supposition in Ockham, see McCord Adams (1976), Loux (1979), Panaccio (1983). Still useful is Boehner (1946).

23) See, e.g., William of Ockham, *Summa logicae*, ed. Boehner *et al.* (1974), 1, 64, 196, 33-37; 1.65, 197-198, 3-10.

reason for distinguishing simple from material supposition, since both suppositions concern terms that are taken not significatively. This difficulty becomes particularly pressing when one deals with mental language. This was one of the reasons why authors such as John Buridan and Albert of Saxony rejected the distinction between material and simple supposition, and in particular denied that such a distinction occurs in mental language.[24]

3. Franciscus de Prato's Criticism of Ockham's Account of the Division of Supposition

Both in his early *Tractatus de suppositionibus*, which dates to 1302, and later on in the *De puritate artis logicae*, composed between 1325 and 1328, Walter Burley explains the division of supposition in a different way.[25] In particular, Burley revives William of Sherwood's core distinction between formal and material supposition.[26] Such a distinction is taken up by Francis of Prato, who incorporates into it the division of supposition derived from Peter of Spain's *Tractatus*. Such a distinction is the basis for understanding Francis's arguments against Ockham's explanation of the division of supposition.

Despite Francis's generally critical attitude towards the philosophy of Ockham, Francis agrees with Ockham that signification is a property that terms have before entering into a proposition, while supposition is a property that terms have only when they are parts of a proposition.[27] For him, this assumption entails two consequences: first, unlike Thomas Aquinas and Peter of Spain, Francis holds that no natural supposition can be attributed to terms, since

---

24) See, e.g., John Buridan, *Summulae. De suppositionibus*, ed. Van der Lecq (1998), 4.3.2, pp. 38-44; Albert of Saxony, *Perutilis logica*, Venice (1522), II, 2, f. 11ra. On this, see Berger (1991) and Kann (1994). Already Walter Chatton noted the problem of the overlap between material and simple supposition in mental language. See Walter Chatton, *Reportatio et lectura super Sententias*, ed. Wey (1992), 1.3.2.

25) See Walter Burley, *Tractatus de suppositionibus*, ed. Brown (1972), §§2.2-2.3, 35-36; *De puritate artis logicae tractatus longior*, ed. Boehner (1955), 1.1.1, p. 2, l. 14-p. 3, l. 25. On this, see Spade (1999).

26) Cf. William of Sherwood, *Introductiones in Logicam*, ed. Brands and Kann (1995), 136: 'Est igitur suppositio quaedam materialis, quaedam formalis. Et dicitur *materialis*, quando ipsa dictio supponit vel pro ipsa voce absoluta vel pro ipsa dictione composita ex voce et significatione. (...) *Formalis* autem est, quando dictio supponit suum significatum.'

27) See Francis of Prato, *Tractatus de suppositionibus*, ed. Amerini (1999/2000), I, 487-488, 46 ff. (also see the following footnote); William of Ockham, *Summa logicae*, ed. Boehner (1974), 1, 63, 193, 2-3. For an introduction to Francis of Prato's logic and philosophy, see Rode (2004) and Amerini (2005).

there is no way of semantically setting apart signification from natural supposition; second, supposition is the property of a term that is already signifying.[28] This second conclusion reveals that, for Francis, supposition must serve the fundamental function of specifying the signification of terms when they occur in a proposition. Although Francis is not as explicit as one would expect him to be, these two conditions seem to entail, further, that neither syncategorematic terms nor non-significant words can be said to supposit in a genuine way (except for materially).[29]

The rationale behind Francis's division of supposition is that each term can stand for itself or for its meaning (*significatum*). Because terms have two meanings, this explains why we can have only three kinds of supposition and not more. Francis does not say that a term can signify itself and its meaning, but a reader can easily infer this, since, for Francis, supposition logically presupposes signification. The strong logical subordination of supposition to signification leads Francis to reject Ockham's division. His criticism develops along to the following lines.

Francis divides supposition into proper and improper supposition. Leaving aside improper supposition, which happens when a term stands for something it has not been imposed to stand for, proper supposition holds when a term stands for something for which it can stand in virtue of its primary imposition.[30] As was said, each significant term can stand either for itself or for its meaning.[31] Consequently, there are two basic kinds of supposition. *Material* supposition occurs in the first case, i.e., when a term stands for itself. When a term stands

28) See Francis of Prato, *Tractatus de suppositionibus*, ed. Amerini (1999/2000), I, 487-488, ll. 46-69: 'Prima conclusio est quod suppositio [...] differt a significatione in tribus. Primo, quia significatio convenit termino, sive ille terminus ponatur in propositione sive non [...]; suppositio vero convenit termino solum quando est in propositione. Secundo, quia significatio est prior ordine naturae vel rationis ipsa suppositione, cum significatio sit rei praeintellectae per vocem repraesentatio; suppositio autem est acceptio termini iam significantis. Tertio [modo], quia significatio convenit termino prout importat absolute rem significatam; sed [...] suppositio convenit termino vel voci prout importat aliquid in comparatione ad alterum. Secunda conclusio est quod falsum dicunt omnes illi qui dicunt quod terminus per se sumptus habet aliquam suppositionem.'

29) Such a consequence, only implicit in Francis, instead is explicitly drawn by John Buridan. See, e.g., John Buridan, *Summulae. De suppositionibus*, ed. Van der Lecq (1998), 4.1.2, p. 9, 20-p. 10, 2.

30) See Francis of Prato, *Tractatus de suppositionibus*, ed. Amerini (1999/2000), VI, 501, 4-7: 'Suppositio propria est illa quando terminus in propositione supponit pro illo pro quo supponere potest non solum ex usu communi seu ex pacto, sed ex natura sua, idest ex primaria sua impositione.'

31) *Ibidem*, lines 7-9: 'Ubi est notandum quod omnis terminus, qui aliquid significat, de se est natus vel potest supponere vel pro se vel suo significato.'

for its meaning, it has *formal* supposition.[32] Francis further subdivides formal supposition: when a term stands for its formal or primary meaning (i.e., a universal form), it has simple supposition; when a term stands for its material or secondary meaning (i.e., any individual bearer of the universal form, that is, the underlying subject of the form), it has personal supposition. In other words, a term has simple supposition when it stands for its meaning as compared to an attribute of reason, while it has personal supposition when it stands for its meaning as compared to a real attribute.[33]

These characterizations clearly show that Francis agrees more with Burley[34] than with Ockham. In particular, Francis argues that it is 'a certain fiction' (*quaedam fictio*) to say—as Ockham actually did—that a term can stand for an intention of the mind even though it was not introduced into language in order to refer to such an intention. The reason seems to be that it is semantically impossible to have an intermediate case in between those illustrated before. A given term can stand either for itself or for its meaning; there is no room for saying that a term can stand for something else that is not itself or its meaning.[35] Presumably, Francis thinks that it is improper to make use of

---

32) *Ibidem*, lines 9-13: 'Et ideo de suppositione propria datur talis divisio, quia aut terminus in propositione supponit pro se ipso, et tunc talis suppositio dicitur suppositio materialis; aut terminus in propositione supponit vel stat pro suo significato, et tunc talis suppositio dicitur formalis.'

33) See Francis of Prato, *Tractatus de suppositionibus*, ed. Amerini (1999/2000), VII, 504-505, 4-5 and 18-20: 'Suppositio autem simplex sic describitur: suppositio simplex est quando terminus in propositione positus stat pro suo significato ad aliquod attributum rationis <comparato>.'; VIII, 511, 4-7: 'suppositio personalis, generaliter loquendo, potest sic describi: suppositio personalis est illa quando in propositione terminus positus supponit pro suo significato solum in comparatione ad aliquod [ad] attributum in ente reali.' Francis explicitly refers to Hervaeus Natalis's theory of the two-fold signification of terms in VII, 505-507, 36-81. From Francis's perspective, there is no significant difference between Peter of Spain's and William of Sherwood's characterizations of simple and personal supposition. Compare Francis's definitions with Peter of Spain, *Tractatus*, V, §5-7 (81, 5-82, 12), and William of Sherwood, *Introductiones in Logicam*, ed. Brands and Kann (1995), 136.

34) Francis deviates from Burley in a few cases. For instance, he disagrees with Burley in explaining the supposition of the term 'Socrates' in the proposition 'Socrates is an individual'. According to Francis, 'Socrates' has simple and not personal supposition, since the predicate makes reference to an attribute of reason. Cf. Francis of Prato, *Tractatus de suppositionibus*, ed. Amerini (1999/2000), VII, 510, 171-184.

35) See Francis of Prato, *Tractatus de suppositionibus*, ed. Amerini (1999/2000), VII, 508, 122-134: 'Secunda conclusio est quod falsa est opinio Guillelmi Ockham, qui ponit quod tunc solum terminus habet suppositionem simplicem quando supponit pro intentione et passione animae, idest quando supponit pro actu intelligendi. Cuius falsitas patet ex hoc: nam quaedam fictio videtur esse quod vox significativa posita in propositione stet pro aliquo alio quam pro se ipsa vel quam

two criteria for distinguishing the various kinds of supposition. If one distinguishes simple from material supposition according to the different objects to which the terms of the proposition refer, one can also differentiate personal from material/simple supposition by way of the same criterion. For example, a term might be said to have personal supposition precisely when it stands for an extra-mental thing (independently of whether it was imposed to signify an extra-mental thing), just like a term can be said to have simple supposition when it stands for an intention of the mind (one could add: independently of whether it was imposed to signify an intention of the mind). But if one invokes a semantic criterion, Francis seems to think that one ought to link supposition to signification in a closer way.

4. Savonarola's Account of Supposition

Although the most-discussed kind of supposition after Ockham is material supposition, Francis assumes that the explanation of simple supposition is the weak point of Ockham's theory.[36] Francis seems to have two problems with Ockham's account of simple supposition. First, Ockham equates intentions of the mind with the acts of understanding, described as items existing subjectively in the mind; this entails that when we pronounce a proposition such as 'Man is a species', we are actually referring to psychological entities; this appears implausible to Francis. Also Ockham's move of distinguishing between two ways of considering an act of understanding—*qua* psychological token and *qua* sign of external things—appears unsatisfactory to Francis. Second, if terms actually maintain their ability to signify when they occur in a proposition, it seems to be incorrect to say that terms have simple supposition when they are taken not significatively. Since Ockham makes use of an ontological criterion for distinguishing simple from material supposition, this means that

pro suo significato; cum ipsa vox significativa non sit ipse actus intelligendi nec significet actum intelligendi, ut patet per superius dicta, sequitur quod vox nullo modo potest supponere pro intentione animae sive pro actu intelligendi, nisi forte sit talis vox quae specialiter sit instituta ad significandum talem actum intelligendi, quia tunc talis vox, sic instituta, poterit supponere pro actu intelligendi.' Here Francis seems to sum up what Burley extensively said in his *De puritate artis logicae tractatus longior*, ed. Boehner (1955), 1.1.3, p. 7, 17-p. 10, 34. As known, Burley called Ockham's explanation of simple supposition a *valde irrationabiliter dictum*.

36) For an introduction to later medieval debates on the nature of supposition, see Ashworth (1974), 77-100.

Ockham leaves unspecified the semantic function that terms serve when they are in simple and material supposition.[37]

More than one hundred years later, Girolamo Savonarola develops a similar line of argument. Savonarola introduces the theory of supposition in the tenth book of his *Compendium logicae*, the book devoted to sophistical syllogisms. Specifically, supposition is introduced to clarify the fallacy of figure of speech.[38] Savonarola's account makes explicit the parallelism between signification and supposition, which was hidden in Francis's treatise. As noted, the relationship that supposition bears to signification is a point of difficulty for the medieval theories of supposition, and medieval logicians express different attitudes towards it. William of Sherwood, for example, argued against the reduction of supposition to signification, stating that terms never signify themselves, but only their meaning. Thus—William holds—only material supposition is able to supplement the signification of terms, for material supposition is the only case in which terms are taken not significatively.[39] By contrast, Girolamo Savonarola defends a full symmetry between supposition and signification. According to Savonarola, since signification fulfills an intensional function (for it is the act of representing something to the mind), it follows that every term can represent to the mind either itself or something else. When a term signifies itself (*significatum materiale*) and stands for the same, it has material supposition, while, when it signifies its meaning (*significatum formale*) and stands for it, it has formal supposition. Savonarola then subdivides formal supposition: when a term stands for its primary meaning (*significatum primarium*), it has simple supposition, while when it stands for its secondary meaning (*significatum secundarium*), it has personal supposition.[40]

---

37) See above, §2.

38) Cf. Girolamo Savonarola, *Compendium logicae*, ed. Garfagnini and Garin (1982), 10. 12, 144, 19-22: 'Et ad hunc modum [*scil.* figure dictionis] omnis deceptio reducitur quae provenit ex varia suppositione terminorum. Ut autem hoc melius cognoscatur aliquid in hoc loco de suppositionibus dicemus.' Another important Dominican, the Spanish Vincent Ferrer († 1419), stresses the relationship between supposition and fallacies. The last chapter of his *Treatise on suppositions* is devoted to the fallacies arising from shift in the supposition of terms. See Vincent Ferrer, *Tractatus de suppositionibus*, ed. Trentman (1977), X, 181-184.

39) See William of Sherwood, *Introductiones in logicam*, ed. Brands nd Kann (1995), dub. 1, 138.

40) Cf. Girolamo Savonarola, *Compendium logicae*, ed. Garfagnini and Garin (1982), 10. 13, p. 144, l. 24-p. 145, l. 7: 'Significare enim est aliquid intellectui repraesentare, unaquaeque autem res seipsam repraesentat et ideo unaquaeque vox significativa aliqualiter duo significat, videlicet seipsam, et hoc significatum dicitur materiale, et illud ad quod significandum est imposita, et hoc dicitur formale, quod aliquando est principale et aliquando secundarium. Ut haec vox homo materialiter seipsam repraesentat, formaliter autem significat naturam humanam secundum se,

Historically speaking, Savonarola is not the first to connect supposition to signification in such a strict way. Among others, Giles of Rome,[41] Vincent Ferrer[42] and, nearer in time to Savonarola, Georgius Rovegnatinus,[43] had already made the same connection explicit. Nonetheless, Savonarola does not seem to be acquainted with their theories or at least interested in taking them into account in his handbook of logic. More generally, it is worth noting that Savonarola does not discuss other theories of supposition or other interpretations of Aquinas's position. If Georgius Rovegnatinus gives an interpretation of Aquinas's doctrine of supposition that is not that far from that of Savonarola,[44] Vincent Ferrer provided a significantly different explanation of supposition and, interestingly enough, he reconnected it to Aquinas's *De ente et essentia*.[45] Savonarola, however, neither discusses Ferrer's explanation nor any other interpretation of Aquinas.

Savonarola's explanation of the division of supposition reveals that in at least one point he diverges from Francis's vocabulary. As noted above,[46]

---

idest hominem, et hoc est eius significatum principale, ex consequenti autem et secundario significat Sortem vel Platonem, quia homo non potest esse nisi in singularibus hominibus. Hoc igitur est significare.'

41) Cf., e.g., Giles of Rome, *Expositio super libros Elenchorum Aristotelis*, ed. Venice (1496), I, ff. 10vb-11ra.

42) Cf. Vincent Ferrer, *Tractatus de suppositionibus*, ed. Trentman (1977), III, 100-103; VII, 163: 'Circa suppositionem autem materialem sciendum quod suppositio materialis non distinguitur a suppositione formali ex eo quod in suppositione formali fiat pro significato termini, in suppositione vero materiali non, ut quidam volunt. Nam omnis suppositio est pro significato termini, sive sit suppositio formalis sive materialis. Sed distinguuntur in hoc, quia suppositio formalis est quando terminus supponit pro illo vel illis quod formaliter significat; suppositio vero materialis est quando terminus supponit illud quod materialiter significat.' Vincent Ferrer's *Tractatus de suppositionibus* dates to around 1372. See Thomas (1952) and Trentman's introduction to the edition of Ferrer's Treatise.

43) Cf. *Ex divo Thoma suppositionum collectula* I, f. a ii^{v}: 'Significant quippe termini per se utique aliquid, at indeterminate sic aut sic. Quare alio indigent ad hunc vel illum significandi modum pertrahente'; f. a iii^{v}: 'Tunc enim supponere terminum dicimus quum determinate sic aut sic suum representat significatum. Nec si quis ispius sancti Thomae locutiones diligenter examinaverit, aliud esse suppositionem quam determinatam termini significationem aut significationis determinationem illum sensisse intelliget. [...] Itaque nihil aliud est suppositio quam determinata termini ad hunc vel ad illum modum significatio vel significationis ad hunc vel illum modum determinatio, vel etiam termini certus et determinatus significandi modus. Quare aperte manifestum est quam absurda sit Pauli Veneti de suppositione diffinitio, quae huiusmodi est: Suppositio est acceptio termini in propositione pro aliquo vel pro aliquibus.'

44) See Georgius Rovegnatinus, *Ex divo Thoma suppositionum collectula* III, f. b i^{r}-b iii^{v}

45) Cf. Vincent Ferrer, *Tractatus de suppositionibus*, ed. Trentman (1977), III, 100-101.

46) See §3.

following Hervaeus, Francis calls 'material meaning' (*significatum materiale*) the secondary or remote meaning of terms, while Savonarola calls 'material meaning' the term itself, only labeling the remote meaning as 'secondary'.[47] This is, however, a minor terminological difference; except for this, in fact, Savonarola entirely agrees with Francis and Burley. For example, like them he relates simple and personal supposition, which exhaust the domain of formal supposition, to the primary and secondary meanings of terms, respectively. Nonetheless, such a difference shows that, unlike Francis, Savonarola regards as fundamental the strong subordination of supposition to signification.

In brief, Savonarola holds that supposition does not serve to enrich signification, but simply to *determine* the signification of a term in one sense or in another.[48] Savonarola argues that this determination can happen only within a propositional context, and this explains why supposition has been traditionally associated with predication. As a matter of fact, terms do not change their signification if they occur in a proposition, so supposition is required in order to specify the semantic spectrum of terms.[49] But how do we establish the

---

47) Savonarola does not dwell on such double usage of the terms 'material' and 'formal'. Attention to this point is instead called by Georgius Rovegnatinus, who traces back the two-fold use of this couple of terms to Aquinas: 'Duplicem autem universaliter esse significationis substantivi termini determinationem, quarum altera ad semetipsum, ad aliud vero quam se effigianda altera trahitur, omnibus dubio procul est. Formalem hanc, materialem vero illam dicimus suppositionem. Haec ex primo *Sententiarum* atque ex commentariis sancti Thomae in librum primum *De interpretatione* Aristotelis elicere possumus. [...] Verumtamen priusquam ad postremam huius partitionis veniamus, sciendum est tam materialem quam formalem ab ipso divo Thoma dupliciter accipi suppositionem, et materialem quidem: primo ut supra expositum a nobis est, aliter vero quum pro naturae supposito communis supponit terminus. Formalis itidem.' (Georgius Rovegnatinus, *Ex divo Thoma suppositionum collectula*, III, ff. a iv^{v}-b i^{r}).

48) Cf. Girolamo Savonarola, *Compendium logicae*, ed. Garfagnini and Garin (1982), 10. 13, p. 145, ll. 8-24: 'Supponere autem est determinare dictionem ad aliquod istorum significatorum, vel etiam ad plura, et hoc non fit nisi in propositione; verbi gratia haec vox homo potest determinari ad significatum materiale, ut cum dicitur: homo est nomen, et tunc stat vel supponit materialiter, vel ad significatum formale principale, idest ad naturam humanam, etsi determinatur ad ipsam prout est in intellectu, ut cum dico: homo est species, tunc supponit pro ipsa natura humana inquantum est intellecta, quam etiam suppositionem quidam vocant materialem. Si autem determinatur ad ipsam secundum se, ut cum dico: homo est animal rationale, tunc supponit pro natura humana simpliciter. Quam suppositionem quidam vocant simplicem. Si autem determinatur ad ipsam prout est in individuis, ut cum dico: homo currit, hanc suppositionem quidam vocant personalem.'

49) Cf. Girolamo Savonarola, *Compendium logicae*, ed. Garfagnini and Garin (1982), 10.13, 146, 8-24: 'Nam in omnibus propositionibus descriptis homo semper idem significat, non enim variat significatum terminus cum ponitur in propositione [...]. Sed tamen homo non eodem modo supponit in omnibus.'

semantic value of a term? Here Savonarola's answer is interesting. Savonarola thinks that there is no intrinsic reason to endow a term with a specific signification and supposition (for instance, personal supposition), but it is the speakers' use of the term, both in a social or in a narrower scientific context, that establishes its primary semantic value.[50]

5. Final Remarks

1. The central question of this paper has been the following: What logical relationship does supposition bear to signification? We have seen that Francis of Prato and Girolamo Savonarola suggest relating supposition strictly to signification. In particular, Savonarola fully embraces the controversial idea, only implicit in Francis, that every term can also signify itself besides its meaning.[51] Since Ockham breaks the parallelism between signification and supposition, nobody should be surprised if Ockham assigns to the theory of supposition a different logical function.[52]

2. Historically speaking, Ockham's theory of supposition has its supporters. Nicolaus Drukken from Denmark, for example, is one of them.[53] But Burley's proposal seems to be the one most followed. Richard Brinkley, for example, is sympathetic with Burley and attacks Ockham by insisting on the same point as Francis: according to Ockham's division, it is difficult to give a semantic basis for simple supposition.[54] Gerald Odonis also emphasizes the strong connection between supposition and signification, explaining the division of supposition

50) Cf. Girolamo Savonarola, *Compendium logicae*, ed. Garfagnini and Garin (1982), 10. 14, 147, 5-10: 'Regula ergo generalis erit quod terminus in propositione supponit et determinatur ad illud significatum, vel ad illum modum significandi quem sequuntur communiter homines vel, si sumus in aliqua scientia, quem sequuntur communiter sapientes in illa.'

51) In this, Girolamo Savonarola differs from Peter of Mantua (*Logica*), who argues for the same interconnection between supposition and signification, drawing however the opposite conclusion: since no term can signify itself, it follows that a term can have only personal supposition. On this, see Paul of Venice, *Tractatus de suppositionibus*, ed. Perreiah (1971), 2.8-2.14, 52-72.

52) For a non-standard interpretation of Ockham's theory of supposition, see Dutilh Novaes (2007) and Dutilh Novaes (2008).

53) Cf. Nicolaus Drukken, *Tractatus de suppositionibus*, ed. Ebbesen (1997), §§1-5, 235, 1-19, esp. §§4-5, 235, 11-16: 'Suppositio simplex est quando subiectum vel praedicatum supponit pro intentione quam non significat. [...] Omnis autem dictio supponens pro intentione quam significat supponit personaliter et significative.'

54) Cf., e.g., Richard Brinkley, *Logica* II (*De universalibus*), ed. Cesalli (2008), §33, 310, 319-325.

in essentially the same way as Burley,[55] and Odonis is perhaps the first to deal extensively with the problem of how to account for supposition when non-signifying terms are involved.[56] Already Giles of Rome at Paris and Richard Campsall at Oxford, however, rejected the position of those who denied that a term can signify itself besides its meaning, and for this reason Campsall in particular stressed the inaccuracy of Ockham's account of simple and material supposition.[57] At the beginning of the fifteenth century, Paul of Venice reappraises the medieval debate concerning supposition, although he himself does not substantially deviate from Burley's explanation of the division of supposition. At the same time, Paul of Venice once more rejects Ockham's explanation of simple supposition with arguments that are not far from those of Francis. Consider the proposition 'Man is a species'. Paul asks the question whether 'man' has meaning or not. The opponent cannot say that it has no meaning, for it does not have material supposition. If instead it has meaning, it signifies a particular man or it signifies an intention of the mind. The first alternative cannot be held because the term 'man' only signifies some thing if it leads the intellect to the knowledge of that thing. But this is not the case if 'man' has material or simple supposition. Therefore, one must concede that it signifies an intention of the mind and that it stands for the same, as Ockham's position states; hence 'man' stands for some thing it signifies.[58]

---

55) See Giraldus Odonis, *Tractatus de suppositionibus*, ed. Brown (1975), §§1.11-1.12, 13. Also see Giraldus Odonis, *Opera Philosophica*. Vol. I: *Logica*, ed. De Rijk (1997), I, 2.

56) Giraldus Odonis, *Tractatus de suppositionibus*, ed. Brown (1975), §§1.2-1.21, 14.

57) See Richard of Campsall, *Logica*, ed. Synan (1955), 1.50 and 51, 198-199, 200-203, esp. 200: 'Est autem suppositio simplex quando terminus supponit pro illo quod significat naturaliter vel quando supponit pro re concepta ut concepta. [...] Ex isto sequuntur duo; primum est quod non semper quandocumque terminus supponit pro conceptu mentis est suppositio simplex, sed aliquando personalis vel significativa. [...] Ex isto sequitur quod illi non intelligunt quid est suppositio simplex qui dicunt quod tunc est suppositio simplex quando terminus supponit pro intentione animae si significative non sumatur.'

58) See Paul of Venice, (*Tractatus de suppositionibus*), ed. Perreiah (1971), 1.2, 1.9, 2.1-2.5, 2.20, 3.1, and 3.8 ff., (pp. 4, 16-20, 42-48, 76-80, 80-84, and 88 ff.); especially see 2.5b-c, 48: 'declaratio istius descriptionis implicat contradictionem. Nam capio illam propositionem 'Homo est species' iuxta modum ipsorum et quaero an li homo significat aut non. Non est dicendum secundum, quia supponit materialiter; igitur significat ab inferiori ad suum superius. Si ergo significat, vel igitur istum hominem et sic de aliis, vel solum intentiones. Non potest dici primum, quia ille terminus 'homo' non significat aliquid nisi ducat intellectum in notitiam illius. Sed certum est quod ille terminus 'homo' sic sumptus materialiter vel simpliciter iuxta modum ipsorum non potest ducere intellectum in notitiam alicuis hominis sed solummodo terminorum aut intentionum. Igitur non significat aliquem hominem. Expedit ergo dicere quod significat intentiones animae et pro eisdem supponit, ut fatetur haec positio; igitur supponit pro suo significato. Confirmatur breviter;

3. At the end of this investigation, one can get the impression that the ontological presuppositions of the theories of supposition considered in this article are indispensable for characterizing and differentiating them correctly. Technically speaking, different theories of supposition, which actually put forward different divisions of supposition, can be constructed in the same way and can work equally well. Francis of Prato, for example, adopts the same technical devices as Ockham, and particularly Ockham's explanation of the *ascensus/ descensus* mechanisms of personal supposition.[59] Generally speaking, Ockham and Francis agree that, in the proposition 'Man is a species', 'man' has simple supposition because it stands for an intention of the mind. But differences emerge when a reader semantically interprets their theories according to their previously established ontologies. Different ontological presuppositions entail different answers to the question of the truth conditions of propositions and this imbues the theories with different philosophical significance.

Francis's and Savonarola's theories of supposition work with a non-parsimonious realist ontology. In spite of Ockham's criticism, they prefer to follow Burley and Hervaeus, so they hold that the things signified formally or primarily by common terms are things *qua* universally understood; accordingly, they prefer to account for mental language as a complex, predicative combination of things *qua* understood rather than of acts of understanding things. From their perspective, when we say that 'Man is a species', we are referring to a thing *qua* cognized rather than to a psychological act of cognition. Likewise, when we say 'Man is an animal', we are referring to a real thing *qua* cognized. More specifically, in the first case, we are *reflecting* on a cognized thing (i.e., man), while in the second case we are *referring* to a real thing by way of some cognized things (i.e., man and animal). Thus, while for Ockham a proposition such as 'Man is an animal' is true (1) if there exists at least one extra-mental singular man and (2) both 'man' and 'animal' refer to that man, for Francis just as for Hervaeus 'Man is an animal' is true even though condition (1) does not hold, since such a

nam nullus terminus supponit pro aliquo, nisi significet idem, cum supponere inferat significare et non econtra. Sed per istos li homo supponit pro intentione vel intentionibus animae; igitur significat idem vel eadem. Sed nihil significat aliquid vel aliqua, quod vel quae non sit suum significatum; igitur supponit pro suo significato, quod est contradictorium huius opinionis.' On Paul of Venice's treatment of supposition, besides some of the contributions mentioned above (note n. 22), see Perreiah (1967).

59) See Francis of Prato, *Tractatus de suppositionibus*, ed. Amerini (1999/2000), VIII-IX, 511-518; Appendix 2, 549-550.

proposition states a relation existing between things *qua* present to the mind.[60] For them, the fundamental kind of supposition, semantically speaking, continues to be simple supposition, since for them, the primary ontological level is given by the formal meaning of terms.

In conclusion, bringing together Burley's logical treatment of supposition and Aquinas's and Hervaeus's ontological and semantic views, Francis of Prato and Girolamo Savonarola seem to suggest that only if signification is taken to express both (i) the semantic relation that written or spoken terms bear to such understood things and, by way of those things, to real things, and (ii) the semantic relation that terms bear to themselves, one can properly explain why the theory of supposition is needed besides the theory of signification.

Bibliography

Primary Literature

Aegidius Romanus, *Expositio supra libros Elenchorum* (Venetiis 1496)

Albertus de Saxonia, *Perutilis logica* (Venice 1522; repr. Hildesheim-New York 1974)

Franciscus de Prato: Francis of Prato, *Tractatus de suppositionibus*, ed. F. Amerini, in: *Medioevo* 25 (1999/2000), 441-550

Georgius Rovegnatinus, *Ex divo Thoma suppositionum collectula*. S.l., post 1504 (http://gallica.bnf.fr/scripts/ConsultationTout.exe?O=n059236)

Gilles of Rome, *vide s.n.* Aegidius Romanus

Giraldus Odonis, *Opera philosophica*. I: *Logica*, ed. L.M. de Rijk (Studien und Texte zur Geistesgeschichte des Mittelalters, LX; Leiden 1997)

——, *Tractatus de suppositionibus*, ed. S.F. Brown, in: id., 'Gerard Odon's De suppositionibus', in: *Franciscan Studies* 35 (1975), 5-45

Girolamo Savonarola, *Compendium logicae*, ed. G.C. Garfagnini and E. Garin (Scritti Filosofici, 1; Rome 1982)

Gualterus Burlaeus, *De Puritate Artis Logicae Tractatus Longior, with a Revised Edition of the Tractatus Brevior*, ed. by Ph. Boehner O.F.M. (Franciscan Institute Publications; Louvain-Paderborn 1955)

——, *Tractatus de Suppositionibus*, in: S.F. Brown, 'Walter Burleigh's Treatise *De Suppositionibus* and its Influence of William of Ockham', in: *Franciscan Studies* 32 (1972), 15-64

60) See Francis of Prato, *Tractatus de suppositionibus*, ed. Amerini (1999/2000), II, 492, 120-123: 'Hoc pro tanto dico, quia ego teneo cum magistro Hervaeo, in suis *Quodlibetis*, quod, nullo homine existente, ista propositio erit vera 'homo est animal', licet Guillelmus Ockham dicat oppositum.'; also see Hervaeus Natalis, *Quodlibet* III, q. 1, ff. 67vb-70ra.

Gualterus Chatton, *Reportatio et lectura super Sententias. Collatio ad libruym primum et prologus*, ed. J.C. Wey (Studies and Texts, 90, Turnhout 1992)
Guillelmus de Ockham, *Summa logicae*, ed. Ph. Boehner, G. Gál and S. Brown (Guillelmi de Ockham Opera Philosophica, I; St. Bonaventure, N.Y. 1974)
Guillelmus de Shireswood: William of Sherwood, *Introductiones in Logicam. Einführung in die Logik*, ed. H. Brands and C. Kann (Hamburg 1995)
Hervaeus Natalis, *Quodlibeta* (Venice 1513; reprinted Ridgewood, New Jersey, U.S.A., 1966)
——, *In quattuor libros Sententiarum commentaria* (Paris 1647)
Johannes Buridanus, *Summulae. De suppositionibus*, ed. R. van der Lecq (Artistarium, 10-4; Nijmegen 1998)
Nicolaus Drukken de Dacia, *Tractatus de suppositionibus*, ed. S. Ebbesen (Copenhagen 1997)
Paulus Venetus: Paul of Venice, *Logica Magna (Tractatus de Suppositionibus)*, ed. and transl. A.R. Perreiah (St. Bonaventure, N.Y. 1971)
Petrus Hispanus: Peter of Spain, *Tractatus, called afterwards Summulae logicales*. First Critical Edition from the Manuscripts with an Introduction, by L.M. de Rijk (Philosophical Texts and Studies, XXII; Assen 1972)
Richard Brinkley, *Summa logicae*, part II (*De universalibus*), in: L. Cesalli, 'Richard Brinkley *contra dialecticae haereticos*: une conception métaphysico-logique de l'universel', in: *Documenti e studi sulla tradizione filosofica medievale* 19 (2008), 277-333
Richard de Campsall, *Logica*, ed. E.A. Synan, in: J.R. O'Donnell, ed., *Nine Medieval Thinkers: A Collection of Hitherto Unedited Texts* (Toronto 1955)
Thomas de Aquino, *Expositio libri Peryermeneias*, ed. altera retractata, cura et studio Fratrum Praedicatorum (Sancti Thomae de Aquino Opera omnia, 1* 1; Roma-Paris 1989)
——, *Summa contra Gentiles*, ed. C. Pera, P. Marc and P. Caramello (Taurini-Romae 1961)
——, *Summa theologiae*, ed. Leonina (Thomae Aquinatis Opera omnia iussu impensaque Leonis XIII P. M. edita; Romee 1888-1906, t. 4-12)
——, *Quaestiones disputate de potentia Dei* (Taurini-Romae 1931)
——, *Expositio libri Metaphysicorum* (Taurini-Romae 1950)
——, *Scriptum super IV Sententiarum* (Paris 1929-1956, t. 1-4)
——, *Quaestio disputata de unione verbi* (Quaestiones disputatae et quaestiones duodecim quodlibetales, 1; Taurini-Romae 1931)
Vincentius Ferrer: Vincent Ferrer, *Tractatus de suppositionibus*, ed. J.A. Trentman (Stuttgart-Bad Cannstatt 1977)
Walter Burley, *vide s.n.* Gualterus Burlaeus
Walter Chatton, *vide s.n.* Gualterus Chatton
William of Ockham, *vide s.n.* Guillelmus de Ockham
William of Sherwood: *vide s.n.* Guillelmus de Shireswood

Secondary Literature

Amerini, F. (2005), *La logica di Francesco da Prato* (Firenze 2005)
Ashworth, E.J. (1974), *Language and Logic in the Post-Medieval Period* (Dordrecht-Boston 1974)
Berger, H. (1991), 'Simple Supposition in William of Ockham, John Buridan and Albert of Saxony', in: J. Biard, ed., *Itinéraires d'Albert de Saxe. Paris-Vienne au XIVe siècle* (Paris 1991, 31-43)

Boehner, Ph. (1946), 'Ockham's Theory of Supposition and the Notion of Truth', in: *Franciscan Studies* 6 (1946), 261-292 (also in Ph. Boehner, *Collected Articles on Ockham*, ed. E.M. Buytaert (Franciscan Institute Publications, Philosophy Series no. 12, St. Bonaventure, Louvain, Paderborn, 1958), 232-267)

Brown, S.F. (1993), 'Medieval Supposition Theory in Its Theological Context', in: *Medieval Philosophy and Theology* 3 (1993), 121-157

Dutilh Novaes, C. (2007), *Formalizing Medieval Logical Theories: Suppositio, Consequentiae, and Obligationes* (Berlin 2007)

—— (2008), 'An Intensional Interpretation of Ockham's Theory of Supposition', in: *Journal of the History of Philosophy* 46 (2008), 365-394

Kann, C. (1994), *Die Eigenschaften der Termini: Eine Untersuchung zur Perutilis Logica Alberts von Sachsen* (Leiden 1994)

Karger, E. (1982), 'La supposition matérielle comme supposition significative: Paul de Venise, Paul de Pergula', in: A. Maierù, ed., *English Logic in Italy in the 14th and 15th Centuries*. Acts of the 5th European Symposium on Medieval Logic and Semantics. Rome, 10-14 November 1980 (Naples 1982, 331-342)

Libera, A. de (1983), *La logique médiévale et la théorie de la supposition* (Paris 1983)

Loux, M.J. (1979), 'Significatio and Suppositio. Reflections on Ockham's Semantics', in: *New Scholasticism* 53 (1979), 407-427

Maierù, A. (1985), 'A propos de la doctrine de la supposition en théologie trinitaire au XIVe siècle', in: E.P. Bos, ed., *Medieval Semantics and Metaphysics. Studies Dedicated to L.M. de Rijk on the Occasion of his 60th Birthday* (Artistarium, Supplementa, II, Nijmegen 1985, 221-238)

McCord Adams, M. (1976), 'What Does Ockham Mean by Supposition?', in: *Notre Dame Journal of Formal Logic* 17 (1976), 5-37

Normore, C.G. (1997), 'Material Supposition and the Mental Language of Ockham's *Summa logicae*', in: *Topoi* 16 (1997), 27-33

Panaccio, C. (1983), 'Guillaume d'Occam: signification et supposition', in: L. Brind'Amour and E. Vance, *Archéologie du signe* (Toronto 1983, 265-286)

Panaccio, C., and E. Perini-Santos (2004), 'Guillaume d'Ockham et la suppositio materialis', in: *Vivarium* 42 (2004), 202-224

Parsons, T. (2008), 'The Development of Supposition Theory in the Later 12th Through 14th Centuries', in: D.M. Gabbay and J. Woods, eds., *Mediaeval and Renaissance Logic* (Amsterdam 2008, 157-280)

Perreiah, A.R. (1967), *Is There a Doctrine of Supposition in the 'Logica Magna'*? (Indiana University 1967) (Ph.D. Dissertation)

Read, S. (1999), 'How is Material Supposition Possible?', in: *Medieval Philosophy and Theology* 8 (1999), 1-20

Rijk, L.M. de (1962-1967), *Logica Modernorum. A Contribution to the History of Early Terminist Logic*. I. *On the Twelfth Century Theories of Fallacy*; II/1: *The Origin and the Early Development of the Theory of Supposition*; II/2: *Texts and Indices* (Assen 1962-1967)

Rode, C. (2004), *Franciscus de Prato: Facetten seiner Philosophie im Blick auf Hervaeus Natalis und Wilhelm von Ockham* (Stuttgart 2004)

Schoot, H.J.M. (1993), 'Aquinas and Supposition. The Possibilities and Limitations of Logic in divinis', in: *Vivarium* 31 (1993), 193-225

Spade, P.V. (1974), 'Ockham's Rule of Supposition: Two Conflicts in His Theory', in: *Vivarium* 12 (1974), 63-73

—— (1999), 'Walter Burley on the Kinds of Simple Supposition', in: *Vivarium* 37 (1999), 41-59

Sweeney, E.C. (1995), 'Supposition, Signification, and Universals: Metaphysical and Linguistic Complexity in Aquinas', in: *Freiburger Zeitschrift für Philosophie und Theologie* 42.3 (1995), 267-290

Thomas, I. (1952), 'St. Vincent Ferrer's *De suppositionibus*', in: *Dominican Studies* V (1952), 88-102

The Role of 'Denotatur' in Ockham's Theory of Supposition

Catarina Dutilh Novaes
University of Groningen

Abstract

In the scholarship on medieval logic and semantics of the last decades, Ockham's theory of supposition is probably the most extensively studied version of such theories; yet, it seems that we still do not fully understand all its intricacies. In this paper, I focus on a phrase that occurs countless times throughout Ockham's writings, but in particular in the sections dedicated to supposition in the *Summa logicae*: the phrase '*denotatur*'. I claim that an adequate understanding of the role of the concept of *denotatur* within Ockham's supposition theory shall yield a deeper understanding of the theory as a whole. Here, I first examine a few uses of the term '*denotatur*' and its variants by other authors. I then turn to Ockham: first I briefly mention some uses of the term in contexts other than his theory of supposition. Following that, I focus on his supposition theory, in particular on how '*denotatur*' allows him to deal with two crucial puzzles, namely the supposition of empty terms and the supposition of terms in false affirmative propositions. The treatment of these two puzzles suggests that Ockham's theory of supposition must be understood as a theory chiefly intended for the generation of the meanings of propositions.

Keywords

Ockham, supposition theory, *denotatur*, empty terms, affirmative false propositions

1. Introduction

Ockham's theory of supposition is probably still the most extensively studied of such medieval theories. Interest in Ockham's theory in particular can be traced back to the pioneering works of E.A. Moody and Ph. Boehner in the 1940s and 1950s; since then, a considerable number of studies have focused on Ockham's theory, including works by P.V. Spade, E. Karger, and C. Panaccio, among many others. Ockham is also among the most widely edited medieval authors, almost all of his writings having received modern critical editions and often even having been translated from the Latin.

And yet, I think it is fair to say that we still do not fully understand Ockham's theory of supposition; there are still several bits and pieces that are not entirely accounted for in the standard interpretations. This is why I propose to examine his theory yet again, in spite of all the scholarship already produced on the topic. Naturally, I do not claim to offer the final word on the matter here, but I do think that paying attention to some of its aspects that have hitherto not received the attention they deserve should bring us closer to a more thorough understanding of this theory. Generally speaking,[1] the main source of confusion seems to be the pervasive and vigorous association of theories of supposition to theories of reference. Instead, I believe that what could be described as an *intensional* interpretation of Ockham's theory of supposition, based on the idea that it is chiefly aimed at the generation of the different possible readings of propositions, does better justice to the textual and conceptual elements present in Ockham's own formulation of the theory.

Here, and in the spirit of my overall intensional approach to Ockham's theory of supposition, I propose to examine an expression that occurs countless times throughout his writings, but in particular in the sections dedicated to supposition in the *Summa logicae*, namely the expression '*denotatur*'. Not that its importance for Ockham's supposition theory has never been noticed before: scholars such as C. Marmo,[2] U. Eco,[3] E. Karger[4] and E. Perini-Santos[5] have duly noted its significance. Nevertheless, I believe that the crucial role played by '*denotatur*' within Ockham's theory is not yet fully appreciated. More generally, and beyond specific uses of the term '*denotatur*', we should also be interested in the conceptual impact that the analysis of this particular aspect of Ockham's theory may have for a wider understanding of theories of supposition in general.

It is, however, absolutely crucial to keep in mind that Ockham's use of the term '*denotatur*' differs from the current meaning of the term 'denotation' in two fundamental aspects. Firstly, the first relatum in the Ockhamist relation of denotation is not a single term or a description denoting an individual, as is usually the case in current uses of the term 'denotation' in the English-speaking

---

1) As I have argued elsewhere—Dutilh Novaes (2007), ch. 1; Dutilh Novaes (2008).

2) Marmo (1984).

3) Eco (1987).

4) Karger (1976).

5) Perini-Santos (2001), (2006).

philosophical tradition (stemming from Mill and Russell); rather, what denote according to Ockham are *propositions*.[6] But he invariably uses the verb in its passive form, often in the following formulation: '*per <istam> propositionem denotatur* [...]'; therefore, it is more accurate to say that, for Ockham, it is not so much propositions that denote, but rather that by means of propositions it is denoted. Secondly, the second relatum in the relation of denotation, that which is denoted, is not an individual object, as is the case in the current tradition. Following the expression '*denotatur*' one usually finds occurrences of '*quod*' clauses (with subjunctive verbal forms) or nominalized sentences with verbs in the infinitive form and nouns in the accusative form. This is of course related to the fact that the first relatum is a proposition and not a simple term; nevertheless, the exact ontological status of what is denoted is not entirely clear. Those '*quod*' clauses may be interpreted as standing for intensional entities such as propositional meanings, as well as for extensional entities such as states-of-affairs; one can also hold a deflationist view of such 'things', i.e., that they are not 'things' properly speaking. The latter seems to be Ockham's case: what is denoted by a proposition is what the proposition says to be the case, but he never reifies that which is denoted by a proposition; as is well known, Ockham does not hint at anything like what was later to be known as '*complexe significabile*'.

Accordingly, and although literal translations are almost always to be preferred, it would be misleading to translate '*denotatur*' simply as 'it is denoted', given the well-entrenched but significantly dissimilar current meaning of the term 'denotation' and its variations. Therefore, I choose to follow the standard English translations of Ockham's *Summa logicae* and to translate '*denotatur*' as 'it is asserted'.

In this paper, I first examine a few uses by other authors of the term '*denotatur*' and its variants. I then turn to Ockham and briefly mention some uses of the term in contexts other than his theory of supposition. Following that, I focus on his supposition theory, in particular on how '*denotatur*' allows him to deal with two crucial puzzles, namely the supposition of empty terms and the supposition of terms in affirmative false propositions. With respect to the latter I also discuss the similarities between the conceptual role of '*denotatur*' within Ockham's theory and Peter of Spain's distinction between supposition and verification (also extensively discussed by Buridan).

[6] Following the meaning of the medieval term '*propositio*', 'proposition' here is understood as a linguistic entity, a sentence, not as the abstract entity which is the meaning of a sentence. Translations of the Latin passages are my own, unless otherwise stated.

2. Other Authors

The notion of *denotatur*, while being of paramount importance for fourteenth-century semantics, is thus far severely understudied. A comprehensive historical study of uses of this terminology, covering different authors at different times, is certainly a pressing task for historians of later medieval logic and semantics, but it falls outside the scope of the present enterprise. Here, I shall only mention a few occurrences of '*denotatur*' and its variants in the first half of the fourteenth century in order to illustrate that Ockham was making use of a widespread terminology, which might explain why he never offered a technical definition of the notion.

In his treatise on insolubles, written roughly at the same time as Ockham's *Summa logicae*, Bradwardine offers the following definition of the signification of a proposition: 'Every proposition signifies or means [*denotat*] as a matter of fact or absolutely [respectively] everything which follows from it as a matter of fact or absolutely.'[7] What is noteworthy for our purposes is of course that he uses the verbs '*significare*' and '*denotare*' as roughly synonymous, and as pertaining to propositions. Bradwardine does not elaborate on what exactly the signification of a proposition amounts to (as I have argued elsewhere),[8] neither concerning the semantic principles determining it nor concerning its ontological status. It is interesting though that he makes a slightly different use of the terminology: while Ockham uses the verb '*denotare*' in the passive form, Bradwardine uses it in the active form. Moreover, as we know, '*significare*' usually has a different meaning for Ockham, one which is incompatible with his uses of '*denotatur*'. Signification for Ockham normally applies to simple terms, not to propositions, and it has a strong extensional sense, as the relation between a term and the things it can be predicated of (although this can include past, future and possible entities as well).[9]

Earlier uses of the '*denotatur*' terminology can be found for example in Burley's commentary[10] and questions[11] on the *Peri Hermeneias* and in his

---

7) '[...] quelibet propositio significat sive denotat ut nunc vel simpliciter omne quod sequitur ad istam ut nunc vel simpliciter.' Text and translation in Thomas Bradwardine, *Insolubilia*, ed. Read (2010), section 6.3.

8) Dutilh Novaes (2009).

9) But here is an exception: 'Unde oratio dicitur vera quia significat sic esse a parte rei sicut est, et ideo sine omni mutatione a parte orationis, ex hoc ipso quod primo significat sicut est a parte rei, et postea, propter mutationem rei, significat sicut non est a parte rei, dicitur oratio primo vera et postea falsa.' William of Ockham, *Expositio in librum Praedicamentorum*, ed. Gál (1978), c. 9. 13, 201, ll. 30-34. I owe this reference to Laurent Cesalli.

10) Brown (1973).

11) Brown (1974).

treatise on supposition.[12] In these texts, occurrences of the term '*denotatur*' are certainly not as numerous as in Ockham's, but the sense attributed to it seems to be the same. For example, in his *Commentary on the Peri Hermeneias*, Burley uses '*denotatur*' when commenting on a proposition that can receive two interpretations, one according to composition and the other according to division.[13] Even though there is no explicit connection to supposition, here the sense of '*denotatur*' is virtually identical to Ockham's usage. However, in a passage from the *Questions*, the sense apparently attributed to '*denotatur*' seems quite different.[14] This would indicate that, at this earlier stage, or in any case for Burley, the term did not yet have the stable, quasi-technical meaning that it later acquired with Ockham.

In authors writing after Ockham, one usually encounters the term '*denotatur*' in the passive form, and essentially with the same meaning as Ockham seems to attribute to it, namely concerning the (possibly multiple) readings of a proposition. One finds it for example in Albert of Saxony in connection with the notion of *ampliatio*,[15] in Heytesbury, commenting on composite and divided senses,[16] and in sophism 16 of Heytesbury's *Sophismata*.[17] As in

[12] Brown (1972).

[13] Walter Burley, *Commentum medium in Aristotelis Peihermeneias*, ed. Brown (1973), l. 819, 94: 'Intelligendum quod haec oratio 'Omne quod est quando est necesse est esse' primo est distinguenda secundum compositionem et divisionem. In sensu compositionis est vera et denotatur quod haec est necessaria 'Omne quod est, est quando est.'

[14] Walter Burley, *Quaestiones in librum Perihermeneias*, ed. Brown (1974), l. 88, 215: 'Quod in ista 'Tantum homo currit' alicui denotatur praedicatum inesse praecise, quia homini stanti pro supposito denotatur praedicatum inesse praecise et ei quod non est Sortes denotatur praedicatum inesse praecise, et cum ista exclusive non denotet nisi unum, concedendum est quod ei quod non est Sortes inest praedicatum praecise.'

[15] Albert of Saxony, *Perutilis Logica*, cited in Pozzi (1987), II, 10, f. 15vb: 'Restat dicere de ampliatione: unde ampliatio est acceptio alicuius termini pro aliquo vel pro aliquibus ultra hoc quod actualiter est: pro quo vel pro quibus accipi denotatur per propositionem in qua ponitur.'

[16] William of Heytesbury, *De sensu diviso et composito*, Venice (1494), f. 4vb: '[E]t denotatur in sensu composito illam compositionem esse possibilem quod tu sis hic et rome: tu pretransis hoc spacium. Et per aliam propositionem denotatur illam compositionem esse possibilem quod album est nigrum. et sic de talibus. Sensus tamen divisus non sic significat nam illa propositio. album potest esse nigrum. significat quod aliquid quod iam est album vel potest esse album potest esse nigrum. Ista propositio. 'tu potes pertransire hoc spatium' verificatur per illam: 'tu habes potentiam vel habere potes qua potest esse quod tu pertransires hoc spatium', et sic de talibus.'

[17] William of Heytesbury, *Sophismata*, sophism 16, p. 3: 'Unde in hac propositione 'omnis homo et duo homines non sunt tres' iste terminus 'duo homines' supponit confuse et non distributive ita quod sub isto termino non contingit descendere copulative sed disjunctive, et denota-

Ockham, it was also widely used in contexts other than logical contexts, in particular in theological contexts, for example in several passages of Gregory of Rimini's commentary on the first and second books of Peter Lombard's *Sentences*[18] and in Robert Holkot's commentary on the fourth book of the *Sentences*.[19] Speaking of Holkot, here is very interesting passage where Holkot applies the notion of *denotatur* not only to linguistic sentences, but also to other kinds of signs having a similar assertive power:

> We in fact see that a barrel-hoop of a certain material signifies as much as will the sentence 'There is wine in this cellar', written on the wall. Hence, 'This barrel-hoop is true' could be conceded, since the case is as is signified [*denotatur*] by it.[20]

Technically, it appears that Holkot is using *denotatur* with respect to the proposition 'This barrel-hoop is true' and not with respect to the barrel-hoop itself, but this passage seems to suggest that that there is wine in the cellar is signified (*denotatur*) by the barrel-hoop itself. Note that here again we encounter the simultaneous use of *denotare* and *significare* that we had observed with Bradwardine, but again in this case pertaining to propositions and other signs with propositional assertive power.

It seems thus that at least as early as in the first decades of the fourteenth century, the notion of *denotatur* was well established and often used in the sense of pertaining to propositional meaning; it is perhaps exactly for this reason that one does not encounter a precise, technical definition of the notion in these authors' writings, as if it were current enough so as not to require an

---

tur per talem propositionem quod omnis homo est talis, qui et [aliud] duo homines non sunt tres homines, et hoc est verum, sed in hac propositione 'nullus homo et duo homines sunt tres homines' supponit iste terminus 'duo homines' confuse distributive, et denotatur per illam quod nullus est homo qui et duo homines sunt tres homines; et ex hoc sequitur quod nulli duo homines et unus sunt tres.'

18) Gregory of Rimini, *Lectura super primum et secundum Sententiarum*, ed. Trapp *et al.* (1979-1984).

19) Robert Holkot, *Commentary on the Fourth Book of the Sentences*, a2v: 'Tertio arguo sic: non est in potestate hominis quod appareat sic esse ex parte rei sicut totaliter per propositionem denotatur. Ergo non est in potestate hominis quod propositio appareat sibi vera, nec per consequens quod credat eam esse veram. Nam homo non credit aliquam propositionem esse veram nisi illa appareat sibi vera, nec aliqua apparet sibi vera nisi appareat sibi sic esse sicut per eam denotatur.'

20) Robert Holkot, *Quodlibetal Dispute*, ed. Courtenay (1971), 7; translation in Spade (1980), n. 614: 'De facto videmus quod unus circulus de certa materia tantum significat, sicut faciet haec propositio scripta in pariete tabernae 'In hoc cellario est vinum'. Unde hoc posset concedi 'Iste circulus est verus,' quia sic est sicut per eum denotatur.'

explicit definition. In these two aspects, i.e., with respect to the specific meaning attributed to the expression '*denotatur*' as well as to the lack of an explicit definition, Ockham's usage was perfectly in accordance with general uses of the term at the time. Note also that it was used in connection with a variety of semantic topics, not only with respect to supposition. It must be noted, however, that (to my knowledge) no other author seems to have made as crucial and wide a use of this notion as Ockham did.

3. Ockham

3.1. *In Other Contexts*

Indeed, Ockham uses the term '*denotatur*' very broadly, not only in his logical writings but also when writing on, e.g., physics or theology; this fact is clearly related to the distinguishably semantic approach that he adopts in these different fields of investigation. In fact, his uses of '*denotatur*' elsewhere (i.e., other than in the logical writings) are again in perfect harmony with what one encounters in the *Summa*, more specifically, and in his other logical writings, more generally. In other words, even if '*denotatur*' lacks some of the features typical of a technical term, in particular that of being explicitly defined, Ockham certainly uses it broadly and consistently. Moreover, '*denotatur*' clearly plays a crucial role within Ockham's semantics (and consequently, within his theoretical system as a whole); these elements together indicate that it can be seen as a quasi-technical term, or in any case as a concept absolutely crucial for Ockham's overall doctrines.

Ockham uses the term '*denotatur*' for example in his *Questions* on the *Physics*[21] and in his commentaries on the *Sentences*.[22] Note, however, that in

---

21) William of Ockham, *Quaestiones in libros Physicorum Aristotelis*, ed. Brown (1984), q. 3, 402, ll. 53-60: 'Et si dicas quod tales propositiones sunt falsae quando subiecta supponunt personaliter, igitur subiecta supponunt pro aliquoo vel aliquibus quae sunt intelligibilia; respondeo quod in talibus propositionibus, quamvis termini supponant personaliter, non tamen supponunt pro aliquo; sed sufficit quod denotetur supponere pro aliquo. Sicut, si nullus homo sit albus, in ista propositione 'homo albus est homo' subiectum non supponit pro aliquo sed denotatur supponere pro aliquo; et nec pro se nec pro voce, et ideo supponit personaliter.'

22) William of Ockham, *In librum secundum Sententiarum, Reportatio*, ed. Gál and Wood (1981), q. 1, 20, 10-14: 'Et hoc non est aliud nisi quod per copulam denotatur quod illud pro quo supponit subiectum est idem cum illo pro quo supponit praedicatum.'
William of Ockham, *Ibid.*, q. 15, 376, 5-9: 'Angelus tamen loquens potest sibi—exprimendo—manifestare quod haec sit notitia unius singularis et non alterius, puta formare aliquod

these passages the concept of supposition is often present, which thus clearly confirms that supposition and '*denotatur*' go hand in hand for Ockham.

3.2. *In Supposition Theory*

But here, given the theme of this volume, I shall focus on Ockham's uses of '*denotatur*' in the formulation of his theory of supposition in the *Summa logicae*. That the concept is central in this context can be inferred from the mere observation of its countless occurrences in the parts of the *Summa logicae* dedicated to supposition; however, surprisingly few scholars have explicitly noted this fact. I attribute this oversight to the reputation of Ockham's supposition theory as purely extensional; the concept of *denotatur* as it is actually used by Ockham in this context simply does not fit in well with the extensional interpretation of this theory.

Indeed, even scholars who do notice the central role occupied by *denotatur* are often still under the spell of the theory's purported extensional character. For example, it is merely on the basis of this purported purely extensional character that Eco[23] attributes to Ockham an extensional reading of the term '*denotatur*'. He seems to reason as follows: '*denotatur*' is crucial for Ockham's supposition theory, and this theory is purely extensional; hence, '*denotatur*' has an extensional meaning for Ockham. But as I have argued elsewhere,[24] the insistence on the purported extensional character of Ockham's supposition theory, and in particular its association to modern theories of reference, is the main source of misunderstanding in these modern interpretations. In fact, I believe that the crucial role played by *denotatur* in Ockham's theory and the evident intensional import that it has in this context are compelling reasons to reassess the purported purely extensional character of this theory, as I have done elsewhere.

As is well known, Ockham does not provide a full-blown, technical definition of supposition. But in the passages that come closest to offering definitions of supposition, we invariably find the term '*denotatur*'. Take for example

complexum et ei assentire, in quo denotatur aliquod praedicatum convenire uni singulari quod non convenit alteri.'

Some other places where the term '*denotatur*' is found in Ockham's works are: *Scriptum in Librum primum Sententiarum. Ordinatio*, ed. Etzkorn and Kelley (1978), d. 27, q. 3, p. 255, ll. 3-8; *Expositio in librum Perihermeneias Aristotelis*, ed. Gambatese and Brown (1978), Prooe., 9; 366, 75-86. I owe these references to F. Perini-Santos.

23) Eco (1987).

24) Dutilh Novaes (2007), ch. 1; Dutilh Novaes (2008).

a passage from *Summa logicae* I, ch. 63 (the very chapter where supposition is introduced), and the famous 'angel' passage from *Summa logicae* II, ch. 2. Both of them illustrate how crucial *denotatur* is for Ockham's concept of supposition.

> More generally, if the suppositing term is a subject, it supposits for that of which, or of the pronoun demonstrating it, it is asserted [*denotatur*] by the containing proposition that the predicate is to be predicated. [...] Thus, by the proposition 'Man is an animal' it is asserted [*denotatur*] that Socrates truly is an animal, so that were the proposition 'This is an animal' (pointing at Socrates) formed, it would be true.[25]
>
> Thus, for the truth of 'This is an angel', it is not required that the common term 'angel' be really identical with what is posited as the subject, or that it be really in the subject, or anything of this sort. Rather, it is sufficient and necessary that the subject and predicate supposit for the same thing. And, therefore, if in 'This is an angel' the subject and predicate supposit for the same thing, the proposition will be true. Thus, it is not asserted [*denotatur*] that this thing has angelhood or that angelhood is in it—or anything of this sort. Rather, it is asserted [*denotatur*] that this thing is truly an angel—not, indeed, that it is the predicate, but that it is that for which the predicate supposits.[26]

The concept of *denotatur* is also called upon to account for the different modes of supposition and the different readings of a proposition that are obtained when there is variation of the modes of supposition of its terms:

---

25) Here I quote the full passage in Latin (Ockham, *Summa logicae*, ed. Boehner *et al.* (1974), I, 63, 194, ll. 16-32: 'Et sic universaliter terminus supponit pro illo de quo—vel de pronomine demonstrante ipsum—per propositionem denotatur praedicatum praedicari, si terminus supponens sit subiectum; si autem terminus supponens sit praedicatum, denotatur quod subiectum subicitur respectu illius, vel respectu pronominis demonstrantis ipsum, si propositio formetur. Sicut per istam 'homo est animal' denotatur quod Sortes vere est animal, ita quod haec sit vera si formetur 'hoc est animal', demonstrando Sortem. Per istam autem 'homo est nomen' denotatur quod haec vox 'homo' sit nomen, ideo in ista supponit 'homo' pro illa voce. Similiter per istam 'album est animal' denotatur quod illa res quae est alba sit animal, ita quod haec sit vera 'hoc est animal' demonstrando illam rem quae est alba; et propter hoc pro illa subiectum supponit. Et sic, proportionaliter, dicendum est de praedicato: nam per istam 'Sortes est albus' denotatur quod Sortes est illa res quae habet albedinem; et si nulla res haberet albedinem nisi Sortes, tunc praedicatum praecise supponeret pro Sorte.'

26) William of Ockham, *Summa logicae* ed. Boehner *et al.* (1974), II, 2, 249-250, ll. 8-21: '[...] sicut ad veritatem istius 'iste est angelus' non requeritur quod hoc commune 'angelus' sit idem realiter cum hoc quod ponitur a parte subiecti, nec quod insit illi realiter, nec aliquid tale –, sed sufficit et requiritur quod subiectum et praedicatum supponant pro eodem. Et ideo si in ista 'hic est angelus' subiectum et praedicatum supponant pro eodem, propositio erit vera. Et ideo non denotatur quod hic habeat angelitatem vel quod in isto sit angelitas vel aliquid huiusmodi, sed denotatur quod hic sit vere angelus; non quidem quod sit allud praedicatum, sed quod sit illud pro quo supponit praedicatum.'

> In the former case [if the subject has simple supposition] we have a true proposition asserting [*denotatur*] that a concept or intention of the soul is a species, and that is true. In the latter case [if the subject has personal supposition], we have the absolutely false proposition asserting [*denotatur*] that some thing signified by 'man' is a species, which is clearly false.[27]

But the crucial function of the notion of *denotatur* comes to the fore even more explicitly in connection with two notorious puzzles for Ockham's theory (as well as for theories of supposition in general), to which I now turn.

3.3. *(Affirmative) Propositions with Empty Names*

Supposition is, of course, the relation between a term in a proposition and the thing or things it stands for. This presupposes, or at least seems to imply, that there are indeed things that a given term that has supposition supposits for. But what if there are no such things? This is most patently the case of terms that signify things that do not and could not ever exist, such as 'chimera' or 'unicorn', but it equally attains terms that at some particular point in time or at a given situation happen to be empty, for example 'man' if there are no men. In such a situation, what is the *suppositum* of 'man' in 'A man is running'? At first sight, it does not seem to supposit for anything, from which one may infer that it does not supposit at all in this situation. But if such empty terms do not have supposition, this means that supposition theory cannot handle cases of terms with empty supposition: properly speaking, empty supposition would be no supposition at all—this would be a contradiction in terms. And if that is indeed the case, then there is an important class of propositions that supposition theory simply cannot handle; this would obviously constitute a severe limitation to the whole framework.

Motivated by the issue of empty supposition, E. Karger[28] identifies two different 'logics' operating in Ockham's system, namely a logic in which categorical propositions indeed have atomic forms, but where empty supposition cannot occur, and a logic where empty supposition can occur but where

---

27) William of Ockham, *Summa logicae*, ed. Boehner *et al.* (1974), I, 65, 198, ll. 17-21: 'Primo modo est propositio vera, quia tunc denotatur quod una intentio animae sive conceptus sit species, et hoc est verum. Secundo modo est propositio simpliciter falsa, quia tunc denotatur quod aliqua res significata per hominem sit species, quod est manifeste falsum.' Another interesting passage is William of Ockham, *Summa logicae*, ed. Boehner *et al.* (1974), I, 70, 210, 21-25: 'Et ideo dicitur suppositio determinata quia per talem suppositionem denotatur quod talis propositio sit vera pro aliqua singulari determinate; quae singularis determinate sola, sine veritate alterius singularis, sufficit ad verificandam talem propositionem.'

28) Karger (1978).

categorical propositions must be represented as molecular assertions in order to make explicit the presupposition of existence of the thing or things supposited for by the subject. She claims that it is only within each of these logics that the relations of ascent and descent between categorical and singular propositions can be preserved. But both logics are not entirely satisfactory, as each of them leaves out an important element of Ockham's overall system: either empty supposition cannot be accounted for, or categorical propositions are not really categorical but rather hypothetical (molecular) propositions. This conundrum leads Karger to conclude that 'le concept de supposition vide est étranger à la doctrine de la supposition personelle dans son ensemble.'[29]

What is surprising in Karger's argumentation is that elsewhere (in her PhD dissertation) she had been one of the few to recognize the importance of *denotatur*: 'The key word that allows recognition of this fact is '*denotatur*' and its grammatical variations: Ockham is defining the supposition a term is 'supposed' or 'intended' to have inside a sentence.'[30] Equally surprising is the fact that Ockham himself explicitly deals with this objection in chapter 72 of *Summa logicae* I (the chapter addressing objections). But what is not surprising is that the solution to this puzzle relies entirely on the notion of *denotatur*, as spelled out in the following passage:

> To the second doubt, it must be said that, according to the proper meaning of the expression,[31] it must be conceded that if no man is white and no man sings the mass and if God does not create, then in the aforementioned propositions, the subject does not supposit for anything. And yet it is taken significatively, since 'taken significatively' or 'supposit personally' can be understood in two ways: either that the term supposits for one of its *significata*, or that it is asserted [*denotatur*] to supposit for something, or that it is asserted to supposit for nothing. For in such affirmative propositions, it is always asserted that the term supposits for something, and therefore if it supposits for nothing, the proposition is false. In negative propositions, however, it is asserted that the term does not supposit for anything, or that it supposits for something of which the predicate is not true, and therefore such negative [propositions] have two causes of truth. [...] In '*homo albus est homo*', if no man is white, the subject is taken significatively and personally, not because the subject supposits for something, but

29) Karger (1978), 55.

30) Karger (1976), 20. See also Karger (1976), 94.

31) Note that here Ockham is applying the notion of *de virtute sermonis* to the very conceptual apparatus used for semantic analysis, and not to a proposition to be interpreted (a meta-use of the concept). He says that, *de virtute sermonis*, for a term to supposit really means to supposit for something (i.e., a strictly extensional interpretation of the term); but secondarily, for a term to supposit means that it is asserted to supposit for something. And it is clearly this second, intensional reading of 'to supposit' that is relevant to solve this puzzle.

> because it is asserted to supposit for something; and since it supposits for nothing, and yet it is asserted to supposit for something, the proposition is simply false.[32]

It seems to me that Ockham's reply to this objection is clear and entirely satisfactory. The notion of supposition can indeed be given a strictly extensional interpretation, as pertaining to terms and actually existing things, in which case there would not even be such a thing as empty supposition. But it can also be given an alternative, intensional interpretation, relying on the notion of *denotatur*, in which case by means of an affirmative proposition it is indeed always asserted that its terms supposit for something. When this condition fails, the proposition is simply false, but the very relation of supposition does not fail. With this device, empty supposition is no longer a threat to the overall effectiveness of supposition theory as a semantic theory, as it is perfectly able to deal with affirmative propositions with empty terms. The theory simply deems such propositions to be false, and this is indeed what they are. By contrast, in the case of negative propositions with an empty subject, they come out as true according to the theory: by means of a negative proposition it is asserted that its terms do not co-supposit, and this occurs in particular if there is no *suppositum* for the subject (or predicate).

The issue that had motivated Karger's reflections in the first place, namely the inferential relations of ascent and descent between categorical propositions and the corresponding singular propositions, is also easily dealt with by my interpretation. Descent corresponds to the inferential relation from a categorical proposition to a conjunction or disjunction of singular propositions, while ascent corresponds to the reverse inferential relation. Now, the hallmark of a valid consequence, according to Ockham himself,[33] is that it is impossible

---

32) William of Ockham, *Summa logicae*, ed. Boehner *et al.* (1974), I, 72, 218-219, ll. 113-130: 'Ad secundum dubium dicendum est quod de virtute sermonis est concedendum, si nullus homo est albus et si nullus homo cantat missam et si Deus non creat, quod in praedictis propositionibus subiecta pro nullo supponunt. Et tamen sumuntur significative, quia 'sumi significative' vel 'supponere personaliter' potest dupliciter contingere: vel quia pro aliquo significato terminus supponit, vel quia denotatur supponere pro aliquo vel quia denotatur non supponere pro aliquo. Nam semper in propositionibus talibus affirmativis denotatur terminus supponere pro aliquo, et ideo si pro nullo supponit est propositio falsa. In propositionibus autem negativis denotatur terminus non supponere pro aliquo, vel supponere pro aliquo a quo vere negatur praedicatum, et ideo talis negativa habet duas causas veritatis. [...] In ista autem propositione 'homo albus est homo', si nullus homo sit albus subiectum sumitur significative et personaliter, non quia supponit pro aliquo, sed quia denotatur supponere pro aliquo; et ideo quia pro nullo supponit, cum tamen denotatur supponere pro aliquo, est propositio simpliciter falsa.'

33) William of Ockham, *Summa logicae*, ed. Boehner *et al.* (1974), III-3, ch. 1.

for the antecedent to be true while the consequent is false. And indeed, in all situations where the antecedent of a descent is an affirmative proposition that is true, there can be no empty supposition, precisely because the proposition is true. By contrast, a situation where there are no *significata* of the subject of a proposition immediately falsifies the antecedent of a descent; but as we know, a situation where the antecedent does not obtain can never be a counterexample to a putative consequence, only a situation where the antecedent obtains but the consequent does not obtain. And this can simply not happen in such cases: if an affirmative proposition is true, automatically its subject is not empty, and thus the appropriate conjunction or disjunction of singular propositions will be verified.

Things might at first sight seem slightly more complicated with negative universal propositions with empty subjects, but this is again not the case. What would invalidate the descent from a negative universal proposition 'No A is B' to a conjunction of singular propositions of the form 'This is not B' (pointing at A-things) is a situation where 'No A is B' is true, for example if there are no A's, and yet the appropriate conjunction of singular propositions would be false, that is, instances of 'This is not B' would be false. But all occurrences of 'This is not B' are true because they are negative propositions with an empty subject: since there are no A's, all occurrences of 'This' or 'That' are empty terms, and thus the negative proposition where they feature are automatically true.

In short, the notion of *denotatur* allows for a very compelling reply to the empty supposition objection. Moreover, the usual formulations of the relations of assent and descent implicitly presuppose that the antecedent is true, as it is not required from a valid relation of consequence that it should guarantee the truth of the consequent also in case of the falsity of the antecedent.

3.4. *False Affirmative Propositions*

A similar puzzle arises concerning affirmative false propositions. Take the classic example of an affirmative false proposition, '*Homo est asinus*' ('A man is a donkey'). It might seem that the two terms '*homo*' and '*asinus*' do not supposit for anything, as there are no men who are also donkeys. This objection could be generalized to all affirmative false propositions, namely those by means of which it is asserted that subject and predicate co-supposit (but this does not happen): presumably, the terms should supposit for objects that verify the proposition, but in the case of false propositions, there are no such objects. If this was indeed the case, it would again be a serious shortcoming of the general suppositional framework: it could only handle true propositions, as it would fail to give a semantic account of false (affirmative) propositions.

I am not aware of passages where Ockham explicitly discusses the puzzle of the supposition of terms in false affirmative propositions, but the objection is a familiar one (if I am not mistaken, it is present even in Geach's *Reference and Generality*). Buridan, for instance, explicitly discusses the case of false affirmative propositions and the supposition of their terms, clearly stating that they do supposit. What fails to occur in such cases is not the supposition of the terms, but the verification of the proposition, as discussed in the following passage:

> (2) Again, it is possible that terms have supposition in a proposition without the verification of the proposition, in affirmatives as well as in negatives, as in 'A man is a donkey' or 'A man is not an animal.' (3) Furthermore, there can be verification without supposition in negatives, as in 'A chimera is not a goat-stag.' But in the case of true affirmatives it is necessary that the proposition be verified of some thing or of some things for which its terms supposit.[34]

In particular, in '*Homo est asinus*', the two terms do have supposition, even though the proposition is not verified. Presumably, terms can supposit even when the proposition is not verified, and thus *a fortiori* for things other than objects that verify the proposition. One may still wonder for what things terms in affirmative false propositions supposit, and an answer to this question can be found in Peter of Spain's *Summulae*. Note that Buridan's passage just quoted is part of the main text of his *Summulae*, that is, the passages purportedly taken from Peter's *Summulae* which Buridan is commenting on (but which are, as we know, heavily modified versions of Peter's own text). And indeed, Buridan takes the distinction between supposition and verification directly from Peter, as the passage below indicates:

> Determinate supposition is the acceptance of a common term taken indefinitely or of a common term taken with a sign of particularity, as in 'A man runs' or 'Some man runs'. The supposition in both these [propositions] is said to be determinate because although in both the term 'man' supposits for every man, both those running and those not running, they

---

34) John Buridan, *Summulae. De Suppositionibus*, ed. Van der Lecq (1998), 10. Translation by Klima in *Summulae. De Dialectica*, p. 224: 'Item, possibile est esse suppositionem terminorum in propositione sine verificatione propositionis, tam in affirmativis quam in negativis, ut 'homo est asinus' vel 'homo non est animal'. Item, potest esse in negativis verificatio sine suppositione, ut 'chimaera non est hircocervus'; sed necesse est in affirmativis veris verificationem propositionis esse pro aliquo vel aliquibus pro quo vel quibus termini supponunt.'

> are nevertheless true with only one man running. That is because suppositing is one thing and it is another thing for a locution to be true of something. Therefore in the propositions mentioned above, the term 'man' supposits for every man, running and not running alike, although it renders a true locution only in connection with the running ones.[35]

Following Peter's reasoning, it is patent that, in '*Homo est asinus*', '*homo*' supposits for all men and '*asinus*' for all donkeys, and thus not only for the men and donkeys that verify the proposition (of which there are of course none).

It is worth noting that Buridan follows Peter of Spain's distinction between supposition and verification, but (to my knowledge) he never uses the term '*denotatur*', in Ockham's sense or otherwise. Indeed, all but one (namely Albert of Saxony) of the authors mentioned in the first section of the present text as those having used '*denotatur*' belong to the English tradition (Gregory of Rimini being something of an in-between case). It is also conspicuous that Albert of Saxony makes use of a notion virtually identical to Ockham's *denotatur* in his treatment of *insolubilia*, but there he uses the other term used by Bradwardine to refer to propositional meaning, namely '*significat*': 'The fourth supposition: every affirmative sentence signifies that its subject and predicate supposit for the same thing(s), and this is clearly shown to us by its affirmative copula.'[36]

But now back to Ockham and *denotatur*: how can this concept help us solve the puzzle of affirmative false propositions? As said above, I am not aware of passages where Ockham explicitly discusses this specific issue, let alone of passages where he tackles the puzzle by means of the concept of *denotatur*, but we can easily do this on his behalf. With an affirmative proposition, as the passages by Ockham already quoted clearly suggest, it is asserted that the subject and the predicate co-supposit: if this does not occur, either because one of the terms is empty or because there is no intersection between the class formed by *supposita* of the subject and the class formed by the *supposita* of the predicate, the proposition is simply false, but supposition occurs nevertheless. We can easily reformulate Ockham's observations concerning 'empty supposition' so

---

35) Peter of Spain, *Tractatus*, ed. De Rijk (1972), 82 (14-22): 'Determinata suppositio dicitur quam habet terminus communis indefinite sumptus vel cum signo particulari, ut 'homo currit' vel 'aliquis homo currit'. Et dicitur utraque istarum determinata, quia, licet in utraque illarum iste terminus 'homo' supponat pro omni homine tam currente quam non currente, tamen uno solo homine currente vera sunt. Aliud enim est supponere et aliud est reddere locutionem veram pro aliquo. In predictis enim, ut dictum est, iste terminus 'homo' supponit pro omni homine, tam currente quam non currente, sed reddit locutionem veram pro currente.'

36) Albert of Saxony, *Sophismata*, in: Pozzi (1987), 316: 'Quarta est: omnis propositio affirmativa significat <esse> idem pro quo supponit eius subiectum et praedicatum, et hoc manifeste ostendit nobis copula in ea affirmativa.'

as to deal with the puzzle of affirmative false propositions (words within brackets are the ones changed from the original passage quoted above):

> In ['*homo est asinus*'], if no man is [a donkey], the subject [and predicate are] taken significatively and personally, not because the subject [and predicate] [co]-supposit for something, but because [they are] asserted to [co]-supposit for something; and since [they do not co-]supposit for [anything], and yet [they are] asserted to [co]-supposit for something, the proposition is simply false.

Hence, in a false affirmative proposition subject and predicate do supposit, even though there is nothing they can co-supposit for, because in fact it is *asserted* that they co-supposit. The absence of a co-*suppositum* again does not imply that they do not have supposition, but simply that what is asserted to be the case is not the case—and this is precisely why the proposition is false.

The similar roles occupied by the notions of *denotatur* and verification in Ockham's and Buridan's respective theories of supposition also suggest that *denotatur* does indeed have some connection with the notion of truth. And in fact, it has been remarked by E. Perini-Santos[37] that what is 'denoted' (asserted) by a proposition is simply its truth-conditions; that is, what is asserted is that things are in such a way so as to verify the proposition. This analogy also indicates that concepts similar to that of *denotatur* may be present in other versions of supposition theory. Thus, the oversimplified motto 'A term supposits, thus it supposits for something' seems to have received more elaborate formulations also in these other versions, in particular in order to deal with the two puzzles that I have treated here, namely that of vacuous supposition and that of affirmative false propositions.

4. The Final Formulation of the Notion of Supposition

In short, the purely extensional interpretation of supposition theory, according to which supposition is only the relation between a word and the (actually existing) thing(s) it stands for, simply cannot account for affirmative false and vacuous propositions. Ockham seemed to be very much aware of these issues, and his response was to formulate his theory of supposition in what may be called intensional terms, relying crucially on the notion of *denotatur*:

---

37) Perini-Santos (2001), section 1.3.1.

One might contend that the notions of 'to supposit' and 'to supposit for nothing' are incompatible since the following is a valid inference: a term supposits, therefore it supposits for something. The response is that the inference is not valid. The following inference, however, is valid: the term supposits, therefore it is asserted either to supposit for something or to supposit for nothing.[38]

5. Conclusion

The foregoing considerations suggest that supposition theory (at any rate, in Ockham's formulation) is essentially a method to establish the readings a proposition may have, i.e., what can be asserted [*denotatur*] through it, rather than a theory of what terms are about, as the received view would have it. This is particularly evident in Ockham's treatment of propositions that must be distinguished, by means of which different assertions can be made.

Besides being fundamentally related to the notion of propositions that must be distinguished (something that I have discussed at greater length elsewhere),[39] we have just seen that the notion of *denotatur* is crucial for the resolution of the two important puzzles discussed in this paper. But some issues concerning this key notion remain. For example, what exactly are these 'things' that are asserted by propositions, that is, their meanings? It is widely acknowledged that Ockham resists any reification of propositional meaning as is later found in the theories of *complexe significabile*; for Ockham, that something is asserted by a proposition does not entail that there is something (in a strong sense) that is asserted by it.

The present contribution was not intended to exhaust the issues surrounding the notion of *denotatur*. But I hope at least to have shown that it is a crucial notion, deserving far more attention than it has received up to now, and that a deep understanding of its role within Ockham's supposition theory is mandatory for a satisfactory account of the latter. Also, it would be important to investigate whether the notion of *denotatur* has counterparts in other theories of supposition: I have suggested that Peter of Spain's and Buridan's distinction between supposition and verification plays a similar role in their theories, but comparisons with other authors are certainly worth undertaking in order to attain a better understanding of supposition theories in general.

---

38) William of Ockham, *Summa logicae*, ed. Boehner *et al.* (1974), I, 72, 219, ll. 135-139: 'Et si dicatur: ista non stant simul 'supponit' et 'pro nullo supponit', quia sequitur 'supponit, igitur pro aliquo supponit', dicendum est quod non sequitur, sed sequitur 'supponit, igitur denotatur pro aliquo supponere, vel denotatur pro nullo supponere'.'

39) Dutilh Novaes (2007), ch. 1; Dutilh Novaes (2008).

Bibliography

Primary Literature

Albertus de Saxonia, *Perutilis logica* (Venice 1522; repr. Hildesheim-New York 1974)

Gregorius Ariminensis, *Lectura super primum et secundum Sententiarum*, ed. D. Trapp, V. Marcolino, W. Eckermann, M. Santos-Noya, W. Schulze, W. Simon, W. Urban and V. Vendland (Spätmittelalter und Reformation Texte und Untersuchungen, 6-11; Berlin-New York 1979-1984)

Gualterus Burlaeus, *Tractatus de Suppositionibus*, in: S.F. Brown (1972), 'Walter Burleigh's Treatise *De Suppositionibus* and its Influence on William of Ockham', in: *Franciscan Studies* 32 (1972), 15-64

——, *Commentum medium in Aristotelis Perihermeneias*, in: S.F Brown, 'Walter Burley's Middle Commentary on Aristotle's *Perihermeneias*', in: *Franciscan Studies* 33 (1973), 45-134

——, *Quaestiones in librum Perihermeneias*, in: S.F. Brown (1974), 'Walter Burley's *Quaestiones in librum Perihermeneias*', *Franciscan Studies* 34 (1974), 200-295

Guillelmus de Heytesbury, *De sensu diviso et composito* (Venetiis 1494)

——, *Sophismata*. Available at http://ddata.over-blog.com/xxxyyy/4/07/38/29/Ttextdx-Philomed/Heytesbury/W_Heytesbury_Soph

Guillelmus de Ockham, *Summa logicae*, ed. Ph. Boehner, G. Gál and S. Brown (Guillelmi de Ockham Opera Philosophica, I; St. Bonaventure, N.Y. 1974)

——, *Expositio in librum Praedicamentorum Aristotelis*, ed. G. Gál (Guillelmi de Ockham Opera Philosophica, II. St. Bonaventure, N.Y. 1978, 133-339)

——, *Scriptum in librum primum Sententiarum. Ordinatio*, ed. G. Etzkorn and F. Kelley (Guillelmi de Ockham Opera Theologica, IV; St. Bonaventure, N.Y. 1978)

——, *Expositio in librum Perihermeneias Aristotelis*, ed. A. Gambatese and S. Brown (Guillelmi de Ockham Opera Philosophica, II; St. Bonaventure, N.Y.: 1978, 341-504)

——, *In librum secundum Sententiarum, Reportatio*, ed. G. Gál and R. Wood (Guillelmi de Ockham Opera Theologica, V; St. Bonaventure, N.Y. 1981)

——, *Brevis summa Physicorum, Summula philosophiae naturalis et Quaestiones in libros Physicorum Aristotelis*, ed. S. Brown (Guillelmi de Ockham Opera Philosophica, VI. St. Bonaventure, N.Y. 1984)

Johannes Buridanus: John Buridan, *Summulae de dialectica*, transl. G. Klima (New Haven 2001).

Johannes Buridanus, *Summulae. De suppositionibus*, ed. R. van der Lecq (Artistarium, 10-4; Nijmegen 1998)

Petrus Hispanus: Peter of Spain, *Tractatus, called afterwards Summulae logicales*. First Critical Edition from the Manuscripts with an Introduction, by L.M. de Rijk (Philosophical Texts and Studies, XXII; Assen 1972)

Robertus Holkot, *Commentary on the Fourth Book of the Sentences*. Available at http://cpricom.bc.edu/siepm/DOCUMENTS/HOLCOT/holkot1497.pdf

——: *Quodlibet*, ed. W. Courtenay, 'A Revised Text of Robert Holkot's Quodlibetal Dispute on whether God is able to Know more than he Knows', in: *Archiv für Geschichte der Philosophie* 53 (1971), 1-21

Thomas Bradwardinus (Thomas Bradwardine), *Insolubilia*, ed. and transl. by S.R. Read (Dallas Medieval Texts and Translations, 10; Leuven 2010)

William of Ockham, *vide s.n.* Guillelmus de Ockham
Wiliam of Heytesbury, *vide s.n.* Guillelmus de Heytesbury

Secondary Literature

Dutilh Novaes, C. (2007), *Formalizing Medieval Logical Theories*—Suppositiones, Consequentia *and* Obligationes (Berlin 2007)

—— (2008), 'An Intensional Interpretation of Ockham's Theory of Supposition', in: *Journal of the History of Philosophy* 46 (2008), 365–393

—— (2009), 'Lessons on Sentential Meaning from Medieval Solutions to the Liar Paradox', in: *Philosophical Quarterly* 59, 682-704

Eco, U. (1987), 'Meaning and Denotation', in: *Synthese* 73 (1987), 549-568

Karger, E. (1976), *A Study in William of Ockham's Modal Logic* (Berkeley, 1976) (PhD dissertation)

—— (1978), 'Conséquences et inconséquences de la supposition vide dans la logique d'Ockham', in: *Vivarium* 16 (1978), 46-55

Marmo, C. (1984), 'Guglielmo di Ockham e il significato delle proposizioni', in: *Versus—Quaderni di Studi Semiotici* 38/39 (1984), 115-148

Perini-Santos, E. (2001), *Modalité et Evidence: la Description Ockhamienne de la Connaissance*, University of Tours. Available at http://tede.ibict.-br/tde_busca/arquivo.-php?codArquivo=107 (dissertation)

—— (2006), *La théorie ockhamienne de la connaissance évidente* (Paris 2006)

Pozzi, L. (1987), *Il Mentitore e il Medioevo* (Parma 1987)

Spade, P.V. (1980), 'Synonymy and Equivocation in Ockham's Mental Language', in: *Journal of the History of Philosophy* 18 (1980), 9-22 (repr. in Spade, *Lies, Language and Logic in the Late Middle Ages* (London 1988))

Ockham and Buridan on Simple Supposition

Claude Panaccio
University of Quebec at Montreal

Abstract
What is at stake, philosophically, in the disagreement between Ockham and Buridan is whether there is simple (or material) supposition in the mental language or not. The key difference is that Ockham's theory allows for the possibility of use/mention ambiguities within mental language, while Buridan's approach, whatever it is exactly, does not.

Keywords
simple supposition, material supposition, ambiguity

1. Where the Disagreement Lies: The Semantics of Simple Supposition

The difference that is most often stressed by commentators between Ockham and Buridan on the topic of simple supposition is that Ockham accepts it while Buridan rejects it. But this is a red herring, a mere matter of terminology. Ockham, it is true, divides supposition into personal, material and simple, while Buridan divides it merely into personal and material.[1] What Ockham calls 'simple supposition', however, is simply included by Buridan within material supposition as a special subdivision of it. Ockham's definition of simple supposition is the following: 'Simple supposition is when a term supposits for an intention of the mind [i.e., a concept], but without being taken significatively',[2] the standard examples being, 'man is a species' or 'man is a concept', where the term 'man' supposits not for what it signifies—not for human beings, that is—, but

[1] See William of Ockham, *Summa logicae*, ed. Boehner *et al.* (1974), I, 64, 195-197, and John Buridan, *Summulae. De suppositionibus*, ed. Van der Lecq (1998), 4. 3. 2, pp. 38-44.

[2] William of Ockham, *Summa logicae*, ed. Boehner *et al.* (1974), I, 64, 196: 'Suppositio simplex est quando terminus supponit pro intentione animae, sed non tenetur significative.' (translations are mine, unless otherwise stated).

for the corresponding concept, the concept of *man*. Now Buridan has no problem at all with accepting this sort of case in spoken language. They are explicitly included under material supposition in the pseudo-Peter of Spain text that he is supposed to be commenting upon in his *Summulae logicales*, but which, in this occurrence, he presumably wrote himself: 'Material supposition occurs when a spoken word supposits for itself or for something similar to itself *or for its immediate significate* [...]'.[3] What Buridan calls the 'immediate significate' of a spoken word he explains right away: the immediate significate of a word, he says, 'is the concept according to which the word was imposed at signifying, like the term 'man' in the proposition 'man is a species'.'[4]

Apart from terminological differences, Ockham and Buridan basically agree on the nominalist thesis that in sentences such as 'man is a species' or 'animal is a genus', the terms 'man' and 'animal' stand not for any universal *in re*, but merely for the concept from which the term receives its meaning. Buridan indeed—with Ockham in mind, most probably—is very explicit that calling this sort of case 'simple supposition' or 'material supposition' is a mere matter of terminology: 'Others', he writes, 'call it simple supposition when the term supposits for the concept according to which it was imposed, and material supposition when it supposits for itself or for something similar to itself. *And this can be allowed*, but I don't care, since I call both material supposition'.[5] Since Buridan says he doesn't mind, we can, therefore, use Ockham's terminology (as I will do in the rest of this paper) and conclude that both Ockham and Buridan agree that cases of this sort—cases of simple supposition in Ockham's sense—do occur in spoken and written discourse, and that 'man' in 'man is a species' or 'man is a concept' yields a paradigmatic case of such simple supposition.

There is, nevertheless, one real and important difference between the two nominalist leaders about simple supposition. It is that Ockham accepts it *within mental discourse*, while Buridan doesn't. Ockham writes the following, for example: 'Just as these different forms of supposition [i.e., personal, simple

[3] John Buridan, *Summulae. De suppositionibus*, ed. Van der Lecq (1998), 4.3.2, p. 38: 'Sed suppositio materialis dicitur quando vox supponit pro se aut sibi simili *aut pro suo significato immediato*. [...]' (my transl. and italics).

[4] *Ibid.*: '[significatum immediatum] est conceptus secundum quem [vox] imposita est ad significandum, ut iste terminus 'homo' in ista propositione 'homo est species'.'

[5] *Ibid.*, p. 39: 'Aliter autem vocant alii suppositionem simplicem quando terminus supponit pro conceptu secundum quem imponitur, et materialem quando supponit pro se ipsa vel consimili. *Et hoc potest permitti*, sed non curo, quia utrumque voco suppositionem materialem' (my italics).

and material] can happen to a spoken or written term, they can also happen to a mental term, since a concept can supposit for what it signifies [this is personal supposition] *or for itself* [simple supposition] or for a spoken or written word [material supposition]'.[6] Ockham here interestingly admits in mental language not only simple supposition, but even (what he calls) material supposition as well. Yet, I will leave that latter thesis aside in the present context—however intriguing it is—and I will focus exclusively on simple supposition: what will concern us is that Ockham explicitly accepts that in the mental proposition corresponding to 'man is a species', the concept 'man' itself occurs as subject, but in simple supposition (rather than personal supposition).

Buridan, on the other hand, holds that only personal supposition can occur in mental discourse. Here is what he says in a famous passage of his *De fallaciis*:

> One should know, as it seems to me, that material supposition [including Ockham's simple supposition] occurs only in spoken significative words. For no term in a mental proposition supposits materially, but always personally, because we do not use mental terms by convention [*ad placitum*], as we do with spoken and written words; *never indeed does the same mental sentence have a variety of significations or acceptations*: for the passions of the soul are the same for all, just like the things which they are similitudes of, as we read in chapter 1 of the *Peri Hermeneias*. I say, therefore, that the mental proposition corresponding to this one 'man is a species', insofar as it is true, is not a proposition in which the specific concept of men is the subject, but it is a proposition in which the subject is the concept by which we conceive of the specific concept of men, and this concept does not supposit for itself, but for the specific concept of men.[7]

[6] William of Ockham, *Summa logicae*, ed. Boehner *et al.* (1974), I, 64, 197: 'Sicut autem talis diversitas suppositionis [= personalis, simplex, materialis] potest competere termino vocali et scripto, ita etiam potest competere termino mentali, quia intentio potest supponere pro illo quod significat *et pro se ipsa* et pro voce et pro scripto.' (my italics). See also *Summa logicae*, ed. Boehner *et al.* (1974), I, 67, 207.

[7] John Buridan, *De fallaciis*, ed. Ebbesen (1976), 7.3.4, 156: 'Sciendum est ergo, ut mihi videtur, quod suppositio materialis non est nisi ratione vocis significativae: nullus enim terminus in propositione mentali supponit materialiter, sed semper personaliter, quia non utimur terminis mentalibus ad placitum, sicut vocibus et scripturis, *numquam enim eadem oratio mentalis diversas significationes vel acceptiones habet*: eaedem enim omnibus passiones animae sunt et [*or*: sic] etiam res quarum ipsae sunt similitudines, ut habetur primo *Peri Hermeneias*. Unde ego dico quod propositio mentalis correspondens huic propositioni prout est vera 'homo est species' non est propositio in qua subicitur conceptus specificus hominum, sed est propositio in qua subicitur conceptus quo concipitur conceptus specificus hominum, et ille iam supponit non pro se, sed pro conceptu specifico hominum. [. . .]' (transl. Klima (2001), 522).

Here is, then, a substantial disagreement between Ockham and Buridan about what can occur in human mental discourse. I want to discuss what is at stake in it, and how these two rival theories can be assessed. Since neither Ockham nor Buridan developed the point in any detail, my interest will be in what each of them is committed to and in how to evaluate these commitments.

2. Ockham's Position Revisited

Most commentators that I know of who have taken a stand on this have favoured Buridan's approach as more philosophically sound than Ockham's. Paul Vincent Spade, for one, famously argued that Ockham's admittance of simple and material suppositions in mental language leads to the possibility of ambiguities within human thought itself, which he—Spade—took to be at odds with what mental language is supposed to be for Ockham, namely a logically ideal language.[8] Ockham, indeed, holds that a sentence with a second-order predicate such as 'man is a species' is always semantically ambiguous between (at least) a personal supposition and a simple supposition reading of its subject term.[9] If both kinds of supposition are admitted in mental language, it follows, then, that some mental propositions are semantically ambiguous. It would follow, moreover, Spade remarks, that 'we do not always know what we are asserting in a mental sentence'.[10] In Spade's view, Buridan's elimination of simple and material suppositions from mental language is the theory 'Ockham *ought* to have had'.[11] And Harald Berger in a 1991 paper on simple supposition, to take another salient example, explicitly agrees with Spade that 'the advantages of Buridan's innovative theory are convincing'.[12]

According to these critiques, then, the problem with Ockham's admittance of simple supposition in mental language is twofold:

---

8) See in particular Spade (1974) and (1980).

9) See William of Ockham, *Summa logicae*, ed. Boehner *et al.* (1974), I, 65, 197-198. In the chosen example, the sentence 'man is a species' would of course turn out to be trivially false if the subject is taken in personal supposition, since no singular man is a species, but it would nevertheless be meaningful: personal supposition readings are *always* semantically acceptable, according to Ockham.

10) Spade (1980), 21.

11) Spade (1980), 22.

12) Berger (1991), 36. Berger eventually concludes, it should be noted, that Albert of Saxony's approach to simple supposition is even better than Buridan's, but I will leave Albert aside here.

- it opens the door to ambiguities in the language of thought;
- we would not always know, consequently, what it is that we are thinking.

The first problem, actually, corresponds to some extent to Buridan's own objection against Ockham in the passage I quoted earlier. Buridan's argument there is a bit contrived, admittedly, but its main point is to be found, no doubt, in the part I have italicized above: '*never does the same mental sentence have a variety of significations or acceptations*'.[13]

Let us note, however, that Ockham does bite this bullett. He does explicitly accept in the third part of the *Summa* that there can be *some* sort of ambiguities in mental language:

> And it must be noted that this third mode of equivocation [ambiguities of supposition, namely] can be found in a purely mental proposition, although the first two modes occur only among conventional signs. This mental proposition 'man is a species', consequently, can be distinguished, since its subject can supposit significatively [i.e., in personal supposition] or for itself [simple supposition].[14]

Ockham's critics, then, cannot rest content with simply pointing out that this consequence follows from his position. They must explain why exactly they deem it unacceptable, why it is, in other words, that such ambiguities cannot be countenanced in mental language. Spade, for one, thinks that Ockham is inconsistent in accepting them because he thinks that Ockham's mental language is meant to be a logically ideal language. But this is not the case: Ockham is quite willing to accept both certain forms of redundancy and certain forms of ambiguity within mental language, and, as I have extensively argued on other occasions, he can consistently do so.[15]

Buridan, on the other hand, seems to have a different background argument against Ockham in the passage quoted above (immediately after the italicized part): it is that concepts 'are the same for all [...] as we read in chapter 1 of the *Peri Hermeneias*'.[16] Ambiguities, in other words, are not to be allowed in

---

13) See above footnote 7.

14) William of Ockham, *Summa logicae*, ed. Boehner *et al.* (1974), III-4, 4, 763: 'Et est notandum quod iste tertius modus aequivocationis potest reperiri in propositione pure mentali, quamvis duo primi modi non habeant locum nisi in signis ad placitum institutis. Unde ista propositio mentalis 'homo est species' distingui potest, eo quod subiectum potest supponere significative vel pro se ipso.'

15) See in particular Panaccio (2004a), ch. 4-6.

16) See above footnote 7.

mental language because concepts are natural signs, not instituted *ad placitum*. But this is off the point, surely. Ockham does agree that concepts are natural signs insofar as their *signification* is concerned. What we are talking about now, however, is *supposition*, not signification. In the mental proposition 'man is a species', the concept 'man', for Ockham, does not lose its natural signification: it still naturally signifies all men. Ockham's idea is that it can, in such a sentence, have either personal or simple *supposition*. That concepts, insofar as they are natural signs, are 'the same for all' is irrelevant. The question is: can a given concept, while retaining its original natural signification, acquire different semantic features in different contexts, according to how the thinking subject decides to use it?

Buridan, let me remark, is himself committed to accepting the possibility of non-natural semantic variations in mental language, although not, in his case, with respect to supposition as in Ockham, but with respect to what he (Buridan) calls 'appellation'. Take a sentence such as:

(1) Miss Scarlett knows Peabody's murderer.

This can mean one of two things, either:

(2) Miss Scarlett knows who murdered Peabody,

or:

(3) Miss Scarlett knows the person who in fact murdered Peabody (although she might not know that this person is Peabody' murderer).

In Buridan's regimented spoken language, this ambiguity does not occur because the order of words disambiguates the sentence:

> With respect to the verbs which signify acts of the soul [such as 'to know', 'to understand', 'to love', and so on], if certain terms follow such verbs and are construed with them as their direct complements, then these terms appellate the concepts according to which they signify what they signify. But if the terms precede the verbs, then they do not appellate these concepts.[17]

[17] John Buridan, *Summulae, De suppositionibus*, ed. Van der Lecq (1998), 4.5.3, p. 84: 'Termini autem respectu verborum significantium huiusmodi actus animae, si sequantur illa verba et construantur cum eis tamquam terminantia transitum eorum, appellant rationes secundum quas significant ea quae significant. Si vero illi termini praecedant illa verba, non sic appellant illas

In (1) above, for instance, the complement 'Peabody's murderer' occurs after a verb signifying an act of the soul ('to know' in this case) and appellates, consequently, the corresponding concept. In Buridan's analysis, the sentence means that Miss Scarlett knows Peabody's murderer under the concept *Peabody's murderer*; she knows, in other words, who killed Peabody. If (3) was meant instead, a Buridanian speaker should have used:

(4) Peabody's murderer is known by Miss Scarlett (where the phrase 'Peabody's murderer' occurs before the verb 'to know').

Word order, however, is a feature of spoken or written discourse. What about the corresponding mental propositions? Buridan, as far as I know, is not explicit about them, but since he apparently wants to exclude all ambiguities from mental language, he is committed to saying that they are *not* ambiguous, because either the mental phrase 'Peabody's murderer' in them appellates its *ratio* or not: if it does, the mental sentence means the same as (1); if it doesn't, it means the same as (4). This is all very well: the order of spoken or written words can be seen as a surface indicator for what is the *real* disambiguating factor, namely the presence or absence of the appellation of reason (*appellatio rationis*) within the corresponding mental proposition. Buridan, then, can successfully avoid admitting an ambiguity in the mental proposition in such cases. In doing so, however, he commits himself to accepting that the *same* mental phrase—'Peabody's murderer' in this case—can be taken in two different uses according to whether one decides to take it with or without the appellation of reason. The natural signification of the phrase, nevertheless, should remain constant. The variation, here, is due to how the cognitive agent takes the phrase, and not to its natural signification. My point, let me insist, is not that Buridan is committed to accept any ambiguity in mental language in such instances. It is that he is committed to admitting that various tokens of the same natural sign can have various uses according to how the cognitive agents

rationes.' What Buridan calls '*rationes*' in such contexts are nothing but concepts as he explicitly states in a number of other similar passages. See for example his *Tractatus de consequentiis*, ed. Hubien (1976), III, II, 3, p. 102: 'Si tamen dicti accusativi aut termini qui ab huiusmodi verbis aut participiis reguntur praecedunt illa verba vel participia, tunc non restringuntur ad appellandum *illas rationes vel illos conceptus* ' (with my italics); and again a little down the same page: 'Similiter etiam exceptio *de appellatione rationis vel conceptus*. [...] ' (with my italics). See also Buridan's *Summulae. De suppositionibus*, ed. Van der Lecq (1998), 4.3.8.4, p. 68, where Buridan directly states that the relevant accusatives 'appellate their concepts': '[...] illi accusativi [...] appellant conceptus suos [...]. '

take the sign in question in context. He cannot, therefore, argue against Ockham's stand on personal and simple supposition in mental language on the sole basis that mental concepts have a natural signification that one cannot toy with, as he seems to be doing.

Buridan's only way out of this predicament would be to deny that the mental phrase 'Peabody's murderer' with an appellation of reason and the mental phrase 'Peabody's murderer' without such an appellation are tokens of the same concept. Since, however, the 'appellation of reason' is always presented by him as a feature that a given term can sometimes have and sometimes not, this would come down to simply abandoning the whole notion of 'appellation of reason' when analyzing mental language. Since he has no other relevant theoretical tool, Buridan would thus be left with no theory at all about indirect contexts in mental language such as those which are introduced by intentional mental verbs like 'to know', 'to love' and so on. This would severely amputate his semantics and deprive it of much of its interest.

The net result of the discussion so far is this: what we have here is a plain disagreement. Ockham admits that there can be suppositional ambiguities in mental language, while Buridan denies it. But Buridan has no good argument against Ockham on the matter. And Spade's argument that if Ockham were right, mental language would not be a logically ideal language does not fly either, since mental language is not meant to be in Ockham's view a logically ideal language of the Frege-Russell type.

What about Spade's other point that if Ockham were right in allowing suppositional ambiguities in mental language, then we would not always know what we are thinking? Well, in my opinion, this is more an argument *for* Ockham's position than against it! It seems highly probable to me that we don't indeed always know what it is exactly that we are thinking. Of course, Ockham must admit that even when we think something like 'man is a species', we do *sometimes*—and perhaps most of the time—know what we are thinking. Ockham, in other words, is committed to accept that there is a possible disambiguating factor in such cases. In a joint paper on material supposition, Ernesto Perini-Santos and I have argued that Ockham's theory of material supposition requires him to acknowledge the possibility of mental disambiguating factors, which, however, are not to be identified with extra-mental terms within mental propositions, a little bit like how the tonic accent can sometimes disambiguate a spoken sentence without having to be counted itself as an extra term within the sentence.[18] The same, obviously, must be said about Ockham's

18) See Panaccio and Perini-Santos (2004), especially section 5, pp. 219-223.

theory of simple supposition in mental language: it does require the possibility of disambiguating elements which are not extra terms within the mental propositions. I would now go further in the same direction and propose that Ockham's best theory on the matter—both for material and for simple supposition in mental language—would be that such disambiguating features should be identified with acts of the will, or *volitions* in Ockham's vocabulary.

Although Ockham does not want to distinguish the will from the intellect as two distinct faculties, he always insists that volitional acts do differ from intellectual acts[19] and that some volitions are usually involved in the production of mental propositions. We might want to say, then, that what can disambiguate a mental proposition such as 'man is a species' for him is the presence not of a special intellectual act within the proposition, an additional conceptual term in other words, but the presence of a distinctive volitional act, the volition, namely, to take the term 'man' in simple supposition in this case (or to take it in personal supposition, as it may happen). Ockham, admittedly, does not elaborate the point, but the intervention of a relevant volitional act in such cases is neatly suggested by him when he says in so many words that a term can be taken in personal supposition in any proposition unless it is meant to be taken otherwise *by a voluntary decision of the users* (*ex voluntate utentium*);[20] and this indeed seems to be his best available theory on the matter.

If this is so, then it is readily understandable that the mental proposition itself, considered as a sequence of intellectual acts, is ambiguous, as Ockham wants it[21]—*semantically* ambiguous, we might say—, but that it can be disambiguated for the cognitive agent himself if it is accompanied by a relevant volition. On this view, it could happen, on the other hand, that the disambiguating volition is simply not produced by the agent: the mental proposition in this case would be ambiguous *for the cognitive agent himself.* If this particular agent neglects to mentally make a decision about whether he takes the term 'man' in

---

[19] See for instance William of Ockham, *Quaestiones in librum quartum Sententiarum. Reportatio*, ed. Wood and Gál (1984), IV, q. 16, pp. 358-359: 'Operatio interior duplex est: una, quae immediate est in potestate voluntatis, sicut volitio; alia, quae non est in potestate voluntatis nisi mediante primo actu [...] sicut intellectio. Et ideo unum actum elicit anima libere et contingenter, et alium naturaliter.'

[20] See William of Ockham, *Summa logicae*, ed. Boehner *et al.* (1974), I, 65, 197: 'Notandum est etiam quod semper terminus, in quacumque propositione ponatur, potest habere suppositionem personalem, nisi *ex voluntate utentium* arctetur ad aliam.'

[21] See William of Ockham, *Summa logicae*, ed. Boehner *et al.* (1974), I, 65, 198: 'Et est propositio [*viz.* 'homo est species'] distinguenda penes tertium modum aequivocationis, eo quod subiectum potest habere suppositionem simplicem vel personalem.'

personal or in simple supposition, then he cannot quite be clear about what he is thinking exactly, because what he is thinking exactly is *in fact* left indeterminate, due to the lack of the required disambiguating volition. That Ockham's theory allows for such cases speaks for it, it seems to me, rather than against it.[22]

So where are we now? Having reassessed Spade's two objections to Ockham's position, as well as Buridan's own distinctive objection, we have found, as it were, no good argument against Ockham's acceptance of simple supposition in mental language, and one argument in favour of it, namely that it allows in principle for certain use/mention ambiguities in thought, and not only in spoken and written discourses.

3. A Problem for Buridan

Let us turn now to Buridan's rival theory, the theory, that is, that no concept can *ever* be taken in simple supposition, and that what does correspond to simple supposition in mental language is actually the occurrence of a second–order concept taken in personal supposition. The mental proposition corresponding to 'man is a species' does not have, in this view, the concept *man* itself as subject-term, but a second-order concept which signifies the concept *man*.[23] What I would like to point out is that there is a problem here, and a quite serious one: what is this second-order concept supposed to be? I can think of three main possibilities in this respect (or families of possibilities), the

---

22) It should be noted that Ockham's point in the passage from the *Summa* quoted above in footnote 20 is not that a term is always to be taken in personal supposition unless otherwise decided by the user. If that was so, the absence of a relevant volitional act would not yield any ambiguity for the cognitive agent himself, since the term under consideration (e.g., 'man' in 'man is a species') would then be in personal supposition, period. The point is rather—just as the text says, actually—that in the absence of such a decision, the term *can* be taken in personal supposition. Just a little further down the same paragraph, Ockham adds that if the predicate is a second-order term (like 'species'), then the subject term *can* also be taken in simple supposition (*ibid.*, 198: 'Sed in ista propositione 'homo est species', quia 'species' significat intentionem animae ideo *potest* habere suppositionem simplicem ' (with my italics)), which is why the absence of a special decision results in ambiguity for a proposition such as 'man is a species'. No such problem occurs, by contrast, in a mental proposition like 'man is an animal': the predicate here being a first-order term, the subject can *only* be taken in personal supposition.

23) Buridan's position in this regard is closely related to Tarski's (and others's) reading of 'man' in, e.g., 'man is a three-letter word' as the *name* of the word 'man' rather than as a token of it. For a critical discussion of this Tarskian approach in comparison with the medieval theory of material supposition, see Panaccio (2004b).

first two based on the idea that this second-order concept is a mental *description* of the first-order concept it purports to pick out, while the third one is that it is some sort of mental intuitive grasping of the first-order concept. All of these possibilities raise delicate—and interesting—problems for Buridan's theory. Let us look at each in turn.

The first possible approach is that the second-order concept corresponding to 'man' in 'man is a species' should actually be a mental description of the first-order concept 'man' based on what this concept is supposed to be a concept *of*. Let us call it the 'semantic description approach'. This is indeed how Buridan himself proceeds in the passage of his *De fallaciis* we have quoted earlier, in which he states his thesis that only personal supposition can occur in mental discourse. He says there that the subject of the mental proposition corresponding to 'man is a species' is the concept by which we conceive of '*the specific concept of men*' (*conceptus specificus hominum*).[24] Instead of mentioning the term 'man' here, Buridan chooses—self-consciously, no doubt—to describe the designated concept with the phrase 'the specific concept of men'. He expressly uses the plural 'men' here (*hominum*) in order to make it clear that he is not thereby mentioning the concept 'man' in simple supposition in this context, but that he is referring to what the concept he speaks about is a concept of: '*hominum*' in the phrase '*conceptus specificus hominum*' is in personal supposition.[25]

What the semantic description '*conceptus specificus hominum*' does in a case like this is that it specifies both what sort of concept it is that we are talking about (a natural-kind concept: *conceptus specificus*) *and* what the significates of this concept are (men, in this example). But what now if two different concepts have the same significates? Think of 'man' and 'rational animal': how can we distinctively describe these two without conflating them? Well, a Buridanian might say, we can be a bit more precise in our descriptions and speak of the *simple* specific concept for men, as opposed to a coextensive, but *complex* concept such as *rational animal*. That is certainly fine as far as it goes, but we are still left in need of a correct semantic description of the complex concept in question. Suppose you want to describe mentally the complex concept *rational animal* without ever mentioning it (simple supposition not being allowed) and you don't want your description to apply to any other

24) See the text quoted above in footnote 7.

25) That Buridan does that on purpose is confirmed by the fact that the phrase '*conceptus specificus hominum*' occurs *three* times in the passage in question, with no occurrence at all of '*conceptus specificus hominis*' where '*hominis*' could have been interpreted as being in simple supposition.

coextensive complex concept such as *rational animal living on Earth*, for instance. You will need, to succeed, a very precise definite description such as 'the complex concept for men which is composed of one simple absolute generic concept signifying all animals, and one simple connotative concept signifying all rational beings and connoting their rationalities'. That might do it, or something like that might. But please note that the method bears with it a *very* strong requirement, namely that you should be able to precisely characterize any concept you want to refer to, and that you should be able, in particular, to semantically *analyze* any complex concept you want to refer to.

This seems to be an implausibly strong requirement. And even for Buridan himself it is, since he holds that a standard old woman (the *vetula*) can second-orderly judge that she believes that, for example, no dog is a horse or that any horse she sees is larger than a dog.[26] As the French medievalist Christophe Grellard has recently shown, the *vetula* regularly reappears in Buridan's works as the figure of a commonsensical uneducated person.[27] But think of all the analytical and technical apparatus such a person would need if she were to mentally describe to herself the conceptual content of her own belief that any horse she sees is larger than a dog, without ever using simple supposition!

A different descriptive method that could be tried would be to mentally refer to a concept by way of mentally referring to a conventional word or phrase which is subordinated to that concept. We could, for instance, use a description such as 'the concept which the English word 'rational' is subordinated to'. This in fact is what we do most of the time in spoken language: we refer to concepts by way of referring to the words that express them. When we say things such as 'Ockham's concept of *supposition*', let's say, what we actually mean is: the concept which is expressed by Ockham's word '*suppositio*' (or the concept which Ockham's word '*suppositio*' is subordinated to). This is not a bad method in principle, to be sure. But it cannot suit Buridan's needs very well, I am afraid, with respect to what is supposed to be going on in mental language. Not only would it require some indirect way of mentally referring to words without ever using *material* supposition (which, of course, Buridan would not want to allow in mental language anymore than simple supposition), but it would, most importantly, make our second-order thoughts much

26) John Buridan, *Quaestiones De anima* III, 9 (*tertia lectura*), ed. Patar (1991), Appendix V, p. 826: '[...] intelligit vetula intellectum suum, quia ipsa experitur et iudicat se scire aut credere quod nullus canis est equus et quod omnis equus quem videt est maior cane; ergo ipsa cognoscit se hoc scientem vel credentem.'

27) See Grellard (forthcoming).

too language-dependent for Buridan's taste, I am sure. And it would fail anyway, in the end, to avoid ambiguities, as Buridan wants. Consider an ambiguous spoken or written word, like 'bank' in English: any mental description like 'the concept which the English word 'bank' is subordinated to' would *ipso facto* be ambiguous as well.

There may be a few more possibilities, or combinations of such possibilities, but the best approach I can see from Buridan's outlook, although he never himself suggests anything of the sort, would be something like the so-called *paratactic* approach that Donald Davidson proposed for analyzing indirect discourse.[28] The method would be the following: you first mentally produce a token of the concept you want to refer to, and then you second-orderly grasp that concept intuitively through a mental demonstrative such as '*that* concept'. Suppose you presently entertain the thought, for example, that man is a species. What is going on, according to this paratactic interpretation, is that you produce a token of the concept *man* and then you mentally—and second-orderly—refer to it in a mental proposition such as 'that concept is a species'. Your tokening of the concept *man* in this case would not involve simple supposition—it would not involve any supposition at all, actually, since it would not itself occur within a proposition[29]—, but it would have, nevertheless, its normal natural signification, and all use/mention ambiguities, therefore, could be avoided, just as Buridan wants. As far as I can see, such an approach—or some variant of it—might turn out to yield a workable theory for Buridan, but it would certainly require some further story about what our second-order intuitive graspings of our own concepts are, and how they are possible, which he does not provide.

On the whole—and this will be my provisional conclusion—, no second-order Buridanian theory, whether descriptional or intuitivistic, seems clearly superior to Ockham's straightforward acceptance of simple supposition in mental language. The key difference, in the end, is that Ockham's theory allows for the possibility of use/mention ambiguities within mental language, while Buridan's approach, whatever it is exactly, does not. My own feeling at this

---

28) The *locus classicus* of the paratactic theory is Davidson (1968).

29) It is common wisdom in fourteenth-century supposition theory that a term has supposition only when it is inserted in a proposition. See William of Ockham, *Summa logicae*, ed. Boehner *et al.* (1974), I, 63, 193: '[…] restat dicere de suppositione, quae est proprietas conveniens termino sed numquam nisi in propositione,' and John Buridan, *Summulae. De suppositionibus*, ed. Van der Lecq (1998), 4.1.1, p. 7: '[…] non est propositionis supponere, sed termini qui est subiectum vel praedicatum [propositionis].'

point is that a philosophical theory of human mental language should *not* rule out *a priori* the possibility of such confusions within our own minds.

Bibliography

Primary Literature

Guillelmus de Ockham, *Summa logicae*, ed. P. Boehner, G. Gál and S. Brown (Guillelmi de Ockham Opera Philosophica, I;. St. Bonaventure, N.Y. 1974)

——, *Quaestiones in librum quartum Sententiarum* (*Reportatio*), ed. R. Wood and G. Gál (Guillelmi de Ockham Opera Theologica, VII; St. Bonaventure, N.Y. 1984)

Johannes Buridanus, *Tractatus de consequentiis*, ed. H. Hubien (Philosophes médiévaux, 16; Louvain, Paris 1976)

——, *The Summulae. Tractatus VII: De fallaciis*, ed. S. Ebbesen, in: *The Logic of John Buridan*, ed. J. Pinborg (Copenhagen 1976, 139-160)

——, *Quaestiones de anima* (*tertia lectura*), ed. B. Patar, in: *Le Traité de l'âme de Jean Buridan (De prima lectura)* (Louvain-la-Neuve 1991, Appendix V, 779-834) (selected excerpts)

——, *Summulae. De suppositionibus*, ed. R. van der Lecq (Artistarium, 10-4; Nijmegen 1998) (Engl. translation by G. Klima in: John Buridan, *Summulae de Dialectica* (New Haven 2001))

John Buridan, *vide s.n.* Johannes Buridanus

William of Ockham, *vide s.n.* Guillelmus de Ockham

Secondary Literature

Berger, H. (1991), 'Simple Supposition in William of Ockham, John Buridan and Albert of Saxony', in: J. Biard, ed., *Itinéraires d'Albert de Saxe. Paris-Vienne au XIV*[e] *siècle* (Paris 1991, 31-43)

Davidson, D. (1984), 'On Saying That', in: *Synthèse* 19 (1968), 130-146 (repr. in: D. Davidson, *Inquiries into Truth and Interpretation* (Oxford 1984, 93-108))

Grellard, C. (forthcoming), 'How is it Possible to Believe Falsely ? John Buridan, the *vetula* and the Psychology of Error', in K. Gosh, D. Denery, R. Copeland & N. Zeeman (eds.), *Uncertainty in the Middle Ages* (Turnhout, forthcoming)

Panaccio, C. (2004a), *Ockham on Concepts* (Aldershot 2004)

—— (2004b), 'Tarski et la *suppositio materialis*' in: *Philosophiques* 31 (2004), 295-309

—— et E. Perini-Santos (2004), 'Guillaume d'Ockham et la *suppositio materialis*', in: *Vivarium* 42 (2004), 202-224

Spade, P.V. (1974), 'Ockham's Rule of Supposition: Two Conflicts in His Theory', in: *Vivarium* 12 (1974), 63-73 (repr. in Spade (1988))

—— (1980), 'Synonymy and Equivocation in Ockham's Mental Language', in: *Journal of the History of Philosophy* 18 (1980), 9-22 (repr. in Spade (1988))

—— (1988), *Lies, Language and Logic in the Late Middle Ages* (London 1988)

Descent and Ascent from Ockham to Domingo de Soto: An Answer to Paul Spade

E. Jennifer Ashworth
University of Waterloo, Canada

Abstract
Paul Spade has attacked the theory of the modes of personal supposition as found in Ockham and Buridan, partly on the grounds that the details of the theory are incompatible with the equivalence between propositions and their descended forms which is implied by the appeal to suppositional descent and ascent. I trace the development of the doctrines of ascent and descent from the mid-fourteenth century to the early sixteenth century, and I investigate Domingo de Soto's elaborate account of how descent and ascent actually worked. I show that although Soto himself shared some of Spade's doubts, including those about the use of merely confused supposition, he had a way of reducing at least some propositions containing terms with such supposition to equivalent disjunctions and conjunctions of singular propositions. Moreover, he gave explicit instructions on how to avoid the supposed problem of O-propositions.

Keywords
ascent, descent, merely confused supposition, modes of personal supposition, O-propositions

1. Introduction

Personal supposition was normally divided into discrete and common supposition, and the latter was said to have at least three modes, determinate, distributive, and merely confused.[1] These modes, as analysed through descent

[1] I shall not be discussing the fourth mode, collective supposition, also called 'merely confused conjoint supposition', as distinguished from 'merely confused disjoint supposition'. Modes are more usually called 'species' by our authors (e.g., Anonymus, *Commentum* [...] *in primum et quartum tractatus Petri Hyspani*, Hagennaw (1495), sig. q 5 v: 'Utrum tantum tres sunt species suppositionis communis?'), if called anything, though Buridan does speak of 'modes and

and ascent, are of central importance for much medieval semantics, and their presentation by Ockham and Buridan has stimulated a good deal of discussion. A number of twentieth-century authors have attempted to formulate them by using the tools of contemporary logic.[2] In turn, their interpretations have been attacked, most notably by Paul Spade. He has argued, not only that the interpretations are misguided, but that the theory itself has significant weaknesses, including a lack of any specific function. In this paper I shall give a brief outline of how Ockham and Buridan presented the modes of personal supposition, and I shall suggest that some of the problems raised by Paul Spade can be dealt with if we consider how the theory was developed by later logicians working within the same tradition. I am assuming that a late fourteenth, fifteenth, or early sixteenth-century logician is a more reliable guide to the intentions of Ockham and Buridan than is a logician who approaches them from the perspective of modern symbolic logic, however interesting and fruitful such an approach may be.

I shall be referring in particular to three works. First, there is an anonymous commentary on Marsilius of Inghen's *Suppositiones* which was first published in Basle in 1487 and received two subsequent editions in Hagenau in 1495 and 1503.[3] I know of no manuscript versions of this commentary, but Maierù has taken it to date back at least to the late fourteenth or early fifteenth century,[4] and I can see no reason in the text to reject his dating. Second, there is John Dorp's commentary on Buridan, written in the last decade of the fourteenth century, of which I shall make some limited use. I say 'limited' because the tract on supposition theory is very puzzling. The lemmata are all taken from Buridan's *Summulae de Dialectica*, but the actual commentary is addressed, not to Buridan, but to Marsilius of Inghen. To take just one example, Dorp's discussion of the common modes and divisions of supposition begins with Buridan's lemma on the distinction between proper and improper supposition, but his commentary on the lemma contains a discussion of discrete and

divisions' (e.g., Johannes Buridanus, *Summulae. De suppositionibus*, ed. Van der Lecq (1998), 4.3.1, p. 37).

2) See, e.g., Priest and Read (1977), Weidemann (1979).

3) For a discussion of syntactic analysis in this work, see Karger (1997).

4) Maierù (1972), 234. One might think that the anonymous author's mention of the special sign 'a' as producing merely confused supposition (Anonymus, *Commentum* [...] *in primum et quartum tractatus Petri Hyspani* (1495), sig. q 5 v) warrants a late dating, but the sign is also found in Johannes Dorp, *Commentary* in: *Johannes Buridanus, Compendium*, Venice (1499) sig. b 4 ra.

common supposition which relates to Marsilius's text.[5] Finally, there is the work of Domingo de Soto, the first version of whose *Summule* was published in 1529. Although there are other elaborate early sixteenth-century discussions by such men as Jerónimo Pardo[6] and Juan de Celaya, I have chosen Soto's work for two reasons. First, it is an intelligent and detailed account of the theory as developed by various logicians from the mid-fourteenth century onwards. Second, there is a surprisingly close relationship between the issues raised by Spade and those discussed by Soto.

2. Spade's Arguments

I begin with a quotation from Spade's seminal article, 'The Logic of the Categorical: The Medieval Theory of Descent and Ascent.' He writes:[7]

> By the early fourteenth-century, the doctrine of modes as a theory of reference had been pretty much defeated. But once this had happened, the theory of modes of supposition was left with *no question to answer*. It was no longer a theory of reference; it was not, I have argued, a theory of truth conditions or analysis. The theory was left with no task to perform. No wonder it has proved so hard for scholars to agree about what it was trying to accomplish. By the early fourteenth-century, it was no longer trying to accomplish anything at all!

In another place, he expands on the last remark by claiming that

> [...] no mediaeval author I know of ever says what the theory is trying to do at all![8]

In short, Spade believes that the theory of modes of personal supposition has no function because it fails to be either a theory of reference or a theory of truth conditions as captured by a certain kind of quantificational analysis.

Spade's first weapon is based on his belief that earlier theories of the modes of personal supposition were theories of reference,[9] and that this had obviously

5) See Johannes Dorp, *Commentary* in: *Johannes Buridanus, Compendium*, Venice (1499), sig. h 4 ra-va. For Buridan's lemma, see Johannes Buridanus, *Summulae. De suppositionibus*, ed. Van der Lecq (1998), 4.3.1, p. 37; for Marsilius, see Marsilius of Inghen, *Suppositiones*, ed. Bos (1983), 52.

6) See Pérez-Ilzarbe (1996).

7) Spade (1988), 212.

8) Spade (2002), 291.

9) Spade (1988), 208-213; Spade (2002), 294-298. Cf. criticisms in Matthews (1997).

changed, given the insistence, implicit in Ockham but explicit in Buridan,[10] that a term with personal supposition supposits for all its significates.[11] Without considering the truth of Spade's beliefs about the thirteenth century, I think it is certainly true that while the theory deals with reference in a loose sense, it does not correspond to theories of reference in any modern sense.[12] It offers no explanation of how it is that a linguistic expression designates one or more particular things. That is, it does not tell us how I can hook proper names on to individuals, or what is going on when I say meaningfully 'Someone in this room is a liar, and I know it's not Tom, Dick or Harry.' What supposition theory, along with the related doctrines of ampliation and restriction, *does* do is tell us how to determine what kind of thing is spoken of, whether a word, a common nature, or an individual thing, and, at least in this third case, what the range or scope of reference is.[13] On the one hand, it determines whether we are speaking of past, present, future or possible things. On the other hand, it tackles the question, 'Do all or some members of the group spoken of have to stand, or fail to stand, in a particular relationship to members of another group in order for our claim to be true?' On the face of it, without a procedure for specifying answers to this question, we cannot handle truth claims. Nor can we judge the soundness of inferences, spot fallacies, or analyse difficult cases. I shall argue that for late medieval logicians the particular function of the theory of modes is to give detailed, practical answers to this question.

However, before we can respond fully to Spade's claim that the theory of modes had no function, we must consider his reasons for rejecting the view that it provides a theory of truth conditions as captured by a certain kind of quantificational analysis. This view stems in part from T.K. Scott, who wrote that the theory of modes is 'a set of rules for the syntactical analysis of propositions containing quantifying words and is thus the quantification theory of terminist logic.'[14] The immediate response to Scott's remark was to reintroduce a semantic element, and to argue that the theory of modes had the dual function of presenting medieval quantification theory by explaining the effect of

---

10) Johannes Buridanus, *Summulae. De practica sophismatum*, ed. Pironet (2004), ch. 3, sophisma 5, p. 53; Johannes Buridanus, *Summulae. De suppositionibus*, ed. Van der Lecq (1998), 4.4.4, 72.

11) Spade (1988), 212. He modifies these remarks slightly in what follows.

12) For more on this point, see Priest and Read (1980), 266-267. They describe supposition theory (267) as 'a theoretical apparatus within which to give the truth conditions of sentences.' For a longer recent discussion, see Dutilh Novaes (2008).

13) Some authors, such as Marsilius of Inghen, applied the theory of modes to material as well as personal supposition.

14) Scott (1966), 30.

quantifier words and their ordering and, at the same time, of giving a complete analysis of the truth conditions of categorical propositions by reducing them to logically equivalent conjunctions and/or disjunctions of singular propositions.

Spade has three arguments against this interpretation.[15] He claims that the theory of modes as presented by Ockham and Buridan cannot be a theory of quantification or of truth conditions because (1) they present their accounts of truth conditions in another context; (2) they do not claim any logical equivalence between propositions and their descended forms, and the fact that later authors seem to have dealt with equivalences is no help, because 'it leaves the origins of the theory of modes of supposition utterly mysterious';[16] (3) the details of the theory are incompatible with such an equivalence. In particular, the attribution of merely confused supposition to the predicates of A-propositions militates against any neat analysis into disjunctions and conjunctions of singular propositions, while the attribution of distributive supposition to the predicates of O-propositions would render such analysis invalid.

3. Accounts of Truth Conditions

I shall consider these arguments in order, starting with the theory of truth. It is true that Buridan's most relevant account of truth conditions is found not in his *Summulae de dialectica* but in his *Summulae de practica sophismatum*, chapter 2, and that no mention is made of descent and ascent.[17] Ockham's equivalent account is found in Part II of the *Summa logicae*, especially chapters 2, 3 and 4. There is a simple reason for Ockham's organization, in that the modes of supposition, discussed in Part I, concern terms, whereas truth conditions in general concern propositions. A feature of medieval texts is that one and the same issue is often discussed in a variety of places, and not brought together in one section or chapter, and so one ought not to place too much emphasis on the presence of separate discussions. What is more important is the absence of reference to the modes of personal supposition, but this too can be explained away if we take it that the general account of truth conditions for basic non-modal categorical propositions with present-tense verbs was intended to be supplemented by the earlier, much more detailed, exposition in the context of supposition theory.

---

15) Spade (1988), 206-207; Spade (2002), 291-292.

16) Spade (1988), 206.

17) Johannes Buridanus, *Summulae. De practica sophismatum*, ed. Pironet (2004), 42-46.

There are two features of Ockham's account of truth conditions which must be emphasized. First, there is the issue of empty subject terms. An affirmative proposition can be true only if the subject term supposits, whereas a negative proposition can be true either because the subject term fails to supposit or because it supposits for something for which the predicate does not supposit. The second issue is that of common terms such as 'sun' and 'phoenix' which have only one suppositum. Ockham raises this issue only in passing, by remarking that if the subject and predicate terms have only one instance, as in 'Every phoenix is every phoenix', universal propositions can be true, and if the predicate has only one instance, as in 'Some animal is every man' (if there were only one man), a particular proposition can be true.[18]

I mention these two issues because they correspond to two places in which the fit between the general theory of truth conditions and the theory of descent and ascent is less than perfect. The issue of common terms with only one instance was touched on, though not normally pursued in detail. For instance, John Dorp says that 'sun' by virtue of its signification has distributive supposition in 'Every sun exists' because if there were many suns, 'sun' would supposit for them, and there would be conjunctive descent. Alternatively, he remarked, one can say that from 'Every sun exists' one can descend to 'this sun exists and this sun exists', pointing to the same sun in each case.[19] The author of the anonymous *Commentum* required that the singular terms used in descent be non-synonymous, and so denied the propriety of using 'this sun' to refer to one and the same object in different singular propositions.[20] He also required an actual plurality of supposita for both determinate and distributive supposition.[21]

---

18) William of Ockham, *Summa logicae*, ed. Boehner *et al.* (1974), I. 4, 266, ll. 192-204.

19) Johannes Dorp, *Commentary* in: *Johannes Buridanus, Compendium*, Venice (1499), sig. i 3 ra-rb. Cf. sig. b 1 vb, on 'sol', 'mundus', and 'deus': '[...] licet illis terminis repugnet supponere pro pluribus; tamen illa repugnantia non provenit ex modo sue impositionis vel naturalis significationis, sed provenit ex parte rerum significatarum, quia ex parte rei repugnat plures esse deos.' Cf. Johannes Eckius, *In Summulas Petri Hispani*, Augustae Vindelicorum (1516), f. lxxxix vb: 'Non requiritur ad terminum supponentem determinate quod sub eo valeat descensus disiunctivus. [...] Sufficit enim quod sub alio consimiliter tento valet descensus disiunctivus.' He gives three examples, including 'mundus est creatus'.

20) Anonymus, *Commentum* [...] *in primum et quartum tractatus Petri Hyspani*, Hagennaw (1495), sig. q 7 r-v.

21) Anonymus, *Commentum* [...] *in primum et quartum tractatus Petri Hyspani.* Hagennaw (1495), sig. q 1 r ('Dicitur 'habente actu plura supposita' propter illos terminos 'sol', 'deus', 'fenix', 'mundus' etc.') and sig. q 3 v. See also sig. q 7 v on the requirement that the singularized common term in descent supposit for fewer things than when it is not singularized: '*Suppositum in supponendo tantum* est terminus communis singularisatus supponens pro paucioribus quam non singulari-

Domingo de Soto, on the other hand, took the position that every common term whose acceptance was not inconsistent with its suppositing for many should, if it supposited at all, supposit as a common term (*communiter*).[22] Thus the subjects of 'God reigns' (*deus regnat*), '[A] phoenix flies' (*phenix volat*) and '[A] sun shines' (*sol lucet*) all have common supposition, but 'chimera' in '[A] chimera exists' has no supposition, while in 'This phoenix is alone in the world' (*hec phenix est sola in mundo*) and '[The] sun which is Phoebus shines' (*sol qui est phebus lucet*) 'phoenix' and 'sun' have only discrete supposition. In these last two cases, the context causes the common term itself to be taken discretely, and not just the phrases 'this phoenix' and 'sun which is Phoebus',[23] but Soto rejected any attempt to distinguish a class of terms which were common in signifying but singular in suppositing, by arguing that signification and supposition must go hand in hand.[24]

As Soto's example of the chimera shows, this general requirement applied only to common terms that do in fact supposit; and this brings us to the more important issue of the special rules for empty subject terms. Ockham himself insisted that universal propositions with empty subject terms have no singulars, and he used the example 'Every white man is white', when there are no white men.[25] More generally, people agreed that terms such as 'chimera' do not supposit, and hence they could not be involved in suppositional ascent and descent, except in very special contexts involving the possibility of ampliation to imaginary and impossible objects.[26] In 'A chimera runs', 'chimera' does not have determinate supposition, but 'A chimera is imagined' (*chimera*

satus, ut ly 'homo' ibi: 'ille homo currit.' Correlarium: ille terminus 'deus' in illa 'deus est' non est suppositum in supponendo, quia non supponit pro paucioribus singularisatus quam non singularisatus. Idem dicatur de illis terminis 'sol', 'mundus', 'fenix'.' Cf. Johannes Eckius, *In Summulas Petri Hispani*, Augustae Vindelicorum (1516), f. lxxxix va: 'sub quolibet termino communi habente actu plura supposita potest fieri descensus, si sit singularisabilis.'

22) Domingo de Soto, *Summule*, Burgos (1529), f. xix rb-va.

23) Domingo de Soto, *Summule*, Burgos (1529), f. xix rb: 'Suppositio discreta est suppositio termini discreti, aut communis discreti tenti [...] non solum illud totum 'hic homo' aut 'homo qui est Petrus', verum ly 'homo' supponit discrete.'

24) Domingo de Soto, *Summule*, Burgos (1529), f. xix rb. '[...] omnis terminus communis in significando [...] est communis in supponendo.' Cf. f. xxiiii rb. 'Cum dicitur 'singularibus', 'universale', satis intelligitur de universalibus et singularibus in supponendo, quia nullum est in significando, quin sit in supponendo.'

25) William of Ockham, *Summa logicae*, ed. Boehner *et al.* (1974), III-3, c. 32, 708.

26) See Ashworth (1977).

imaginatur) might be a special case,[27] as might 'Every chimera is able to exist (*potest esse*)'.[28] Both Dorp and the author of *Commentum* agreed that one can say that the term 'chimera' is distributed in such contexts as 'No chimera is running', for otherwise the proposition would not be universal.[29] However, one cannot say that it has confused and distributive supposition, because the term does not supposit at all. Dorp also makes plain an important point about the difference between *acceptio* and *suppositio*. In Buridan supposition in the wide sense (which includes parts of subjects and predicates) is taken to be synonymous with *acceptio*,[30] but in authors such as Albert of Saxony and Marsilius of Inghen, and in the *Commentum*, supposition is defined as the acceptance of a term for some thing or things of which the term is verified by means of the copula of the proposition in which the term appears.[31] This allows one to say that acceptance is wider than supposition. Thus Dorp in his discussion of the difference between distribution and distributive supposition indicates that there can be *acceptio* whether a term is verified or not, whereas supposition requires both *acceptio* and verification.[32] This point is expanded on by Soto with reference to terms whose supposition varies through time. He offers an elaborate example whereby a term signifies more than it is accepted for, and is accepted for more than it supposits for. He remarks that in 'Some man disputes', 'man' signifies all humans, male and female, who are imagined to exist, it is accepted only for males, even those who are imagined to exist, but it supposits only for males who currently exist.[33] At the end of his list of syntactical

27) Anonymus, *Commentum* [...] *in primum et quartum tractatus Petri Hyspani*, Hagennaw (1495), sig. q 1 r, sig. q 1 v, sig. r 1 v.

28) Domingo de Soto, *Summule*, Burgos (1529), f. xxiiii rb. For discussion of this case, see below.

29) Johannes Dorp, *Commentary* in: *Johannes Buridanus, Compendium*, sig. i 2 ra. Dorp uses the example 'nulla chimera est chimera' and makes the point about the proposition being universal. Anonymus, *Commentum* [...] *in primum et quartum tractatus Petri Hyspani*, Hagennaw (1495), sig. q 4 r has 'Ly 'chimera' in ista 'nulla chimera currit' [...] stat distributive sed non supponit distributive.' See also George of Brussels, *Expositio Georgii super summulis magistri Petri Hyspani*, Parisius (1491), sig. A 7 rb and Johannes Eckius, *In Summulas Petri Hispani*, Augustae Vindelicorum (1516), f. xc rb: 'Distributio est in plus quam suppositio distributiva.' On the difference between distribution and its origins and distributive supposition, see Parsons (2006), especially p. 65.

30) Johannes Buridanus, *Summulae. De suppositionibus*, ed. Van der Lecq (1998), 4.4.1, p. 70.

31) Albert of Saxony, *Perutilis logica*, Venice (1522), II, c. 4, f. 11 ra; Marsilius of Inghen, *Suppositiones*, ed. Bos (1983), 52; Anonymus, *Commentum* [...] *in primum et quartum tractatus Petri Hyspani*, Hagennaw (1495), sig. o 4 v.

32) Johannes Dorp, *Commentary* in: *Johannes Buridanus, Compendium*, Venetiis (1495), sig. i 2 ra.

33) Domingo de Soto, *Summule*, Burgos (1529), f. xvi va.

rules governing what mode of personal supposition a term will have, he adds that the rules must be understood conditionally, by means of the added clause 'if such terms supposit'.[34]

The reactions to the treatment of propositions with empty subject terms have been varied. Gareth Matthews found it very problematic, writing:[35] 'Yet, since, according to Ockham, the E proposition has no existential import, there will be cases of true E propositions such that there will be no singulars for us to descend to.' He added that something similar was true of O-propositions, and suggested that the problem should be solved by rejecting the view that the theory of ascent and descent is supposed to give truth conditions for categorical propositions. Elizabeth Karger has suggested that Ockham had two alternative logical systems, only one of which held for universal terms and their singulars, and presumably she would be willing to extend this claim to the other logicians who refused to descend from empty terms.[36] Her suggestion casts doubt on her other theory, that the modes of personal supposition have the function of providing a basis for inference in general,[37] but it does make sense if we take it that the theory of modes of personal supposition tells us how to deal with propositions whose common terms have a plurality of supposita. That is, it tells us how to handle those propositions that have a fighting chance of being both true and informative. Propositions about chimeras, and possible things that we know not to exist, may enter into valid inferences, but they will not lead us to truths that are important in practice, and they can be ignored. Nor do propositions whose common terms have only one suppositum normally need careful analysis, so they too can be ignored.

This approach fits in with what various medieval logicians show us about the uses of suppositional descent, and with what some of them make explicit. I take my first two examples from Ockham. He argues that in the proposition: 'At every time after Adam some man existed (*omni tempore post Adam aliquis homo fuit*)', 'man' must supposit merely confusedly, because if it were

---

[34] *Ibid.*, f. xx rb-vb. See f. xx vb: 'Has omnes regulas intelligito conditionaliter, id est, si tales termini supponant.'

[35] Matthews (1997), 38-39.

[36] Karger (1978), 49-51.

[37] Karger (1984), 103, said 'One important function which must consequently be recognised to TM (= Theory of Modes) is that of unifying the otherwise fragmented logic of formal immediate inferences.' In a later paper (Karger (1993), 407-408), she argued for the claim that standard categorical inferences are part of a generalized theory which requires modes of personal supposition.

distributive or determinate, every singular would be false.[38] We know the proposition to be true, and the rules of descent enable us to show this. Similarly, we know that 'Socrates begins to be literate (*Sortes incipit esse grammaticus*)' can be true, and the failure of all types of descent show that the proposition must be construed as a conjunction of two propositions, one affirmative and one negative.[39] In turn, John Dorp uses the modes of personal supposition and the process of descent to explain various features of both contradiction and conversion.[40] He also remarks that we can show 'Every planet the sun is not (*omnis planeta sol non est*)' to be false by descending under 'planet' until we reach a false singular.[41] The author of the *Commentum* is explicit about what is going on. He writes that supposition theory was developed in order to save the truth or falsity of propositions which are commonly granted or denied, and that descent is probative of the truth or falsity of propositions. If the descendents are true, then the original proposition is true.[42] In the late fifteenth century, George of Brussels said that supposition theory was invented in order to inquire into the truth and falsity of propositions, the values of consequences, and the truth of complex contradictories. He also mentioned the subtlety and propriety of speech.[43] In the sixteenth century, Domingo de Soto tells us that descent to singulars is used to show the falsity of propositions, and ascent from singulars is used to show their truth.[44] He also gives a long and elaborate

38) William of Ockham, *Summa logicae*, ed. Boehner *et al.* (1974), I. 73, 227, ll. 24-27.

39) William of Ockham, *Summa logicae*, ed. Boehner *et al.* (1974), I. 73, 231-233.

40) Johannes Dorp, *Commentary* in: *Johannes Buridanus, Compendium*, Venetiis (1499), sig. b 4 ra-rb on contradiction, sig. b 6 va-sig. c 3 ra on conversion.

41) Johannes Dorp, *Commentary* in: *Johannes Buridanus, Compendium*, Venetiis (1499), sig. c 1 rb. For more examples of the use of the modes of personal supposition, see Ashworth (1974), (1976) and (1978).

42) Anonymus, *Commentum* [...] *in primum et quartum tractatus Petri Hyspani*, Hagennaw (1495), sig. n 7 r-v; sig. q 7 r.

43) George of Brussels, *Expositio Georgii super summulis magistri Petri Hyspani*, Parisius (1491), sig. A 1 vb-sig. A 2 ra. He asks why supposition pertains only to terms within propositions, and answers that this has to do with the reasons for which supposition was found: 'Prima est, quia suppositio est inventa ad inquirendum veritatem vel falsitatem propositionum. [...] Secunda causa: quia suppositiones sunt invente propter propriam locutionem et subtilem, propter valores consequentiarum et propter veras contradictiones complexas.'

44) Domingo de Soto, *Summule*, Burgos (1529), f. xxiiii ra. Cf. Juan de Celaya, *Magne Suppositiones*, Parisiis (1526), sig. B 1 rb. 'Preterea descensus fit ut manifestetur veritas vel falsitas propositionis.' See also Hieronymus de Sancto Marco, *Compendium*, sig. B 3 r. Writing about supposition, ampliation and appellation, he said: 'sine quibus (si verum fateri velimus), nemo de quantitate, oppositione, ac de veritate et falsitate propositionum expeditus esse poterit. Immo ipsis ignoratis expedit artem veritatem et falsitatem propositionum cognoscendi ignorare.'

reduction of an actual example to singulars in order to show how one can test the truth of a proposition by using all the tools of logic, including not only supposition theory but *exponibilia* and modal logic.[45] He moves in an orderly way from 'Only a creature is contingently a substance' to 'This is this', a proposition which he takes to be epistemologically basic.

In short, the modes of personal supposition give us a testing device, to be employed in cases where there is doubt about the truth value of a proposition and its proper interpretation. It is used within an already-established framework of logical equivalences and inferences, and within a wider theory of truth that embraces propositions with non-suppositing terms.

4. Logical Equivalence

Now that we have some idea about the purpose served by the theory of modes of personal supposition, we can turn to Spade's other arguments, namely that Ockham and Buridan do not claim any logical equivalence between propositions and their descended forms, and that the details of the theory are incompatible with such an equivalence.

Let us first remind ourselves of the modes of personal supposition[46] and how they applied to standard categorical propositions. I shall ignore collective supposition, and I shall also ignore all the special cases, such as 'I promise you a horse', or 'Of every man some horse is running' and so on. Determinate supposition applies to the subjects and predicates of particular affirmative or I-propositions and to the subjects of particular negative or O-propositions. For both Ockham and Buridan there is a full descent to a disjunction of singular propositions, but there is only a partial ascent.[47] That is, they both speak of the fact that from 'This man is running' we can infer 'Some man is running', but they do not contemplate the case in which we move from 'Either this man is running, or this man is running, or this man is running, and so for all the singulars' to 'Some man is running.' Distributive supposition applies to the subjects

45) Domingo de Soto, *Summule*, Burgos (1529), f. xxiii va-vb.

46) For summaries of Ockham and Buridan see Spade (1988), 194 and Spade (2002), 281-286. Note that there are two approaches. One is more purely semantic, with reference to descent and inference, and one is more heavily syntactic, with reference to the place of terms in propositions and their relation to syncategorematic terms, as well as to exponible terms, terms signifying acts of mind, and so on. See Dutilh Novaes (2004), for discussion of Buridan's semantic approach, and Karger (1993), for discussion of the syntactic approach.

47) William of Ockham, *Summa logicae*, ed. Boehner *et al.* (1974), I. 70, 210; Johannes Buridanus, *Summulae. De suppositionibus*, ed. Van der Lecq (1998), 4.3.5, pp. 49-50.

of universal affirmative or A-propositions, to both the subjects and the predicates of universal negative or E-propositions, and to the predicates of particular negative or O-propositions.[48] For both Ockham and Buridan there is a full descent. That is, from 'Every man is running' one can descend to the conjunction of all the relevant singular propositions. Buridan also has partial descent, which allows descent to just one of the singular propositions. Spade remarks that ascent fails for Ockham and Buridan, but he is speaking only of partial ascent from an arbitrary singular rather than of full ascent from 'This man is running and this man is running and so for all the singulars'.[49] However, as we shall see, full ascent does appear for both authors under the rubric of induction. Finally, there is merely confused supposition which applies to the predicate of A-propositions and allows descent is to a disjoint term, as in 'Every man is this animal or that animal or the other animal, and so for all the singulars.'[50] In one place, Ockham allows the partial ascent from 'Every man is this animal' to 'Every man is an animal',[51] but this seems to be a special case, especially when one considers the distinction between saying 'Every man is one and the same animal' and 'Every man is some animal or other.'[52] Soto has the same inference, which he regards as valid, and he uses it to cast doubt on the claim that a descent is valid if and only if the corresponding ascent is valid.[53] The inference is significant in another way for Spade, who organizes Ockham's account of the modes of personal supposition by saying that in each case we get descent to a singular, or ascent from a singular, but never both at once.[54]

---

48) William of Ockham, *Summa logicae*, ed. Boehner *et al.* (1974), I. 70, 211; Johannes Buridanus, *Summulae. De suppositionibus*, ed. Van der Lecq (1998), 4.3.6, p. 50.

49) Spade (1988), 194.

50) William of Ockham, *Summa logicae*, ed. Boehner *et al.* (1974), I. 70, 211; Johannes Buridanus, *Summulae. De suppositionibus*, ed. Van der Lecq (1998), 4.3.6, p. 50.

51) William of Ockham, *Summa logicae*, ed. Boehner *et al.* (1974), I, 70, 211. See also I, 75, 231 (one of the two indexed places where he uses the word 'ascendere'. For the other place, see I, 76, 235, l. 69).

52) Later logicians brought out the distinction between these two cases by using the special sign 'b': see Ashworth (1978), 602. Domingo de Soto, *Summule*, Burgos (1529), f. xx rb, writes: 'Etenim si dicas 'omnis homo est aliquod animal', predicatum supponit confuse, adminus ut sophistis huius temporis placet, sed si dicas 'omnis homo est .b. animal', ly 'animal' supponit determinate.'

53) Domingo de Soto, *Summule*, Burgos (1529), f. xxiiii ra: 'Haud enim scio quid obstet bonitati huius ascensus: 'omnis homo est hoc animal vel omnis homo est hoc animal etc., ergo omnis homo est animal', quod econtrario non valet descensus.' Cf. f. xxv vb.

54) Spade (2002), 303-306.

The assignment of modes of personal supposition is perfectly standard and will be found in subsequent authors. However, it does seem quite clear that Spade is right in claiming that neither Ockham nor Buridan was trying to construct elaborate equivalences between propositions and their fully descended form. This approach only evolved rather gradually during the fourteenth century, although it was to be standard by the late fifteenth century. To begin with, there is the insistence on full descent in every case which we find in Albert of Saxony, Marsilius of Inghen, and John Dorp, as well as in the *Commentum*.[55] Ascent is another matter. There is no discussion of ascent as such in Albert of Saxony, Marsilius of Inghen, or Dorp. Nor does ascent appear in Paul of Venice's *Logica parva*, and in the *Logica magna* he merely mentions the converse of disjunctive descent without using the word 'ascent'.[56] However, full descent paired with full ascent for terms with determinate and distributive supposition did appear in the 1360s in the work of Ralph Strode. His vocabulary is particularly illuminating. When presenting distributive supposition, he writes 'And the first mode is called *probatio vel inductio*, as in 'This man runs and this one and so for all the singulars, and these are all the men there are, therefore every man runs.' He adds that the converse is called descent.[57] The reference to *probatio* links ascent to the proofs of terms, and we find that Paul of Venice has what amounts to ascent in Tract 4 of the *Logica Parva*, which is devoted to that subject, when he discusses how to prove an E-proposition from its singulars.[58] Paul of Pergula has the same material in his own section on the

---

[55] Albert of Saxony, *Perutilis logica*, Venice (1522), II, c. 4, f. 11 va, II, c. 6, f. 12 vb; Marsilius of Inghen, *Suppositiones*, ed. Bos (1983), 56, 58; Johannes Dorp, *Commentary* in: *Johannes Buridanus, Compendium*, Venetiis (1499), sig. h 6 ra, sig. h; f. 6 vb, sig. i 1 ra.

[56] Paul of Venice, *Logica Magna* (*Tractatus de suppositionibus*), ed. Perreiah (1971), 88.

[57] See the quotation in Spade (1988), 223-224. Strode begins with ascent, which he calls proof: 'Terminus supponit determinate cum propositio in qua ponitur contingit disjunctive probari per ejusdem supposita, sicut ista 'Homo currit' potest verificari vel quia iste homo currit vel quia iste homo currit vel quia ille homo currit et sic de singulis.' He then describes the converse, the inference to a disjunction of singulars which is possible 'cum debita constantia'. He describes distributive supposition in terms of both ascent and descent: 'Et primus modus dicitur probatio vel inductio, ut 'Iste homo currit et iste et sic de singulis, et isti sunt omnes homines; ergo, omnis homo currit'. Et e converso dicitur descensus, ut 'Omnis homo currit, et isti sunt omnes homines; ergo, iste homo currrit et iste homo currit et sic de singulis'.' See also extracts in Maierù (1972), 273, note 189, on determinate supposition, where Strode says that he is speaking of formal inference, and 275, note 194, on distributive supposition.

[58] Paul of Venice, *Logica parva*, ed. Perreiah (2002), 71.

proofs of terms.[59] More importantly, we find ascent where A and E-propositions are concerned in the discussions of induction by Ockham, Buridan and, especially, Dorp. Ockham gives as a rule that if all the singulars of some proposition are true, the universal is true, and he remarks that this rule holds for both affirmative and negative universal propositions.[60] In Dorp's commentary on Buridan's definition of induction,[61] he describes induction in the broad sense as a valid consequence inferring a universal proposition from its singular or singulars with the addition of the clause 'and so for all the singulars' or 'there are no more of the same sort', or some similar locution.[62] These clauses are, he says, necessary for the consequence to be formally valid, though he notes that formality also requires the use of 'this man' rather than proper names.[63] He explicitly presents induction as the counterpart of descent.[64]

In the anonymous *Commentum*, descent and ascent are presented in their full forms, involving all the singulars, and a necessary condition for the validity of a descent is said to be the validity of the corresponding ascent. Moreover, ascent is described as a subspecies of induction, the difference between them being that induction allows the use of proper names, whereas ascent requires demonstrative phrases.[65] The intent here is to safeguard the link between the common term and its singulars, for one might be tempted to substitute such names as 'Favellus' and 'Brunellus', traditionally used of horses or donkeys, when instantiating the word 'man'. Another requirement is the addition of an extra premise, whether in the form 'these are all the As there are', which we have already seen in the quotation from Strode, or in the forms 'these As exist'

59) Paul of Pergula, *Logica*, ed. Brown (1961), 48.

60) In *Summa Logicae*, ed. Boehner *et al.* (1974), III-3 c. 32, 708-714, William of Ockham discusses induction and on 709, ll. 21-23 he writes 'Alia regula est: si omnes singulares alicuius propositionis universalis sint verae, universalis est vera. Et ista regula est generalis tam de affirmativa quam de negativa.' Buridan's lemma appears in Johannes Dorp, *Commentary* in: *Johannes Buridanus, Compendium*, sig. o 5 va as: 'Inductio est a singularibus ad universalem progressio probandam: ut 'Sortes currit, Plato currit, et sic de singulis; ergo omnis homo currit'.'

61) Johannes Dorp, *Commentary* in: *Johannes Buridanus, Compendium*, Venice (1499), sig. o 5 va-6 ra.

62) Johannes Dorp, *Commentary* in: *Johannes Buridanus, Compendium*, Venice (1499), sig. o 5 va: 'Inductio est consequentia bona illativa propositionis universalis ex sua singulari vel ex suis singularibus cum additione istius clausule 'et sic de singulis' vel 'non sunt plura talia' vel alterius equivalentis.'

63) Johannes Dorp, *Commentary* in: *Johannes Buridanus, Compendium*, Venice (1499), sig. o 5 va-vb.

64) Johannes Dorp, *Commentary* in: *Johannes Buridanus, Compendium*, Venice (1499), sig. o 5 va.

65) Anonymus, *Commentum (...) in primum et quartum tractatus Petri Hyspani*, Hagennaw (1495), sig. q 8 r.

or 'this A exists and this A exists and so on'.[66] The purpose of such a clause was to guarantee validity, not in the cases where A itself is known to be an empty term, but in the case where the demonstrative phrase 'this man' points to someone already dead, or perhaps to a donkey. These additions allowed descent and ascent to be formally valid inferences, and thus to capture logical equivalences, provided, of course, we remember that they are only available once we know (or have grounds to assume) that we are dealing with suppositing terms. Not everyone accepted this proviso. Hieronymus de Sancto Marco, for instance, mentioned the alternative view that one could descend either to singulars in supposition alone (*in supponendo tantum*), or to singulars in acceptance alone (*in accipiendo tantum*), which would allow the descent from 'Every chimera runs' to 'this chimera runs and this chimera runs and so on.'[67]

Another alternative is found in Juan de Celaya, who read the proviso as not guaranteeing what it was supposed to guarantee, for he claimed that one could have a consequence such as 'Every man is running, and these are all the men, therefore this man is running, and this man is running, and so on,' which is a valid descent today and not tomorrow, owing to the death of one of the men picked out by the singular term 'this man'.[68] Domingo de Soto claimed that descent was a formally valid consequence[69] and, in his consideration of arguments similar to that of Celaya, made two points. First, he denied that one could claim that an argument in whose conclusion 'this man' referred to Antichrist, who does not currently exist, was similar in form to an argument in whose conclusion 'this man' refers to an existent man. He insisted that there was a relationship of synonymy between *constantia* in the antecedent and the list of singulars in the consequent, without which there could be no similarity of form.[70] Here it is important to note that Soto insisted that similarity of form

66) For Strode, see note 57 above, and see also Maierù (1972), 274, note 190, for his definition of *constantia*. Strode and Domingo de Soto, *Summule*, Burgos (1529), f. xxiiii ra, call the first form 'constantia' and the second form 'antiquum medium'. For Anonymus, *Commentum (…) in primum et quartum tractatus Petri Hyspani*, Hagennaw (1495), sig. q 8 r-v and Johannes Eckius, *In Summulas Petri Hispani*, Augustae Vindelicorum (1516), f. lxxxix rb, the reverse is the case.

67) Hieronymus de Sancto Marco, *Compendium*, Cologne (1507), sig. C 1 v.

68) Juan de Celaya, *Magne Suppositiones*, Parisiis (1526), sig. B 1 rb: 'Sequitur tertio quod aliqua est consequentia que hodie est bonus descensus et crastina luce non erit, et habebit precise illa significatione quam modo habet. Patet de descensu dato pro exemplo, et posito quod cras essent aliqui homines mortui de illis qui demonstrantur.'

69) Domingo de Soto, *Summule*, Burgos (1529), f. xxiii vb.

70) Domingo de Soto, *Summule*, Burgos (1529), f. xxiiii ra-rb. He writes (f. xxiiii rb): 'Habet enim esse inter subiectum constantie et subiecta singularium talis connexio, ut si constantia ponatur per singularia determinata, singula eorum et non alia sint subiecta in singularibus, <et> quod si

should encompass signification, in that the parallel arguments both required strict synonymy between the *constantia*-phrase and the series of singular terms in the conclusion, and that this in turn required that the *constantia*-phrase have exactly the same set of referents as the series of singular terms.[71] This condition is not met when 'this man' comes to refer to Antichrist; nor would it be met if through an individual's death 'this man' ceased to supposit. Secondly, he cast doubt on the requirement that there should be actual supposition in all cases, saying that this requirement allowed some *moderni* to argue that some consequence counted as an ascent although it could cease to be an ascent while retaining exactly the same signification if one of the significates were to die.[72] Soto replied that the requirement was simply that it was sufficient that it should not be inconsistent for the superior term to supposit for many, and that it should not be inconsistent for the inferior term to supposit for one, while being inconsistent for it to supposit for many. His main example brings us back to the issue of chimeras, for it was 'This chimera is able to exist, and this chimera is able to exist, and so on; therefore every chimera is able to exist.'[73] Soto's claim that actual supposition is not necessary for correct ascent seems to rest on two principles. First, so long as a term can have supposition, the death of a given singular will simply mean that the conjunctive antecedent of a correct ascent becomes false, and this will not affect validity. Second, there are contexts involving ampliation in which even a term such as 'chimera' is admissible.[74]

ponatur per pronomen demonstrativum collective sumptum, nihil habet demonstrari in singularibus quod non demonstretur in constantia; et illa synonimitas inter subiecta illa singularium et constantie est ita intrinseca ut si in alia non servetur, non erit similis forme.'

71) Domingo de Soto, *Summule*, Burgos (1529), f. xxiii vb.

72) Domingo de Soto, *Summule*, Burgos (1529), f. xxiiii rb: 'Et dicunt (quod peius apparet) ad hoc quod termini quipiam dicantur superior et inferior in supponendo, exigi quod actualiter supponant, unde inferunt aliquam consequentiam esse ascensum et tamen posse manente eadem omnino significatione desinere esse ascensum, utpote si termini desinant supponere.'

73) Domingo de Soto, *Summule*, Burgos (1529), f. xxiiii rb: 'Sed memento superiorum ex quibus habes ad terminum communem aut singularem non requiri quod actualiter supponat, sed sufficere quod illi non repugnet supponere pro pluribus, et huic non repugnet supponere pro uno; repugnet autem supponere pro pluribus. Et proportionaliter ad terminum superiorem sufficit quod non stet inferiorem mediante aliqua copula supponere pro aliquo pro quo non supponat superior; stet tamen ediverso. Ex quo sequitur hanc consequentiam esse ascensum: 'hec chimera potest esse, et hec chimera potest esse, etc., ergo omnis chimera potest esse'.'

74) Soto allowed ampliation to impossibles: Domingo de Soto, *Summule*, Burgos (1529), ff. xxviii va-xxix ra.

5. A-propositions

We must now consider the final claim that the details of the theory are incompatible with equivalence and indeed are problematic even when no equivalence is claimed. First of all, there is the claim that the attribution of merely confused supposition to the predicates of A-propositions militates against any neat analysis into disjunctions and conjunctions of singular propositions. Some authors, such as Swiniarski, have tried to get rid of merely confused supposition altogether.[75] While Spade recognizes that it did have a distinctive role in the discussion of special cases such as promising,[76] he writes:[77]

> I think this appeal to disjoint terms is something of a mark of desperation. I suspect it was done only because people felt a need to be able to make some kind of descent in every case. And since with merely confused supposition descent by means of the familiar conjunctive or disjunctive propositions was not possible, another kind of descent was contrived to fill the gap.

He also pointed out that complex disjoint terms are always available, and said[78] 'If it is legitimate to appeal to complex terms in these special cases, it was also legitimate to appeal to them all along, even in cases where other kinds of descent were available as well. Yet they were *not* appealed to except when other kinds of descent failed.' These remarks are not without some justification. Dorp and the author of the *Commentum* recognized that disjoint descent was always available. They pointed out that when a fully descended proposition has been obtained, if it is a conjunctive proposition, it will imply both a disjunctive proposition and a proposition with a disjoint predicate, and if it is a disjunctive proposition, it will imply one with a disjoint predicate.[79] Strode

---

75) See references in Ashworth (1973), 38.

76) Spade (2002), 286-287. See Ashworth (1973), 38-39 for a list of special contexts.

77) Spade (2002), 286.

78) Spade (2002), 307.

79) See Johannes Dorp, *Commentary* in: *Johannes Buridanus, Compendium*, Venice (1499), sig. i 3 va: 'Nam sub omni termino sub quo potest fieri descensus copulative etiam potest fieri descensus disiunctive et disiunctim, et non e contra. Et sub omni termino sub quo potest fieri descensus disiunctive, potest fieri descensus disiunctim, et non e contra.' See Anonymus, *Commentum* [...] *in primum et quartum tractatus Petri Hyspani*, Hagennaw (1495), sig. q 3 v: 'Et ille terminus dicitur stare pro suis significatis copulative sub quo vel sub alio consimiliter accepto habente actu plura supposita et singularisabili valet descensus copulativus gratia forme. Et non debet addi 'sic quod non disiunctivus nec disiunctus' quia, sub quocumque termino valet descensus copulativus, sub eodem etiam valet descensus disiunctivus vel disiunctus.' See also George of Brussels, *Expositio Georgii super summulis magistri Petri hyspani*, Parisius (1491), sig. A 5 va. Johannes Buridanus,

spoke of the disjoint term being convertible with the common term it replaced;[80] and the author of the *Commentum* required that the disjoint term have the same mode of supposition as the original term. Thus from 'No man is every animal' we descend to 'No man is every this animal or this animal and so for all the rest.'[81] Soto suggests that one can always insert a disjoint term, provided that one keeps the original supposition, so that in 'Every man is an animal' one can carry out a double replacement, which results in 'Every this man or this man or so for all the singulars is this animal or this animal or so for all the singulars.'[82] Moreover, his third rule stipulates that one may always descend from a term with disjoint supposition, just as one may always ascend from its replacement.[83] However, since terms with merely confused supposition allow

---

Summulae. De suppositionibus, ed. Van der Lecq (1998), 4.3.6, p. 51 adds some remarks about the relationship between modes of personal supposition when he notes the order that holds between propositions in which a given term exhibits various types of supposition. From distributive supposition one can infer determinate or even merely confused supposition, and from determinate supposition one can infer merely confused supposition. His examples have to do with propositions exhibiting these types of supposition. Thus from 'animal est omnis homo' we can infer 'omnis homo est animal.' However, care has to be taken to alter or remove syncategorematic signs, as in the inference from a distributed term to a determinate term, in 'omnis homo est animal; ergo animal est homo.'

[80] Read (1991), 72, note 68 (and translation in text) quotes Strode's passage on merely confused supposition (my punctuation): 'Terminus dicitur supponere confuse tantum quando respectu istius suppositionis pro pluribus non contingit eius propositionem nec copulative probari nec descendere, nec disiunctive, sed forte coniunctim vel disiunctim convertibiliter, sicut non sequitur 'omnis homo est animal, ergo omnis homo est aliquod animal vel omnis homo est hoc animal', nec copulative ut patet. Sequitur tamen disiunctim cum medio in ista de disiuncto extremo, scilicet 'omnis homo est animal et ista sunt omnia animalia, ergo omnis homo est hoc animal vel hoc animal vel hoc animal et sic de singulis'.'

[81] Anonymus, *Commentum* [...] *in primum et quartum tractatus Petri Hyspani*, Hagennaw (1495), sig. r 2 r: From 'nullus homo est omne animal' we go to 'nullus homo est omne hoc animal vel hoc animal et sic de aliis.'

[82] Domingo de Soto, *Summule*, Burgos (1529), f. xxv rb: 'Ascensus aut descensus copulativus debetur immediate aut mediate termino supponenti distributive; disiunctivus, debetur termino supponenti determinate. Copulatus <debetur> termino collective sumpto; sed disiunctus cuicunque termino communi potest contingere dummodo disiunctum per quod descenditur eodem modo supponat sicut terminus ipse sub quo descenditur, quod in hoc genere descensus apprime observandum est. Idem enim est dicere 'omnis iste homo vel iste homo vel sic de singulis est hoc animal vel hoc animal vel sic de singulis' (signo distribuente totale subiectum) ac si diceretur 'omnis homo est animal.' Sed quia termino supponenti confuse tantum nullus alius convenit descensus quam disiunctus, dicitur esse peculiaris terminorum supponentium confuse tantum.'

[83] Domingo de Soto, *Summule*, Burgos (1528), f. xxv vb: '*Tertia <regula>*. Sub quocumque termino supponente confuse copulatim aut disiunctim, licet quandocunque copulatim aut disiunctim pro libito ascendere vel descendere. Et ratio est quia tale copulatum aut disiunctum communiter

nothing other than disjoint descent, this kind of descent is said to be peculiar to them; and, according to Soto, because the type of supposition remains the same, disjoint ascent and descent are not as useful for showing truth and falsity as are other kinds of ascent and descent.[84]

On the other hand, there is a possibility of getting rid of disjoint terms, at least for standard A-propositions, through a rule governing the priority of analysis. The author of the *Commentum* does give a rule for when a term with distributive supposition appears with a term having determinate or merely confused supposition, but in fact the rule pertains only to the priority of determinate supposition over distributive.[85] We have to wait for Domingo de Soto in the sixteenth century to find a full account of how descent from the predicate of an A-proposition works in combination with descent from the subject, and how it is that, in this context at least, the disjoint predicate can legitimately give way to a disjunction of propositions, so that for standard categorical propositions all descent is in the end productive of conjunctions and disjunctions of singular propositions.[86] He writes:

> *Second rule*: From a term with merely confused disjoint supposition in relation to universality, one cannot descend disjunctively before the said universality is resolved, for otherwise one will argue from a single distribution to many [distributions] with respect to the parts of the particularity, which is a defect opposite to the defect of the preceding rule. For instance, this does not follow: 'Every man is an animal and these animals are all the animals, therefore every man is this animal or every man is this animal, or so for all the singulars.' However, the corresponding ascent is valid. And so in any proposition whatsoever, one should first ascend or descend from the term with determinate supposition, and then from the term with distributive supposition, and then disjunctively from the term which did supposit confusedly

---

et eodem modo habet supponere in descendenti sicut terminus ipse communis supponebat in descensa.' Note that the rule also applies to merely confused conjoint supposition, that is, to collective supposition.

84) Domingo de Soto, *Summule*, Burgos (1529), f. xxv rb.

85) See below, note 91.

86) Domingo de Soto, *Summule*, Burgos (1529), f. xxv vb: '*Secunda regula*. Sub termino supponente confuse tantum disiunctim in ordine ad universitalitatem, non licet descendere disiunctive priusquam dicta universalitas resolvatur, nam alias argueretur ab unica distributione ad plures respectu partium particularitatis, qui est defectus oppositus defectui precedentis regule. V.g., non sequitur 'omnis homo est animal et hec animalia sunt omnia animalia, ergo omnis homo est hoc animal, vel omnis homo est hoc animal, vel sic de singulis'; recte tamen ediverso valeret ascensus. Igitur in quavis propositione prius ascendendum aut descendendum est sub termino supponente determinate, deinde sub termino distributo, deinde disiunctive sub termino qui supponebat confuse in ordine ad distributionem, qui videlicet dempta distributione remanet pure determinate.'

> in relation to the distribution, for when the distribution has been taken away, [the term] supposits purely determinately.

Given that distributive supposition has precedence over merely confused supposition, from 'All A is B' we can descend to 'A_1 is B and A_2 is B or so for all the singulars.' We now have a sequence of singular propositions whose predicates have determinate supposition, and so for any number i we can descend correctly to 'A_i is B_1 or A_i is B_2 or so for all the singulars.'[87] As a result, any standard proposition can be reduced to logically equivalent conjunctions and/or disjunctions, either immediately or after some antecedent reduction. On the other hand, if we are dealing with simple supposition, or with merely confused supposition in the context of collective terms or such special signs as 'promise', we will not obtain this desirable result.[88]

6. O-propositions

Finally, we must consider the claim that the attribution of distributive supposition to the predicates of O-propositions renders analysis in terms of descent and an equivalent ascent invalid. The problem is this.[89] If we assume that there are two men (Socrates and Plato) and two Greeks (Socrates and Plato once more) and we take the false proposition 'Some men are not Greek' we can apparently descend to 'Some man is not Socrates and some man is not Plato', and then to '(Either Socrates is not Socrates or Socrates is not Plato) and (either Plato is not Socrates or Plato is not Plato).' That is, we get a truth from a falsehood, while the corresponding ascent yields a falsehood from a truth, which renders it invalid. It seems, then, that one cannot claim any equivalence between propositions and their descended forms, any more than one can claim that descent is intended to provide an analysis of truth-conditions. Neither Ockham nor Buridan discussed this problem, but it certainly did not pass unnoticed by later logicians, though perhaps sometimes with a sense of mischief. Hieronymus de Sancto Marco in the early sixteenth century suggested that if you want to show that propositions have an appropriate truth value, you

87) See Priest and Read (1980), 288-289, on my misreading of this rule in Ashworth (1973), 40.

88) Domingo de Soto, *Summule*, Burgos (1529), f. xxv rb-va.

89) Spade (1988), 206; Spade (2002), 292.

should start with one order of descent, and if you want to cast sophistical doubts, you should start with the other order.[90]

There were two solutions, both found in the anonymous *Commentum*, and both preserving all the appropriate truth-relationships.[91] The first solution was that any descent from the predicate must involve the term 'the same (*idem*)' in the second singular proposition and all those that followed.[92] Thus in the above case, descent from the predicate term would have to start in this way: 'Some man is not Socrates and the same man is not Plato', which is as false as the original proposition. No further descent was offered, but one might think of it as going like this: '(The first man is not Socrates and the same man is not Plato) or (the second man is not Socrates and the same man is not Plato).' This first solution is also found in Dorp, though in a rather obscure formulation.[93] Soto discussed it, but argued (unconvincingly) that counter-examples could be found if one considered an infinite series of supposita (see below).

The second solution was to give determinate supposition precedence over distributive, so that the first descent is to 'Socrates is not a Greek or Plato is not a Greek.' We can then descend to '(Socrates is not Socrates and Socrates is not Plato) or (Plato is not Socrates and Plato is not Plato)', which is as false as the original proposition. Soto adopted this second solution, presenting it most clearly in his discussion of the rule quoted in relation to A-propositions (see

---

90) Hieronymus de Sancto Marco, *Compendium*, Cologne (1507), sig. C 2 v: 'Quandocumque sunt duo termini sic se habentes quod unus stat confuse et distributive, alter vero supponit determinate, tunc ad veritatem propositionis clare aspiciendum ex parte suppositionis, fiat descensus sub termino supponente determinate. Sed ad apparenter impugnandum veritatem, fiat descensus sub termino supponente confuse et distributive.' On sig. C 3 r, he gave the same kind of advice about propositions with a term standing distributively in relation to one with merely confused supposition. Here for truth one should start with the distributed term. 'Si autem volueris apparenter veritatem impugnare, facias descensum sub termino supponente confuse tantum.'

91) Anonymus, *Commentum* [...] *in primum et quartum tractatus Petri Hyspani*, Hagennaw (1495), sig. r 1 v: '*Secunda regula*: quando cum termino stante distributive ponitur terminus stans determinate vel confuse tantum, tunc prius debet descendi sub termino stante determinate, et postea in una singularium termini stantis determinate debet descendi sub termino stante distributive. Vel potius prius descendi sub termino stante distributive sic tamen quod in qualibet singularium preterquam in prima, termino stante determinate addatur relativum identitatis, nisi terminus stans distributive fuerit determinatio termini stantis determinate secum non unica acceptione accepta.'

92) Hieronymus de Sancto Marco, *Compendium*, Cologne (1507), sig. C 2 v, suggested that if, when descending first from the term with distributive supposition, you refrain from adding a relative of identity to the term with determinate supposition, you are arguing sophistically.

93) Johannes Dorp, *Commentary* in: *Johannes Buridanus, Compendium*, Venetiis (1495), sig. i 1 ra-rb.

section 5). However, he approached the problem in much greater detail in his discussion of his first rule, where he focuses on the issue of what happens when you have assumed the mistaken form of descent which gives priority to the distributed term, and then perform the corresponding ascent. He writes:[94]

> *Rule one.* One cannot ascend from a term which has distributive supposition in relation to some determinate before one ascends from that determinate, for otherwise one would argue from many determinates related to the parts of a multitude to a single determinate related to the multitude as a whole. For instance, given the predicate of this [proposition] '[Some] man is not an animal', this ascent is invalid: '[Some] man is not this animal and [some] man is not this animal, and so on; therefore some man is not an animal.' And to argue from many determinates [related to the parts of a multitude] is to argue from a term often taken conjunctively in relation to many singulars to the same [term] suppositing determinately in relation to a distribution through which such singulars are numbered. This indeed is considered to be a defect in [the argument] because the antecedent indicates that some suppositum or other of the term suppositing determinately agrees with singular supposita of the distributed term, or that some [suppositum] or other is removed from [such] singulars, while the consequent indicates that all the supposita of the distributed term agree

94) Domingo de Soto, *Summule*, Burgos (1529), f. xxv va-vb: '*Prima* <*regula*>. Sub termino supponente distributive in ordine ad aliquam determinatam non potest ascendi priusquam sub tali determinata. Nam alias argueretur a pluribus determinatis respectu partium multitudinis ad unicam determinatam respectu totius multitudinis. V.g., sub predicato huius 'homo non est animal' non valet ascensus, 'homo non est hoc animal et homo non est hoc animal etc., ergo homo non est animal.' Et arguere a pluribus determinatis etc. est arguere a termino pluries copulative sumpto in ordine ad plures singulares ad ipsum determinate supponentem in ordine ad distributionem sub qua tales singulares numerantur, quod quidem eo reputatur defectus, quia in antecedenti denotatur singulis suppositis termini distributi aliquod suppositum termini supponentis determinate convenire aut a singulis aliquod removeri, et in consequenti eidem supposito termini supponentis determinate omnia supposita termini distributi convenire. Sed sub tali distributione recte licet prius descendere quam sub determinata. Sunt qui velint sub termino distributo ascendere prius dummodo addatur relativum idemptitatis termino supponenti determinate. Itaque hec consequentia est bonus ascensus 'homo non <est> Petrus et idem homo non est Paulus et sic de singulis animalibus, ergo homo non est animal. Nam si non essent, v.g., ex animalibus nisi tres homines, ultima singularis esset falsa, scilicet idem homo qui nec est Petrus nec Paulus non est Thomas. Sed certe modus hic ascendendi non est securus neque in affirmativis neque in negativis. Si enim essent infiniti homines et infiniti equi, itaque daretur primus et non ultimus, et Petrus videret omnes equos et Paulus omnes dempto primo et Thomas omnes demptis duobus primis et sic ordinatim. Tunc non sequitur 'equum Petrus videt et eundem equum, quem scilicet Petrus videt, Paulus videt, quia secundum, et eundem, quem scilicet Petrus et Paulus vident, Thomas videt, quia tertium, et sic in infinitum; <ergo etc.>' Consequens tamen est falsum, si inferas 'ergo equum quilibet homo videt.' Item, neque negative sequitur in eodem casu: 'homo non est Petrus, et idem homo non est Paulus, et sic de singulis, ergo homo non est homo'.'

with one and the same suppositum of the term suppositing determinately [or are removed from it]. But given a distribution of this sort it is allowable to descend before one descends from the determinate.[95] There are those who wish to ascend first from the distributed term provided that a relative of identity is added to the term suppositing determinately. And so this consequence is a valid ascent: '[Some] man is not Peter and the same man is not Paul, and so for all singular animals, therefore [some] man is not an animal', for, if there were only three men among the animals, the last singular [proposition] would be false, namely that the man who is neither Peter nor Paul is not Thomas. But clearly this way of ascending is not safe in either affirmatives or negatives. For if there were infinitely many men and infinitely many horses, and if there were a first but not a last [of each set], and if Peter sees all the horses and Paul [sees] all except the first and Thomas [sees] all except the first two, then this would not follow: 'There is a horse that Peter sees, and Paul sees the same horse that Peter sees (namely the second), and Thomas sees the same [horse] that Peter and Paul see (namely the third), and so on in an orderly manner, [therefore there is a horse that every man sees].'[96] However, if you do infer 'therefore there is a horse that every man sees', you have a false consequent. Again, it does not follow negatively in the same [sort of] case: '[Some] man is not Peter, and the same man is not Paul, and so for all the singulars, therefore [some] man is not a man.'

Let us consider the first part of this rule in more detail, and let us assume that there are only two individuals. Soto is arguing as follows. If we take a false proposition of the form 'Some A is not B' and first perform a descent on the distributed term B we will get the true proposition 'Some A is not B_1 and some A is not B_2.' In this proposition, there are two clauses containing a term with determinate supposition, and in each clause some suppositum or other of that term is said to be related to a suppositum of B through negation, but it need not be the same suppositum of A in both clauses. Nor does any clause relate a single suppositum of A to all the supposita of B. However, if we now perform the ascent back to 'Some A is not B' we will get a false proposition in which at least one suppositum of A is said to be related to all the supposita of B through negation. This ascent is invalid, and so we have a clear example of the so-called problem of O-propositions. If we now adopt Soto's preferred solution to this problem, we will take the false proposition 'Some A is not B' and begin by performing a descent from A. This will give us 'A_1 is not B or A_2 is not B', an analysis which successfully relates at least one single suppositum of A to all the Bs, and which allows a valid ascent back to the original proposition.

95) I am not sure what case Soto has in mind here.

96) As G. Klima has pointed out in correspondence, the supposed counter-example involves a shift of reference.

6. Conclusion

In this paper, I have ignored all the difficulties involved in giving a precise account of how the modes of personal supposition were actually applied to the analysis of complex propositions not in standard categorical form. I have also touched very lightly on the undoubted tension between the later requirement of formal equivalence and the limitations implied by the view that descent and ascent properly involve only those common terms which not only supposit, but supposit for more than one individual. Nonetheless, it is clear that Domingo de Soto and some of his predecessors were aware of the issues raised by Paul Spade, and that they did have a view about what the theory was trying to accomplish.

Bibliography

Primary Literature

Albertus de Saxonia, *Perutilis logica* (Venice 1522; repr. Hildesheim-New York 1974)

Anonymus, *Commentum emendatum et correctum in primum et quartum tractatus Petri Hyspani Et super tractatibus Marsilii de Suppositionibus; ampliationibus; appellationibus et consequentiis* (Hagennaw 1495; reprinted as: Marsilius von Inghen, *Commentum in primum et quartum tractatum Petri Hispani*. Frankfurt am Main 1967)

Domingo de Soto, *Summule* (Burgos 1529)

Georgius Bruxellensis, *Expositio Georgii super summulis magistri Petri hyspani* (Parisius 1491)

Guillelmus de Ockham, *Summa logicae*, ed. P. Boehner, G. Gál and S. Brown (Guillelmi de Ockham Opera Philosophica, I; St. Bonaventure, N.Y. 1974)

Hieronymus de Sancto Marco, *Compendium preclarum quod parva logica seu summule dicitur* (Impressum in alma Coloniensi universitate 1507)

Johannes Buridanus, *Summulae. De suppositionibus*, ed. R. van der Lecq (Artistarium, 10-4; Nijmegen 1998)

——, *Summulae. De practica sophismatum*, ed. F. Pironet (Artistarium, 10-9; Turnhout 2004)

Johannes Dorp, Commentary in: Johannes Buridanus, *Compendium logice doctissimi viri Joannis Buridani cum expositione acutissimi Ioannis Dorp* (Venetiis 1499; repr. Frankfurt/Main 1965)

Johannes Eckius, *In Summulas Petri Hispani* (Augustae Vindelicorum 1516)

Juan de Celaya, *Magne Suppositiones* (Parisiis 1526)

Marsilius ab Inghen: in: *Marsilius of Inghen, Treatises on the Properties of Terms. A First Critical Edition of the Suppositiones, Ampliationes, Appellationes, Restrictiones and Alienationes* with Introduction, Translation, Notes and Appendices by E.P. Bos (Dordrecht-Boston 1983)

Paulus Venetus: Paul of Venice, *Logica Magna* (*Tractatus de Suppositionibus*), ed. and transl. A.R. Perreiah (St. Bonaventure, N.Y. 1971)

Paulus Venetus: *Logica Parva*, ed. A.R. Perreiah (Leiden 2002)

Paulus Pergulensis: Paul of Pergula, *Logica*, ed. Sister M.A. Brown (St. Bonaventure, N.Y., Louvain-Paderborn 1961)
William of Ockham, *vide s.n.* Guillelmus de Ockham

Secondary Literature

Ashworth, E.J. (1973), 'Priority of Analysis and Merely Confused Supposition', in: *Franciscan Studies* 33 (1973), 38-41
—— (1974), ''For Riding is Required a Horse': A Problem of Meaning and Reference in Late Fifteenth and Early Sixteenth Century Logic', in: *Vivarium* 12 (1974), 146-172 (repr. in Ashworth, 1985)
—— (1976), ''I Promise You a Horse': A Second Problem of Meaning and Reference in Late Fifteenth and Early Sixteenth Century Logic (1)', in: *Vivarium* 14 (1976), 62-79. Continued: ibid. 139-155 (repr. in Ashworth, 1985)
—— (1977), 'Chimeras and Imaginary Objects: A Study in the Post-Medieval Theory of Signification', in: *Vivarium* 15 (1977), 57-79 (repr. in Ashworth, 1985)
—— (1978), 'Multiple Quantification and the Use of Special Quantifiers in Early Sixteenth Century Logic', in: *Notre Dame Journal of Formal Logic* 19 (1978), 599-613 (repr. in Ashworth, 1985)
—— (1985), *Studies in Post-Medieval Semantics* (Variorum Reprints, London 1985)
Dutilh Novaes, C. (2004), 'The Buridanian Account of Inferential Relations between Doubly Quantified Propositions: a Proof of Soundness', in: *History and Philosophy of Logic* 25 (2004), 225-243
—— (2008), 'An Intensional Interpretation of Ockham's Theory of Supposition', in: *Journal of the History of Philosophy* 46 (2008), 365-393
Karger, E. (1978), 'Consequences et inconsequences de la supposition vide dans la logique d'Ockham', in: *Vivarium* 16 (1978), 46-55
—— (1984), 'Modes of Personal Supposition: The Purpose and Usefulness of the Doctrine within Ockham's Logic', in: *Franciscan Studies* 44 (1984), 87-106
—— (1993), 'A Theory of Immediate Inferences Contained in Buridan's Logic.' in: K. Jacobi, ed., *Argumentationstheorie: Scholastische Forschungen zu den logischen und semantischen Regeln korrekten Folgerns* (Leiden 1993, 407-429)
—— (1997), 'Some 15th and Early 16th Century Logicians on the Quantification of Categorical Sentences', in: *Topoi* 16 (1997), 65-76
Maierù, A. (1972), *Terminologia logica della tarda scolastica* (Rome 1972)
Matthews, G.B. (1997), 'Two Theories of Supposition?', in: *Topoi* 16 (1997), 35-40
Parsons, T., 'The Doctrine of Distribution', in: *History and Philosophy of Logic* 27 (2006), 59-74
Pérez-Ilzarbe, P. (1996), 'The Doctrine of Descent in Jéronimo Pardo: Meaning, Inference, Truth', in: I. Angelelli and M. Cerezo, eds., *Studies on the History of Logic: Proceedings of the III. Symposium on the History of Logic* (Berlin, New York 1996, 173-186)
Priest, G., and S. Read (1977), 'The Formalization of Ockham's Theory of Supposition', in: *Mind* 86 (1977), 109-113
—— (1980), 'Merely Confused Supposition: A Theoretical Advance or a Mere Confusion?', in: *Franciscan Studies* 40 (1980), 265-297
Read, S. (1991), 'Thomas of Cleves and Collective Supposition', in: *Vivarium* 29 (1991), 50-84

Scott, T.K. (1966), *John Buridan: Sophisms on Meaning and Truth*. Translation and Introduction (New York 1966)

Spade, P.V. (1988), 'The Logic of the Categorical: The Medieval Theory of Descent and Ascent' in: N. Kretzmann, ed., *Meaning and Inference in Medieval Philosophy* (Synthese Historical Library, 32; Dordrecht-Boston-London 1988, 187-224)

—— (2002), *Thoughts, Words and Things: An Introduction to Late Mediaeval Logic and Semantic Theory*. Version 1.1. August 9, 2002. http://www.pvspade.com/Logic

Weidemann, H. (1979), 'Wilhelm von Ockhams Suppositionstheorie und die moderne Quantorenlogik', in: *Vivarium* 17 (1979), 43-60

When the Inference 'p is true, therefore p' Fails: John Buridan on the Evaluation of Propositions

Ernesto Perini-Santos
Departamento de Filosofia, UFMG, Brazil

Abstract

For John Buridan, truth-bearers are assertions. This fact explains why the inference 'p is true, therefore p' may fail. On the one hand, the tense of the verb plus the time of utterance do not determine the time about which a sentence is intended to be true: the intention of the speaker is needed. On the other hand, since the meaning of vocal and written words is conventional, it may seem that they can be used with different meanings on each side of the inference. While the antecedent may talk about a situation different from the present one, this doesn't make it the actual situation of utterance, and the words have meanings only in the actual situation of use. Although such situations are different, in both of them we are asked to see the importance of identifying features of which the disquotational schema doesn't keep track and that can only be specified in the context of use.

Keywords

John Buridan, sophism, truth, truth-bearer, proposition, disquotational schema

1. Introduction

The biconditional 'p is true, if and only if p' is not valid for John Buridan: further conditions must obtain for the inference to be valid. The required precisions are not the same for each side of the biconditional. In this paper, I will be concerned mostly with 'p is true, therefore p'. However, let us take a look first at the other half, 'p, therefore p is true'.

For John Buridan, the bearers of truth-values are proposition-tokens,[1] that is, written, vocal, and, primarily, mental occurrences of sentences. Probably,

[1] The term 'proposition' here should be understood as equivalent to 'sentence', following the medieval use of '*propositio*'.

the first motivation for this position is his nominalism: proposition-types are simply not available in his ontology, not being individual entities. Truth-bearers are singular occurrences of propositions in the mind, in speech or in writing. As a matter of fact, this is not entirely correct—as we will see, in a Buridanian framework, it seems more appropriate to say that the bearers of truth-values are assertions, rather than sentences, which doesn't change the ontological commitments of the theory.

It is important to have proposition-tokens, or assertions, as the bearers of semantic values, to cope with certain semantic phenomena, such as to explain how a sentence can be true about a situation in which it doesn't exist, and in which its existence would make its truth impossible—e.g., 'no proposition is negative'.[2] Different frameworks can accommodate the same phenomena, the crucial point being the distinction between the context in which the content of an assertion is determined, and the circumstance of evaluation of this content, whatever it is that is deemed to be the content bearer. Be that as it may, we need parameters that are only determined for sentence-tokens.

Some such situations are discussed in the important eighth chapter of the *Sophismata*, on self-reflexive propositions. It is well known that the inference 'p, therefore p is true' is crucial in Buridan's examination of the Liar paradox.[3] This inference is invalid if a certain requirement is not fulfilled—the antecedent must exist for the consequent to be true.[4] Things can be as signified by the antecedent, without being as signified by the consequent, for the consequent has a proposition in material supposition, and therefore is true only about a situation in which there is a proposition for which its subject can suppose.

---

[2] See Johannes Buridanus, *Summulae. De practica sophismatum*, ed. Pironet (2004), VIII, 2: 'Nulla propositio est negativa, ergo quaedam propositio est negativa', which is the title of the sophism. Conversely, the truth of a sentence can be guaranteed by its existence: it may be the case that a sentence is true in every situation in which it is formed, even though it is not a necessary proposition—e.g., a sentence exists.

[3] For a recent assessment, see Read (2002). For a general view of Buridan's approach to the Liar Paradox, see Pironet (1993); for recent overviews of medieval treatments of this paradox, see Dutilh Novaes (2008) and Yrjönsuuri (2008), for an interpretation of Buridan's solution the the liar paradox, see Perini-Santos (2011).

[4] Johannes Buridanus, *Summulae. De practica sophismatum*, ed. Pironet (2004), VIII, 2, 144, 22-28: '[...] illa regula, sicut ponebatur, non est vera de virtute sermonis, scilicet quod ad quamlibet propositionem sequatur quod ipsa sit vera; immo non sequitur 'homo est; igitur haec est vera 'homo est' ', quia homo possest esse, licet nulla propositio esset, et etiam quia ita esse est possibile sicut per istam propositionem 'homo est' significatur, vel significaretur si proponeretur, non existente ita sicut per istam significaretur 'haec est vera 'homo est' '. Ita enim esset sicut per primam significatur, si homo esset et nulla propositio esset.'

Not so for the antecedent, which can be true about a situation in which no proposition exists. Indeed, both sides of the consequence are true only in situations in which they exist, but only the left-hand side can be true about a situation in which no proposition exists. It is precisely this difference that makes the consequence invalid, if stated without further conditions.[5]

In this paper, I will be concerned mostly with the other half of the biconditional, 'p is true, therefore p'. We will go backwards in the *Sophismata*, and as we go backwards in the text, we will find Buridan concerned with more primitive conditions: the determination of truth-conditions of assertions, in the 7th chapter, and the signification of propositions, in the 6th chapter. The conditions won't be the mere existence of propositions, but features that can be specified only for tokenings.

2. The Role of the Speaker's Intention

You say:

1) Socrates is not running

and during the precise time of your utterance, he was running: your proposition is false. To contradict you, I say:

2) Socrates is running.

Since these propositions are contradictories, and your proposition is false, my proposition is true, and therefore Socrates is running. As it happens though, as soon as I begin to talk, he sits down. Annoyingly, it seems that my proposition is false. Does that mean that two contradictories are false? Not so. If each utterance concerns the minimal interval of time containing the totality of the utterance event, then they are both false, but they are not contradictories. If they concern the same slice of time, they are contradictories, but only one of them can be false. The description above that suggests that 1 and 2 are both false presupposes that each utterance is true about the exact slice of time in which it is uttered. In this case, they are indeed both false, but not contradictories. Obviously, for sayings of 1 and 2 to be contradictories, they have to be about the same slice of time, and if they are not exactly simultaneous—as is the case in

[5] On this point, see Klima (2004).

our story, another determination of the time about which they are said to be true is needed.

This distinction is possible because the tense of the verb plus the time of the utterance are not sufficient to determine the time about which it is true. If the tense of the verb plus the time of the utterance does not determine the time for which a proposition is true, what brings about the further determination? If I intend to contradict you, I intend to talk precisely about the same slice of time you talked about, and that establishes the time *pro quo* the proposition is true:

> [...] in intention I have to refer the verb of my proposition to the same time as that to which you referred the verb of your proposition, so that the intention should be to deny what you affirmed, or conversely [to affirm what you denied], for the same time, even if that time coexisted with your proposition and not with mine.[6]

A present-tense sentence can be evaluated strictly at the time of its utterance, or at the minute, the hour, the day, the year of the utterance, or further.[7] The extension is determined by the intention of the speaker, who can choose the slice of time he wants to use as present:

> [...] it is not determined for us how much time we ought to use as present, but we may use as much as we want.[8]
> [...] in our speech we can use as present a slice of time sometimes bigger, and sometimes smaller, for instance, sometimes a whole year, and sometimes a single day or a single hour, or only the time that is exactly coincident with our speech.[9]

6) Transl. Klima (2001), 943; Johannes Buridanus, *Summulae. De practica sophismatum*, ed. Pironet (2004), VII, 2, p. 130, 14-17: '[...] secundum intentionem ego debeo referre verbum meae propositionis ad idem tempus ad quod tu referebas verbum propositionis tuae, ita quod intentio sit pro eodem tempore negare pro quo tu affirmabas aut e converso, licet illud tempus coexisteret propositioni tuae et non meae.'

7) Johannes Buridanus, *Summulae. De practica sophismatum*, ed. Pironet (2004), VII, 1, 128, 22-129, 8.

8) Transl. Klima (2001), 942; Johannes Buridanus, *Summulae. De practica sophismatum*, ed. Pironet (2004), VII, 1, 129, 1-2: '[...] non est nobis determinatum quantum sit tempus quo debeamus uti tamquam praesenti, sed licet nobis uti quanto volumus.'

9) Johannes Buridanus, *Questiones longe super librum Perihermeneias*, ed. Van der Lecq (1983), I, 9, p. 44, 12-14: '[...] in nostra locutione possumus tamquam presenti uti tempore aliquando maiore parte aliquando minore parte, ut aliquando uno toto anno et aliquando una sola die vel hora vel isto solo tempore quod adequate coexistit locutioni nostre.'

It may seem that the time for which a proposition is true is strictly the time of its utterance:

> [...] some say that literally speaking we should use as present only exactly the time in which a proposition is uttered. However, from this thesis some unwelcome consequences follow. Firstly, there would not be a present year, nor a present day. Secondly, every conjunction of contradictory propositions about contingent matters would be possible. For instance, this would be possible: 'Socrates is running and Socrates is not running', for referring to the time in which the affirmative was uttered, Socrates is running, and [the affirmative] is true, [and] referring to the time in which the negative proposition was uttered, Socrates would not be running, and the negative would be true; therefore both would be true, and the conjunction would be true.[10]

To avoid this consequence, we have to abandon the idea that, literally speaking, a present-tense utterance should be evaluated exactly at the time of the utterance event. The intention of the speaker determines the slice of time about which the utterance is said to be true, and the exact time of the utterance event is no more natural as to the time about which it is true than any other slice of time including the time of the utterance event.

When I pronounce 'Socrates is running', our philosopher is sitting; therefore, in a part of any extension of time that includes the time of the utterance of this sentence, to say that he is running is false. Buridan says that, in this case, I speak as if the time of my utterance was the same time as yours.[11] There is an intentional shift of what is to be taken as the time of evaluation of the utterance—I do not intend that the very time of the utterance of the sentence be part of the time at which the utterance is evaluated. This sophism exploits a tricky case (we are dealing with a sophism, after all!), but this shouldn't make us miss the general point.

It seems to me that we can explain in this framework an interesting case of a radical extension of the time at which a present tense sentence is intended to

---

[10] '[...] aliqui dicunt quod de virtute sermonis nos debemus uti pro presenti tempore illo solo tempore in quo adequate propositio proponitur: sed ad illud dictum sequuntur inconvenientia: primo quod non esset annus presens nec dies presens; secundo quod omnis copulativa ex contradictoriis in materia contingenti esset possibilis, verbi gratia: ista esset possibilis 'Sortes currit et Sortes non currit': quia pro tempore in quo affirmativa proponeretur Sortes currit: et sic est vera, pro tempore in quo negativa proponeretur Sortes non curreret et sic negativa esset vera: ergo ambe essent vere et sic illa copulativa esset vera.' Johannes Buridanus, *In Metaphysicam* IV, Parisiis (1518), q. xv, f. 25vb. I thank Ria van der Lecq for some minor corrections to this text.

[11] Johannes Buridanus, *Summulae. De practica sophismatum*, ed. Pironet (2004), VII, 2, 130, 17-19: 'Ego enim secundum intentionem loquor, ac si loquerer in illo tempore, et hoc licitum est facere ut possimus conferre ad eandem intentionem.' On this point, see Perini-Santos (2008).

be evaluated: the case of scientific propositions, such as 'every thunder is a sound made in the clouds' (*omne tonitruum est sonus factus in nubibus*), one of the senses of Buridanian natural supposition. In his treatise on suppositions, Buridan says that such sentences are not literally true,

> [...] the verb 'is' by virtue of its proper meaning was imposed to signify only the present time, whereas there may be no thunder or lunar eclipse at the present time.[12]

These sentences are used for the sake of brevity, instead of 'every thunder, whenever it is, was, or will be, etc.'—thus stated, they are true, but, if they were considered according to their proper meaning (*ad sensum proprium*), they would be false. Since we do not have a new word imposed to signify such mental copula,

> [...] we can use the verb 'is' by convention (*ad placitum*) to signify such a copula by which the present time will no more be signified than is the past or the future: indeed, [it will signify] no time at all.[13]

This seems to be a conventionalist solution to the determination of the extension of the time about which a proposition is denoted to be true. What is missing in this case is a contrast between the present, the past and the future; in other cases, even if the extension of the time considered as present may vary from one utterance to another, according to the intention of the speaker, there always seems to be room for this contrast. It doesn't seem, however, that the vocal copula '*est*' is ambiguous between a 'temporal' and an 'eternal' reading. Wouldn't it be more economical to consider this case an extreme extension of the time of evaluation due to the intention of the speaker?

In his questions on *Metaphysics*, Buridan simply associates both cases. How much time must we use as present? There is no rule.

---

12) Transl. Klima (2001), 261; Johannes Buridanus, *Summulae. De suppositionibus*, ed. Van der Lecq (1998), 4.3.4, p. 47, 10-12: [...] hoc verbum 'est' de proprietate sermonis non sit impositum ad significandum nisi praesens tempus; et tamen forte nec est tonitruum nec eclipsis lunae in praesenti tempore.'

13) Transl. Klima (2001), 261; Johannes Buridanus, *Summulae. De suppositionibus*, ed. Van der Lecq (1998), 4.3.4, p. 47, 25-27: '[...] possumus ad placitum hac voce 'est' uti ad significandum talem copulam per quam non significabitur magis tempus praesens quam praeteritum vel futurum, immo nullum.'

> If you say that we must use as present one whole hour, and not more, you cannot give a reason why one hour, and not more. And for any period of time you choose, you cannot give a reason why this time should be used instead of a bigger or smaller slice of time, unless you say 'I want to use this much [as present], and not more nor less'. I approve this thesis, that anyone can use as present a slice of time as extended as one wants, bigger or smaller, even if the utterance is more appropriate when a smaller slice of time is used as present. Therefore, Aristotle says that the sky always moves, believing that the sky is always eternally moving, and in this thesis uses an infinite time as present.[14]

The contrast between the past, the future and the present is still missing, but the role of the intention of the speaker is acknowledged. Be that as it may for scientific propositions, in any other situation the intention of the speaker is essential in the determination of the time at which a present-tense proposition should be evaluated.[15]

The importance of the distinction between the time *at which* [*in quo*] the *propositio* is true and the time *for which* [*pro quo*] it is true has been shown by Arthur Prior in his classic paper 'The possibly-true and the possible'.[16] In a 1968 paper, 'Fugitive Truth', he examines Buridan's solution to the *sophisma* 'Socrates is sitting and Socrates is not sitting'.[17] Following Buridan, he says that this conjunction 'may well be true, since Socrates may be sitting down while

---

14) Johannes Buridanus, *In Metaphysicam*. Parisiis [1518], IV, q. xv, 25va: 'Si enim dicas quod debemus uti pro presenti una totali hora et non pluri tempore tu non potes dare rationem quare magis una hora et non pluri: et ita quodcumque tempus tu assignes tu non potes dare rationem quare magis illo quam pluri vel pauciori nisi dicendo quod tanto volo uti et non pluri nec pauciori et illud dictum ego approbo, videlicet quod cuilibet licet uti pro presenti tempore quantocumque uti voluerit sive parvo sive magno; quamvis locutio est magis propria quando parvo tempore utimur presenti. Unde diceret Aristoteles quod celum semper movetur; credens tamen quod semper celum perpetue moveri et in isto dicto infinito tempore utitur pro presenti.'

15) Albert of Saxony, when examining whether two contradictory propositions can be true or false simultaneously ('Quaeritur utrum propositiones contradictoriae possint esse simul verae vel simul falsae etc.', q. XXIV, pp. 311-318), comes to similar conclusions: for some propositions, we can only determine whether they are contradictory or not once contextual parameters are determined. For instance, '*ego sum Sortes*' and '*ego non sum Sortes*' may be both true, if pronounced simultaneously by Socrates and Plato—their truth-conditions can only be established once contextual parameters are fixed. He doesn't extend this conclusion, however, to the role of the speaker's intention; see Albertus de Saxonia, *Quaestiones Circa Logicam*, ed. Fritzgerald (2003), q. xxiv, 315, 12-15.

16) Prior (1969).

17) Prior (1968). For the sophism '*Haec copulativa est vera 'Socrates sedet et Socrates non sedet*', see Johannes Buridanus, *Summulae. De practica sophismatum*, ed. Pironet (2004), VII, 4, 131, 17-133, 10.

we utter the first part of it but may stand up while we utter the second part'.[18] The conjunction may well be true, and it certainly may be false. How can we tell one situation from another? Prior suggests a conventionalist solution. He, however, seems to realize how hard it is to accept conventions associated with each occurrence of each sentence.[19] It is not a good solution, and anyway it is not Buridan's. This determination is due to the intention of the producer of the sentence token, and as she does that, she does not create a new word, but determines what is insufficiently determined by the meaning of the word.

The reason why the disquotational schema is incomplete is that the intention of the speaker is needed for the sentence to have determined truth conditions, and the inference by itself does not keep track of the intentions associated with the antecedent and with the consequent.

> And to the authority of Aristotle I reply that 'the proposition 'Socrates is sitting' is true, therefore, Socrates is sitting' is indeed valid, but only as long as in the consequent we take the time consignified by the verb 'is sitting' for the time *for* which the proposition is true, and not for the time *at* which it is true, unless the former and the latter [times] are the same. Therefore, although the proposition 'Socrates is sitting' was true *at* time B, this was [the case] *for* the time A.[20]

The inference is valid, as long as we take the antecedent and the consequent to be about the same slice of time. We can fix that both sides are true about the same slice of time. This amendment is indeed possible, but it is no longer a disquotational schema. We cannot formulate this amendment without stating the time about which both sides are true—e.g., 'and both sides are intended to be true about the same slice of time'—, and we will have the consequent taken in material supposition in a complete formulation of the inference. We can simply take both sides to be true about the same slice of time, but as the argument for the sophism shows, it is not a property of the sentence, but of the

18) Prior (1968), 6. It is interesting to note that Buridan does not use 'we', but 'I' or 'you', which are better suited to descriptions of utterances.

19) 'It is clear that we need to make our conventions more explicit at this point', Prior (1968), 6.

20) Transl. Klima (2001), 944; Johannes Buridanus, *Summulae. De practica sophismatum*, ed. Pironet (2004), VII, 3, 131, 11-16: 'Et ad auctoritatem Aristotelis ego dico quod optime sequitur 'haec propositio 'Socrates sedet' est vera; ergo Socrates sedet', dum tamen in hoc consequente, scilicet 'Socrates sedet', accipiamus tempus consignificatum per hoc verbum 'sedet' pro illo tempore pro quo propositio est vera, et non pro illo in quo est vera, nisi sint idem hoc et illud. Unde licet ista propositio 'Socrates sedet' esset vera in tempore B, tamen hoc erat pro tempore A.'

assertion with a determining intention of the speaker. In order for the inference 'p is true, therefore p' to be valid, we need to have assertions, and not sentence tokens: read as an inference linking sentence tokens, it is incomplete, that is, it is not valid.

3. The Situation of the Use of a Word and the Situation Signified by Words

In the sixth chapter of the *Sophismata*, John Buridan refuses an inference close to the disquotational schema, for a reason not entirely unlike that of the seventh chapter. The first sophism of this chapter is 'You will be a donkey' (*Tu eris asinus*). The argument for the *sophisma* goes as follows:

> Proof: tomorrow this will be true: 'You are a donkey'; therefore, today this is true: 'You will be a donkey'. The consequence is clear from the order of present, past, and future: for what is now present, afterwards will be past, and before was future, and this also holds for propositions. For if it is true that you will run, then sometime it will be true that you are running and conversely.[21]

This last inference, obviously, is not an example of a disquotation—we have that-clauses in both sides of the inference; that-clauses wouldn't justify the antecedent 'tomorrow this will be true: 'You are a donkey''. Buridan comes back to an instance of a disquotation in the justification of the antecedent:

> [...] positing the case that you and others want to change your name by convention and want to give you the name 'donkey'; then it is obvious that tomorrow this will be true and to be conceded by you: 'You are a donkey'.[22]

Since tomorrow the sentence 'You are a donkey' will be true, today it is true 'You will be a donkey'. Buridan leaves to the reader the inference from 'today this is true: 'You will be a donkey'' to 'you will be a donkey', which is the

---

21) Transl. Klima (2001), p. 930; Johannes Buridanus, *Summulae. De practica sophismatum*, ed. Pironet (2004), VI, 1, 116, 3-8: 'Probo: quia cras ista erit vera 'tu es asinus'; ergo hodie haec est vera 'tu eris asinus'. Consequentia est manifesta ex ordine temporum praesentis, praeteriti et futuri ad invicem: quia quod modo est praesens, post erit praeteritum et ante fuit futurum, et etiam ita est de propositionibus. Nam si est verum quod tu curres, aliquando erit verum quod tu curris et e converso.'

22) Transl. Klima (2001), 930; Johannes Buridanus, *Summulae. De practica sophismatum*, ed. Pironet (2004), VI, 1, 116, 9-11: '[...] posito casu quod tu et alii ad placitum velitis mutare nomen tuum et imponere tibi hoc nomen 'asinus', tunc apparet quod haec cras erit vera et a te concedenda 'tu es asinus'.

sophism. This last inference is valid; the problem lies in the passage from 'tomorrow this will be true: 'You are a donkey' ' to 'today this is true: "You will be a donkey" '.

It is certainly possible to change one's name voluntarily, and, more generally, since written and vocal terms signify *ad placitum*, there are different situations in which the same written or vocal forms (or different tokens of the same written or vocal type) signify different mental terms. This isn't enough, however, to justify the inference. Why? Because a proposition is always used according to its present signification:

> And when it is said 'tomorrow 'You are a donkey' will be true', I concede this concerning a similar utterance. But that [utterance] will not hold in terms of the signification that the term 'donkey' now has and according to which the sophism is at present propounded. Therefore, [the conclusion] 'You will be a donkey' does not follow.[23]

The antecedent may be true *secundum vocem*, that is, considering the mere sounds '*tu es asinus*', and in that case, it signifies nothing, and there is no possible disquotation. The sentence '*tu es asinus*' can be disquoted only if it signifies something, and, in that case, according to the signification it has in the situation in which it has been proposed, it is false.

The next sophism, '*Ba baptizabitur*', deals with an analogous problem: a child will be baptized tomorrow, and will receive the name 'Ba', therefore Ba will be baptized.[24] We have again the same alternative. If it is taken materially, then the proposition is evidently false (*'Ba baptizabitur' est una propositio manifeste falsa*): this written form won't be baptized tomorrow. If 'Ba' is taken personally,

> [...] in the case posited the utterance or inscription 'Ba will be baptized' is neither true nor false, for it is neither a proposition nor a sophism, nor is it even an expression, since an expression is defined as being a conventionally significative utterance whose parts are separately significative. Therefore, when you say 'This boy tomorrow will be Ba', I say that if the utterance 'Ba' supposits materially, then this [proposition] is to be denied as false; and if it does not supposit materially, then the whole of what is said, namely, 'This boy tomorrow

[23] Transl. Klima (2001), 933; Johannes Buridanus, *Summulae. De practica sophismatum*, ed. Pironet (2004), VI, 1, 119, 10-13: 'Et quando dicitur 'ista cras erit vera 'tu es asinus' ', concedo de ista vel de simili secundum vocem. Sed hoc non erit secundum significationem quam modo habet iste terminus 'asinus' secundum quam sophisma nunc proponitur. Ergo 'tu eris asinus' non sequitur.'

[24] Johannes Buridanus, *Summulae. De practica sophismatum*, ed. Pironet (2004), VI, 2, 119, 17-18.

> will be Ba', is neither true nor false, and is neither to be conceded nor denied, just as if you were to say 'bu ba'.[25]

A proposition can only be true according to the signification it has at the moment it is uttered:

> [...] a proposition, unless its terms are taken materially, is never to be said to be true or false except according to the signification it and its terms have when it is propounded, and not according to the signification it may have or perhaps will have but not yet has.[26]

Both sophisms depend on the illusion that a signified situation of evaluation can play the role of the utterance situation. We can refer to the signification a term will have in a future situation, or to the signification it may have in any other situation different from the present one, but we cannot use the term except with the signification it has in the actual situation of utterance. In both cases, the antecedent states that certain *voces* will have a certain signification distinct from their present signification—e.g., '*asinus*' will signify you, and '*Ba*' will be the name of a child –, and in the consequent, through a disquotational inference, this *vox* is used as if it had this different signification. This is precisely what can't be done: one can use a *vox* only with its present signification.

The sophisms in this chapter depend on a pretended displacement of the utterance situation of the consequent to a non-present situation signified in the antecedent. As a matter of fact, the story can be a little bit more complicated. In the sophism 'It is within our power that a man should be a donkey', he asks us to imagine a disputation in which we arbitrarily decide to change the meaning of the words 'man' and 'donkey', so that 'a man is a donkey' signifies a

---

25) Transl. Klima (2001), 934; Johannes Buridanus, *Summulae. De practica sophismatum*, ed. Pironet (2004), VI, 2, 120, 14-20: '[...] in casu posito ista vox vel scriptura 'Ba baptizabitur' non est vera nec falsa, quia non est propositio nec sophisma, immo nec est oratio, cum oratio diffiniatur 'quod est vox significativa ad placitum, cuius partes sunt significativae separate' Quando ergo dicis 'iste puer cras erit Ba', dico quod si haec vox 'Ba' supponat materialiter, illa est neganda tamquam falsa; et si non supponat materialiter, tunc illud dictum totale 'iste puer cras erit Ba' nec est verum nec falsum, nec concedendum nec negandum, sicut si diceres 'bu ba'. [...]'

26) Transl. Klima (2001), 933; Johannes Buridanus, *Summulae. De practica sophismatum*, ed. Pironet (2004), VI, 2, 120, 6-9: '[...] numquam propositio, si termini eius non sumantur materialiter, debet dici vera vel falsa nisi secundum significationem quam ipsa et sui termini habent quando proponitur, et non secundum significationem quam potest habere vel forte habebit sed nondum habet.'

true mental proposition. We certainly can do that. However, we cannot use the words 'man' and 'donkey' except with their actual signification:

> [...] when we use the terms 'man' and 'donkey' significatively, we use them according to the signification that they now have.[27]

If a vocal proposition *p* is proposed in a situation *s*, and in *s* its words mean the mental proposition *A*, *p* cannot be evaluated in *s* except as signifying *A*. Indeed, I can arbitrarily change the meaning of the vocables I use; I can say, for instance, that I will use from now on the *vox* 'donkey' to name philosophers. But if I start to use this *vox* in this unorthodox way, it doesn't retain its actual signification, and the mental proposition signified by 'Socrates is a donkey' in this imaginary situation is the proposition signified now by the proposition 'Socrates is a philosopher'. However, I cannot use the vocable 'donkey' with a given meaning only by describing a possible situation in which it would have this meaning: again, 'we use [words] according to the signification that they now have', not according to the signification we can imagine them having.

Indeed, Buridan is not disputing the arbitrary character of *voces*, but making a distinction between the situation of the use of a word, in which we presuppose it to have a specific meaning, and a situation that can be signified by words already endowed with meanings. We can change what *voces* mean according to our will, but we cannot make it the case that the actual situation of use of words is an imaginary situation described by our words.

4. Conclusion

We have dealt with two different sets of phenomena. Firstly, we talked about the slice of time *about which* a proposition is true. In the second case, we were dealing with the situation *in which* a proposition is true. What do these topics have in common? Let us start with the apparent cause of truth of the sophisms. The arguments for the sophisms miss the fact that some semantic features of words are only fixed in a context of use, and they appeal to a disquotational inference (or to a modified version of it) because this inference does not keep track of such features. Moreover, in both cases, there is an intentional element that, in the first case, determines the slice of time about which a proposition is

[27] Transl. Klima (2001), 937; Johannes Buridanus, *Summulae. De practica sophismatum*, ed. Pironet (2004), VI, 4, 124, 3-5. '[...] quando utimur istis terminis 'homo' et 'asinus' significative, utimur eis secundum significationem quam habent nunc.'

true, and, in the second case, may change the meaning of a vocable. However, they don't miss exactly the same fact.

The determination of the slice of time about which a proposition is true is determined by the intention of the speaker in making such and such assertion—the time of utterance plus the tense of the verb are not enough. The intention is not determined by the mere instantiation of a propositional type, but only in the use of a sentence in a speech act. The inference 'p is true, therefore p', if read as linking sentence tokens, doesn't keep track of the intention of the speaker. Even granted the existence of both sentences in the situation in which they are evaluated—and, as it is exploited by one of the sophisms above, the slices of time in which they exist are not the same—, nothing guarantees that the times for which both sides are intended to be true are one and the same, and if they are not the same, the inference is invalid.

The determination of the situation at which the assertion is to be evaluated requires the intention of the speaker; for the inference to be valid an extra element is needed. We can fix a parameter that is only determined by the use of the sentence, but this extra clause guaranteeing that both the antecedent and the consequent are intended to be true about the same time cannot be read in the mere tokening of sentences.

Our second type of argument aims at another kind of phenomenon: not that the meaning of words underdetermines what is said, but that the same sounds or written marks may have totally different meanings (if any) in different situations. An inference of the form "'You are a donkey' will be true, therefore today it is true 'You will be a donkey'" tries to shift the situation that determines the mental proposition signified by some words by signifying, in the antecedent, a situation different from the actual one—or uses some other device to the same effect. The antecedent may talk about a situation different from the present, be it a past situation, a future situation, or a possible situation. But merely talking about a situation doesn't make it the actual situation of utterance, and the words have meanings only in the actual situation of use. This is not a semantic fact of which the disquotational schema, or rather a variant of it, doesn't keep track, but rather a presemantic fact, concerning the signification of the words used, borrowing a distinction from John Perry.[28]

In the first case, we are asked to distinguish the situation in which a proposition is true from the situation about which it is true. In the second case, a distinction is made between the mental proposition actually expressed in a given situation and the mental proposition the same *voces* would have signified in a

28) On the distinction between semantic and presemantic, see Perry (1998).

different situation, be that different situation signified by the mental proposition actually expressed. In both cases, we are asked to see the importance of identifying features that can only be specified in the context of use.

Bibliography

Primary Literature

Albertus de Saxonia, *Quaestiones circa logicam*, ed. M.J. Fitzgerald, in: id., *Albert of Saxony's Twenty-five Disputed Questions on Logic. A Critical Edition of his Quaestiones circa Logicam* (Leiden 2002)

Johannes Buridanus, *Quaestiones in Aristotelis Metaphysicam* (Paris 1518, repr. Frankfurt am Main 1964)

——, *Questiones longe super librum Perihermeneias*, ed. R. van der Lecq (Artistarium, 4; Nijmegen 1983)

Johannes Buridanus: John Buridan, *Summulae. De dialectica*, transl. G. Klima (New Haven 2001)

Johannes Buridanus, *Summulae. De practica sophismatum*, ed. F. Pironet (Artistarium, 10-9; Turnhout 2004)

Secondary Literature

Dutilh Novaes, C. (2008), 'A Comparative Taxonomy of Medieval and Modern Approaches to Liar Sentences', in: *History and Philosophy of Logic* 30 (2008), 227-261

Klima, G. (2001), see John Buridan, *Summulae. De dialectica*, transl. G. Klima (New Haven 2001)

——. (2004), 'Consequences of a Closed, Token-based Semantics: the case of John Buridan', in: *History and Philosophy of Logic* (25) 2004, 95-110

Perini-Santos, E. (2008), 'John Buridan on the Bearer of Logical Relations', in: *Logica Universalis* 2 (2008), 69-79

——. (2011), 'John Buridan's Theory of Truth and the Paradox of the Liar', in: *Vivarium* 49 (2011), 184-213.

Perry, J. (1998), 'Indexicals, Contexts, and Unarticulated Constituents', in: A. Aliseda, R. van Gabeek and D. Westerståhl, eds., *Computing Natural Language* (Stanford 1998, 1-11)

Pironet, F. (1993), 'John Buridan on the Liar Paradox: Study of an Opinion and Chronology of the Texts', in: K. Jacobi, ed., *Argumentationstheorie. Scholastische Forschungen zu den logischen und semantischen Regeln korrekten Folgerns* (Leiden 1993, 293-300)

Prior, A.N. (1968), 'Fugitive truth', in: *Analysis* 29 (1968), 5-8

——. (1969), 'The possibly-true and the possible', in: *Mind* 78 (1969), 481-492

Read, S. (2002), 'The Liar Paradox from John Buridan back to Thomas Bradwardine', in: *Vivarium* 40 (2002), 189-218

Yrjönsuuri, M. (2008), 'Treatments of Paradoxes of Self-reference', in: D. Gabbay and J. Woods, eds., *Handbook of the History of Logic, 2: Mediaeval and Renaissance Logic* (Amsterdam 2008, 579-608)

XV-XVI-XVIITH CENTURIES

Logic in Salamanca in the Fifteenth Century The *Tractatus Suppositionum Terminorum* by Master Franquera

Angel d'Ors†*
Universidad Complutense, Madrid

Abstract

This paper looks into the contents of the *Tractatus suppositionum terminorum* by Master Franquera, in the context of the teaching of logic in Salamanca (and elsewhere in Spain) in the fifteenth century. Franquera's work is characterised by its explicit realist bias and its rejection of Ockhamist theses, i.e., by its recognition of the existence of a *natura communis* or a *universale in re*, which is evident in all discussions related to *suppositio simplex* and the theory of *significatio*. But, apart from this, Franquera's discussion of the theory of *suppositio* stands out as a strange mixture of different doctrines: some of them are derived from thirteenth-century (or earlier) analyses, others from fourteenth-century developments; realist goals are reached by means of instruments that, while not being nominalist, are definitely inspired by terminism. While upholding the theses of realist schools, Franquera adopts the definitions and rules of nominalist authors.

Keywords

logic in Salamanca in the fifteenth century, commentaries on Peter of Spain's *Tractatus*, '*Glose Salamantine*', Master Bartholomew, Master Franquera

1. Introduction

In his article on logical works attributed to Richard Billingham,[1] Professor De Rijk drew our attention to a number of manuscripts in Spanish libraries (or of Spanish origin) that contain such works, pointing out that these manuscripts

*) Angel d'Ors passed away before he received proofs of this paper, and not all references had been completed by the time of his death.

1) De Rijk (1976), 125-132. See also De Rijk (1975), 118-120 and 135-137.

also preserve various logical treatises by Spanish authors that are related in one way or another to Billingham (and the '*Logica Cantabrigiensis*' or the '*Logica Oxoniensis*'). These manuscripts are as follows:[2]

i) Oxford, *Bodleian Library*, Ms. Misc. lat. E. 100 (which can probably be identified with the manuscript from Zaragoza, *Biblioteca del Cabildo Metropolitano*, 15-82, now lost);
ii) Segovia, *Archivo de la Catedral*, Ms. B-293 (olim 31);
iii) and iv) Toledo, *Biblioteca Capitular*, Mss. 94-27 and 94-28;
v) and vi) Salamanca, *Biblioteca Universitaria*, Mss. 1735 and 1882;
vii) Roma, *Biblioteca Casanatense*, Ms. 5445;
viii) and ix) Barcelona, *Arxiu de la Corona d'Aragó*, Mss. Ripoll 141 and Ripoll 166;
x) Gdańsk, *Biblioteka Gdańska Polskiej Akademii Nauk*, Cod. 2181;
xi) Sevilla, *Biblioteca Capitular y Colombina*, Ms. 5-I-14;
xii) Paris, *Bibliothèque nationale de France*, Ms. lat. 258.

To these twelve manuscripts we should add:

xiii) Cambridge, *Corpus Christi College Library*, Ms. 378;
xiv) and xv) Segovia, *Archivo de la Catedral*, Mss. B-283 (olim 30) and B-355 (olim 33);
xvi), xvii) and xviii) Salamanca, *Biblioteca Universitaria*, Mss. 2002, 2080 and 2107;
xix) and xx) Barcelona, *Biblioteca de Catalunya*, Mss. 71 and 768;
xxi) Pamplona, *Biblioteca de la Iglesia Catedral*, Ms. 6;
xxii) Sevilla, *Biblioteca Capitular y Colombina*, Ms. 7-3-13;
xxiii) Tortosa, *Archivo Capitular*, Cód. 25;
xxiv) Madrid, *Biblioteca Nacional de España*, Ms. 2166;

as well as the family of manuscripts of the '*Glose Salamantine*' on Peter of Spain's *Tractatus* ('*Ut ait Philosophus in primo Posteriorum*'), associated with the name of Bartholomew.[3]

---

2) In presenting the list I have followed the order used by De Rijk (1976). The order is neither chronological nor geographical.

3) Toledo, *Biblioteca Capitular*, Ms. 94-27, ff. 1r-61r (14th c. (?), commentary on the first 4 treatises of the *Tractatus*); Seville, *Biblioteca Capitular y Colombina*, Ms. 7-7-7, ff. 2ra-165ra (14th c. [?], on all 12; Paris, *Bibliothèque de France*, Ms. lat. 6433, ff. 153ra-258vb (15th c., on the first 7); Paris, *Bibliothèque de France*, Ms. lat. 258, ff. 77r-126v (15th c., on the first 4); Segovia, *Archivo de la Catedral*,

In these manuscripts we find a range of treatises belonging to various genres: '*Suppositiones*', '*Ampliationes*' and '*Appellationes*'; '*De probationibus propositionum (sive Terminus est in quem sive Speculum puerorum sive Exponibilia)*'; '*Consequentiae*' and '*Contraconsequentiae*', and '*Obligationes*'. There are also works of the '*summulae*' variety, as well as commentaries on Peter of Spain's *Tractatus* and on various works from Aristotle's *Organon*, which encapsulate the way logic was taught and the reception of the '*logica modernorum*'[4] in Spain in the second half of the fourteenth century and the first seventy years of the fifteenth. Through these manuscripts, we perceive the presence of three major groups of works and authors: a Castilian group, centering on Salamanca (but to which Valladolid also belongs); an Aragonese group, whose focal points are Huesca and Zaragoza; and a Catalan group with its main base in Lérida (but in which Barcelona is also important). However, we have hardly any information about the scholars working in those centres, or the relations between them.

Many of these treatises are presented without the author's name, and in others they are attributed to figures who are unknown today, for example Pedro Fajardo, Guillermo Ferrer, Salvador de Terradis, Benedicto de Undis, Nicolás Surrina, Master Bartholomew, Juan de Pastrana, Guillermo de Osma, Master Franquera, García de Castello, Juan de Medina, Antonio Vallarono, Juan de Santa Cruz and Cerdo. Their lives span the century between the activities of Guillermus Arnaldi, Antonio Andrés, Pedro Tomás, Nicolás Eymerich and Vicente Ferrer, on the one hand, and those of Fernando de Córdoba, Pedro de Castrovol, Andrés Limos and Angel Stanyol, whose works were subsequently printed and circulated more widely.

Ms. B-355 (olim 3) (14th/15th c.; this manuscript is preserved in incomplete form, but the commentaries on 7 of the 12 treatises are extant, as far as *De relativis*, omitting the tract *De fallaciis*; the manuscript also contains Thomas Aquinas' *De fallaciis* plus a treatise on *De proprietatibus terminorum*, which would seem to replace the treatises of the *Tractatus* on which no commentaries are offered, even though they are written in a different hand from the commentary '*Ut ait Philosophus*'); Salamanca, *Biblioteca Universitaria*, Ms. 1882, i) ff. 104v-106r (15th c., on the initial fragment of the first treatise as far as '*nomen* [...]'- and ii) ff. 106v-115r (15th c., on the first treatise). See De Rijk (1970), 41-49.

4) Though associated mainly with British figures (Burley, Heytesbury, Billingham, Strode, etc.), the '*logica modernorum*' seems to have reached Spain via Italy.

2. Logic in Salamanca in the Fifteenth Century

My purpose in this paper is to examine the *Suppositiones* attributed to Master Franquera, in the framework of the logical literature of Salamanca in the mid-fifteenth century, where three names seem to dominate the scene: Master Franquera, García de Castello and Cerdo. These three authors seem to be closely connected with each other and to depend on the teaching of Master Bartholomew. We should perhaps also see them as related in some way to Pedro Fajardo, Juan de Pastrana, Guillermo de Osma and Juan de Santa Cruz, as well as to some anonymous treatises.

a) *Juan de Pastrana*

Of all these names, the only one that is not completely unfamiliar to us today is Juan de Pastrana, the author of a popular *Compendium grammaticae* ('*Thesaurus pauperum sive Speculum puerorum*') that was printed several times and to which Nebrija refers in his *Diccionario*, the reason for Pastrana's fame.[5] Salamanca, *Biblioteca Universitaria*, Ms. 1882, ff. 1r-36v, contains an incomplete *Logica* attributed to Pastrana. A work in the genre of the *summulae*, the *Logica*, includes brief sections on terms, '*Suppositiones*'[6] and '*Consequentiae*', as well as a treatise *De obligationibus*.

We have no precise information about Juan de Pastrana's biography and academic activity. One manuscript copy of Pastrana's *Compendium grammaticae* is dated 1462,[7] and Pastrana and this work are mentioned in the '*Libro de*

---

5) Antonio de Nebrija, *Diccionario latino-español* (Salamanca, 1492), f. 2ra: 'Et quod ex universa prope modum Hispania Alexandros, Petros Helias, et adhuc nomina duriora Galterios, Evrardos, Pastranas et nescio quos indignos qui nominentur grammatistas ac litteratores funditus eradicavi'. The *Compendium grammaticae* is preserved in Salamanca, *Biblioteca Universitaria*, Ms. 2107; Madrid, *Biblioteca Nacional*, Ms. 9748; and Sevilla, *Biblioteca Capitular y Colombina*, Ms. 7-2-22. It was printed several times in the 15th and 16th centuries: Salamanca, c. 1485; Tolosa, c. 1492; Lisboa, 1497; Valencia, 1533; Mallorca, 1545; Mallorca, 1554; Mallorca, 1559; and Barcelona, 1578. A facsimile of the edition c. 1485 is now available: *Grammatica latina, Juan de Pastrana. Materies grammaticae, Ferdinandus Nepos*, Servicio de Publicacións e Intercambio Científico, Universidad de Santiago, Santiago de Compostela (2001), as well as a critical edition: Codoñer (2000). See also Casas Homs (1949), 1-16, and Lozano Guillén (1998), 344-354.

6) The division of *suppositio* into species which we find in Pastrana's *Logica* is atypical, which makes it hard to determine its possible relationship with works by other authors. The arrangement of the divisions as '*singularis/communis*', '*materialis/formalis*', '*propria/impropria*' is anomalous, and the division of '*suppositio communis*' into '*indistributa/distributa*' is also unusual.

7) Madrid, *Biblioteca Nacional de España*, Ms. 9748: 'Explicit compendium grammaticae breve et utile sive tractatus In-/-titulatus thesaurus pauperum sive expeculum (*sic*) puerorum edi-/

Claustros' of the University of Salamanca in 1467.[8] More useful for dating purposes is Toledo, *Biblioteca Capitular*, Ms. 94-27, which gathers together various anonymous treatises on which Master Franquera's *Suppositiones* seem to rely, as well as what appears to be the oldest copy of Franquera's treatise *Terminus <est> in quem*. Master Franquera seems to have written these works in the 1440s, so the anonymous treatises in the manuscript must have been written earlier. One of these anonymous texts, the *Notabilia* on Billingham's *Consequentiae* on ff. 75r-90v,[9] refers to Pastrana's *Logica*, giving us a *terminus ante quem* the latter work, which must date between the late fourteenth century and ca. 1430.

In the '*Accesus*' that accompanies the incunabular editions of Pastrana's *Compendium grammaticae* he is mentioned as '*Sacrarum Literarum magister*' and a member of the Dominican Order.[10] Although his name might appear to indicate that he was from Castile,[11] various authors have regarded him as being an early fifteenth-century scholar from Mallorca, although I do not know what their justification for this might be.[12] The printings of Pastrana's *Compendium grammaticae* indicate that it was well received in Mallorca and, in general, on the Mediterranean coast;[13] the work also contains the example '*vado Barchinone*'. The *Logica* seems to show the influence of Lull's thought,[14] which might

-tum a devoto Johanne de pastrana. / Laus tibi christe liber explicit iste. Qui fuit perfectus anno / domini millesimo ccccºlxºiiº fernandus perfecit immaculata christi / virgo maria oret semper pro eo. Amen'.

8) *Libro de Claustros* of the University of Salamanca, 29-VIII-1467 (1, 212-213).

9) Edited in Weber (2003), 79-125.

10) Juan de Pastrana, *Compendium grammaticae*, ed. c. 1485, aiiir; Codoñer (2000), 77: 'Efficiens ille est qui opus conficit, ut fuit hic Johannes de Pastrana, Sacrarum Literarum magister ac professor et, ut quidam dicunt, Ordinis Predicatorum frater, qui videns scolasticos per septennos decennosque annos <...>, placuit dolori pauperum hoc breve ac utile compendium conficere, quo deo duce duobus in annis, unoque si velint elaborare, in arte grammatica optime valeant erudiri'. The expression '*ut quidam dicunt*' seems to indicate that some time has elapsed, perhaps in addition to a shade of doubt.

11) 'Pastrana' is the name of a town in the province of Guadalajara, close to the border with the province of Madrid. This is the best known and most important 'Pastrana', but not the only one. There is also a 'Pastrana' in the province of Murcia near the border with the province of Almería (and therefore near the Mediterranean), and there may well be others, so the name 'Juan de Pastrana' does not enable us to determine his origin with any degree of certainty.

12) Bover (1976), 65; *Gran Enciclopedia Catalana*, vol. 11 (1978, reprinted 1981), 355a; *Gran Enciclopedia de Mallorca*, vol. 13 (1989), 15b-c.

13) See notes 5 and 11.

14) Juan de Pastrana, *Logica*, Salamanca, *Biblioteca Universitaria*, Ms. 1882, f. 4v: 'Ut autem planius distinctionis huiusmodi terminorum / notitia possit haberi, subiungitur exemplaris figura que /

be consistent with the notion that he came from Mallorca. Nevertheless, Juan de Pastrana seems also to have exercised an equally powerful influence in Portugal.[15] I know of no document which can enable us to situate Juan de Pastrana more precisely in space and time.[16] Although his *Compendium* was used and his *Logica* had some influence in Salamanca, he does not seem to have been chiefly active in Castile, but rather in the Mediterranean area.

b) *Master Bartholomew*

If Juan de Pastrana is the best known author in this group, Master Bartholomew is probably the most influential. We know the name 'Master Bartholomew' from the family of commentaries on Peter of Spain's *Tractatus*, '*Ut ait Philosophus in primo Posteriorum* [. . .]', which De Rijk called '*Glose Salamantine*'. Various copies or versions of these are extant, some of which are thought to date from the second half of the fourteenth century, others from the first half of the fifteenth century.[17] As De Rijk pointed out, on the grounds of the way in which the example of 'example' in the *Tractatus* is phrased ('*Zamorenses contra Taurenses pugnare malum est; ergo Salamantinos contra Albenses pugnare malum est*'), it would seem to be beyond question that this commentary is linked to

---

Arbor Logicalis dicitur. In qua primo ponuntur septem transcen-/-dentia velut [strips] <stirps> et radices essendi. [Strips] *<Stirps> enim dicitur ens, / cui radices convenientes (?) dicuntur 'magnitudo', 'duratio', 'potestas', 'veritas', / 'virtus', 'bonitas' et hiis plereque (?) similes.* Logice tamen indeginis (?) / exordio sufficiunt iam predicte cum suis oppositis convenientibus / cum non ente, tum quia iste sunt potissime per se note, tum quia / obiective transcendentis (?) sub harum rationibus sua [~~posit~~] potius apprehen-/-dunt obiecta, *cum spiritualis potentia (?) rationis per se principalis(?) sit apprehensiva / entis ut ens, et per memoriam ut durans, et per intellectum (?) ut verum (?) et per / voluntatem ut bonum, disponente vero magnitudine ipsum (?) durans / ad esse recolibile, et potestate verum ad esse intelligibilem, virtute / bonum ad esse volibile.* Sub transcendentibus ponitur substantia, sub qua / ponitur substantia corporea. Sub substantia corporea substantia elementa-/-lis, sub substantia elementali ponitur substantia vegetativa. Sub substantia / vegetativa substantia sensitiva, sub qua ponitur homo. Sub / homine ponitur ego, tu, illa'. See also f. 6r.

15) See Codoñer (2000), 13, n. 2; Verdelho (1995), 89-122.

16) In his *Logica*, in the example of the *suppositio impropria*, Pastrana adapts the paradigmatic example '*Anglia pugnat*' in the form '*Gallee pugnat*', but I do not know how this adaptation should be interpreted. On the other hand, in 1474 we have evidence of one Juan Ruiz de Pastrana, linked to the chapter of Toledo Cathedral, but there is no reason to think that he is the same person as the author of the *Compendium Grammaticae*, who would seem to have belonged to a previous generation. See Lop Otín (2003), 492.

17) See note 3.

the world of Salamanca.[18] But the attribution to Bartholomew is problematic, since his opinion is cited in at least two of the versions of these '*Glose*':

> Modo dicit magister quod ista ars quam tradidit a<u>ctor, satis bona est <ad> reducendum entimema ad sillogismum; [cum] <tamen> *magister Bartolomeus* hic ponit aliam, et dicit ipse quod eadem est cum ista quam ponit a<u>ctor, nisi quod est magis declarata. Et dicit ipse quod in entimemate quandoque debet supleri maior, quandoque minor. Unde si tu volueris sci[e]re quando supletur maior et quando minor, debes notare quod ille terminus vel illa extremitas que sumitur bis, aut subicitur in utraque aut predicatur in utraque. Si subicitur in utraque, tunc suplenda est maior, verbi gratia 'omnis homo currit, ergo omnis homo movetur', et hic idem terminus subicitur in utraque (quia 'homo'), ergo debet supleri maior; et supletur sic: 'omne illud quod currit movetur et omnis homo currit, ergo omnis homo movetur'; modo sillogismus est perfectus. Si predicatur in utraque, tunc suplenda est minor, verbi gratia 'omne animal currit, ergo omnis homo currit'; hic idem terminus predicatur in utraque (quia 'currit'), ergo debet supleri minor sic: 'omnis homo est animal', et facias sillogismum perfectum sic: 'omne animal currit, omnis homo est animal, ergo omnis homo currit'.

This reference to '*magister Bartolomeus*' is found in Segovia, *Archivo de la Catedral*, Ms. B-355 (olim 33), and in the extant Sevilla, *Biblioteca Capitular y Colombina*, Ms. 7-7-7, thought to be from the second half of the fourteenth century (?). The *explicit* of the Seville manuscript provides the basis for attributing these '*Glose Salamantine*' to Master Bartholomew: '*Et in hoc terminatur lectio, et per consequens / totus liber Bartholomey supra primam / partem magistri Petri Ispani de ordine / Predicatorum. Deo gracias. Amen*'.[19] Thus, on the basis of the Seville *explicit* we can conclude that Master Bartholomew was actually the author of a commentary on Peter of Spain's *Tractatus*, but because of the reference to him in the third person in the passage quoted above from the same Seville codex and the Segovia witness, the commentary preserved in these two manuscripts cannot be comfortably ascribed to him. As De Rijk pointed out, the various manuscripts offer different versions of these '*Glose Salamantine*', which raises the question as to whether they are different versions of the work of a single author, in which case the '*Ut ait Philosophus* [...]' would not be Master Bartholomew's commentary, or whether they are the work of several

18) The analysis of the notion of method, as De Rijk also indicated, also brings out this bond with the Salamanca area: '*et stricte est via brevis, ut / si quis vult ire Zamoram, illa via que est re-/-ctior et compendiosior dicitur esse methodus*' (f. 1vb).

19) I have not had the opportunity to consult the manuscript Paris, *Bibliothèque nationale de France*, Ms. lat. 6433; in the other manuscripts, the commentary does not reach as far as the fifth treatise of the *Tractatus*.

authors, which would mean that Master Bartholomew might be the author of one of the versions.

The rule that this commentary attributes to Master Bartholomew is also found, without any attribution to him, in various commentaries on the *Tractatus*, for example, in the commentary '*Ut vult Philosophus in tertio De anima*', attributed to Guillermus Arnaldi (Madrid, *Biblioteca Nacional*, Ms. 1070, f. 42ra), or in the anonymous commentary '*Viderunt eam filie Syon*' (Salamanca, *Biblioteca Universitaria*, Ms. 2080, f. 105rb). This rule therefore does not provide us with a reliable way of identifying Bartholomew's commentary. The adaptation of the example of 'example' from the *Tractatus* shows that the first of these commentaries was written in the south of France ('*ut si dicam narbasonenses contra carcasonen-/-ses pugnare malum est, ergo tolosanos contra aux-/-itanos pugnare malum est*' [f. 42rb]) and the second in an area of Spain which is difficult to locate precisely ('*ut burgenses contra toletanos pugnare malum est, ergo ilerdenses contra barchi-/-nonenses pugnare malum est*' [f. 105va]).[20]

An examination of the seven manuscripts in which the commentary '*Ut ait Philosophus*' is preserved, enables us to determine, first, that the *incipit* of two of these is not, in a strict sense, '*Ut ait Philosophus*', but '*Ut ait Aristotiles*' (Toledo 94-27 and Salamanca 1882), and, second, that only three of these (the three of the family '*Ut ait Philosophus*') include the commentary on the fifth treatise of the *Tractatus*, in which enthymeme and example are discussed and the reference to Master Bartholomew is included, pointing to his link with the area of Salamanca (Sevilla 7-7-7, Paris 6433, and Segovia B-293). It is therefore not tenable to attribute to Master Bartholomew the commentary included in one of these three manuscripts. In turn, this casts doubt both on the idea that Master Bartholomew wrote a commentary on the twelve treatises of the *Tractatus* (although it does not rule this out) and that he had links with Salamanca.

The *explicit* of the Seville manuscript ('*Et in hoc terminatur lectio, et per consequens / totus liber Bartholomey supra primam / partem magistri Petri Ispani*') would seem to indicate that Bartholomew only commented on the first treatises of the *Tractatus*, a commentary which one or several other authors later

---

20) Given that the rule attributed to Master Bartholomew can be found both in the commentary '*Ut vult Philosophus*', attributed to Guillermus Arnaldi, and in the anonymous commentary '*Viderunt ea*', we cannot rule out the possibility that this could be Bartholomew's commentary. I consider that there are good reasons to think that this is Bartholomew's commentary, but I shall not argue in favour of this hypothesis here (it would be necessary to analyse the relations between this commentary and the family of commentaries '*Ut ait Philosophus*'; we can say that the first is much more directly subordinate to Peter of Spain's *Tractatus* insofar as the latter already shows the influence of Walter Burley and Albert of Saxony).

completed.[21] Since the Toledo 94-27 ('*Ut ait Aristotiles*') and Seville 7-7-7 ('*Ut ait Philosophus*') manuscripts seem to be the earliest, it might be reasonable to think that the '*Ut ait Aristotiles*' is Bartholomew's commentary on the first part of the *Tractatus*, while the '*Ut ait Philosophus*' is the work of another author, who finished off the partial commentary by Bartholomew. Both commentaries could have met with different fates, which would explain the existence both of copies containing partial commentaries and copies containing commentaries on the twelve treatises of Peter of Spain. The attribution to Bartholomew of the rule concerning the reduction of enthymemes would be unexplained, because, if this hypothesis is true, Bartholomew did not write a commentary on the fifth treatise. However, this attribution may have arisen from his oral teaching rather than any written commentaries, which would again tend to back up the notion that he was active in or around Salamanca.

We know nothing about Bartholomew. The only statement we can make is that he wrote a commentary on the *Tractatus*, probably in the late fourteenth century, and that he may also have been the author of the *Communis tractatus* and the *Tractatus aureus*[22] found in Cambridge, *Corpus Christi College*, Ms. 378. However, we do have reports of two Bartholomews linked in one way or another to logic and the Salamanca environment in the late fourteenth century. One of these was Bartolomé López, '*cler. Hispalen.* (Seville) *studenti in logicalibus de benef. ad collationem archiepo. Hispalen.*' (1381), who reappears a few years later with a degree in law, as a reader at the *studium* of Montpellier, and then as a doctor of laws with a chair in Avignon, and who was given a benefice in Salamanca Cathedral in 1387.[23] The other was Bartolomé Rodríguez de Caracena, '*cler. Seguntin.* (Sigüenza) *dio. studenti in logicalibus, de benef. vel archipresb. in civ. et dio. Usamen.* (Osma)', who was studying in Valladolid in 1394. Bartolomé López's links with Montpellier and Avignon make it likely that he could have imported Peter of Spain's *Tractatus* to Salamanca, as well as the

21) The significant differences between the commentaries in the Seville and Segovia manuscripts lead us to think that the commentaries belonging to the '*Ut ait Philosophus*' family had several authors, all of whom were apparently linked to Salamanca. The way the author of this commentary refers to Master Bartholomew seems to support the hypothesis of multiple authorship: '*auctor*' (Peter of Spain), '*magister Bartolomeus*', '*magister*', and the author of the commentary.

22) Edited in De Rijk (1975), 118-135.

23) Beltrán de Heredia (1967), vol. I, 487: 'Archidiaconatum de Alba in eccl. Salamantin. cum praestim. usque ad valorem ducentorum florenorum de Aragonia, vacantes per obitum, posset conferri Bartholomaeo Lupi, legum doctori, Avinion. legenti, qui obtinet praeb. cum [...]' See also vol. I, 448 and 468, and vol. III, 364. Bartolomé López later held the archdiaconates of Zamora and Jaén.

commentary by Guillermus Arnaldi. Master Bartholomew seems to follow Arnaldi's model on the commentary. However, apart from the fact that the names and dates are the same, there is no further reason for assuming that Master Bartholomew was Bartolomé López. Further research is needed to establish the truth.

c) *Master Franquera, García de Castello and Cerdo*

If we try to reconstruct the teaching of logic in Salamanca in the mid-fifteenth century, the most precise information available is that provided in a false colophon to the *Tractatus contraconsequentiarum* by García de Castello, an autograph copy of which is in Segovia B-283:

> Grates redo humani generis redemptori, qui mihi, cum inmenso / labore, dedit huic operi finem imponere, quod die veneris, / in nocte tertia die mensis septembris, Anno domini milesimo quadracente-/-simo quinquagesimo quarto, ego frater garsias de castello in artibus / bachalarius, ordinis sancti agustini, dum studerem in estudio sa-/-lamantino copilavi et illud originale manu propria scripsi et hoc / agi[or: egi ?] ut valetudo permisit. AMEN (f. 100v).

Thus on 3 September 1454, García de Castello, '*Ordinis Sancti Augustini*', introduces himself to us as '*in artibus bachalarius*', '*in studio salamantino*'. Regarding García de Castello, we only know that he declares himself to be a pupil of Master Franquera (Fancaria or Franquaria)[24] and that he subsequently wrote an untitled treatise of the genre '*Terminus est in quem*', which is also in Segovia B-283, and a series of very short '*Notabilia*' (*Super suppositiones ad veram notitiam terminorum, Super suppositiones de quia ignorantibus, Super ampliationes* and *Super tractatum appellationum*), collected, together with a copy of his *Tractatus contraconsequentiarum*, in Salamanca 1735.

The widespread customs of calling students and masters from Salamanca by their place of origin, Latinizing their Spanish names, and adopting new names on entering religious life hinder our task of finding out more information about the biography and academic career of García de Castello and other figures of this period. Thus, it is only possible to devise hypotheses, for which we possess the following three pieces of information:

---

24) 'lege *suppo-/-<sitio>nes reverendi mei magistri de franquaria*' (Segovia B-283, f. 92r).

i) In Segovia B-293, there is a *Tractatus consequentiarum* attributed to 'Cerdo' or 'Cerdone', which is closely related to, seemingly dependent on, and often coinciding literally with García de Castello's *Tractatus contraconsequentiarum*.

ii) Between 1465 and 1479, in documents concerning the academic life of Salamanca, there are frequent references to Juan de Quintanapalla, first as a bachelor in charge of '*Repetición*' of logic (from which post he resigned on 2 April 1468), and then as bachelor in charge of '*General de lógica*' (from 18 January 1469 to 10 October 1475), figuring as a graduate after 24 September 1477. During these years he worked as a substitute in the chairs of Logic (13 July 1465), Rhetoric (27 July 1468, 18 October 1469) and the Bible (3 August 1468). He was '*Consiliario*' (7 January 1469), '*Síndico*' (between 23 August 1471 and 8 February 1479), '*Visitador*' (10 November 1472), '*Diputado*' (24 May 1473) and 'Vice Rector' (7 December 1476). He aspired to a chair in theology on 15 February 1475 (unsuccessfully), and finally left Salamanca, on 8 February 1479, to become a canon of Segovia Cathedral (where these two manuscripts are kept). In two of the documents he is referred to as Juan de la Cerda de Quintanapalla.[25] Juan de la Cerda de Quintanapalla reappears years later as a graduate in '*Decretos*', canon of Toledo Cathedral (between 1491 and 1495) and '*Arcediano*' of Cuéllar.[26]

iii) In a document dated 2 November 1448, a reference is made to one Juan Rodríguez de Franqueira as '*cler. Tuden.* (Túy) dio. et *magister in artibus et cathedram ordinariam in studio salamantino regens*'. However, I have been unable to obtain any further information about this.[27]

On the basis of this information, it would not seem too hazardous to advance the hypothesis that Juan de la Cerda de Quintanapalla, first master of logic at Salamanca, then canon of Segovia Cathedral and, later, of Toledo Cathedral, could be the 'Cerdo' or 'Cerdone' to whom the *Tractatus consequentiarum* preserved in Segovia B-293 is attributed and who was responsible for taking to

---

[25] See Marcos Rodríguez (1964).

[26] See Lop Otín (2003), 450, 533, 537, 538 and 540.

[27] 'Supplicat s. v. Joannes Roderici de la Franqueira, cler. Tuden. dio. et magister in artibus et cathedram ordinariam in studio Salamantin. regens, quatenus sibi specialem gratiam facientes, de quadam integra portione in eccl. Salamantin. cuius fructus <...>. Dat. Romae quarto nonas novembris anno secundo' (Beltrán de Heredia (1967), vol. III, n. 1101, 30-31). Academic documents in Salamanca refer frequently to various people named 'Juan Rodríguez', but they do not provide precise information that can enable us to identify any of these as Juan Rodríguez de la Franqueira.

Segovia both this manuscript and B-283, which contains the work of the man who may have been his master, García de Castello (and probably also the other manuscripts on logic conserved there). Nor would it seem to be too venturesome to put forward the hypothesis that Juan Rodríguez de Franqueira, '*magister in artibus et cathedram ordinariam in studio salamantino regens*' in 1448, could be identified as Master Franquera. Given that García de Castello calls himself a pupil of Master Franquera and says that he wrote his *Tractatus contraconsequentiarum* in Salamanca in 1454, we can reconstruct this sequence: Juan Rodríguez de Franqueira (c. 1440-1450)—García de Castello (c. 1450-1460)—Juan de la Cerda de Quintanapalla (c. 1465-1475).[28]

I have found no references to García de Castello under this name in any of the extant documents about academic life in Salamanca in this period. We do not know whether 'García' is his first name or his surname, or whether 'Castello' is his surname or his place of origin. We do know, on the one hand, that in 1464 Diego de Navalmarcuende succeeded Fray Pedro de Padilla in the chair of '*Texto viejo de lógica*' (and that he was first substituted, in 1465, and then succeeded, in 1469, by Juan de Quintanapalla) and, on the other, that in the documents of these years there are frequent references to Diego García de Castro, 'Vice Rector' on 15 May 1464 (twelve years before Juan de Quintanapalla), who is not known to have taught logic at Salamanca (but no documents on the academic life of Salamanca are known to survive concerning the years before 1464, when he might well have done so). Diego de Navalmarcuende and Diego García de Castro are found together in some academic documents concerning Salamanca, which means that despite the similarity of their first names, they cannot be the same person. However, given the similarity between 'Castello' and 'Castro', we cannot rule out the possibility that García de Castello might be the same person as Diego García de Castro. Whatever the case, García de Castello may have taught logic in Salamanca around 1454, that is, in the period between Master Franquera, on the one hand, and Fray Pedro de Padilla and Diego de Navalmarcuende, on the other.

As far as our sequence is concerned, Juan Rodríguez de Franqueira (c. 1440-1450)—García de Castello (c. 1450-1460)—Juan de la Cerda de Quintanapalla (c. 1465-1475), we still have to situate i) the author of the *Notabilia*

---

28) During these years, Fr. Pedro de Padilla is known to have held the chair of the '*Logica magna*', which he relinquished on 4 July 1464, when he was succeeded by Diego de Navalmorcuende, who occupied the chair from 3 August 1464 to 13 May 1469, and who was in turn followed by Juan de Quintapalla; at the same time, on 1 May 1464 Martín de Espinosa was given the chair of the '*Summulae*', but no work on logic attributed to any of these three Salamanca professors is known.

about the '*Consequentie*' preserved in Toledo 94-27, which quotes Juan de Pastrana and on which the *Suppositiones* of Master Franquera seem to depend, which means that he must therefore predate Master Franquera,[29] and ii) Guillermo de Osma, the author of the *De Consequentiis* in Gdańsk, *Biblioteka Gdańska Polskiej Akademii Nauk*, Cod. 2181, 68r-71v,[30] about whom we know nothing, but who has to be placed in the same intermediate period between Master Bartholomew and Master Franquera, if not in the period immediately before that. On the other hand, Juan de Santa Cruz, the author of the '*Terminus in quem' tractatus minor*[31] that is found in Roma, *Biblioteca Casanatense*, Ms. 5445, ff. 157r-174r and Oxford, Oxford, *Bodleian Library*, Ms. Misc. lat. E. 100, ff. 64r-69r, though active in Zaragoza and not Salamanca, seems also to have some relationship to Master Franquera, which leads us to regard him as his contemporary, or, perhaps, a contemporary of one of his pupils, García de Castello or Juan de la Cerda. We have no information about Pedro Fajardo, whose commentaries on Aristotle I have not had the opportunity to examine.

3. Master Franquera and His Works

a) *Master Franquera: Juan Rodríguez de Franqueira (?)*

The name 'Master Franquera' has reached us via two manuscripts, Segovia, *Archivo de la Catedral*, Ms. B-293, and Salamanca, *Biblioteca Universitaria*, Ms. 1735,[32] in which two works are attributed to him: a *Tractatus vocatus 'Terminus <est> in quem'*[33] and a *Suppositiones*.[34] In the *Suppositiones*, the author refers

29) Since Juan de Pastrana seems to have worked not in the Salamanca area but in the Mediterranean area, the fact that these '*Notabilia*' quote Pastrana does not allow us to draw the conclusion that they are linked with Salamanca circles. However, the genre of these '*Notabilia*' seems to reflect the way in which the teaching of logic was organized in Salamanca ('*lecture*', '*notabilia*', '*questiones*') and appears to resemble other works of Salamancan origin.

30) Edited in Schupp (1991).

31) Edited in De Rijk (1975), 135-153.

32) See the description of this manuscript in Lilao and Castrillo (2002), 74-76.

33) Segovia, *Archivo de la Catedral*, B-293 (ff. 113v-128v), f. 113v: 'Sequitur tractatus vocatus terminus <est> in quem compositus a vene-/-rabili magistro de franquera salamantice catedram regenti'.

34) Salamanca, *Biblioteca Universitaria*, 1735 (ff. 96r-114v), f. 96r: 'Incipiunt Suppositiones a reverendo magistro de Franquera copilate'; f. 114r: 'Expliciunt supposiciones quas reverendus magister de franquera copilavit ac petrus poncius studens / amantissime oculavit et ora tertie die sabati / valuit patrare et manu propria scripsit'.

to the *Tractatus vocatus 'Terminus <est> in quem'*,[35] and García de Castello refers to both as works by Master Franquera (whose pupil he declares himself to be) in his *Tractatus contraconsequentiarum*,[36] written in Salamanca in 1454. García de Castello also refers to the *Suppositiones* of Master Franquera in his *Notabilia super Ad veram notitiam terminorum*.[37] All of this evidence confirms the attribution we find in these manuscripts and enables us to state that the *Tractatus vocatus 'Terminus <est> in quem'* is earlier than the *Suppositiones*. We have no information about any other work attributed to Master Franquera.

Regarding Master Franquera himself (García de Castello also calls him 'Fancaria' and 'Franquaria'), the only solid information we have (apart from the fact that he wrote these two works) is i) that he had a chair at the University of Salamanca ('*salamantice catedram regenti*'), and ii) that in 1454 García de Castello referred to his works and stated that he had been his pupil ('*reverendi mei magistri de franquaria*'), from which we can deduce that Master Franquera taught at Salamanca and wrote his work before this date, although probably not much earlier.[38]

However, as I indicated above, V. Beltrán de Heredia published a document dated 2 November 1448, in which reference is made to Juan Rodríguez de Franqueira as '*cler. Tuden. dio. et magister in artibus et cathedram ordinariam in studio salamantino regens*'.[39] Juan Rodríguez de Franqueira thus figures as '*magister in artibus*' and '*cathedram ordinariam in studio salamantino regens*' in 1448, that is, in the period in which Master Franquera must have held his chair in Salamanca, which makes the hypothesis that 'Master Franquera' was Juan Rodríguez de Franqueira very likely, if not absolutely certain. ('Franquera' would be his birthplace: 'La Franqueira', a small town in the Galician province of Pontevedra, belonging to the diocese of Tuy.) If we accept this identification, we could say that the *Tractatus* and the *Suppositiones* were probably written in the 1440s.

---

35) *Ibid.*, f. 109v: 'dico quod fallit ista regula in exceptiva, ut videbis in *Termino in quid*'.

36) Segovia, *Archivo de la Catedral*, B-283, f. 71v: 'Item si plura queras de istis expositis et causatis superius dictis, lege *Terminum in quem quam reverendus magister de fancaria* copilavit'; f. 92r: 'Multa relinquo que non posui in litera precedenti propter prolixitatem, sed <si> magis vis scire <de> istis proprietatibus terminorum, lege *Suppo<sitio>nes reverendi mei magistri de franquaria* et ibi ista lacius invenies'.

37) *Ibid.*, f. 77v: 'Nota quot modis dicitur hoc nomen suppositio. Quere *Suppositiones magistri Franquera* et ibi lacte reperies'.

38) In the Segovia B-293 manuscript, in an endnote, it is stated that this codex was purchased in Toledo in April 1453, which means that it must predate this. See note 44.

39) See note 27.

b) *The Works of Master Franquera*

Master Franquera's *Tractatus vocatus Terminus <est> in quem* ('*Nobili efluens*...'), in addition to the copy preserved in the Segovia manuscript,[40] comes down to us in a further three copies, in which no author is named: Toledo, *Biblioteca Capitular*, Ms. 94-27, ff. 101r-122v,[41] Roma, *Biblioteca Casanatense*, Ms. 5445, ff. 60r-103v,[42] and Salamanca, *Biblioteca Universitaria*, Ms. 1735, ff. 114r-116r[43] (only initial fragment). I have not seen the Rome manuscript, but the version in Toledo 94-27 seems to be the oldest.

Master Franquera's *Suppositiones* ('*Exaudiende sunt preces* [...]'; *incipit*: '*Ad cuius tractatus evidentiam pleniorem*...'), besides the copy in Salamanca 1735,

40) *Titulus*: 'Sequitur tractatus vocatus terminus <est> in quem compositus a vene-/-rabili magistro de franquera salamantice catedram regenti' (113v). *Prologus*: 'Nobili efluens a radice latere ex utroque quem / moralium virtutum colegium laudabiliter comen-/-dat pulchriter et decorat exiguum munus quod ego tue sincere / dilecionis radio inflamatus ad tuam procul intellectus caliginem / propulsandam iusta (*sic*) veritates vel falsitates propositionum in logica / pulchriter discernendas solicite copilavi' (113v). *Incipit*: 'Quem tractatum iuvenum (?) / in duas partes divido principales in prima premitam difinitiones et / divisiones multifarias terminorum in secunda vero probaciones propositionum ad-/-iungam contemplacionem eorum. / Quantum ad primum quero quid est terminus' (113v). *Explicit*: 'et sic satis est evidens veritas / eiusdem propositionis. diceres tu eadem est ratio in alia. respondeo [...]' (121v) (the following pages are missing).

41) *Titulus*: 'Incipit terminus in quem, scilicet speculum puerorum' (f. 105r). *Prologus*: '<N>obili efluens a rradice latere ex utroque / colegium quem moralium virtutum / comendat pulchriter et decorat exi-/-guum munus quod ego tue sincere / dilectionis radio inflamattus ad tuam / procul intellectus caliginem propulsandam iuxta veritates et / falsitates propositionum in logica pulchriter discernendas soli-/-cite copilavi' (f. 105r). *Incipit*: 'Quem tractatum iuvenis in duas partes divido principales in prima premitam difinitiones et divisiones multifarias terminorum in secunda vero probaciones propositionum adiungam contemplacionem eorum. Quantum ad primum quero quid est terminus' (f. 105r). *Explicit*: 'Item neutrum oculum habendo tu potes videre equivalet / huic dum neutrum oculum habes tu poteris videre. Et hec / est falsa, ideo prima est falsa. Explicit tractatus' (f. 126v).

42) I have not had the opportunity to consult this manuscript.

43) *Prologus*: 'Nobili efluens a radice latere ex utroque / quem moralium <virtutum> colegium laudabiliter comen-/-dat pulchriter et decorat exiguum munus / quod ego tue sincere dilectionis radio inf(*corr. superl.* l)amatus ad tuam / procul [c] intelec[c]tus caliginem propulsandam iuxta ve-/-ritates et falsitates propositionum in logica pulchriter / discernendas solicite copilavi' (f. 114v). *Incipit*: 'Quem tracta-/-tum iuvenis (?) in duas partes divido principales: in prima / tibi premitam difinitiones et divisiones multifarias terminorum / in secunda <vero> probationes propositioum adiungam contemplationem eorum. / Quantum ad primum quero [?] primo quid est terminus' (f. 114v). *Explicit*: 'si vero talis terminus mediet inter acusativum et infinitivum causatur sensus / divisus ut hominem posibile est currere. Alius est terminus oficialis qui habet oficium privativum [...]' (f. 116r) (the treatise breaks off abruptly).

is preserved in Segovia, *Archivo de la Catedral*, Ms. B-293, ff. 94r-114r,[44] as well as a simplified adaptation in Oxford, *Bodleian Library*, Ms. Misc. lat. E. 100, 91r-102v[45] ('*Ad evidentiam tractatus suppositionum pleniorem*...'), which is a copy adapted to an Aragonese setting (the example of *suppositio impropria*, '*Castella pugnat*', is replaced by '*Aragonia pugnat*'), in which the prologue is omitted, as are the passages explaining the controversies between Ockham and the realists, but which can otherwise be considered identical to the work by Master Franquera.[46] In Segovia B-293, this work is entitled '*Tractatus suppositionum*', and in the prologue the author declares that his intention is to write a '*Tractatus suppositionum terminorum*', which we can use as its name.

Of the three copies that exist of Master Franquera's *Tractatus suppositionum terminorum*, the oldest and most complete is that preserved in Segovia B-293, which was written before 1453. The one in Salamanca 1735, must postdate this, since the *Tractatus contraconsequentiarum* by García de Castello, which we know to have been written in 1454, is also reproduced here. We know nothing about the possible date of the adapted copy in Oxford E. 100. The Segovia copy, on the other hand, after the treatise proper, which is the same as in the other two copies, includes two appendices: one in which eleven rules are formulated

[44] In this manuscript, in the title of the treatise, the author's name is scratched out: '*Sequitur tractatus de suppositionibus magistri ???*' (f. 94r) (perhaps this name was '*Cerdo*', that of the author of the treatise *In Consequentias Berlingani* in the same manuscript, Segovia B-293, ff. 5r-93v). On the other hand, in an end note to this manuscript it says: '*Iste codex est alfonsi de ortega qui furatus fuerat (?) in pa-/-tibulo sui (?) peccatum (?), et emit eum anno domini M. cccc° liii / mensse aprilis toleti conditus in eadem urbe emptus, et / continentur in eo consequentie de cerdo et supposiciones de vene-/-rabili berlingano et terminus in quem de franquera* (add. al. m. *et suppositiones, ampliationes et apelationes*).' This note confirms the attribution of the *Tractatus vocatus 'Terminus <est> in quem'* to Master Franquera, but the *Tractatus suppositionum terminorum* seems to be attributed to Billingham; however, since this treatise debates his teaching, it is clear that this attribution cannot be correct; this treatise is the same as that ascribed to Master Franquera in Salamanca 1735; Billingham is only the author of the brief treatises on the '*suppositiones*', '*ampliationes*' and '*apellationes*' on the first four pages of the manuscript.

[45] See the description of this manuscript in *Bodleian Library Records*, 5 (1956), 332 and in De Rijk (1976), 125-126. Folios 103v-104r, however, do not form part of the *Controversie inter Ocam et Scotum*, but deal with a different '*questio*': '*Utrum suppositio sit possibilis*'.

[46] The same manuscript contains a copy of the treatise *Terminus in quem tractatus minor* (ff. 64r-69r), which in Rome, *Casanatense*, 5445, ff. 157r-174r, is attributed to Johannes de Sancta Cruce, and in which it is stated that the manuscript was written in Zaragoza. We may therefore consider whether this same Johannes de Sancta Cruce, or one of his pupils, may have been the author of this adaptation of Master Franquera's *Suppositiones* to the milieu of Aragon.

'per quas practica colligitur efluens ex principiis sursum traditis huius artis',[47] and another *nota 'ad comprehendendum suppositionem* (sic) *materialem* (sic) *in generali breviter et in suma'*,[48] in which the main ideas of the treatise are summarised. These appendices are not reproduced in either of the other two manuscripts. The Segovia manuscript, however, contains some features (reading mistakes, jumped lines which break the flow of the arguments, which can be rectified by reference to the other two manuscripts)[49] that lead us to think that this is neither the original manuscript of this work nor the one from which the other two are copied. The doubt therefore arises as to whether these two appendices were part of the original text or added by the scribe. The copy in the Salamanca manuscript contains a tree diagram in which the different species of *suppositio* (f. 99r) are set out.[50] This does not appear in the other two copies, but may have been present in the original. My purpose below is to analyse the contents of this treatise.

4. Master Franquera's Tractatus Suppositionum Terminorum

a) *General Remarks*

In his article on logical works attributed to Billingham,[51] Professor De Rijk reported the existence of this work by Master Franquera (preserved in Salamanca 1735 and Segovia B-293), as well as the *Notabilia Ad veram notitiam terminorum* by García de Castello (in the same Salamanca 1735). De Rijk regarded

---

[47] These eleven rules are not the same as those in the body of the treatise; they are all common rules, but I have been unable to identify a source from which they could have been taken with this order and wording. The first six rules are from Billingham's *Suppositiones*, while the last five, about exclusive and exceptive propositions, seem to be a reworking of the *Perutilis logica* by Albert of Saxony.

[48] This 'note' is a mere extract from the definitions of the different species of *suppositio* examined in the main part of the treatise.

[49] For example, in the title to the second appendix, we read '*suppositionem materialem*', where we ought to read '*suppositionum materiam*'; in the last part of the prologue, the Segovia manuscript omits some parts which interfere with the meaning of the phrase: '*sileant detractores invidi et mordaces qui <sagita propria feruntur> (sagita enim non ledit lapidem, sed resiliens sepe vulnerat dirigentem); <nihil enim aliud peragunt detractores quam ut dum suflant in pulverem pulvis in eorum oculis incitetur>*'; in the example of *suppositio simplex* it says '*homo est resibilis*', instead of '*homo est species*', etc.

[50] This diagram would seem to have been inspired by the one illustrating the copy of the '*Suppositiones Quia ignorantibus* [...]' in Toledo 94-27, ff. 91r-92v.

[51] See note 1.

Franquera's work as 'a rather extensive' commentary on Billingham's treatise on the '*Suppositiones*' ('*Ad habendum veram notitiam terminorum . . .*').[52] Nonetheless, although the *Notabilia* by García de Castello are indeed a commentary on Billingham's work, Franquera's work cannot be regarded as such. In formal terms, it does not have the form of a commentary, but rather that of an independent treatise. From a theoretical point of view, even though it takes Billingham's work into account, it offers an analysis of the doctrine of *suppositio* which for the most part differs substantially from his.

From the formal point of view, Franquera follows the structure which is common to all the treatises '*De suppositionibus*': after looking at some previous issues of a general nature, he offers a definition of *suppositio*, presents its divisions into species, and examines each one of these in turn (giving their definitions, distinguishing their subspecies, providing examples, analysing sophisms, devising rules). This examination gives him the opportunity to analyse the virtues of the different species of syncategorematic terms and to explain the different types of descent (which allow us to resolve a proposition according to the type of *suppositio* of its terms). In this perspective, perhaps the most characteristic feature of Franquera's treatise is its focus on the doctrinal controversy between the followers of Ockham and the realists, taking in both the theory of *significatio* and, particularly, the issue of *suppositio simplex*, which leads him to examine various definitions and divisions and to offer a parallel presentation, at least partially, of the doctrines of each side.

Another feature which can be regarded as characteristic of this treatise is the writer's limited understanding (if not total incomprehension), sometimes acknowledged explicitly, of some of the doctrines being explained, which leads him to express his disagreement (or his perplexity), and to declare that he does not teach them to his students because they are unclear.[53] This seems to show that Franquera knew about these doctrines as a reader, through the manuscripts that had reached Salamanca, which would appear to rule out the possibility that Franquera had studied at another university or had heard of the

[52] I know of two commentaries on this tract. A fifteenth-century Spanish master named De Franquera of Salamanca wrote a rather extensive one which is found in Salamanca, *Univ.* 1735, ff. 96r-114r <. . .>. The same commentary is found anonymous in Segovia (Spain), *Cabildo de la Catedral*, vitrina pars media (*sign. antiqua 19-67-82*), 94r-113r. The Salamanca manuscript (Univ. 1735) contains also some *notabilia* from the hand of a *frater Garsia de Castello* (ff. 77v-78v)'. See De Rijk (1976), 127.

[53] 'Opositum sustinens et illum descensum diformiter doceat suos iuvenes, quia non bene illum intelligo'; 'qui autem voluerint partem opositam sustinere, doceant apertissimis documentis quam ego non consuevi meis iuvenibus edocere.'

new developments in these doctrines via some master who was already directly familiar with them. This limited familiarity with the doctrines in question is also apparent in the work of his pupil García de Castello. In this context, the works of Franquera and García de Castello are a great contrast to those of the Spanish logicians of the Mair School, like Pardo, Lax and Encinas, who knew these doctrines thoroughly from their time in Paris, and contributed to their subsequent development.

From the theoretical point of view, Franquera's work can be characterised by his explicit declaration in favour of realism, and his rejection of Ockham's theses. That is, he acknowledges that a *natura communis* exists, as is shown particularly by the way he handles *suppositio simplex*. This rejection of Ockham's doctrines is also shown as far as the theory of *significatio* is concerned. However, leaving the realist theses aside, in the particular characteristics of his explanation of the doctrine of *suppositio*, Franquera's work is a curious mixture of doctrines, some typical of thirteenth-century authors or even earlier ones, and others reflecting fourteenth-century developments. It also contains a strange mixture of realist claims and instruments which, if not nominalist, are at least terminist in their inspiration. In his *Tractatus suppositionum terminorum*, Master Franquera, while holding and defending theses which are characteristic of the realist schools, for the sake of greater clarity, also adopts definitions and rules by authors working within a nominalist framework.

The sources of Master Franquera's *Tractatus* are clear and explicit: Richard Billingham's *Suppositiones* ('*Ad habendum veram notitiam terminorum* [...]'); the *Logica Cantabrigiensis* ('*Quia ignorantibus suppositiones* [...]'); Walter Burley's *De puritate artis logicae*; William of Ockham's *Summa logicae* (though apparently only indirectly); Albert of Saxony's *Perutilis Logica*; Peter of Spain's *Tractatus*, and probably, though not explicitly, the commentary on Peter of Spain's *Tractatus*, '*Ut ait Philosophus primo Posteriorum* [...]' and the *Notabilia* to the '*Suppositiones Quia ignorantibus*' in Toledo 94-27, ff. 92v-99v. It is from Burley, whom he regards as the greatest interpreter of Aristotle ('*Aristotelis maximus insequitor*'), that Franquera receives everything that has to do with the division of *suppositio* into species; and it is from Peter of Spain and the commentary '*Ut ait Philosophus*...' and the *Notabilia* to the '*Suppositiones Quia ignorantibus*' that he takes everything concerning the theory of *significatio* and the relations between *significatio* and *suppositio*. Franquera bases his realist doctrines on his reading of these authors. On the other hand, he adopts Albert of Saxony's definitions, which are simpler and clearer, when they do not specifically affect realist doctrines. Billingham and the '*Logica Cantabrigiensis*' provide the formal model for the work's organisation in some respects. Finally, Franquera owes to Ockham the nominalist contrast to the realist theses which he defends.

b) *Some Preliminary Issues. 'Significatio' and 'suppositio'*

In Billingham's *Suppositiones Ad veram notitiam terminorum*, three initial questions are posed: i) '*quid est suppositio*'; ii) '*quid est supponere*'; and iii) '*ubi termini supponunt et ubi non*'. This is parallel to the '*Logica Cantabrigiensis Quia ignorantibus*', which also begins with three questions: i) '*quid est supponere*'; ii) '*qui termini supponunt et qui non*'; and iii) '*ubi <termini> supponunt et ubi non*'. The point of these questions is the same in both treatises: to restrict *suppositio* to the sphere of the terms of the proposition considered as such: *suppositio* is regarded as a property which belongs to the subject and predicate of a proposition when these are compared to each other within that proposition. In the '*Logica Cantabrigiensis*' it is stressed that the terms have to be complete extremes (not their parts), but in both, *suppositio* is restricted to the terms of the proposition: '*Suppositio est proprietas termini secundum quod unus terminus comparatur ad alium in propositione*' (Billingham); '*Suppositio est proprietas termini secundum quod unum extremum comparatur ad aliud in propositione*' (*Logica Cantabrigiensis*). It is presupposed that terms, considered as parts of the proposition, have signification, but *significatio* is not the object of consideration.

Following this model, but under Burley's influence and, apparently, that of the commentary '*Ut ait Philosophus*', Master Franquera begins his treatise by posing six preliminary questions: i) '*quot modis sumitur hoc nomen suppositio*'; ii) '*quid sit suppositio prout a logico est intenta*'; iii) '*ubi termini supponunt et ubi non*'; iv) '*que differentia est inter significationem, suppositionem, copulationem, ampliationem, appellationem et connotationem*'; v) '*qui termini supponunt et qui non*'; and vi) '*an termini communes pro quibuscumque supponant, illa significent*'. The reason for lengthening the list of questions is to broaden the perspective taken in the earlier treatises in order to tackle the issue of the relations between *suppositio* and *significatio*.

In answer to the first question, Franquera points to four senses of the word '*suppositio*':[54] a physical sense ('*unius rei sub alia positio*'), a grammatical sense

[54] In the commentary '*Ut ait Philosophus* [...]', three of these senses of the word '*suppositio*' are given, and in the *Notabilia* to the '*Suppositiones Quia ignorantibus*' we can find the same four senses of the word '*suppositio*'.

(*'actus verbi supportatio'*), an epistemological sense (*'propositio per se nota que non est demonstrabilis'*), and a logical sense (in which the *'comparatio'* of Burley, Billingham or the *Logica Cantabrigiensis* is completed by the *'acceptio pro aliquo'* of Peter of Spain and Albert of Saxony). The preciseness of this logical sense forms the subject of the second question in which, after his examination of the definitions by Billingham and Burley and using as a basis Burley's distinction between the acceptation of *suppositio* when taken broadly (*'large'*) and its acceptation when taken strictly (*'stricte'*) (according to which *suppositio* belongs to the subject term and not to the predicate), Franquera rescues Peter of Spain's definition in order to link *suppositio* not to subject terms but to substantives, and to provide a definition of *suppositio* in which all the definitions come into play: *'Suppositio stricto modo est proprietas termini substantivi in subiecto positi ad predicatum comparati qui accipitur pro aliquo in propositione'*. Franquera thereby links *suppositio* to the way of signifying which is proper to a substantive in its function as subject (*'per modum per se stantis'*). *Suppositio* is thus understood as a property which belongs to the substantive in its function as subject of the proposition. It does not belong to the adjective or the substantive in its function as predicate, since the predicate has *ratio forme*, and the *ratio forme* is *inherere*. The predicate signifies *per modum inherentis*, and so the *ratio appositi* is more appropriate to it than is the *ratio suppositi*.

Franquera's examination of Peter of Spain's definition (as *'acceptio termini substantivi pro aliquo'*) provides him with an opportunity to look at the distinction between *suppositio naturalis* and *suppositio accidentalis* as types of *suppositio* which may belong to a common term, the first *per se sumptus*, outside the proposition, through a relationship with its *significatio*; the second *in propositione*, through its relationship to the verb to which it is joined. Nevertheless, both the focus on the acceptation of *'suppositio'* taken strictly and the distinction between *suppositio naturalis* and *accidentalis* seem to be meant merely as a historical note. The only result Franquera obtains from this examination of the senses and definitions of *'suppositio'* is his consideration of it as *'acceptio pro aliquo'*, which leads him finally to accept Albert of Saxony's definition of it as *'acceptio seu usus termini cathegorematici pro aliquo vel pro aliquibus in propositione'*. That is, a definition of *suppositio* taken *'large'*, which allows him to analyse the terms by their relationship to 'extra-propositional' entities and which facilitates the subsequent division of *suppositio* by attending to the nature of the entities 'for which the terms stand' in the proposition. *Suppositio* thus belongs to categorematic terms, whether they are adjectives or substantives, and whether they exercise the function of the subject or the predicate.

The analysis of the senses of '*acceptio pro aliquo*' ('*pro voce vel pro proprietate vocis*', '*pro conceptu (ocanice loquendo)*', '*pro re particulari vel pro re universali*'),[55] provides him with the basis for later dividing *suppositio* into species.

Regarding the third question ('*ubi termini supponunt et ubi non*'), Franquera turns once more to Peter of Spain, to indicate that if we consider *suppositio naturalis*, the terms supposit '*extra propositionem*', but if we consider *suppositio accidentalis*, the terms only supposit '*in propositione existentes*'. The fifth question ('*qui termini supponunt et qui non*') has no other purpose than to sharpen the meaning of Albert of Saxony's definition, which restricts *suppositio* to the categorematic terms: *suppositio* belongs to everything that can be the extreme of a proposition, be it a proposition or a term, categorematic or syncategorematic. Supposition does not belong only to categorematic terms proper, but to everything that can be used '*cathegorematice*' as an extreme of a proposition.

However, it is the fourth and sixth questions, in which the relations between *significatio* and *suppositio* (and other *proprietates terminorum*) are examined, which offer the greatest interest. Franquera examines the relations between *significatio* and the *proprietates terminorum* in terms of the sequence *vox-dictio-terminus-subiectum* and within the framework of the hylomorphic theory.[56] A *vox* is the matter which, when formed by a '*primaria impositio*', is constituted as a *dictio*. From this form, which is thought as a '*ratio significandi*', comes the *significatio*, which is considered to be a specific property not of the *vox*, but of the *dictio* insofar as it is already composed.[57] As a result, this composite which is the *dictio* can be used to replace the thing it signifies (*suppositio naturalis*). In particular, it can be used in the place of what it signifies in a proposition (*suppositio accidentalis*), and this confers on the *dictio* (which is now behaving as matter) a new form (a *ratio terminandi*), which constitutes it as a *terminus*. This *ratio* changes according to the function that the term has in the proposition ('*ratio subiiciendi*', '*ratio predicandi*'), constituting it as subject, predicate or copula.[58] *Suppositio* is thus conceived of as the specific property deriving from the use considered as a new form of the *dictio*, and is multiplied according to whether we are thinking about its use in general, its

[55] In the commentary '*Ut ait Philosophus* [...]' we can find an analogous analysis of these three senses of the '*pro aliquo*', but set out with a different order and form: '*pro significato vel pro suppositis significati*', '*pro modo significandi vel pro contentu eius*', '*pro voce vel pro proprietate vocis*'.

[56] Here Franquera reviews the analyses developed by the *modistae*. See Marmo (1994), 109-136.

[57] 'Significatio est proprietas vel operatio dictionis per impositionem vel rationem significandi.'

[58] 'Suppositio est proprietas termini iam constituti ex dictione ut ex parte pro materia et ex ratione terminandi vel subiiciendi vel predicandi pro forma.'

use as part of a proposition, or, in the most specific sense, its use as subject of a proposition. Considered in this strict sense, the predicate should have a different property, *appellatio*, and the copula another, *copulatio*.[59] *Suppositio* would thus be a property of the *terminus* (in the strictest sense, of the subject), which presupposes as an earlier, more basic property, the *significatio* of the *dictio*.

His analysis of the relations between *suppositio* and *significatio* also serves as the occasion for analysing *significatio*, within the framework of the ideas set out by Aristotle in the *Peri hermeneias*, and for distinguishing it from *connotatio*: '*Significare est intellectum constituere*', '*significare est, primo et principaliter, aliquid intellectui representare*'; '*connotare est aliquid secundario designare*'. For his analysis of *significatio*, Franquera contrasts verbs and adjectives to substantives, be they concrete or abstract, and maintains that the substantive signifies '*formam precisam a subiecto*' (or '*non concernendo subiectum cui inheret*'), and that the substantive name '*primo et principaliter*' signifies this form. For its part, the adjective signifies '*formam per modum inherentis substantie*', while the verb signifies an action '*respectu temporis*', but the form or action is what the adjective or verb signify '*primo et principaliter*'. The subject in which the form is inherent, or the time during which the action is performed, is not regarded by Franquera as part of what the *dictio* signifies, but as something that it '*secundario designat*' or '*connotat*'. Franquera uses the 'temporal consignification' of the verb, on the basis of *Peri hermeneias*, to extend this notion and apply it to the adjective and substantive as well, and to maintain that neither the subject in which the form signified by the adjective is inherent, nor the individuals in which the form signified by the substantive are multiplied, are signified '*primo et principaliter*' by the *dictio*, but that they are connoted or designated in a secondary fashion.[60] This is why Franquera defends the notion that a term can supposit for something that it does not signify.

---

59) Master Franquera distinguishes two uses of '*copulatio*', one '*large*', which he takes from Peter of Spain, as '*acceptio termini adiectivi pro aliquo*', and another '*stricto modo*', which he defines as '*relatio quedam qua duo extrema formaliter uniuntur*', which only belongs to the copulative verb. When he examines *copulatio*, Franquera explains the distinction between the use of the verb 'to be' '*de secundo adiacente*', according to which the latter should be considered a categorematic term which predicates '*esse realis existencie*', and its use '*de tertio adiacente*', according to which it is a '*purum sincathegorema*', which predicates being imported by the predicate (either '*esse existencie accidentale*', or '*esse essentiale essentie*'). Such analyses would seem to be based on '*Ut ait Philosophus*', where other uses of the word '*copulatio*' are also considered.

60) 'Pro cuius evidentia est sciendum quod *duplex distinguitur significatum* huius termini 'homo': unum est *primarium et adequatum*, quod est *quid commune et non particulare nec individuum* (teste Philosopho, qui dicit quod 'nomen secunde substantie non significat *hoc aliquid*, id est, individuum et unum numero, sed magis *quale quid*, id est, quid commune'); alia sunt signifi-

To determine the 'first and principal' significate of a *dictio*, Franquera does not pay attention to the intended end of the imposition of the *ratio significandi* which constitutes the *vox* as *dictio*, but to the intellectual intermediary, which is a condition of the imposition. One can only signify what is understood, which is the form, and this is why the form is the primary and principal significate. This form is, at the same time, the *natura communis* and the concept, which is its *similitudo*. Franquera thus tries to reject Ockham's theory of signification, and lays the foundations for his realist analysis of *suppositio*.

The connotation or secondary significate of a *dictio*, however, is not considered by Franquera to be a transferred signification deriving from use ('*ex usu loquendi*'), but something that arises from the primary signification ('*ex sua primaria significatione*'). In the analysis of signification, the intended end of the imposition of a *ratio significandi* on the *vox* thus makes its presence felt. In parallel to this, alongside *significatio* and the issues arising from it, the different levels of matter involved in the formation of the *dictio* and the *terminus* also play a part. For this reason, their presentation of themselves by the *dictio* and the *terminus* is similarly not regarded as a transferred signification, quite apart from the form that they signify.

Nonetheless, despite the apparent subordination of *suppositio* to *significatio* which seems to derive from the foregoing analyses, Franquera gives primacy to *suppositio personalis* over and above any other kind of *suppositio*, which forces him to pay attention not to *significatio*, which comes from the *ratio significandi* imposed on the *vox*, but to the reason or intended end of the *impositio* of this *ratio significandi*, which is the designation of things (not the signification of their forms). He thus introduces the notion of a '*primaria impositio*', on which is based the reason why any term can, in any proposition, have *suppositio personalis*[61] (which seems a strange addition to the theory set out above).

---

cata huius termini 'homo' que *non dicuntur adequata nec prima, nec etiam propria significata*, sed magis *supposita vel connotata*. Significat enim iste terminus 'homo' naturam humanam concretive, concernendo Sortem vel Platonem, vel connotando ista in quibus illa natura communis existit formaliter (humanitas vero significat illam formam abstractive, non concernendo supposita). Unde sicut 'album' significat formam concernendo subiectum in quo subiective illa forma dicitur existere [et] sicut concretum concernendo subiectum, sic 'homo' dicitur concretum concernendo supposita, in quibus non est subiective, sed formaliter et quiditative, et sic Sortes et Plato dicuntur *supposita et connotata*, et non prima significata.'

61) 'Et ratio est quoniam conveniens est semper accipere terminos secundum naturam et rationem formalem; sed ratio formalis dictionis est eius primaria impositio; igitur semper conveniens est accipere terminos secundum suam primariam impositionem; igitur cum dicitur: 'homo est bisilabum', li 'homo', secundum convenientiam illius predicati et secundum usum et communem modum loquendi, eo quod predicatum non potest verificari de subiecto nisi materialiter accepto,

c) *Species of suppositio: 'Suppositio propria' and 'impropria'*

When Franquera analyses the relations between *suppositio* and *significatio* he pays attention to the distinction between *suppositio naturalis* and *accidentalis*, but this distinction is consigned to oblivion when it comes to dividing *suppositio* into species. Here, Franquera reproduces literally the schema of dyadic divisions from Burley's *De puritate artis logicae*. He divides *suppositio*, first into '*propria*' and '*impropria*', and the latter into '*antonomatica*', '*methonomatica*' and '*sinecdochica*'. The *propria* type is divided into '*materialis*' and '*formalis*'; the latter into '*simplex*' and '*personalis*'; the last of these into '*discreta*' and '*communis*'; and this in turn is divided into '*determinata*' and '*confusa*'; '*confusa*' is divided into '*confusa tantum*' and '*confusa et distributiva*'; and the last of these into '*mobilis*' and '*immobilis*'. He not only takes this general division of *suppositio* into species from Burley, but he also borrows the divisions into subspecies: the five types of *suppositio materialis* and, above all, the division of *suppositio simplex* into '*absoluta*' and '*comparata*', and that of the latter into '*generalis*' and '*specialis*'. On the other hand, he rejects Albert of Saxony's division between a descent which is '*uniformis*' and one which is '*diformis*'.

Nonetheless, Franquera distances himself from Burley in some aspects concerning the order of discussion and the definitions of some of the types of *suppositio*. Unlike Burley, Franquera discusses *suppositio impropria* before *propria*, and superposes Burley's handling on that of Billingham. He presents Billingham's general definition of *suppositio impropria* ('*aliter supponit ex usu loquendi quam ex sua primaria significatione*') and that of Burley ('*ex usu loquendi supponit pro illo pro quo de virtute sermonis supponere non permittitur*') as equivalent, using Billingham's type of example ('*Castella pugnat*'), but Burley's divisions, even though he places the *suppositio metonomatica* before the *sinecdochica*.[62] As far as *suppositio propria*, which Billingham does not define, is concerned, Franquera modifies Burley's definition to make it parallel to his

habet suppositionem materialem; sed cum inspicitur virtus et ratio dictionis que est eius impositio primaria, supponit personaliter.' The different manuscripts of Franqueras's *Suppositiones* present an uncomfortable oscillation between '*impositio*' and '*significatio*' in this context, which makes it hard to interpret them exactly. In the treatise, there does not seem to be an explicit, coherent account of the way that *acceptio* and *suppositio* are constructed on *impositio* and *significatio*.

62) Franquera defines *suppositio antonomatica* as that in which the term stands '*pro eo cui maxime convenit per excellentiam*', and adds to Burley's example, '*Apostolus*', the new examples '*Philosophus*', '*Magnus*'. He defines *suppositio metonomatica* as that in which '*continens supponit pro contento*', expands Burley's example, '*cyphum*' into '*bibi cyphum vel lagenam*', and adds the example '*comedi ollam*'. He expands Burley's defininiton of *suppositio sinecdochica* as '*pars*

definition of *suppositio impropria*: the term '*supponit ex usu loquendi pro illo pro quo de virtute sermonis supponere permittitur*' (introducing the clause '*ex usu loquendi*', which no longer signifies anything; if the use of a term is justified '*de virtute sermonis*', there is no point in justifying it '*ex usu loquendi*' as well; its inclusion in this definition seems to have something to do with the problem of the terms which '*ex usu loquendi*' have *suppositio materialis*, but which '*ex sua primaria impositione*' can also be interpreted with *suppositio personalis*).

d) '*Suppositio materialis*'

As far as *suppositio materialis* is concerned, Franquera initially poses five questions: i) '*quid est*'; ii) '*quot modis terminus materialiter dicatur supponere*'; iii) '*quare materialis suppositio apellatur*'; iv) '*quibus modis cognoscatur*'; and v) '*an terminus supponens materialiter possit supponere formaliter*'.

To answer the first question, Franquera examines Billingham's definition ('*non capitur pro illo ad quod fuit impositus ad significandum, sed capitur pro se ipso scripto vel voce prolata*'), which he discards because it is incomplete, as well as that of Burley ('*capitur pro se vel pro alia voce que non est inferior ad eam*'), and ends up by accepting that of Albert of Saxony, which he considers to be equivalent, but which facilitates the subsequent division into species ('*capitur pro se solum vel pro sibi simili vel dissimili, eodem modo vel aliter supponente, cui non fuit impositus ad significandum proprie, nec illud pro quo supponit naturaliter et proprie representat*'). Burley's interest in the different types of language (oral and written) or the different uses of the same term (use of an adjective as a substantive or as an adjective) seems to Franquera to be unnecessarily complex, which is why he prefers to adopt Albert of Saxony's analysis, which is more straightforward, but which shares what he regards as the essential aspects of Burley's explanation (although it omits everything related to mental language which is present in the *Perutilis logica*).

Following Burley, Franquera wants *suppositio materialis* not to be restricted to the *suppositio* of a term for itself (his reason for rejecting Billingham's definition), but he wants it to be extended to any kind of *suppositio* of one term for another. What characterises *suppositio materialis* is not 'that for which the term supposits' (a term) but 'the way in which it supposits for that thing', and this cannot be both '*significative*' and derived from imposition, for in that case there would be *suppositio personalis*.

---

supponit por toto' by adding the final phrase '*vel totum pro parte*'. Like Burley, he illustrates it with the example '*prora est in mari*', and adds the new example '*Petrus est in paradiso*'.

Regarding the species of *suppositio materialis*, Franquera distinguishes the five species which are virtually anticipated in Albert of Saxony's definition, which he illustrates partly using the same examples as this author.[63] The contrast of '*pro se ipsa / pro alia*', which is characteristic of Burley, is replaced by the contrast '*pro se / pro sibi simili / pro sibi dissimili*' from Albert of Saxony; and Burley's restrictive clause '*quae non est inferior ad illam vocem*' is replaced by the clause '*cui non fuit impositus ad significandum proprie, nec illud pro quo supponit naturaliter et proprie representat*'. Franquera seems not to notice that the first part of this (which is also present in Billingham's definition) contradicts his theory of signification, or that the second part refers to mental language, to which he has paid no attention. Nor does he seem to be aware that these five species do not correspond to the five species distinguished by Burley. Burley's fifth, which refers to the term taken in itself '*eo modo quo non potest supponere nec suppositionem habere*'—illustrated by the statement '*albus non potest supponere*', in which '*albus*' is an adjective taken materially as a substantive in order to say of itself, taken as an adjective, that it cannot supposit—, loses its specific character.

As far as the third question, about the reason behind the name '*suppositio materialis*', is concerned, Franquera returns to his hylomorphic analysis of the *dictio*, to indicate that '*suppositio materialis*' is that in which the terms supposit '*pro materia vel parte materiali dictionis*', and '*suppositio formalis*' is that in which the terms stand for what they signify by virtue of the '*ratio significandi*', which is their form; this is why they have their respective names. Franquera appears to forget that it is not only the *vox* which counts as matter but that the *dictio* also counts as matter in relation to the term, as does the term in relation to the subject, predicate, and copula. Accordingly, we can see that '*suppositio materialis*' will apply to several different cases, and '*homo est bisilabum*', '*homo est nomen*', '*homo est subiectum*' illustrate the different levels of matter.

To answer the fourth question, he gives, as was usual, two criteria for recognising *suppositio materialis*: the use of a '*signum materialitatis*' ('*li*', '*hec dictio*', etc.), or the presence in the proposition of a term of second intention, predicated about a term of first intention ('*excluso omni impedimento*'). Franquera

[63] The five species are (1) '*pro se*' ('*homo est bisilabus*'); (2) '*pro sibi simili eodem modo supponente*' ('*homo est terminus prolatus a Sorte*', which Plato said in reference to Socrates's utterance of the first example); (3) '*pro sibi simili aliter supponente*' ('*cuius casus est li homines?*', referring to the statement '*video homines*', the example in which he differs from Albert of Saxony most); (4) '*pro sibi dissimili eodem modo supponente*' ('*de homine predicatur bisilabum*', in reference to the first example); (5) '*pro sibi dissimili aliter supponente*' ('*animal predicatur de homine*', in reference to '*homo est animal*').

uses this opportunity to present a threefold classification of the predicates ('*predicata realia pertinentia ad supposita / predicata intentionalia que competunt rebus universalibus / predicata intentionalia que tantum ad voces pertinent et ad scripta*'), whose purpose is none other than to restrict the scope of predicates of second intention which allow us to recognise the *suppositio materialis* of the subject, in order to leave space for *suppositio simplex*.[64] However, he seems not to notice that the '*predicata realia que competunt rebus universalibus*', required for his realist theses, have been left out.

Regarding the fifth and final question, Franquera admits Burley's rule according to which '*subiectum respectu diversorum predicatorum potest habere diversas suppositiones, tamen respectu cuiuscumque predicati pertinentis ad quamcumque suppositionem semper potest habere suppositionem personalem*'. From this rule one can infer that any proposition in which the subject has either *suppositio materialis* or *suppositio simplex* is equivocal. This is the place where Franquera introduces the notion of '*primaria impositio*', to which I referred above, by virtue of which *suppositio personalis*, in which the term is taken not for what it signifies, but for what it connotes or designates, comes to take primacy over any other kind of *suppositio*.

e) '*Suppositio simplex*'

Neither Billingham, nor Albert of Saxony, nor Peter of Spain pays attention to *suppositio formalis*. Burley does, but without giving a general definition, for he proceeds immediately to consider its species, '*personalis*' and '*simplex*'. Franquera, however, before going on to analyse its species, feels that he must formulate a general 'description' of *suppositio formalis*, which, as in the case of the definition of *suppositio materialis*, anticipates the subsequent division into species: '*suppositio formalis est quando terminus accipitur pro re vel pro rebus quas formaliter, id est, ex sua primaria impositione, representat*'. Franquera turns here to the notion of 'representation' as a genus which encompasses *significatio* (of the form or nature) and *connotatio* (of the subjects of the form or the inferiors of the nature), both of which follow the '*primaria impositio*', either

64) 'Respondetur (meliori iudicio salvo) quod in predicatis secunde intentionis est differentia, quoniam *quedam pertinent ad suppositionem materialem*, ut 'bisilabum', 'trisilabum' <...>, et talia trahunt subiecta ad suppositionem materialem <...>; *alia sunt predicata intentionalia que atribuuntur rebus universalibus prime intentionis et oriuntur a modis essendi repertis in primis intentionibus per actum comparativum intellectus*, et talia trahunt subiecta ad supponendum pro rebus universalibus, ut 'homo est species', 'animal est genus'. Et sic intelligitur regula quod *predicata pertinentia ad voces vel ad scripta, quales sunt partes orationis et sua accidentia, trahunt subiecta secundum usum et <communem modum loquendi> ad suppositionem materialem.*'

because of the imposed *ratio significandi*, or because of the intended end of the imposition. Once the definition has been formulated, however, he goes on immediately to examine each of the species individually.

In his discussion of *suppositio simplex*, Franquera first explains Ockham's doctrine, in a summarised, schematic form, in order to proceed to a more detailed explanation of the 'peripatetic' doctrine following Burley's model. Analysis of Ockham's doctrine is confined to setting out the doctrine of three languages (oral, written and mental), and to formulating the definition of *suppositio simplex* as '*acceptio termini vocalis vel scripti pro simplici conceptu mentis, cui non fuit impositus ad significandum proprie*'. This definition seems to have been taken from Albert of Saxony, although Franquera replaces '*intentio*' with '*conceptus*' (as in the '*Suppositiones Quia ignorantibus*') and introduces some other small modifications. Also following Albert of Saxony, Franquera uses this definition to extract three corollaries: that *suppositio simplex* cannot belong to a mental term (as it would supposit for itself, and would therefore have *suppositio materialis*); that the term must supposit '*pro simplici conceptu mentis*' and not '*pro re*'; and (analogously to the requirements imposed on *suppositio materialis*) that it should not have been imposed to signify this *intentio*.

To explain the doctrine of the realists, Franquera first asks three questions: i) '*quid sit*'; ii) '*quot modis fit*'; and iii) '*quotuplex sit*'. To answer the first, he points to the controversy between 'Ockhamists' and 'realists' concerning the *natura universalis*, which the latter accept and the former reject, which is the reason why they handle the *suppositio simplex* differently, and he then introduces the realist definition by Peter of Spain: '*suppositio simplex est quando terminus communis capitur pro re universali significata per ipsum*', a definition which he reinforces by indicating that the '*res universalis*', unlike the concept, '*est de esse individuorum*'. But once he has committed himself to the realist thesis, since he cannot turn to the Ockhamist Albert of Saxony for help, he instead has recourse to Burley.

The complexity of *De puritate artis logicae*, however, seems to be too much for Franquera, who ends up by expressing his perplexity at Burley's thinking. In *De puritate*, Burley offers us two definitions of *suppositio simplex*, one short, another more complex, in which all its different species are brought together:

i) '*suppositio simplex est quando terminus communis vel singulare agregatum supponit pro eo quod significat*' and
ii) '*suppositio simplex est quando terminus communis supponit pro suo significato primo vel pro omnibus contentis sub suo significato primo vel quando terminus singularis concretus vel terminus singularis compositus supponit pro suo significato totali*'.

Franquera shares Burley's theory of signification, which is why it is not difficult for him to understand and accept that, when the common term stands for the *natura communis*, it is there '*pro suo significato primo*'. Similarly, he seems to understand the problem of the 'singular accidental compounds' ('*Sortes albus*') to which Burley refers, and to allow that, if they are taken '*pro suo significato totali*', they could also have *suppositio simplex* (despite the fact that they are not *naturae communes* or forms, but accidental complexes). But he would seem not to understand either the problem of the '*terminus singularis concretus*',[65] or the problem of the acceptation of the common term '*pro omnibus suis contentis sub suo significato primo*'.

Franquera, however, adopts a definition of *suppositio simplex* which is almost a literal copy of Burley's definition, which he does not understand: '*suppositio simplex est quando terminus communis capitur pro suo significato primo vel pro [conceptis] <contentis> sub significato primo vel quando terminus [simplex] <singularis> aggregatus supponit pro suo significato totali*' (omitting the consideration of the '*terminus singularis concretus*').

Once he has devised this definition, Franquera goes on to answer the second question, that concerning the species of *suppositio simplex*, in which he distinguishes three species or degrees which are virtually contained in the definition, though modified in certain respects: Burley's '*terminus singularis concretus*' (omitted in his definition) is replaced by the '*concretum de genere accidentis*' ('*album*', to which Burley assigned *suppositio personalis*) and the accidental complexes ('*album dulce*', which Burley does not consider and which hint at a problem which is apparently beyond the scope of the theory of *suppositio*), which are assimilated to the '*terminus singularis aggregatus*' (relative to entities *per accidens*). But the complex structure of the *De puritate artis logicae*, in which the explanation of *suppositio simplex* is interrupted by the controversy against Ockham, prevents Franquera from noticing that the explanation of the third species, when the common term is taken '*pro contentis sub significato primo*', is what Burley explains through the different species of the *suppositio simplex comparata*. Franquera thinks that this is a different issue (which accounts for why he separates the question '*quot modis fit*' from the question

---

65) Burley's stance towards singular terms is ambiguous, since the singular term does not signify a nature, but a singular, which has no inferiors, which is why insofar as the singular is its significate, and *suppositio* for the significate is what characterises *suppositio simplex*, it would be expected to have this type of *suppositio*. This is the position which Burley seems at first to be defending: '*Suppositio simplex est quando terminus communis <. . .> vel quando terminus singularis concretus vel <. . .>*'. But since *suppositio* for a singular is what characterises the *suppositio personalis*, Burley seems finally to incline towards assigning the latter type to it.

'*quotuplex sit*'), and is disconcerted when confronted with this third species. He interprets '*pro contentis*' as '*pro suppositis*', which is what defines *suppositio personalis*, and, inexplicably, links this third species of *suppositio simplex* to the problem of quantifying the predicate (the example he provides of this third species is '*homo est omnis homo*'). However, he does acknowledge that he has failed to understand why *suppositio simplex* should be assigned to one term or other of this proposition.[66]

The third question serves no purpose other than to take in the other divisions of *suppositio simplex* which he finds in the *De puritate*. Franquera fails to notice that these are only the continuation and explanation of the previous question. Following Burley, he distinguishes the two *rationes* of the universal, on the one hand, insofar as it is '*aptum natum esse in multis*', and on the other, insofar as it is '*predicabile de multis*'. If the term obeys the first ratio, it has *suppositio simplex absoluta* ('*quando capitur pro significato primo secundum esse quod habet in suppositis*', or '*cum precisione a suppositis*'). If it obeys the second, it has *suppositio simplex respectiva* or *comparativa* ('*quando capitur pro significato primo secundum quod est predicabile vel dicibile de multis*' or '*sub habitudine ad inferiora*'), species which are exemplified, respectively, by the propositions '*homo est dignissima creaturarum*' and '*homo est species*'. But when he looks at the subdivision of the latter into '*generalis*' and '*specialis*', he again comes up against his own failure to understand the meaning of the clause '*pro contentis*', and offers a definition of *suppositio comparata generalis* which contradicts the general definition of the *suppositio comparata:* '*generalis est quando terminus capitur pro significato primo non habendo comparationem ad inferiora*', which is a long way from Burley's definition: '*Quando talis terminus generalis habet suppositionem simplicem generalem, tunc supponit pro suo significato absoluto, ita quod non pro aliquo inferiori*'. It is not a question of whether

---

[66] 'Sed huius tertii modi non bene sentiens fundamentum sub dubio sic relinquo, nam si terminus communis caperetur pro contentis, potius personalis suppositio censeretur quam simplex (ut patet intuenti definitionem doctoris definientis suppositionem personalem). <...> Et si aliquis dicat quod in ista: 'homo est omnis homo', li 'homo' in predicato positus supponeret simpliciter, istud esse non potest, quia predicatum distribuitur; ergo est ibi suppositio personalis; ergo non simplex. Non ergo intelligo quomodo subiectum, nec predicatum supponant simpliciter'. We can find the same interpretation and problem in the commentary '*Ut ait Philosophus*', though it offers an argument which is hard to understand, without the author displaying the least perplexity: 'alio modo quando terminus communis supponit pro suis suppositis, ut 'homo est omnis homo', quod probat sic quidam doctor: 'omne nomen secunde substantie est substantia [substantia] secunda; sed homo est nomen secunde substantie; igitur homo est substantia secunda', sed *nomen est secunda substantia quia est genus, igitur est secunda substantia que est species.*' (Sevilla 7-7-7, f. 116ra).

there is a 'comparison' or not, which the predicate '*genus generalissimum*' obviously requires, but of whether the subject is taken for the inferiors or not. *Suppositio comparata specialis* is defined as follows: '*specialis vero est quando genus generalissimum pro aliquo inferiori capitur quod non est individuum*', which is closer to Burley: '*quando habet suppositionem simplicem specialem, tunc supponit pro speciebus, ita quod non pro individuis*'. These species are exemplified, respectively, by these propositions from *De puritate*, '*substantia est genus generalissimum*' and '*substantia est definibilis*'.

f) '*Suppositio personalis*'

Franquera approaches his analysis of *suppositio personalis* by contrasting Ockham's definition with the realist one. His Ockhamist definition seems to have been inspired by Albert of Saxony's definition: '*acceptio termini pro illo quod ex impositione vel naturaliter representat*', which is neutral concerning the theory of signification (which is explained in relation to the distinction between the three languages, oral, written and mental). The realist definition is a reworking of Burley's.

In the *De puritate*, however, Burley offers various definitions of *suppositio personalis*:

i) Suppositio personalis est quando terminus supponit pro supposito vel suppositis vel <pro> aliquo singulari de quo terminus accidentaliter praedicatur (quod dico pro terminis singularibus aggregatis vel concretis);

ii) Suppositio personalis est quando terminus communis supponit pro suis inferioribus (sive illa inferiora sint singularia sive communia, sive sint res sive voces), vel quando terminus concretus accidentalis vel terminus compositus supponit pro illo de quo accidentaliter predicatur;

and

iii) Suppositio personalis est quando terminus singularis simplex supponit vel terminus communis pro singulari vel pro singularibus, vel terminus communis pro omnibus suis inferioribus copulative vel disiunctive (sive illa inferiora sint singularia sive non).

Obviously, these three definitions are not strictly equivalent. In the first two, there is no reference of any kind to the '*terminus singularis simplex*'[67] ('*Sortes*'), which is explicitly discussed in the third; on the other hand, in the third there is no reference of any kind either to the '*terminus singularis aggregatus*' ('*Sortes albus*') or to the '*terminus concretus accidentalis*' ('*album*'), which are explicitly

---

67) See note 65.

discussed in the first two. Yet, Burley runs into major difficulties when he tries to handle the four types of terms that may have *suppositio personalis* in a unified way, which forces him to introduce a range of notions ('*pro suppositis*', '*pro singularibus*', '*pro inferioribus*') that partly overlap, but none of which can perform all the functions. Singular terms have no inferiors, but common terms may have inferiors that are not singular; and the singulars for which a concrete accidental term or an aggregate singular term stand are neither their inferiors nor their *supposita*.

Franquera tries to synthesise and order Burley's series of definitions into a definition in which virtually all the species are anticipated, so that their subsequent division is easier (leaving out the problem of concrete accidental terms, to which he has already assigned *suppositio simplex*). The definition he proposes is as follows:

> Suppositio personalis est quando terminus singularis [simpliciter] <simplex> supponit pro suo significato simplici, vel terminus singularis aggregatus pro parte significati, vel quando terminus communis capitur pro suis inferioribus, copulative vel disiunctive, copulatim vel disiunctim (sive illa inferiora sint singularia sive non, sive voces sive res).

Franquera constructs his definition on the contrast between the singular (simple or aggregate) term and the common term. He does without the notions of *suppositum* and singular, and in their place, he introduces the notion of significate (in relation to singular terms), contrasted with the notion of inferior (in relation to common terms). He thus distinguishes three forms of *suppositio personalis*.

Franquera goes straight on to divide *suppositio personalis* into '*discreta*' and '*communis*'. He presents Burley's definition of *supposito discreta* (which he enriches by introducing personal pronouns as discrete terms) and, by way of definition of *suppositio communis* (which Burley does not define), the corresponding fragment of the general definition of *suppositio personalis*. Regarding *suppositio discreta*, he confines himself to considering the problem of '*hec herba crescit in horto meo*' ('*falsa de rigore sermonis / vera de bonitate intellectus*').

Franquera also adopts Burley's definitions of *suppositio determinata* ('*quando terminus communis capitur pro suis inferioribus, ita quod licitum est sub tali termino sic supponente ad sua supposita descendere disiunctive*') and *suppositio confusa tantum* ('*quando terminus communis capitur pro suis inferioribus, ita quod infertur ex quolibet eorum et ad nullum eorum contingit fieri descensum copulative nec disiunctive, (sed suple copulatim vel disiunctim)*', but since he does not find a general definition of *suppositio confusa et distributiva* in the *De puritate*, he adopts that of Albert of Saxony ('*quando terminus vel dictio supponit pro illis que ex impositione vel naturaliter representat (sive sint singula generum sive genera singulorum) secundum exigentiam signi, dum tamen sub illo termino distributo licitum sit fieri descensum ad omnia sua contenta copulative*'), although, concerning the species of the latter, *mobilis* ('*quando terminus vel dictio accipitur pro suis inferioribus omnibus et sub termino sic supponente licitum est fieri descensum ad omnia sua contenta copulative*') and *immobilis* ('*quando terminus vel dictio distribuitur pro omnibus suis inferioribus virtute alicuius sincathegorematis habentis naturam mobilitandi et non licet fieri descensum ad sua contenta copulative propter aliquem terminum exponibilem vel resolubilem vel oficialem, antecedentem vel subsequentem talem terminum distributum, ratione cuius termini talis descensus variatur et impeditur*'), Franquera seems to adapt Burley's definitions (though he eschews the subdivision '*absoluta/respectiva*').

In the course of his explanation of the different kinds of *suppositio personalis*, Franquera examines the four forms of descent, and the virtues of the different types of syncategorematic terms, which provides him with an opportunity to analyse some of the sophisms which are *topoi* in such works, and to formulate some rules. Franquera seems to follow Burley's *De puritate artis logicae* and Albert of Saxony's *Perutilis logica* alternately, though his account of *suppositio personalis* includes the '*copulatim*' and '*disiunctim*' forms of descent, which are absent in these two authors' works. Furthermore, when discussing one species of *suppositio* or another, he analyses the examples '*omnis homo differt ab homine / ab omni homine*', '*videns omnem hominem est animal*', '*omnis homo habet caput*', '*promitto tibi pomum*', '*Sortes indiget oculo ad videndum*', '*ter comedisti panem*', '*omnis homo preter Sortem currit*', etc., paying attention to topical issues concerning the position of the syncategorematic term, its categorematic use as part of the extremes, etc.

The most significant point here is his rejection of Peter of Spain's thesis about the *suppositio simplex* of the predicate of universal affirmative propositions (to which he assigns, as was usual in the fourteenth century, *suppositio confusa tantum*),[68] or of the term which follows the exceptive clause (to which

[68] 'Secundo infero falsum dixisse predicatum universalis affirmative habere suppositionem simplicem asserentes, cum tale predicatum habeat suppositionem confusam tantum, et per consequens personalem (que contra simplicem est ex oposito distincta).'

he also assigns *suppositio confusa tantum*),[69] and also the limitation of the value of a term's position through the requirement of dependence (as in Burley, that is, certain syncategorematic terms do not confuse any term that follows them, but only those that depend on them)[70] and the recognition of a certain type of syncategorematic term that has '*virtutem mobilitandi*' but not '*virtutem immobilitandi*', in such a way that, contrary to Albert of Saxony, he assigns the same species of *suppositio* to '*homine*' in the propositions '*omnis homo differt ab homine / omnis homo differt ab omni homine*'.[71] In some cases, Franquera accepts Burley's analysis, as in the case of the numerical syncategorematic terms ('*bis*', '*ter*', etc.), whereas in others he seems to follow that of Albert of Saxony, as in the case of comparative or superlative propositions. Finally, Franquera rejects the *diformis* form of descent which Albert of Saxony uses to resolve terms which have a *suppositio confusa et distributiva immobilis*, because he regards it as contrary to their condition of *immobilis*.[72]

5. By Way of Conclusion

Our exploration of the life and work of Master Franquera has shed some light on logical activity in Salamanca in the first half of the fifteenth century. It has

---

[69] 'Ex quo infero falsum dixisse dicentes quod terminus communis post dictionem exceptivam positus non supponit confuse tantum, sed simpliciter. Quod patet non esse verum, quoniam ista convertitur: 'omne animal preter hominem est brutum' et 'de animalibus tantum homo non est brutus' (per regulam 'exceptiva affirmativa et exclusiva negativa de eisdem terminis, si exceptio et exclusio sint equales, sine dubio convertuntur'). Sed per secundam regulam supra scriptam, in ista: 'inter animalia tantum homo non est brutus', li 'homo' supponit confuse tantum, ergo in alia: 'quodlibet animal preter hominem est brutum', li 'hominem' stat confuse tantum (ex quo talia convertuntur), et per consequens supponit personaliter; ergo non simpliciter (quod erat intent[io]<um>).'

[70] 'Sed si dicatur: 'a lapide differt homo', li 'homo' immobiliter supponit, quia determinate, nam, ut dictum est, iste dictiones confundunt terminos communes subsequentes, si tales sint eorum dependentie determinantes (et quia nominativus non determinat dependentiam istarum dictionum, ideo sequitur intentum).'

[71] 'Ad propositum revertens, dico quod dictio exclusiva non habet naturam immobilitandi. Quod probatur, nam si dicatur: 'tantum animal est homo', ut dictum est li 'homo' mobiliter supponit; sed si dicatur: 'tantum animal est omnis homo', li 'homo' ad idem mobiliter supponit sicut in prima; ergo neutra earum dictionum habet naturam immobilitandi (quod erat intentum).'

[72] 'Sic ergo a significatione probabili vocabuli non [descendens] <devians>, dico *ergo suppositione distributivam mobilem ab immobili esse distinctam*, non penes descensum copulative *uniformiter et diformiter*, sed potius penes descensum et non descensum, quia sub mobili fit descensus copulative simpliciter et absolute, sub immobili vero nullo modo.'

also enabled us to show that, under the title of *Terminus <est> in quem* or *Speculum puerorum*, that is, in the shadow of Billingham's name, contrary to what might be expected, nominalist teaching was not taken up in Spain. Instead, paradoxically, there was a reaction to it in a realist vein, which sought its justification in Burley's work, while also drawing, surprisingly enough, on Albert of Saxony.

Bibliography

Primary Literature

Albertus de Saxonia: Alberto de Sajonia, *Perutilis logica o Lógica muy útil (o utilísima)*, ed. A. Muñoz García (México 1988)

Gualterus Burlaeus, *De Puritate Artis Logicae Tractatus Longior, with a Revised Edition of the Tractatus Brevior*, ed. Ph. Boehner O.F.M. (Franciscan Institute Publications; Louvain-Paderborn 1955)

Walter Burley, *vide s.n.* Gualterus Burlaeus

Secondary Literature

Beltrán de Heredia, V. (1967), *Bulario de la Universidad de Salamanca* (1219-1544), 3 vols. (Salamanca 1967)

Bover, J.M. (1868), *Biblioteca de Escritores Baleares* (1868) (Barcelona-Sueca 1868, facsimile edition 1976)

Casas Homs, J. (1949), 'El '*Thesaurus pauperum*' de Juan de Pastrana. Un manuscrito catalán', in: *Analecta Sacra Tarraconensia* 25 (1949), 1-16

Codoñer, C. (2000), *Gramáticas latinas de transición. Juan de Pastrana y Fernando Nepote* (Salamanca 2000)

Lilao, O., and C. Castrillo. (2002), *Catálogo de Manuscritos de la Biblioteca Universitaria de Salamanca. II Manuscritos 1680-2777* (Salamanca 2002)

Lop Otín, M. J. (2003), *El Cabildo Catedralicio de Toledo en el siglo XV. Aspectos institucionales y sociológicos* (Madrid 2003)

Lozano Guillén, C. (1998), 'El 'Compendium' de Juan de Pastrana, una gramática de la España prerenacentista', in: C. Leonardi, ed., *Gli Umanesimi Medievali.* Atti del II Congresso dell' Internationales Mittellateinerkomitee (Firenze 1998, 344-354)

Marcos Rodríguez, F. (1964), *Extractos de los libros de claustros de la Universidad de Salamanca. Siglo XV (1464-1481)* (Salamanca 1964)

Marmo, C. (1994), *Semiotica e linguaggio nella scolastica. Pariggi, Bologna, Erfurt 1270-133): la semiotica dei Modisti* (Nuovi Studi Storici, 26; Roma 1994)

Rijk, L.M. de (1970), 'On the Genuine Text of Peter of Spain's *Summule*. Part V. Some Anonymous Commentaries on The *Summule* Dating from the Thirteenth Century', in: *Vivarium* 8 (1970), 10-55

—— (1975), 'The Place of Billingham's '*Speculum puerorum*' in 14th and 15th Century Logical Tradition with the Edition of some Alternative Tracts', in: *Studia Mediewistyczne* 16 (1975), 99-153

—— (1976), 'Richard Billingham's Works on Logic', in: *Vivarium* 14 (1976), 121-138

Schupp, F. (1991), Wilhelm von Osma, De Consequentiis. *Über die Folgerungen, textkritisch hrsg., übersetzt, eingeleitet und kommentiert* (Hamburg 1991)

Verdelho, T. (1995), *Las origens da gramaticografia e da lexicografia latino-portuguesa* (Aveiro 1995)

Weber, S. (2003), *Richard Billingham 'De Consequentiis' mit Toledo Kommentar* (Bochumer Studien zur Philosophie; Amsterdam 2003, 79-125)

The Hermeneutical Rehabilitation of Supposition Theory in Seventeenth-Century Protestant Logic

Stephan Meier-Oeser
Freie Universität, Berlin

Abstract

The paper focuses on some aspects of the early modern aftermath of supposition theory within the framework of the protestant logical tradition. Due to the growing influence of Humanism, supposition theory from the third decade of the sixteenth century was *the* object of general neglect and contempt. While in the late sixteenth-century a number of standard textbooks of post-Tridentine scholastic logic reintegrated this doctrine, although in a bowdlerized version, it remained for a century out of the scope of Protestant logic. The situation changed when the Strasburg Lutheran theologian J.C. Dannhauer, who in 1630 developed and propagated the program of a new discipline which he called 'general hermeneutics' (*hermeneutica generalis*), accentuating the importance of supposition theory as an indispensable device for the purpose of textual interpretation. Due to Dannhauer's influence on later developments of hermeneutics, which in the seventeenth and early eighteenth centuries was regarded as a logical discipline, supposition theory is still present in several logical treatises of the eighteenth century. The explication of the underlying views on the notion of supposition and its logico-semantic function may give at least some clues as to how to answer the question of what supposition theory was all about.

Keywords

suppositio, exponibilia, humanism, hermeneutics

An average early-sixteenth-century humanist would probably have opened his conference talk on supposition theory with the remark that this doctrine consists of nothing but 'delirious fictions of dreamers' (*somniantium deliria*) which are the 'plague and palsy of any rational philosophy' (*pestis et lues rationalis philosophiae*),[1] or rather 'a disease the human race was infested with ever since

[1] Johannes Arboreus, *Compendiaria in dialecticen elementa introductio*, ed. Lyons (*s. a.*), 51.

the days of Albert the Great'.[2] The doctrine of supposition as well as all the other more truly grammatical additions to Aristotle's dialectics, called *parva logicalia*, being only an 'immoderate accumulation of precepts and inextricable labyrinths without any utility',[3] should, therefore, not be taught to the youth (*quas non consulo iuvenibus ediscendas*).[4]

Obviously, Humanism has been a most hostile medium for any further cultivation of supposition theory. Hence, at least since the third decade of the sixteenth century it was, as most parts of the *Parva logicalia*, the object of general neglect and contempt.[5] And yet, as E.J. Ashworth has shown in her pioneering study on *The Doctrine of Supposition in the Sixteenth and Seventeenth Centuries*, there are quite a number of logical texts from this period that 'offer a detailed analysis which is neither a slavish nor an inept echo of what the mediaeval logicians had said', so that 'the claim that with the end of the fifteenth century all interest in the more sophisticated doctrines of the mediaeval logicians also came to an end, is clearly without foundation'.[6] This, of course, is true, and it is so for two reasons. First, nothing came to an end with the end of the fifteenth century—except the fifteenth century. The medieval form of logic was continued on a very high level up to the second or even early third decade of the sixteenth century, when the growing institutional success of humanism caused a rapid decline of scholastic logic, which, however, was only a temporary one. Second, the humanist overreaction against some late medieval oversophistications was itself revised when, after an interruption of about five decades, the logical content of the *Summulae* and thus supposition theory, too, became—even though in some sort of light version—reintegrated into the logical writings of the Post-Tridentine Catholic reform.

In what follows, however, I shall focus on another aspect of the early modern aftermath of supposition theory, namely on its fate in the Protestant logical tradition. Here the situation was different insofar as the combination of humanism and reformation had created an intellectual climate that was even more hostile to scholastic logic, so that in this context supposition theory as

2) Augustinus Nifus, *Dialectica ludrica*, ed. Venice (1521), f. 86v-87r.

3) Philippus Melanchthon, *Erotemata dialectices*, ed. Brettschneider (1846), col. 750: 'Addita est Aristotelis Dialecticae, doctrina verius Grammatica quam Dialectica, quam nominarunt Parva logicalia, in qua dum praecepta immodice cumularunt, et labyrinthos inextricabiles, sine aliqua utilitate finxerunt.'

4) Johannes Arboreus, *Compendiaria in dialecticen elementa introducttio*, ed. Lyons (*s.a.*), 48.

5) There are, of course, a few exceptions, such as the *Summulae* of Domingo de Soto (1494-1560) or Chrysostomo Javelli's (ca. 1470-1558) *Logicae Compendium*.

6) Ashworth (1969), 285.

well as the entire doctrinal content of the *parva logicalia* remained excluded from the scope of logic for nearly a whole century.

Philipp Melanchthon's Ambivalent Dismissal of Supposition Theory

Philipp Melanchthon (1497-1560), the important reformer and so called '*praeceptor Germaniae'*, took recourse to supposition theory at the very end of his influential *Erotemata dialectices* (1547). However, his remarks mark as well the very end of supposition theory in the Protestant tradition—at least for quite a long time. For the way in which Melanchthon presented the *doctrina suppositionum* apparently was too ambivalent to stimulate any further engagement with this doctrine. And yet, his short remarks already point in the direction from which a new interest in this theory was to come.

Melanchthon's hostility to scholastic logic, which he considered as the garrulous archnemesis of his own rhetorical ideal of logic, is notorious. Already in his 1518 Wittenberg inaugural address *On the Reform of University Study* and other works of that time he 'substantiates' his criticism of the scholastic 'Sophistry' by referring to the endless quarrels over futile sentences and inane questions such as '*Parisiis et Romae venditur piper*' or '*utrum ad equitandum requiratur equus*',[7] which were, as is well known, commonly used examples to discuss the issue of supposition. Thus, in Melanchthon's view the *parva logicalia*, and supposition theory, in particular, epitomise the scholastic '*nugae*'.

Later in his *Erotemata dialectices* the tone is slightly different, at least as far as supposition theory is concerned. Melanchthon devotes the very last two pages of this work to make some notes on the harshly criticized *parva logicalia*, which he views as a grammatical rather than a logical addition to Aristotelian dialectics, and to review briefly some of their 'useful instructions'.[8] In some sort

[7] Cf. Philippus Melanchthon, *De corrigendis adolescentiae studiis*, ed. Stupperich (1961), 34 ff.: '[...] de dialecticis videamus, quae adhuc ex Tartaretis, Bricot, Perversore [!], Eckiis, [...] et aliis huius farinae petimus. Licet hac parte mihi liberius agere, nam et iisdem ego annos iam sex perpetuos paene detritus sum. [...] [35] Complexorum doctrinae, quas tenebras non offunderunt argutiis? Quales sunt: genus est species, nullus et nemo mordent se in sacco, Parisiis et Romae venditur piper.' Cf. Melanchthon, *Adversus furosum Parisiensium Theologastrorum decretum... pro Luthero Apologia* (1521), f. A3r: 'Vidi Joannis Maioris commentarios [...] in Sententias Longobardicas, quem nunc inter Lutetiae theologos regnare aiunt: Bone Deus, quae plaustra nugarum! Quot paginis disputat, utrum ad equitandum requiratur equus. [...]'

[8] Philippus Melanchthon, *Erotemata dialectices*, ed. Brettschneider (1846), col. 750: 'Addita est Aristotelis Dialecticae, doctrina verius Grammatica quam Dialectica, quam nominarunt Parva logicalia, in qua dum praecepta immodice cumularunt, et labyrinthos inextricabiles, sine

of productive misunderstanding, Melanchthon presents the different types of supposition as '*modi significandi*' and compares them to the diverse gesticulations through which the various affections of the soul such as hate, love or fear are expressed by the human body. Thus, any word, by its composition with other words, takes on, as it were, a special gesture or mode of signifying so that one and the same word may have a more extended or restricted meaning depending on the sentential context.[9] Melanchthon gives some examples: 'If I say 'man is a species', I mean the idea; if I say 'man is running' I mean some individual.'[10] Except for the use of the term 'idea'—the scholastics would have rather spoken of concept—these are traditional examples for *suppositio simplex* and *personalis*. The notion of 'idea', however, which unlike 'concept' implies for Melanchthon the connotation of perfection, becomes important in the following instance: for when I say (with Paul): 'charity is the fulfilment of the law' (*dilectio est impletio legis*), it is, as Melanchthon claims, true only regarding the 'idea' of charity, viz. the 'integer and incontaminated charity' which we human beings do not have due to the corruption of our nature. Thus, the context whence the acceptation of the term '*dilectio*' has to be determined in the case given is not the logical context of the proposition itself, but

aliqua utilitate finxerunt, ut: Nullus et nemo mordent se in sacco, etiam illas admonitiones, quarum aliquis est usus, tenebris involverunt. Recensebo igitur breviter ex his quasdam utiles admonitiones'.

9) Cf. Philippus Melanchthon, *Erotemata dialectices*, ed. Brettschneider (1846), col. 750 : 'Ut gestus dissimiles in homine significant dissimiles animorum motus, alius est irati, alius amantis, alius timentis: ita omnia vocabula diversos quasi gestus in compositione habent, qui vocantur modi significandi. Idem vocabulum alias significat communem speciem, alias individua, alias latius patet, alias angustius est.' The Louvain logician and theologian Augustinus Hunaeus (1521-1577), who tried to revive the doctrine of supposition as a propedeutic to the art of disputation, conceding 'that the manyfold acceptions of the words which usually appear in disputations may be easily known and judged' ('ut multiplex vocabulorum acceptio, quae passim in disputationibus [...] obvia est, facile dignosci et diiudicari queat'; Augustinus Hunaeus, *Logices fundamentum* [...], Antwerp (1568), 7), seems to have taken the end of Melanchthon's *Erotemata dialectices* as a starting point. For he opens his *Logices fundamentum* with the remark that 'just as one and the same person expresses and reveals the different affections of his soul solely through a modification of his facial expression and gesticulation, so one and the same word, too, may denote solely on grounds of a modified rationale of signifying now and then otherwise utmost different things' ('... quemadmodum idem homo sola vultus atque gestus mutatione varias animi sui affectiones prodit atque patefacit: ita quoque unum idemque vocabulum, sola significandi ratione mutata, nunc hanc rem, nunc aliam a priori logissime dissitam nobis designat'; *ibid.*, 1568, 1)

10) Philippus Melanchthon, *Erotemata dialectices*, ed. Brettschneider (1846), col. 750: 'Cum dico, 'Homo est species', intelligo ideam; cum dico, 'Homo currit', intelligo individuum aliquod.'

rather the much larger theological context of the doctrine of original sin and justification.

Melanchthon, however, does not differentiate between quantificational modification and this form of textual interpretation which, even though expounded in terms of supposition theory, seems to transcend the boundaries of its traditional understanding. He rather asseverates: 'this was the doctrine of suppositions, as they called it' (*Haec fuit doctrina suppositionum, ut vocarunt*) and claims that it would be sufficient to teach the youngsters (*iuniores*) to observe the different modes of signifying and to look where species and where individuals are signified.[11] Quantification is concerned in the immediately following examples of biblical sentences like 'exibat ad Baptistam omnis Iudaea', 'omne animal fuit in arca Nohae' and 'decet nos implere omnem iusticiam'.[12] But even though Melanchthon expounds these sentences by using traditional technical terms like 'restrictio', 'exclusive' and 'distributive', he claims that cases like these may be judged from the grammatical figures or figures of speech alone, so that there is obviously no need for the more detailed and obscure precepts of supposition theory. For the collation of the passages and circumstances of a sound text will make evident to anyone who is sufficiently educated and not absolutely moronic (*qui non sunt prorsus hebetes*) whether a

---

11) Philippus Melanchthon, *Erotemata dialectices*, ed. Brettschneider (1846), col. 750: 'Haec animadversio necessaria est, ut cum dico: 'Dilectio est impletio legis'. Id verum est de idea, id est, de integra dilectione, et non contaminata. Nam lex vere haec duo postulat, ardentem amorem Dei, et purum amorem proximi, congruentem voluntati Dei. Sed nos in hac imbecilli natura non habemus integram dilectionem. Nondum igitur nostra dilectio legi satisfecit. Ita si dicam: Eloquentia est facultas de rebus omnibus recte et ornate dicendi. Haec propositio vera est de idea. Sed nullius oratoris eloquentia par est huic ideae. Haec fuit doctrina suppositionum, ut vocarunt. Sed satis est monere iuniores, ut observent diversos modos significandi, ac videant, quando species, quando individua significentur. Actiones enim competunt individuis, ut 'Homo currit.'

12) Philippus Melanchthon, *Erotemata dialectices*, ed. Brettschneider (1846), col. 750 ff.: 'Deinde illa attentio admodum necessaria est, videre, quando latius vagetur vocabulum, quando sit angustius. *Exibat ad Baptistam omnis Iudaea.* Hic particula universalis restringenda est. Non enim significat, omnes homines in Iudaea accessisse ad Baptistam, sed multos passim ex tota Iudaea. Grammatico facilis est explicatio, qui dicit esse synecdochen: nam et Grammatici ad hunc usum figuras tradiderunt, ut aliquo modo ostenderent, diversos modos significandi. Ita si dicas synecdochen esse, *Omne animal fuit in arca Nohae.* Haec brevis explica-[751-tio solvet multas quaestiones, ut Gordium nodum Alexander non longo labore evolvit, sed subito dissecuit. Ex plenitudine eius omnes accepimus, exclusive dictum est. Contrahitur enim universalis ad certam speciem, et significatur aliorum exclusio: Quotquot accepimus gratiam, a solo Christo omnes accepimus. Matth. 3: Sic *decet nos implere omnem iusticiam* [= Matth. 3.15], distributive intelligitur, id est, te tuam, me meam, id est, te decet parere tuae vocationi, me meae.'

certain word has to be understood in a weak or strong, wide or narrow sense.[13] As is clear from the examples given, Melanchthon in his somewhat strange report on the *doctrina suppositionum* is obviously focusing on its function for textual, especially biblical interpretation. But at the same moment that this function is brought out and presented to the reader it is transformed into grammar, with the result that from then on supposition theory was virtually absent in Protestant logic for about 80 years.

That there might have been already in the Middle Ages certain connections between supposition theory and the business of interpreting authoritative theological texts has already been shown in Pinborg's studies[14] and has been suggested in Sten Ebbesen's article on *The Semantics of the Trinity according to Stephen Langton and Andrew Sunesen.*[15] More recently and more explicitly Catarina Dutilh Novaes has claimed that because 'many difficult passages of the bible, or problematic dogmas of the Christian faith, would be clarified by means of the distinction of different kinds of supposition [....] it is arguable that the hermeneutic activity of biblical interpretation would have had significant influence in the development of logical methods.'[16] The historical fate of supposition theory within the context of early modern Protestant logic provides at least some support for such a connection. For it is a matter of fact that here the rehabilitation of supposition theory and the introduction of hermeneutics as a discrete discipline historically coincide in one and the same author.

---

13) Philippus Melanchthon, *Erotemata dialectices*, ed. Brettschneider (1846), col. 751: 'Hoc modo multa iudicantur ex figuris grammaticis, de quibus regulae in parvis logicalibus obscuriores traditae sunt. Afferant igitur studiosi cognitionem figurarum ad Dialecticen, et adiungant rhetorica, et in obscuris dictis quaerant non prodigiosas interpretationes, sed veras, concinnas et congruentes iis materiis, quarum explicationem instituant. Collatis enim membris et circumstantiis in recte scriptis, scientía nativa ab iis, qui non sunt prorsus hebetes, et sunt recte eruditi et amantes veritatis, et in iudicando dexteritatem et candorem adhibent, deprehendi potest et cerni, quae vocabula quasi laxanda sint dilatanda, quae restringenda, quae pars propriis verbis, quae figurate dicta sit. 'Meretrices praecedent vos in regno coelorum', laxandum est, videlicet, quae fuerunt tales, sed conversae sunt ad Deum, et mores emendaverunt. Hanc commonefactionem de illa varietate modorum significandi sufficere hic indico, et adhortor studiosos, ut omnes Dialecticae partes diligenter exerceant, in iudicandis omnibus materiis, quas discunt et in scribendo, et in familiaribus disputationibus.'

14) As early as Pinborg (1968).

15) Ebbesen (1987).

16) Dutilh Novaes (2007) 35. See also Dutilh Novaes (2008).

Johann Conrad Dannhauer's Hermeneutical Rehabilitation of Supposition Theory

The author in question is the Strasburg professor of logic, rhetoric, and theology Johann Conrad Dannhauer (1603-1666), who in 1630 for the first time explicitly propagated the program of a discipline he called 'general hermeneutics' (*hermeneutica generalis*), which was meant to be a logical art providing the adequate methodology for textual interpretation. And it is the very same author who, in this very context, after nearly a century of complete disregard for supposition theory, accentuated its importance for that purpose. While earlier Protestant authors had understood the final remarks of Melanchthon's *Erotemata* in such a way that supposition theory was to be banned from logic, Dannhauer seems to have taken them as a hint to its useful and, as he sees it, necessary function for expounding the sense of what has been said or written by others.

Regarding supposition theory and its relation to hermeneutics, two of his numerous writings are particularly important: the *Idea boni disputatoris*, in which he presents the logical tools that shall enable the students of philosophy and the higher faculties 'to proceed in a methodologically correct way, to dispute well and to emasculate any sophistical argumentation' (recte progredi, bene disputare, omnem [...] Sophisticam emasculare); and closely related to the former, the *Idea boni interpretis*, or, as the telling Baroque title reads: 'Presentation of the good interpreter and the malicious calumniator, that teaches, obscurity having been dispelled, to discern the true from the false sense in all writings and speeches of the authoritative writers, and plainly answers the question: whence do you know that this is the sense and no other?'[17]

Historically seen, this text represents the emergence of hermeneutics as a separate logical discipline.[18] The systematic place within the framework of the Aristotelian *Organon* provided for the new *hermeneutica generalis* is, as one may easily imagine, marked by the two books of *Peri hermeneeias* or *De interpretatione*. With these, however, Dannhauer claims, Aristotle had just commenced hermeneutics without working out its details or finishing it, so that the task of adding a new province to the realm of the Aristotelian *Organon* was left to him.[19] Even though at first sight this systematic allocation seems to

[17] Idea boni interpretis et malitiosi calumniatoris, quae obscuritate dispulsa, verum sensum a falso discernere in omnibus auctorum scriptis ac orationibus docet, et plene respondet ad quaestionem unde scis hunc esse sensum, non alium?

[18] This emergence was, of course, no *creatio ex nihilo*. Cf. Meier-Oeser (2011).

[19] Dannhauer, *Idea boni interpretis et malitiosi calumniatoris* ... (1630 / 2004), 4: 'Hermeneuticam [...] in libris *peri hermeneias* Aristoteles incepit, non tractavit aut absolvit.'

suggest itself quite naturally, it nevertheless rests upon a radical revision of the traditional logical notion of '*interpretatio*'. For all participants in the medieval debates about the adequate understanding of the title of this text agreed that according to Aristotle the term *hermeneia* or *interpretatio* does not denote the act of understanding or expounding the expressions of others but, quite to the contrary, the expression of one's own thoughts. The controversial point was rather the question of whether the term *interpretatio* in Aristotle covers, as Boethius had claimed, the *vox significativa* in general, comprising both the *vox singularis* and the *oratio*, or, as for instance Thomas Aquinas stated, solely the '*oratio enuntiativa, in qua verum et falsum invenitur*', i.e., the *logos apophantikos*.[20] Insofar as in these discussions, if at all, the 'modern' understanding of interpretation had been taken into consideration—which had not been the case before the early sixteenth century—, it was explicitly denied that it was interpretation in this sense Aristotle's text was concerned with.[21]

Dannhauer, however, disapproves of this traditional view, also shared by Ramus,[22] Zabarella[23] and most of his contemporaries. He rather claims that

---

20) Cf. Thomas Aquinas, *Expositio libri peryhermeneias, Prooemium*, ed. Leonina (1989), 5 ff.: '[...] dicitur [6] autem interpretatio, secundum Boethium, vox significatiua que per se aliquid significat, sive sit complexa sive incomplexa [...] Set tamen nomen et uerbum magis interpretationis principia esse uidentur quam interpretationes: ille enim interpretari videtur qui exponit aliquid esse uerum vel falsum; et ideo *sola oratio enuntiatiua, in qua verum et falsum invenitur, interpretatio vocatur.* cetere vero orationes, ut optatiua et imperatiua, magis ordinantur ad exprimendum affectum quam ad interpretandum id quod in intellectu habetur.' (italics mine)

21) Cf. Gerard Frilden, *Exercitium veteris artis*, Rostock (1507), f. v3va: 'Notandum quod [...] interpretatio derivatur a verbo interpretor communis generis quod capitur pro expono[...]: et ita interpretatio proportionabiliter tantum valet sicut expositio. Verum in proposito interpretatio non debet accipi pro quacunque expositione [...]: nec ut vulgo accipitur pro expositione unius idiomatis per aliud aut pro declaratione alicuius obscuri per vocabula manifestiora sed in proposito interpretatio est illud quo mediante exponitur et velut per signum declaratur quod in mente conceptum est [...]'; cf. Bartholomaeus Arnoldi of Usingen, *Exercitium veteris artis*, Erfurt (1514), f. S3r: 'Notandum est Interpretationem capi tripliciter. Scilicet primo pro expositione minus noti per magis notam. [...] Secundo pro expressione conceptus mentalis per vocem significativam, non curando an hoc fiat complexe vel incomplexe et sic non solum enuntiatio sed etiam nomen vel verbum dicitur interpretatio. [...] Tertio pro propositione vocali quae dicitur enuntians illud vocaliter quod sibi correspondet mentaliter. Et ab illa fit inscriptio praesentis libri.'

22) Cf. Petrus Ramus, *Scholae in artes liberales*, Basle (1569), 159: [...] *hermeneia* igitur, symbolum et signum pro arbitrio positum, quo sensa mentis explicamus. [...] Titulus igitur operis, materiam operi propositam ostendit *ten hermeneian*: artem videlicet, qua possumus sensa mentis et cogitata significare alteri et explicare.'

23) Cf. Jacopo Zabarella, *De natura logicae*, Cologne (1597), 62 ff.: '[...] est enim interpretatio vox articulata alicuius rei significatrix, et amplectitur tam simplices terminos, nomen et verbum, quam enunciationem [...][...] quid sit interpretatio tunc declarat [sc. Aristoteles] quando inquit conceptus animi esse imagines rerum, voces autem esse signa conceptuum: per voces enim

the title of *peri hermeneias* was not meant to announce what he calls a 'synthetic' art, teaching how to express one's own thoughts to others, but rather an 'analytic' science providing the right method for the interpretation of what has been said or written by others.[24]

On the basis of this conceptual inversion of the traditional logical notion of interpretation,[25] Dannhauer defines the purpose of hermeneutics (*finis hermeneuticae*) as the 'expounding of any kind of speech and the infallible discrimination between the true and the false sense' (*expositio omnium orationum et veri sensus a falso infallibilis discretio*).[26] Accordingly, the subject matter of hermeneutics is the arbitrary 'doctrinal signs' or, as Dannhauer says, the '*signa* [...] *doctrinalia ad logicam accomodata et supposita pro rebus ipsis*'.[27] But in order to make the *objectum hermeneuticae* precisely explicit this determination still requires further specification. For general hermeneutics is concerned with such signs only insofar as they are obscure (*obscura*) but apt to be expounded, so that its proper object (*objectum formale*) is nothing but the 'oratio quatenus obscura sed exponibilis'.[28]

Thus, already in Dannhauer's definition of the purpose and the proper object of hermeneutics, the terminology of *suppositio* and *exponibilia* is present. This is not by chance, for it turns out that the application of these two scholastic doctrines constitutes a characteristic feature of Dannhauer's concept of hermeneutics. The task of this discipline, as he stresses, is not to decide whether a given sentence is true or false, but rather to establish the true sense

significantur res per medios animi conceptus: talis igitur vox dicitur interpretatio [63], quum sit veluti interpres animi nostri.'

24) Johann Conrad Dannhauer, *Idea boni interpretis* [...], Strasburg (1630 / 2004), 24 ff.: '... certum est non aliud hermeneuticae objectum esse, quam in quo libri Aristotelis *Perihermeneias* sunt occupati: quos ego sic dictos existimo, non ratione synthéseos, quasi doceant sensa mentis oratione exponere, sed ratione *analyseos*, quia tradunt modum interpretandi orationes jam dum ab alio seu voce seu scriptura prolatas.'

25) When, for instance, Rudolph Goclenius in his *Lexicon philosophicum graecum* notes: '*Hermeneia* (interpretatio) in genere est explicatio, explanatio, expositio, sive sit vocis phraseosve, sive sententiae legisve [...]' (Marburg 1615, 56b), he refers to the general and unspecific meaning of 'interpretatio' which, as Johann Philipp Ebel remarks, is an extra-logical use of the term ('extra usum logicum'; *Hermes logicus Diterichianus*, Giessen (1620), 137).

26) Johann Conrad Dannhauer, *Idea boni interpretis* [...], Strasburg (1630 / 2004), 11.

27) *Ibid.* 25: 'Sunt autem objecta Hermeneuticae Aristotelicae, non quaevis signa, sed voluntaria et doctrinalia: [...] doctrinalia quia ad logicam accomodata et *supposita* pro rebus ipsis.' (italics mine)

28) *Ibid.* 28 ff.: 'Haec signa quatenus obscura, sunt obiectum Hermeneuticae, [29] ut ita signum voluntarium ac doctrinale, sit materiale, quatenus obscurum, formale. Paucis, omnis oratio foras prolata quatenus obscura sed *exponibilis* est hujus tractatus objectum.' (italics mine)

even of a false proposition. Admittedly in this regard the aim of hermeneutics seems to be in contrast to what late medieval authors claimed to be the aim of supposition theory, namely that 'suppositio est inventa ad inquirendum veritatem vel falsitatem propositionum [...]'.[29] And yet, these procedures are by no means at cross-purposes. For on the basis of what Quine, Davidson and others have called the 'principle of charity', i.e., the presumption of the author's rationality and coherence—which in fact was already a prominent principle of the hermeneutic tradition of the seventeenth and eighteenth centuries[30]—the analysis of the sense in which the terms of a given proposition have to be understood in order to render it true will be the natural starting point for establishing the author's intention, and both procedures will run parallel as long as there is no evidence for his misapprehension or untruthfulness.

Both supposition theory and the doctrine of exponibles therefore have a certain affinity with the hermeneutical task, and it is precisely in this respect that Dannhauer stresses their methological importance. While supposition theory contributes to disambiguating the meaning of a term by taking into account the propositional context, the doctrine of *exponibilia* provides a device for transforming any non-predicative speech (*oratio non enuntiativa*) into predicative speech (*oratio enuntiativa seu logica*).[31] Dannhauer, severely complaining about the general neglect of the *exponibilium doctrina*, considers this procedure as fundamental to all hermeneutical operations. For it is only with regard to a logical speech (*oratio logica*) consisting of a logical noun and verb that we can infallibly ascertain what is affirmed or denied by it and thus bring out its sense.[32] While Thomas Aquinas claimed that solely the declarative

---

29) Petrus Tartaretus, *Expositio* [...] *in summulas Petri Hispani*, ed. Venice (1514), f. 51v; cf. Bartholmaeus Arnoldi of Usingen, *Summa compendiaria*, 1507: f. 6 r: '[...] finis suppositionis est per eam cognoscere an oratio sit vera vel falsa [...].'

30) It is to be found already in Johann Conrad Dannhauer, *Idea boni interpretis* [...] Strasburg 1630 / 2004, 196: 'Cum [...] unius loci ex eventu duo sunt sensus, ille non est intentus qui invehit absurditatem.' See Scholz (1999).

31) It is understandable that the scholastic doctrine of exponibilia must have attracted Dannhauer's attention, considering that an exponible proposition was defined as a 'proposition that has an obscure sense requiring exposition in virtue of some syncategorema occurring either explicitely or included within some word' ('Propositio exponibilis est propositio habens obscurum sensum expositione indigentem propter aliquod syncategorema explicite positum vel in aliqua dictione inclusum.'). Cf. Kretzmann (1982), 215.

32) Johann Conrad Dannhauer, *Idea boni interpretis* [...], Strasburg (1630 / 2004), 100 ff.: 'Exponibilium doctrinam tanquam scholasticam et intricatam multi repudiant. [...] Sed male et periculose hos agere, demonstrabo 1. ratione. Quicunque de auctoribus vel Doctoribus [auctoris vel Doctoris—ed. 1630] quem vult intelligere, textu tenetur ostendere verum sensum, affirmantem

speech in which truth and falsity are to be found is called 'interpretation' ('*sola oratio enuntiativa, in qua verum et falsum invenitur, interpretatio vocatur*'), Dannhauer holds that only speech of this form can be understood. For to unterstand a sentence is to know what it says. Thus, in order to establish the true sense of any non-declarative sentence it is first of all required to render it truth-sensitive.

In this regard Dannhauer refers to his *Idea boni disputatoris*, where he had treated at some length both supposition theory and the doctrine of *exponibles*.[33] Already in this text he emphazises the utility and necessity of the interpreter's acquaintance with both doctrines. True enough, the late medieval logicians had transgressed, also in Dannhauer's view, the limits of the necessary by elaborating these doctrines into a 'bewildering obscurity' (*perplexa obscuritas*). The later logicians, however, were no less mistaken to let this deter them from any further engagement in these useful doctrines, so that, as he remarks, 'our logicians have written nothing or just very scarcely on this subject matter' (*nostri logici vel nihil vel parcissime hac de re scripserunt*).[34] With talk of 'our logicians' (logici nostri) reference is made not to coevality but rather to confessional affiliation. This becomes evident when Dannhauer, a severe exponent of Lutheran controversial theology, praises the more recent Jesuits (*Jesuitae Juniores*), especially Petrus Fonseca (1528-1599) and Petrus Hurtado de Mendoza (1578-1651), for not losing sight of this subject matter, in spite of their criticism of the '*inculta horrida*' and the '*chaos*' of the late medieval *Summulae*.[35] And they were right in doing so. For directly contradicting Melanchthon's attempt to replace supposition theory with grammar, Dannhauer notes: 'due to the fact that it is well possible that even a person who is sufficiently acquainted with the grammatical rules and the peculiarities of language may err regarding the adequate understanding of the words within the context of

ac negantem, absolutum condicionatum, proprium tropicum etc: is tenetur orationes non logicas convertere in logicas, et quasi ad liquidum redigere. Omnis bonus Interpres haec hujusmodi tenetur praestare [...] Ergo etc.: Major inde probatur, quia de sola oratione logica nomine et verbo constanti, expedite certo ac infallibiliter dicere possumus, affirmet ac neget? et omnem ejus sensum liquidissime exsolvere [exolvere—ed. 1630].'

33) Cf. Johann Conrad Dannhauer, *Idea boni interpretis* [...], Strasburg (1630 / 2004), 101. 118. Cf. Dannhauer (1629 / 2009), 53-65: 'De suppositionis ac descensuum, status et ampliationis praenoscenda ratione'; *ibid.* 66-80: 'De non negligenda exponibilium doctrina.'

34) Dannhauer, *Idea Boni Disputatoris* [...] (1629 / 2009), 53 ff.

35) Cf. Dannhaue, *Idea Boni Disputatoris* [...] (1629 / 2009), 53.

a proposition, it was fruitful and necessary to engage in considerations on the *doctrina de nominum suppositione*'.[36]

Dannhauer exemplifies this claim by presenting a sophism which, as he maintains, could hardly be solved by someone who is not well acquainted with supposition theory:

> Verus Deus neque gignit, neque gignitur, neque procedit.
> Unus in essentia Deus gignit, gignitur et procedit. E.<rgo>
> Unus in essentia Deus non est verus Deus.
> At est verus Deus E.<rgo> neque gignit, neque giginitur, neque procedit.

On the basis of supposition theory, however, the answer is easily at hand: due to a change of supposition there is a '*quaternio terminorum*' underlying that syllogism.[37]

The fact that within the Protestant tradition, after almost a century of neglect, the rehabilitation of supposition theory was closely linked to its function as a logical device for textual interpretation may provide—if only *in obliquo*—since support for the thesis that in the Middle Ages as well the theory of supposition was not only used as a means for detecting the truth-value of propositions and conclusions within the context of disputations, but also as a 'good interpretive device'.

Supposition as Context Meaning

Due to Dannhauer's influence, some of the main features of supposition theory and other parts of the *parva logicalia* became integrated into the hermeneutical part of the Protestant logical tradition of the later seventeenth and eighteenth centuries. This is the case for instance in Johannes Clauberg's (1622-1665) important *Logica vetus et nova*[38] and Christian Weise's (1642-1708)

---

36) Dannhauer, *Idea Boni Disputatoris* [. . .] (1629 / 2009), 54 ff.: 'Nihilominus quia fieri potest, ut in vocum acceptione erret etiam, qui praecepta grammatica, et rationem loquendi ac Idiotismos linguarum sufficienter novit, idque in omnibus disciplinis; non prorsus sane abs re [55] fuerit et fruge, aliquam et quidem quantum necesse est, hac de re considerationem instituisse.'

37) *Ibid.* 55: 'Ex doctrina de suppositione in promtu est responsum dare, subesse scilicet quatuor terminos. Nam in majori verus in essentia Deus vel supponitur simpliciter pro immediato significato quod est essentia, ita major vera est: essentia enim divina praecise sumta nec gignit, nec gignitur, nec procedit; minor vero in eadem suppositione accepta est falsa. Vel vox Dei supponitur personaliter pro mediato significato, videlicet pro personis divinis, hoc sensu, is, qui verus Deus est, nec gignit, nec generatur, nec procedit. ita major falsa est, minor vera.'

38) Johann Clauberg, *Logica vetus et nova quadripartita*, Amsterdam (1691), 852 ff.

Doctrina logica.[39] On account of its hermeneutic relevance in the eighteenth century supposition theory even appeared in some prominent logic texts of the then leading school of Christian Wolff, as for instance the *Systema logicum* of Johann Peter Reusch (1691-1758) where it is praised as a most useful device 'for establishing the ways in which the signification of vocal terms may vary due to their connection with other terms'.[40] In the same vein, Joachim Ehrenfried Pfeiffer (1709-1787) in his *Elementa hermeneuticae universalis* presents the topic of supposition after his treatment of metaphors and other rhetorical figures.[41] Pfeiffer, who also devotes some pages to the *genera exponibilium*,[42] considers supposition theory as a necessary supplement to the distinction of proper and figurative speech insofar as it provides further insight into the varieties of signifying (*significandi varietates*).[43]

Another specimen of this tradition is the *Introductio in artem inveniendi* published in 1742 by Joachim Georg Darjes (1714-1791), as well as the German version which appeared in 1776 under the title *Weg zur Wahrheit*. With Darjes, again noting that 'the doctrine of supposition of words [is] of the utmost importance for the art of interpretation as well as regarding the consequences',[44]

---

39) Christian Weise, *Doctrina logica* (. . .), Leipzig-Frankfurt (1690), 440 ff. Even if Weise does not mention Dannhauer, it is obvious that he was the most important source regarding the part '*De interpretatione in genere*'. The same holds for the additional notes on supposition of Johann Jacob Stübelius (1652-1721) in his edition of Johannes Scharff's (1595-1669) *Manuale Logicum* (1701), 319-322.

40) Johann Peter Reusch, *Systema logicum*, Jena (1732), 266: 'ad perspiciendos modos, quibus vocum significationes per nexum cum aliis variari queant'.

41) Joachim Ehrenfried Pfeiffer, *Elementa hermeneuticae* [. . .], Jena (1743), 108-111.

42) Cf. *ibid.* 132-134.

43) Cf. *ibid.* 108 ff.: 'Etsi interpres assecutus sit, utrum proprie terminus an improprie accipiatur, nondum tamen omnem significandi varietatem ita profligavit, ut nihil supersit reliquiarum. Termini enim communes tam proprii, quam improprii aut ideam abstractam, qua talem denotant, et simpliciter supponuntur, aut applicationem eius ad inferiora sua inferunt, et compositam habent seu personalem suppositionem: Novum hinc oritur examen, cui ut satisfiat, contextus in subsidium vocandus indeque [109] perpendendum est, utrum determinatio inseri potest, quae ideam vocabuli ut abstractam spectari innuat. Nisi id procedat, tanto minus suppositio simplex admitti debeat, quanto luculentius patet, communissimam esse et vulgari usu approbatam, quae eidem opponitur, personalem.' Further distinctions mentioned are: suppositio collectiva—distributiva; suppositio completa—incompleta, and suppositio absoluta—restricta.'

44) Joachim Georg Darjes, *Weg zur Wahrheit*, Frankfurt an der Oder (1776), 159: 'Die Lehre von der Supposition der Worte [ist] sowohl in der Auslegungskunst, als auch bey den Schlüßen von der größten Wichtigkeit.'

supposition theory is basically spelled out in terms of context meaning.[45] So, he remarks that

> In the face of the great variety of possible significates that a term is apt to express the conclusion can be drawn immediately that the meaning of a word in most cases depends on its connection with other words. And it was from here that the doctrine of supposition originated with the Aristotelians, so that supposition is the acceptation of a word for its significate resulting from its connection with other words.[46]

Or, in short: 'A term stands in a certain supposition insofar as its meaning is determined from its connection with certain other terms'.[47] These remarks of Darjes are in a sense paradigmatic of the approach to supposition within the framework of seventeenth- and eighteenth-century hermeneutics and logic. For here, seen on the whole, an increasing amalgamation or fusion—and perhaps in a certain sense also a confusion—of the concepts of supposition and meaning is apparent.

I am far from claiming that the treatment of supposition as it appears in the framework of this tradition provides any substantial contribution to the logical aspects of supposition theory in the proper sense, or that it was on par with the medieval discussions on that issue. And yet, the amalgamation of *significatio* and *suppositio*, as it was reinforced by the hermeneutical approach to supposition, brought out a certain 'collateral gain' with regard to the concept of meaning or signification. For the traditional clear-cut distinction between

---

45) Cf. Joachim Georg Darjes, *Introductio in artem inveniendi* [...], Jena (1742), 257: 'Si dicas: A. 'Voluptatis comes dolor' et B. 'Summum hominis bonum est voluptas', facile intelligimus, quod termini voluptatis significationes per nexum cum aliis varientur. Nexus enim requirit, ut in propositione A in malam, et in propositione B in bonam partem. Terminus itaque voluptatis in diversa hic accipitur suppositione, seu diversae huius termini hic supponuntur significationes.'

46) Cf. *ibid.* 256: 'Praemissa explicatione varietatis significatuum, quos terminus quidam exprimere potest, [...] immediate inferri potest conclusio, quod vocis significatio saepissime a coniunctione cum aliis vocibus dependeat. Et inde oritur Aristotelicorum doctrina de suppositione, quae est acceptio vocis pro significatu ex coniunctione cum aliis quibusdam vocibus determinando. CORROLARIUM: Diversae igitur termini suppositiones ex diverso eorum significatu, quae [quem—ed. 1742] vi coniunctionis cum aliis vocibus exprimi possunt, cognoscendae atque deducendae sunt.' On supposition theory see 256-269.

47) *Ibid.* 267: 'Terminus est in suppositione quadam, quatenus eius significatio ex coniunctione cum aliis quibusdam terminis determinanda.' Cf. Joachim Georg Darjes, *Weg zur Wahrheit*, Frankfurt an der Oder (1776), 159: 'Ein Wort stehet alsdenn in einer Supposition, wenn es sich in einer solchen Verknüpfung befindet, welche die Gränzen der Gedanken bestimmet, die wir, woferne keine Unwahrheit entstehen soll, nicht überschreiten können.'

the signification a word has outside or inside a proposition and the supposition a term adopts within or outside a propositional context seems to have consolidated for centuries the assumption that there was something like a determinate meaning inherent in single words due to their imposition. It is, in a way, quite astonishing that within the long-lasting scholastic '*magna altercatio*' (as Scotus labeled it), i.e., the debate over 'whether words signify thoughts or things', the notion of supposition was virtually absent. For along the lines of supposition theory the question of whether for instance the term '*homo*' signifies real men or rather the mental concept of man could have been easily answered with the remark that this question could not be decided in general—and therefore was put in the wrong way. Whether a term has to be taken in one sense or another simply depends on the propositional context in which it occurs.

As it seems, it was particularly the suspension of that clear-cut distinction between supposition and signification through which the 'contextual' or rather 'propositional approach' of terminist logic[48] could result in a more common insight into the context dependency of meaning. It is, however, questionable whether this insight has to be spelled out by completely reducing the notion of meaning to the functional role of *suppositio*, as is to be seen for instance in Wittgenstein, claiming that words have no meaning save in propositions and that they function only in propositions, like levers in a machine.[49] For what is problematic with the assumption of a determinate extra-propositional meaning is not meaning as such—what happens in my mind by reading a page of a dictionary of language x will differ significantly depending on whether or not I am familiar with x—but the acknowledgement of a *determinate* meaning. Here, I think, a notion of meaning that differentiates between the indeterminate or potential lexical meaning of a word and its context meaning, as it has become commonly accepted within the hermeneutical adaptation of supposition theory, is more adequate. In this vein Christian August Crusius (1715-1775), in the hermeneutical chapter of his logic, distinguishes between a 'grammatical meaning' (*grammatischer Verstand*), i.e., 'the totality of meanings a word may ever have in one language', and a 'logical meaning' (*logischer Verstand*) or the 'totality of what a word can mean at a certain place and in a certain context'.[50] Although, however, the hermeutical approach seems to have reinforced such a

---

48) De Rijk (1967), 123-125; cf. *ibid.* 552.

49) Cf. Ludwig Wittgenstein, *Wittgenstein's Lectures*, ed. King and Lee (1989), 2. Cf. *Tractatus* 3.3: '[. . .] nur im Zusammenhange des Satzes hat ein Name Bedeutung.'

50) Cf. Christian August Crusius, *Wege zur Gewissheit* [. . .], Leipzig (1747), 1080: 'Der grammatische Verstand der Worte bestehet in dem gantzen Inbegriffe der Bedeutungen eines Wortes, welche es nur in einer Sprache haben kan. [. . .] Der logische Verstand der Worte ist der Inbegriff desjenigen, was dieselben an einem gewissen Orte und ein einem gewissen Contexte bezeichnen.'

view, it was by no means its unique source. For a certain tendency to override the distinction of supposition and signification is also to be seen in the late scholastic tradition itself. The Milanese Dominican Giorgio Rovegnatino (fl. ca. 1500), for instance, instead of claiming (as Wittgenstein does) that each term by itself signifies nothing, prefers to say that each term by itself signifies only in an indeterminate way (*significat indeterminate sic aut sic*), so that it requires another term to be determined to this or that mode of signifying. If someone just says 'man' without any further addition, he notes, we may understand that something is signified, but we cannot know whether it is the human nature in general or individual men, whether only one particular or a plurality of individuals, existing or not, or whether the word is only meant to denote itself. For words, being apt to express in many ways what they represent, can only mutually determine themselves to this or that mode of signifying.[51]

Bibliography

Primary Literature

Augustinus Hunaeus (Huens), *Logices fundamentum seu prodidagmata de dialecticis vocum affectionibus et proprietatibus, iuxta laudatissimum celeberrimae Lovaniensis Academiae morem* (Antwerpiae 1568)

Augustinus Nifus, *Dialectica ludicra* (Venitiis 1521)

Bartholomaeus Arnoldi of Usingen, *Exercitium veteris artis* (Erfurdiae 1514)

[51] Georgius Rovegnatinus, *Ex divo Thoma suppositionum collectula*, s.l., s.a., ca. 1504, f. a2v: 'Significant [...] termini per se utique aliquid: at indeterminate sic aut sic. Quare alio indigent ad hunc vel illum significandi modum pertrahente. Nam si proferat aliquis terminum hunc ‚homo' simpliciter nullo addito: intelligimus quidem aliquid significari: at naturamne hominis in communi: an supposita ipsa naturae: unumne singulariter: an plura universaliter: existentia an non existentia: aliudne formaliter: an semetipsum potius materialiter significet: scire minime possumus. Quod si protulerit: "homo est nomen" iam manifeste intelligo: ipsammet vocem materialiter designari. [...] a3r : Sicque de singulis arbitrandum est vocibus: quod repraesentant: pluribus id modis esse aptas natas effingere: si in oratione collocentur: nec per se quamlibet ad hunc vel ad illum determinare se posse significandi modum, determinari autem a se invicem terminos alterum ab altero.' Still more explicit is the identification of supposition and signification in Juan Caramuel y Lobkowitz, *Praecursor logicus complectens Grammaticam audacem*, Frankfurt 1654, 11, who claimed: 'Significatio, suppositio, et acceptio tria sunt nomina, sed una res. Scio a multis distingui, sed eorum subtilitas minime placet. Idem omnino est significare ad placitum, quod poni pro obiecto: Idem poni pro illo, quod pro illo accipi. [...] Dicendum [...] est, omnem omnino vocem eo sensu, quo significat, supponi, et accipi. Omnis enim vox seorsim sumpta habet multas significationes potentia (vox Homo est aequivoca, et potest significare materialiter, formaliter, primo intentionaliter, et secundo intentionaliter etc.), nullam actu: et item pro multis rebus potest supponi, pro multis accipi.'

——, *Summa compendiaria totius logicae* (Basiliae 1507)

Christian August Crusius, *Der Weg zur Gewißheit und Zuverlässigkeit der menschlichen Erkenntniß* (Lipsiae 1747)

Christian Weise, *Doctrina logica, duabus partibus sic comprehensa, ut prior terminorum simplicium, propositionum et syllogismorum notitiam; posterior ipsam praxin, in definitione, divisione, demonstratione, methodo, interpretatione et disputatione necessariam exhibeat: exemplis ut plurimum politicis illustrata, rebusque oratoriis sedulo accomodat* (Lipsiae-Francofurti 1690 ([1]1680)).

Georgius Rovegnatinus, *Ex Divo Thoma suppositionum collectula* (ca. 1504, *sine loco, sine anno*)

Gerard Frilden, *Exercitium veteris artis* (Rostock 1507)

Jacopo Zabarella, *De natura logicae*, in: *Opera logica* (Coloniae [3]1597, reprint Hildesheim 1966)

Joachim Ehrenfried Pfeiffer, *Elementa hermeneuticae universalis veterum atque recentiorum et proprias quasdam praeceptiones complexa* (Ienae 1743)

Joachim Georg Darjes, *Introductio in artem inveniendi seu Logicam theoretico-practicam qua Analytica atque Dialectica* [...] *proponuntur* (Jena 1742)

——, *Weg zur Wahrheit* (Francofurti 1776)

Johann Conrad Dannhauer, *Idea Boni Disputatoris Et Malitiosi Sophistae: Exhibens Artificium, Non Solum Rite Et stratagematice disputandi; sed fontes solutionum aperiens, e quibus quodvis spinosissimum Sophisma dilui possit* (Coloniae Agrippinae 1656 ([1]1629); repr. Hildesheim-Zürich-New York 2009)

Johann Conrad Dannhauer, *Idea boni interpretis et malitiosi calumniatoris, quae obscuritate dispulsa, verum sensum a falso discernere in omnibus auctorum scriptis ac orationibus docet, et plene respondet ad quaestionem unde scis hunc esse sensum, non alium? Omnium Facultatum Studiosis perquam utilis* [...] (Coloniae Agrippinae [4]1652 ([1]1630); repr. Hildesheim-Zürich-New York 2004)

Johann Peter Reusch, *Systema logicum* (Ienae 1732)

Johann Philipp Ebel, *Hermes logicus Diterichianus* (Gissae Chatorum 1620)

Johannes Arboreus, *Compendiaria in dialecticen elementa introductio* (Lyons *sine anno*)

Johannes Clauberg, *Logica vetus et nova quadripartita* (Johannes Clauberg, Opera omnia philosophica, Amsterdam 1654, 1691)

Johannes Scharffius, *Manuale logicum, novis* [...] *annotationibus instructis cura ... studioque M. Joh. Jacobi Stübelii* (Lipsiae-Francofurti 1701)

Juan Caramuel y Lobkowitz, *Praecursor logicus complectens Grammaticam audacem* (Francofurti 1654)

Petrus Ramus, *Scholae in artes liberales* (Basiliae 1569)

Petrus Tartaretus, *Expositio magistri Petri Tartareti in summulas Petri Hispani* (Venice 1514)

Philippus Melanchthon, *Adversus furiosum Parisiensium Theologastrorum decretum ... pro Luthero Apologia* (Basiliae 1521)

——, *De corrigendis adolescentiae studiis*, in: Melanchthon, Studienausgabe, 3, ed. R. Stupperich (Gütersloh 1961)

——, *Erotemata dialectices* ed. C.G. Brettschneider (Philippi Melanchthon Opera quae supersunt omnia, Corpus Reformatorum, 13; Halle 1846)

Rudolph Goclenius, *Lexicon philosophicum graecum* (Marchioburgi 1615)

Thomae de Aquino, *Expositio libri Peryermeneias*, ed. altera retractata, cura et studio Fratrum Praedicatorum (Sancti Thomae de Aquino Opera omnia, 1* 1; Roma-Paris 1989)

Wittgenstein, Ludwig: Wittgenstein's Lectures, Cambridge, 1930-1932, from the Notes of John King and Desmond Lee (Chicago 1989)

Secondary Literature

Ashworth, E.J. (1969), 'The Doctrine of Supposition in the Sixteenth and Seventeenth Centuries', in: *Archiv für Geschichte der Philosophie* 51 (1969) 260-285

Ebbesen, S. (1987), 'The Semantics of the Trinity according to Stephen Langton and Andrew Sunesen', in: J. Jolivet and A. de Libera, eds., J. Jolivet and A. de Libera, eds., *Gilbert de Poitiers et ses contemporains aux origines de la* Logica Modernorum. Actes du septième symposium européen d'histoire de la logique et de la sémantique médiévales, Poitiers, 17-22 Juin 1985 (Napoli 1987, 401-435) (repr. in Ebbesen, 2009)

Dutilh Novaes, C. (2007), *Formalizing Medieval Logical Theories*—Suppositiones, *Consequentiae and* Obligationes (Berlin 2007)

—— (2008), 'An Intensional Interpretation of Ockham's Theory of Supposition', in: *Journal of the History of Philosophy* 46 (2008), 365-393

Kretzmann, N. (1982), *Syncategoremata, exponibilia, sophismata,* in: N. Kretzmann, A. Kenny and J. Pinborg, eds., E. Stump, ass. ed., *The Cambridge History of Later Medieval Philosophy. From the Rediscovery of Aristotle to the Desintegration of Scholasticism 1100-1600* (Cambridge-London-New York-New Rochelle-Melbourne-Sydney 1982, 211-245)

Meier-Oeser, S. (2011), 'Brüche und Kontinuitäten in der Genese der frühneuzeitlichen Hermeneutik', in H. Busche, ed., in collaboration with S. Hessbrügen-Walter, Departures for Modern Europe / Aufbruch in das moderne Europa. A Handbook of Early Modern Philosophy (1400-1700) (Essen 2007); Hamburg 2011, 740-753)

Pinborg, J. (1968), [Review of] L.M. de Rijk, *Logica Modernorum,* 2, in: *Vivarium* 6 (1968), 155—158

Rijk, L.M. de (1962-1967), *Logica Modernorum. A Contribution to the History of Early Terminist Logic.* I. *On the Twelfth Century Theories of Fallacy*; II/1: *The Origin and the Early Development of the Theory of Supposition* II/2: *Texts and Indices* (Assen 1962-1967)

Scholz, O.R. (1999), *Verstehen und Rationalität. Untersuchungen zu den Grundlagen von Hermeneutik und Sprachphilosophie* (Frankfurt am Main 1999)

LOGIC: MEDIEVAL AND MODERN

A Quantified Temporal Logic for Ampliation and Restriction

Sara L. Uckelman
Ruprecht-Karls Universität Heidelberg

Abstract

Temporal logic as a modern discipline is separate from classical logic; it is seen as an addition or expansion of the more basic propositional and predicate logics. This approach is in contrast with logic in the Middle Ages, which was primarily intended as a tool for the analysis of natural language. Because all natural language sentences have tensed verbs, medieval logic is inherently a temporal logic. This fact is most clearly exemplified in medieval theories of supposition. As a case study, we look at the supposition theory of Lambert of Lagny (Auxerre), extracting from it a temporal logic and providing a formalization of that logic.

Keywords

temporal logic, appellation, ampliation, restriction, supposition, quantification, necessity, Lambert of Lagny

1. Introduction

Many aspects and branches of medieval logic are essentially pragmatic in nature, developed to be tools for the analysis of properties of natural language (even if that natural language was sometimes the semi-formalized Latin of the later (thirteenth-fourteenth century) Scholastics). Natural language occurs in written, spoken, and mental utterances, and it is these utterances, not the propositions expressed by them, which bear the truth values. Since every utterance has either a past-, present-, or future-tensed verb, the analysis of the truth conditions for an utterance essentially involves time. Thus, there was no such differentiation of logic into tensed and untensed. As a result, no analysis of the truth conditions of natural language sentences can be done without some reference to the tense of the main verbs. In a sense, for the medieval logicians, *all* logic was temporal logic.

The medieval approach to logic contrasts with the development of the modern discipline of temporal logic, which is treated separately and as distinct from ordinary logic, whether predicate or propositional. Modern propositional logic is essentially timeless, dealing with timeless properties and relations. The addition of a temporal structure involves the addition of a more complex semantics, usually involving possible worlds or states of affairs, and an accessibility relation between those worlds.

The fact that medieval logic is essentially temporal in nature is well illustrated in the theories of supposition that developed in the twelfth century and later. While the present-tense fragments of various theories of supposition have been studied from a formal perspective in recent years, the parts of these theories that deal with future- and past-tensed sentences have not been so studied.

Our purpose in this paper is to use the techniques of modern temporal logic to provide a formal analysis of the future- and past-tensed parts of supposition theory. Such a formal analysis has many benefits: it will show that supposition theory, far from being mere scholastic wrangling with innumerable rules and definitions (as later scholars tended to complain), was an implicit solution to various philosophical problems which was lost to modern logicians when supposition theory was lost, in the post-medieval period. Since many of these problems still arise in philosophical discussions today, we will have yet another reason why it is important for contemporary philosophical logicians to be familiar with the developments of their medieval predecessors, as there is much to be gained from a close investigation of their theories. Further, we will provide a way that these medieval solutions can be made accessible to modern practitioners by placing the solutions in a context familiar to people working in philosophical logic.

The techniques that we introduce in this paper will be general in application, but because we cannot create a formal model without a particular theory to formalize, we have picked one text on which to focus. This text is the final chapter, *De suppositionibus et de significationibus*, of Lambert of Lagny's logic textbook, the *Summa Lamberti* or *Logica*.[1] The chapter starts with defi-

[1] The *Logica* is edited in Lambert of Auxerre, *Logica* (*Summa Lamberti*), ed. Alessio, (1971), with a new and better edition of the final chapter in De Libera (1988), with a translation of the final chapter in Lambert of Auxerre, 'Properties of Terms', Kretzmann and Stump, transl., in: *Cambridge Translations of Medieval Philosophical Texts* vol. 1: Logic and the philosophy of language (1988). Because the English translation in 1988 was based on the Latin edition of 1971, all references are made to the earlier edition. The author of this work was previously identified as Lambert of Auxerre by, e.g., Alessio, Kretzmann and Stump. The author is now generally identified as Lambert of

nitions of signification and supposition, followed by the standard division of supposition into its types,[2] and then concludes with a discussion of *appellatio, restrictio, ampliatio, distributio*, and *relatio*. Lambert's supposition theory is not especially remarkable, and so serves as a good 'generic' theory that we can use to exemplify different parts of our formal model.

We thus begin in §2 with a discussion of Lambert's basic notions of signification and supposition, as we need to understand these before we introduce, in §3, the parts of supposition theory which are connected to analyses in terms of time and modality. These analyses we will formalize in a quantified modal-temporal framework in §5 after we have first given a general introduction to Kripke semantics for temporal logic in §4. Once we have our formal system, we apply it to Lambert's theory and prove some results about it in §6, and then make some conclusions in §7.

2. Basic Notions and Definitions

Even though, as we said above, the techniques that we present will have general application to ampliation and restriction in a variety of theories of supposition, we make Lambert's theory our case study, and in order to show how the formal tools we develop can be applied, we must give Lambert's theory in sufficient detail.

We begin with the properties of terms. It is important to study the properties of terms because sentences (utterances) are made up out of terms, and the properties of the terms will induce the properties of the sentences in which they occur. The most basic concept is that of signification. Both sentences and terms signify, with the signification of terms being prior to the signification of

Lagny (see De Libera (1981)). Little is known about the life of Lambert, but we know that he was a Dominican friar living in the middle of the thirteenth century. His *Summa* was most likely written between 1253-1257 at Troyes (or possibly Pamplona), and published in Paris probably around 1260 (De Rijk, 'Some Thirteenth Century Tracts on the Game of Obligation, III', in: *Vivarium* 14 (1976), 39). It is similar in content and style to those of his contemporaries William of Sherwood, *Introductiones in Logicam*, which is edited with a translation into German by Brands and Kann (1995), and translated into English in: William of Sherwood, *Introduction to Logic*, translated with an introduction and notes by N. Kretzmann (1966); Peter of Spain, *Tractatus*, edited by De Rijk (1972) and translated by F.P. Dinneen, in: Peter of Spain, *Language in dispute* (1990); Roger Bacon, *Sumule dialectices*, ed. Steele (1940) (see also the new edition by De Libera (1987) and (1988)), and indeed to most of the mid-thirteenth century logical compendia.

[2] The fact that supposition is so divided is standard; the division itself differs from author to author.

sentences. The signification of a term is, according to Lambert, the concept on which an utterance is imposed by the will of the person instituting the term.[3] Four things are required for signification:

- A thing
- A concept of the thing
- An utterance
- A union of the utterance and the concept

A thing is any extra-mental thing, such as a substance (e.g., Socrates the man), an accident (e.g., the whiteness which is in Socrates), or an activity (e.g., the running which Socrates is doing right now). These extra-mental things are presented to the soul by means of a concept. A term gains signification when it is used in an utterance and the utterance is connected to a concept by the will of the speaker. The concept imposed upon the term in the speaker's utterance is the signification of the term.

Concepts are concepts of things, and terms signify concepts. Signification hence gives us an indirect way to speak of things, but it does not give us a direct way. In order to be able to speak not just of concepts, such as 'man', but of things which fall under those concepts, such as 'Socrates', we need a more sophisticated mechanism. That mechanism is supposition.

The supposition of a term is the 'acceptance of a term for '(*pro/per*)' itself, for its signified thing, for some suppositum contained under its signified thing, or for more than one suppositum contained under its signified thing'.[4] For example, the term 'man' signifies the concept of man, but it can supposit either for the word 'man', the concept man, or individual or multiple men. Supposition is posterior to signification; a term can have signification in isolation, but a term only has supposition within the context of a complete utterance, and it is the context of the sentence that determines what the term supposits for.

The different types of supposition are divided into the tree in Figure 1. Lambert's definitions of these types of supposition are by and large orthodox. We present them here briefly. Natural supposition is 'what a term has on its

[3] Lambert of Auxerre, 'Properties of Terms', 104; idem, *Logica* (*Summa Lamberti*) VIII, 205. 'Significatio termini est intellectus rei ad quem intellectum representandum [representandum *om.* Alessio] rei vox imponitur ad voluntatem instituentis.'

[4] Lambert of Auxerre, 'Properties of Terms', 106; Lambert of Auxerre, *Logica* (*Summa Lamberti*), VIII, 206: 'Quarto modo dicitur suppositio acceptio termini pro [pro: S.L. Uckelman, per: Alessio] se sive pro re sua, vel pro aliquo supposito contempto sub re sua vel pro aliquibus suppositis contemptis sub re sua.'

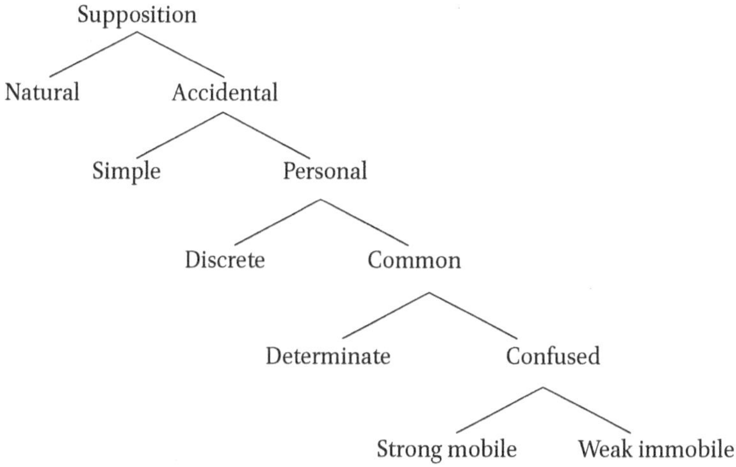

Figure 1: Lambert's division of types of *suppositio*

own and by its nature'.[5] A term with natural supposition supposits 'not only for the things that share its form, but instead for all the things that share, [have shared, and will share] its form—i.e., for present, past, and future things [of that form]';[6] and here we see how time is brought into the definition of supposition.

Accidental supposition is 'what a term has from what is adjoined to it'.[7] This type of supposition is 'in keeping with the requirement of that to which it is adjoined',[8] and the example given is of a term which is conjoined to a tensed being verb (e.g., *est, erit, fuit*). Accidental supposition is divided into two types. The first type is simple, which is 'the kind according to which a term is interpreted for itself or for its [signified] thing, without relation to the supposita contained under it'.[9] This is contrasted with personal supposition, which is '[the sort] according to which a term is interpreted for a *suppositum* or for

---

5) Lambert of Auxerre, 'Properties of Terms', 109; Lambert of Auxerre, *Logica* (*Summa Lamberti*), VIII, 208: 'Naturalis suppositio est quam habet terminus a se et a natura se.'

6) Ibid.: 'Non solum supponit pro hiis que participant formam suam, immo pro omnibus hiis supponit qui participant formam suam, scilicet pro presentibus, preteritis et futuris.'

7) Ibid.: 'Accidentalis suppositio est quam habet terminus ab adiuncto.'

8) Ibid.: '[...] et in hac supponit terminus secundum exigentiam illius cui adiungitur.'

9) Lambert of Auxerre, 'Properties of Terms', 110; idem, *Logica* (*Summa Lamberti*), VIII, 209: 'Simplex suppositio est illa secundum quam tenetur terminus pro se vel pro re sua, non habito respectu ad supposita sub se contempta.'

supposita'.[10] The reason why this latter type is called personal is that 'in the case of rational substance a suppositum or individual is the same as a person'.[11]

Personal supposition is likewise divided into two types, discrete and common. Discrete personal supposition is 'what a discrete term has in itself [...] as when a common term is taken together with a determinate pronoun'.[12] In *Socrates currit* and *iste homo currit*, both 'Socrates' and 'homo' have discrete supposition. Common supposition is, as expected, 'the kind that is appropriate to a common term'.[13]

Common supposition is divided into determinate and confused supposition. Determinate supposition is 'what a common term has when it can be taken equally well for one or for more than one'.[14] The typical example of determinate supposition is *homo* in *homo currit*; the supposition of *homo* here is determinate because it is true 'if one man is running or if more than one are running'.[15] Confused supposition is 'what a common term has when it is interpreted necessarily for all its supposita or for more than one',[16] it is called 'confused' because 'where there is plurality there is confusion'.[17]

The last division is of confused supposition into strong mobile and weak immobile. Strong mobile supposition is 'what a common term has when it is interpreted for all its supposita necessarily and a [logical] descent can be made under it'.[18] This happens when a term with confused supposition is preceded by a universal affirmative or universal negative quantifier (e.g., *omnis* or *nullus*). Weak immobile supposition is 'what a common term has when it is

[10] Lambert of Auxerre, 'Properties of Terms', 110; idem, Logica (*Summa Lamberti*), VIII, 209: 'Personalis suppositio est secundum quam terminus tenetur pro supposito vel pro suppositis.'

[11] Ibid.: 'Suppositum vel individuum in substantia rationali idem est quam persona.'

[12] Lambert of Auxerre, 'Properties of Terms', 111; idem, *Logica* (*Summa Lamberti*), VIII, 209: 'Discreta est illa quam habet terminus discretus in se, [...] ut quando sumitur terminus communis cum pronomine determinato.'

[13] Lambert of Auxerre, 'Properties of Terms', 111; idem, *Logica* (*Summa Lamberti*), VIII, 210: 'Communis est illa que termino communi convenit.'

[14] Ibid.: 'Determinata est illa quam habet terminus communis quando indifferenter potest sumi pro uno vel pro pluribus.'

[15] Ibid.: 'Est enim ista vera uno homine currente, vel pluribus.'

[16] Ibid.: 'Confusa est illa quam habet terminus communis quando de necessitate tenetur pro omnibus suis suppositis vel pro pluribus.'

[17] Ibid.: 'Ubi enim est multitudo, ibi est confusio.'

[18] Lambert of Auxerre. 'Properties of Terms', 112; idem, *Logica* (*Summa Lamberti*), VIII, 210: 'Vehemens mobilis est illa quam habet terminus communis quando tenetur pro omnibus suis suppositis de necessitate, et potest fieri descensus sub eo.'

interpreted necessarily for more than one suppositum contained under it but not for all, and a descent cannot be made under it'.[19] (The terms 'mobile' and 'immobile' refer to whether or not it is possible to make descents from the term to its *supposita*.)

Our focus in the succeeding sections will be on common terms, terms that can apply to more than one object, because discrete terms 'are not able to be restricted or ampliated',[20] and as we'll see in the next section, restriction and ampliation of terms are what give Lambert's logic its temporal character. But before we turn to restriction and ampliation, we must make a general point about signification. Signification is a conventional and use-based notion. A term gains signification when a concept is imposed upon it by the will of a speaker. There are no restrictions on what concepts can be imposed on what terms. It is a conventional fact that speakers of English almost always imposed the same concept upon the term 'dog', but that this convention exists is not a logical fact. In order to have a tractable logic, we are forced to gloss over some of the volitional and psychological aspects of signification. We do so by the following stipulation:

> Antecedent to any logical investigation, we set out that we will always impose on a term its standard (that is, dictionary or conventional) signification.

This stipulation is not strictly required, since we could build agents and signification functions for these agents into our logic, but doing so would not add any clarity or indeed any further expressivity for modeling the medieval theory. Lambert, for example, has no discussion on what to do about the fact that signification can vary from person to person, or how it is that the hearer knows or understands which concept is imposed upon a term by the speaker. These are interesting questions, but they are not wholly logical ones, so we feel no compunction in leaving them aside here. Instead we take the common sense view that when we use the terms 'dog', 'cat', 'donkey', we intend to pick out the expected concepts.

---

19) Lambert of Auxerre, 'Properties of Terms', 112; idem, *Logica* (*Summa Lamberti*), VIII, 211: 'Exilis immobilis est illa quam habet terminus communis quando de necessitate tenetur pro pluribus suppositis sub se contemptis, non tamen pro omnibus, nec sub ipso potest fieri descensus.'

20) Lambert of Auxerre, 'Properties of Terms', 116; idem, *Logica* (*Summa Lamberti*), VIII, 213: 'Ad explanationem istius regule secundum quod terminus communis ibi ponitur ad differentiam termini discreti, qui non potest restringi nec ampliari.'

3. Appellation. Ampliation, and Restriction

After introducing the divisions of types of supposition, Lambert defines five further modifications of supposition: appellation, restriction, ampliation, distribution, and relation. Of these five, we will focus on the first three, since the way each of the three affects the supposition of a term is defined in temporal or modal terms.

Appellatio is 'the acceptance of a term for a suppositum or for supposita actually existing'.[21] The connection between existence and appellation is stressed a few sections later when Lambert says:

> It is important to know, however, that *appellata* are not properly so-called unless they are actual existents; for what is, and not what is not, is properly appellated. And so it is right to say that appellation is for existent supposita, or for an [existent] suppositum.[22]

In the classification of supposition types given in the previous section, appellation is a type of personal supposition.[23] Specifically, it is common terms, not discrete terms, which have appellation. The appellation of a term is, informally speaking, its ordinary reference. If we speak of 'men' without any modifiers, we are speaking of currently existing men. However, there are times when we want to be able to speak of men, but not all currently existing men, or to speak of not just currently existing donkeys, but all donkeys that have ever existed or will ever exist. We can change the supposition of a term by restricting or ampliating it.

Lambert gives the following rule which connects appellation to supposition:

> *Rule* 3.1: 'A substantial or accidental common term that is not restricted by any other means and that serves as the subject or the predicate of a present-tense verb that has no ampliating force of its own or from anything else is restricted to suppositing for present things

---

[21] Lambert of Auxerre, 'Properties of Terms', 114; idem, *Logica* (*Summa Lamberti*), VIII, 212: 'Quarto modo dicitur appellatio acceptio termini pro supposito vel pro suppositis actu existentibus.'

[22] Lambert of Auxerre, 'Properties of Terms', 115; idem, *Logica* (*Summa Lamberti*), VIII, 213: 'Sciendum autem quod proprie loquendo non dicuntur appellata nisi sint actualiter existentia; appellatur enim proprie quod est, et non quod non est, et ideo bene dicitur quod appellatio est pro existentibus suppositis vel pro supposito.'

[23] 'Appellation is always personal supposition' (Lambert of Auxerre. 'Properties of Terms', 114); idem, *Logica* (*Summa Lamberti*), 212: 'Appellatio semper est suppositio personalis.'

if it has *appellata*; but if it does not have *appellata*, it reverts to non-existents (*existentes/ existentiam*).[24]

Appellation allows us to speak of currently existing objects of a certain kind. However, we often want to speak of possible objects, or future or past objects, or a subset of objects of a specific kind. This is done through ampliating and restricting the supposition of a term.

Restriction and ampliation are each other's opposites. *Restrictio* is 'a lessening of the scope of a common term as a consequence of which the common term is interpreted for fewer supposita than its actual supposition requires'.[25] One way that a common term can be restricted is through the addition of an adjective. For example, we can restrict the supposition of 'man' by adding to it the adjective 'white'; 'white man' has fewer supposita than 'man' unmodified. This type of restriction is called *naturalis*. This is distinguished from *usualis* or 'use-governed' restriction, that is, restriction made by convention and not by the addition of a modifying word. An example of use-governed restriction is 'the queen is coming', where 'queen' is not modified explicitly in the statement, but is, by convention, taken to mean the queen of the country in which the sentence is spoken.[26]

It is important to note that not all additions of modifying words or phrases to a common term will result in natural restriction: If the added word or phrase 'destroys' the term, then the result is not called restriction, but diminution. An example of this is the modifying clause *qui non est*. Restriction requires that the common term still have *appellata*, but as defined in Rule 3.1, *appellata* must exist. There is no existing object which falls under the modified common term 'man who does not exist'.

If the modifying word or phrase does not restrict the supposition of a common term but rather expands it, then we are dealing with the opposition of restriction, which is called ampliation. *Ampliatio* is an 'extension of the scope

---

24) Lambert of Auxerre, 'Properties of Terms', 116; idem, *Logica* (*Summa Lamberti*), VIII, 213: 'Terminus communis substantialis vel accidentalis non restrictus aliunde supponens vel apponens verbo presentis temporis non habenti vim ampliandi a se nec ab alio, restringitur ad supponendum pro presentibus, si appellata habeat; si vero non, recurrit ad non existentes [existentiam]'.

25) Lambert of Auxerre, 'Properties of Terms', 134; idem, *Logica* (*Summa Lamberti*), VIII, 226: 'Restrictio est minoratio ambitus termini communis, secundum quam pro paucioribus suppositis teneter terminus communis quam exigat sua actualis suppositio.'

26) Lambert of Auxerre, *Logica* (*Summa Lamberti*), VIII, 227: 'Cum dicitur: 'rex venit'; iste terminus 'rex' restringitur ad supponendum pro rege patrie in qua sermo iste dicitur.'

of a common term as a consequence of which the common term can be interpreted for more supposita than its actual supposition requires'.[27] As an example, Lambert offers *homo potest esse Antichristus*;[28] in this sentence, *homo* is ampliated to stand not only for current men but for future men.

Ampliation is caused by the addition of names (such as 'possible'), verbs ('can'), adverbs ('potentially'), or participles ('praised'):

> For there are certain names that have the power of ampliating—e.g., 'possible', 'necessary'—and certain verbs likewise—e.g., 'can' [....]—similarly also certain adverbs—e.g., 'potentially', 'necessarily' [...].[29]

Whichever type the ampliating term is, ampliation can be divided into two types: 'ampliation [...] brought about because of supposita' and that brought out 'because of times'.[30] Ampliation by reason of *supposita* is caused by 'verbs whose corresponding action is related to the subject and said of the subject but is not in the subject—such as 'can', 'is thought', 'is praised''.[31] A term ampliated in this way stands for both actual and non-existent *supposita*. Ampliation by reason of times is caused by modifiers which 'cause a term to be extended to all the differences of time'.[32] Example of this kind of modifier are temporal operators such as 'always', modal operators such as 'necessarily' and 'possibly', and changes in the tense of the verb. We'll call these two types of ampliation '*supposita* ampliation' and 'tense ampliation', respectively.

The interaction of restriction and ampliation with supposition is partly determined by whether the terms in question are accidental. In future- or past-tensed propositions, the following two rules are relevant:

---

27) Lambert of Auxerre, 'Properties of Terms', 137; idem, *Logica* (*Summa Lamberti*), VIII, 228: 'Ampliatio est extensio ambitus termini communis secundum quod teneri potest terminus communis pro pluribus suppositis quam exigit sua actualis suppositio.'

28) Lambert of Auxerre, *Logica* (*Summa Lamberti*), VIII, 228.

29) Lambert of Auxerre, 'Properties of Terms', 138; idem, *Logica* (*Summa Lamberti*), VIII, 228: 'Quedam enim sunt nomina que habent virtutem ampliandi ut possibile, necessarium; et similiter quedam verba ut potest [...]; similiter et quedam adverbia ut potentialiter necessario [...]'

30) Ibid.: 'Ampliationum alia fit ratione suppositorum, alia ratione temporum.'

31) Lambert of Auxerre, 'Properties of Terms', 138; idem, *Logica* (*Summa Lamberti*), VIII, 229: 'Verba quorum actus comparatus ad subiectum de subiecto dicitur, in subiecto tamen non est, ut sunt ista: potest, opinatur, laudatur.'

32) Lambert of Auxerre, 'Properties of Terms', 138; idem, *Logica* (*Summa Lamberti*), VIII, 229: 'Illa dicuntur ampliare ad tempora que faciunt terminum extendi ad omnes differentias temporis.'

> *Rule* 3.2: A common term pertaining to accident that is not restricted by any other means and that serves as the subject of a past-tense verb can supposit for present and past things even though the term serving as the predicate supposits only [*tantum*] for past things; but if a term pertaining to substance serves as the subject or predicate of a past-tense verb, it always supposits for past things.[33]

> *Rule* 3.3: A common term pertaining to accident that is not restricted by any other means and that serves as the subject of a future-tense verb can supposit for present and future things even though when serving as the predicate it is interpreted only for future things; but if a term pertaining to substance serves as the subject or the predicate of a future-tense verb, it is always interpreted for future things.[34]

These rules and the definitions of appellation, restriction, and ampliation provide the basis for our formal analysis of the temporal elements of Lambert's theory of supposition. In the next section, we move towards developing our formal model.

4. Kripke Semantics for Modal and Temporal Logic

In this section we switch our attention to modern logic and give a general introduction to Kripke semantics, with a view towards our application of these ideas in the next section.

Kripke semantics can formalize our informal ideas about both modality and tense with the same type of structure, which is called a *Kripke frame*. A Kripke frame is a pair $F = \langle W, R \rangle$, where W is a set whose elements can be interpreted as times or as possible worlds depending on the context, and R is a binary relation on W, called an accessibility relation. The accessibility relation indicates which instants of time are past or future with respect to the current instant or which worlds are considered 'possible' with respect to a given world. We can represent Kripke frames schematically, as in Figure 2; the nodes are the worlds and the arrows indicate the accessibility relations.

---

33) Lambert of Auxerre, 'Properties of Terms', 129; idem, *Logica* (*Summa Lamberti*), VIII, 223: 'Terminus communis accidentalis non restrictus aliunde supponens verbo preteriti temporis, supponere potest pro presentibus et preteritis; apponens vero solum [Kretzmann and Stump, terminum Alessio] supponit pro preteritis; si vero fuerit terminus substantialis supponens vel apponens verbo preteriti temporis, semper pro preteritis supponit.'

34) Lambert of Auxerre, 'Properties of Terms', 129; idem, *Logica* (*Summa Lamberti*), VIII, 223: 'Terminus communis accidentalis non restrictus aliunde supponens verbo futuri temporis, supponere potest pro presentibus et futuris; apponens vero solum tenetur pro futuris; si vero fuerit terminus substantialis supponens vel apponens verbo futuri temporis, semper tenetur pro futuris.'

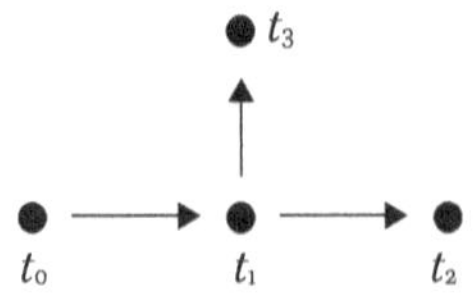

Figure 2: An example of a Kripke frame

Kripke frames are turned into models by the addition of a valuation function, *V*. This function maps propositions (which we indicate by lower case Roman letters *p*, *q*, *r*, etc.) to subsets of *W*. If *w* is in *W* (written formally as $w \in W$) and $w \in V(p)$, then we say that *p* is true at *w* (written formally as $w \models p$). While *V* only gives information about the truth values of simple propositions in worlds or at instants, we can easily extend this to give the truth values of more complex propositions, as follows:

$w \models \sim p$	iff	$w \notin V(p)$
$w \models p \wedge q$	iff	$w \models p$ and $w \models q$
$w \models p \vee q$	iff	$w \models p$ or $w \models q$
$w \models p \rightarrow q$	iff	$w \models \sim p$ or $w \models q$

Note that these definitions all involve only one world; these are exactly the classical truth conditions for the various connectives. The extension into non-classical logic comes when we introduce the temporal operators, *F* (at some time in the future), *P* (at some time in the past), *G* (at all times in the future), and *H* (at all times in the past):

$w \models Fp$	iff	there is a *v* such that *wRv*, $v \models p$
$w \models Pp$	iff	there is a *v* such that *vRw*, $v \models p$
$w \models Gp$	iff	for all *v* such that *wRv*, $v \models p$
$w \models Hp$	iff	for all *v* such that *vRw*, $v \models p$

We call *P* and *H* *past-tensed* operators and *F* and *G* *future-tensed* operators. Note that *P* and *H* are just the symmetric versions of *F* and *G*, and that *H* and *G* can be defined as $\sim P \sim$ and $\sim F \sim$, respectively. To give an example of the temporal operators, suppose *V* assigns *p* to t_0 and t_1, and *q* to t_1, t_2, and t_3 in Figure 2. Then, $t_0 \models Fp$, but $t_1 \models G\sim p$. Also, both $t_0 \models Gq$ and $t_1 \models Gq$, but $t_3 \models G\sim q$ (this latter fact is because t_3 is not related by *R* to any other world, so the antecedent of the condition for $G\sim q$ is vacuously satisfied. On these same grounds, $t_3 \models Gq$). Looking at the past-tensed operators, $t_1 \models Hp$ and $t_2 \models Pq \wedge P\sim q$.

In modern modal logic, the truth conditions of the modal operators $\Box$ 'necessarily' and $\Diamond$ 'possibly' are defined as:

$w \models \Diamond p$	iff	there is a v such that wRv and $v \models p$
$w \models \Box p$	iff	for all v such that wRv, $v \models p$

However, this does not adequately capture Lambert's view of modality. Lambert follows one of the traditional medieval interpretations of modality, the statistical interpretation,[35] where 'necessarily' and 'possibly' are defined with reference to times, not to possible worlds. As he says:

> Thus when one says 'A man is an animal necessarily', it is the same as if one had said 'That is, in every time being an animal applies to a man: in the present, the past, and the future.[36]

Formally, this means the truth conditions of Lambert's modal operators are:

$w \models \Diamond p$	iff	there is a t, $t \models p$
$w \models \Box p$	iff	for all t, $t \models p$

That is, we drop any reference to the accessibility relation. If we are dealing with linear time, it is possible to define these modal operators in terms of the temporal operators:

$\Diamond p := p \vee Fp \vee Pp$
$\Box p := p \wedge Gp \wedge Hp$

Before we discuss adding quantification to what we've set up so far, we introduce a few more definitions and formal notions. A formula which is true in every world in a frame or a class of frames all sharing a certain property regardless of the valuation function V is called *valid* on that class of frames. The validity of certain modal formulas on a frame or class of frames can be connected to properties of the R-relation on these frames. A list of some common properties of R, their names, and formulas which correspond to them can be found

35) Cf. Knuuttila (1993), *passim*.

36) Lambert of Auxerre, 'Properties of Terms', 138; idem, *Logica* (*Summa Lamberti*), VIII, 229: 'Unde cum dicitur: 'home est animal necessario' aliud est ac si diceretur: 'id est, in omni tempore convenit homini esse animal', scilicet in presenti preterito et futuro.'

Table 3: Formulas which correspond to frame properties

Property	formula	name
$\forall x(xRx)$	$\Box p \to p$	reflexivity
$\forall x,y(xRy \to yRx)$	$\Box\Diamond p \to p$	symmetry
$\forall x,y,z(xRy \land yRz \to xRz)$	$\Box p \to \Box\Box p$	transitivity
$\forall x,y,z((xRz \land yRz) \to (xRy \lor yRx \lor x=y))$	$FPp \to (Pp \lor p \lor Fp)$	backwards linearity

in Table 3. The proofs of the correspondence between the properties and the formulas can be found in any standard modal logic text book.[37]

What we have outlined above is basic propositional modal logic. Extensions of this are widely studied by contemporary logicians, but when we turn our attention to quantified modal logic, which allows one to speak of predicates applying to individuals, the situation becomes much more murky. Few logicians consider the subject at all, with many following W.V.O. Quine who, according to Fitting and Mendelsohn, 'does not believe that quantified modal logic can be done coherently'.[38] Fitting and Mendelsohn's book is the standard text on the subject, and we follow their presentation below.

When we introduce quantification into our models, we need a domain of things to quantify over. There are two ways that we can add such a domain. The first is to fix a domain of quantification which remains constant no matter which world you are evaluating formulas at. Fitting and Mendelsohn call such models *constant-domain* models. The second is to allow the domain to differ from world to world, so that the scope of the quantified formulas changes depending on the world in which they are being evaluated. Such models are called *varying-domain* models. These two types of models each capture a different type of quantification, the *possibilist* view of quantification and the *actualist* view, respectively. The question immediately arises which of these two types of models should be preferred. As Fitting and Mendelsohn point out, the answer is: 'It doesn't matter. We can formalize the same philosophical ideas either way, with a certain amount of care'. Two pragmatic reasons to prefer constant-domain models are that they are formally simpler to define and to use, and they also 'model our intuitions about modality most naturally if we take the domain

37) Such as Chagrov and Zakharyaschev (1997), §3.5.

38) See Fitting and Mendelsohn (1998), 89. One exception to this view is Arthur Prior (1957), who believes that a coherent philosophical treatment of quantified modal logic can be given.

to consist of possible existents, not just actual ones, for otherwise we would be required to treat every existent as a necessary existent'.[39] If one wishes to use varying-domain models, it is necessary, in order to address this problem, to introduce an existence predicate, and to relativize all the quantifiers to only the objects which exist at a certain world. It can be shown that if you introduce a relativization of this kind, if a formula is a theorem of a varying-domain model, its relativization is a theorem of a constant-domain model.[40]

Once the domain of and the truth conditions for quantification have been established, the next important question is how the quantifiers interact with the modal or tense operators. The most interesting questions arise from what are called the Barcan formula and the Converse Barcan formula:

Barcan formula	$\forall x \Box \varphi x \rightarrow \Box \forall x \varphi x$
Converse Barcan formula	$\Box \forall x \varphi x \rightarrow \forall x \Box \varphi x$

(Their name comes from the person to first investigate them in detail, Ruth Barcan Marcus. Technically speaking, these are not formulas but formula schemes, with s being substitutable for any formula, though we will continue to speak of the singular 'the Barcan formula'.) These formulas, like the ones in Table 3, correspond to frame properties which express the idea that as we move from one world to another, nothing passes out of existence (the Converse Barcan formula) and nothing comes into existence (the Barcan formula). Put this way, we can see immediately the problems that these formulas present in the context of temporal logic, where we want objects to be able to come into and go out of existence as time passes.[41] As a result, if we work in a varying-domain model, then we must ensure that neither the Barcan formula nor its converse is valid.

5. Constructing a Formal Model

In this section we adapt the general system introduced in the previous section to our specific purposes. We specify our language and formal model, and give definitions of appellation, ampliation, and restriction.

Our base language is that of predicate logic with temporal and modal operators, but we extend this with four copulae for the different types of categorical

39) Fitting and Mendelsohn (1998), 94.

40) Ibid., prop. 4.8.2.

41) For an extensive discussion of the problems, see Prior (1957).

sentences. That is, for any predicates S and Q, the following are well-formed formulas:

$$SuaQ \mid SunQ \mid SpaQ \mid SpnQ$$

We also stipulate that we have a name for every object in our domain; since we are only working with finite domains, this means that our language will still be finite.

Recall that we saw in §4 that Lambert's view of modality is statistical, that is, it is defined in terms of temporal instants just as the tense operators are. This simplifies our system somewhat in that we do not need to specify a possibility relation in addition to a temporal ordering on the possible worlds. Formally, our models are 5-tuples $M=\langle T, <, O, E, V\rangle$, where:

- T is a set of temporal instants. We let variables t with and without subscripts and superscripts, w, x, y, and z range over T.
- $<$ is a transitive, irreflexive, and backward linear relation on T; if $t<t'$ we say that t' is (temporally) later than t and t is (temporally) earlier than t'.
- O is a finite set of objects. We let the variables $a, b, c \ldots$ range over O.
- E is a function from O to subsets of T; if $t \in E(a)$, then we say that *a exists at t* and write $t \models \hat{a}$. We require that E be such that for every $t \in W$, there is an $a \in O$ such that $t \in E(a)$; that is, at every point in time, at least one object exists.
- V is a function from predicate-world pairs to subsets of O; if $a \in V(Q,t)$, then we say that a is Q at t.

On variable-domain models, the Barcan formula is valid iff the domains are anti-monotonic, and the Converse Barcan formula is valid iff the domains are monotonic.[42] Because we wish neither of these to be valid, we require that there exist $t, t', w, w' \in T$ such that $t<t'$ and $E(t) \not\subseteq E(t')$, and $w<w'$ and $E(w') \not\subseteq E(w)$.

We call propositions of the form $\hat{a}$ 'existence propositions'. We also extend V to a global function V' by setting $V'(Q) := \cup t \in \mathrm{T}\ V(Q,t)$. The function V' picks out all the objects in the model that a particular predicate is true of at some point in the timeline; we will use V' in the definition of one type of ampliation.

Finally, if we wish to formalize the medieval concept of existential import, then we can do so as follows:

---

42) Cf. ibid., props. 4.9.6, 4.9.8.

Definition 5.1 (Existential import). A model M has *existential import* iff for all predicates Q and for all $t \in T$, $V(Q,t) \neq \varnothing$, that is, V is a function from predicate-world pairs to *non-empty* subsets of O.

From now on we will only work with models which have existential import.

Giving truth conditions for the four types of present-tensed categorical statements is straightforward.

Definition 5.2 (Truth conditions for categorical propositions)

UA:	$t \models$ Sua$Q$	iff	$V(S,t) \subseteq V(Q,t)$
UN:	$t \models$ Sun$Q$	iff	$V(S,t) \cap V(Q,t) = \varnothing$
PA:	$t \models$ Spa$Q$	iff	$V(S,t) \cap V(Q,t) \neq \varnothing$
PN:	$t \models$ Spn$Q$	iff	$V(S,t) \not\subseteq V(Q,t)$

It should be immediately clear from these definitions that the truth conditions for the four types of categorical propositions respect the relationships in the traditional square of opposition. We prove this just for the case of the subaltern relation:

Lemma 5.3. $t \models$ Sua$Q$ implies $t \models$ Spa$Q$.

Proof. Suppose $t \models$ Sua$Q$. Then by UA of Def. V.2, $V(S,t) \subseteq V(Q,t)$. Since we are working in models with existential import, $V(S,t) \neq \varnothing$; it follows that $V(S,t) \cap V(Q,t) \neq \varnothing$, and hence $t \models$ Spa$Q$.

Note that we have not defined the categorical propositions in terms of universally quantified conditionals or existentially quantified conjunctions. This is because we want to reserve the quantifiers for making statements about objects which actually exist at a certain point in time. We will give truth conditions for such statements below, once we have defined the appellation function.

The appellation, ampliation, and restriction functions are defined in terms of V and the existence propositions. These functions, like V itself, will be indexed to worlds.

Definition 5.4 (Appellation). Ap$(Q,t) = V(Q,t) \cap \{a{:}t \models \hat{a}\}$.

That is, given a predicate and a world, the function Ap selects all the objects which V assigns to the predicate in the world which actually exist in that world. We use the appellation function to define the truth conditions for simple present-tensed predications:

Definition 5.5 (Truth conditions for simple predications).

$t \models \exists xQx$	iff	$\mathrm{Ap}(Q,t) \neq \varnothing$
$\underline{t} \models \forall xQx$	iff	$\mathrm{Ap}(Q,t) = \{a{:}t \models \hat{a}\}$

Boolean combinations of these simple predications are formed in the expected way, e.g., for any infinite predicate $\sim Q$, $t \models \exists x \sim Qx$ iff $\mathrm{Ap}(\sim Q,t) = \{a{:}t \models \hat{a}\} \setminus \mathrm{Ap}(Q,t)$. Note that these definitions allow for the possibility that $V(Q,t) \neq \varnothing$ but that $t \not\models \exists xQx$, namely when all of the things which are Q at t do not exist. This allows us to satisfy Lambert's requirement that 'when no man is in existence "Every man exists' is false, and so its contradictory 'Some man does not exist' will be true".[43]

Defining the truth conditions for the quantifiers and the categorical propositions in this fashion means that our models are a sort of hybrid between constant-domain models and varying-domain models, as these were defined in the previous section. Because $\forall x$ and $\exists x$ are defined in terms of objects which actually exist at a given world, we can say that the models have varying-domains (since it is allowed that $E(a) \neq E(b)$ for $a \neq b$). But if we consider just the categorical operators ua, un, pa, and pn, then we have a constant-domain model, since the range of V does not vary.

We are now in a position to define restriction and ampliation. These functions modify the supposition of a term and are defined in terms of the appellation function. The formalization of natural restriction is straightforward:

Definition 5.6 (Restriction). If Q is (naturally) restricted by S, then $\mathrm{Res}(Q,S,t) = \mathrm{Ap}(Q,t) \cap \mathrm{Ap}(S,t)$

We omit from consideration here use-governed restriction, as this falls under pragmatics, and not the formal theory itself.

Moving on to ampliation, recall from §III that there are two kinds of ampliation, ampliation because of *supposita* and ampliation because of time. We give separate definitions for each. The definition of tense ampliation is straightforward, because it can be defined in a structural fashion.

Definition 5.7 (Tense ampliation). We have four cases, one for each of the temporal operators:

---

43) Lambert of Auxerre, 'Properties of Terms', 123; idem, *Logica* (*Summa Lamberti*), VIII, 219: "Nam nullo homine existente hec est falsa: 'omnis homo est', ergo sua contradictoria erit vera, hec scilicet: 'aliquis homo non est'."

$$\text{Amp}^F_t(Q,w) = \bigcup_{t>w} \text{Ap}\,(Q,t) \qquad \text{Amp}^G_t(Q,w) = \bigcap_{t>w} \text{Ap}\,(Q,t)$$

$$\text{Amp}^P_t(Q,w) = \bigcup_{t<w} \text{Ap}\,(Q,t) \qquad \text{Amp}^H_t(Q,w) = \bigcap_{t<w} \text{Ap}\,(Q,t)$$

We will see in the next section how these definitions can be used to give formal analyses of Rules 3.2 and 3.3.

Giving a definition of *supposita* ampliation is much more difficult. Tense ampliation was straight forward because the different cases have clear syntactic definitions. On the other hand, *supposita* ampliation is caused when a term is 'related to the subject and said of the subject but is not in the subject', and this does not immediately lend itself to a nice syntactic characterization. To address this difficulty, note that, conceptually, tense ampliation is a subset of *supposita* ampliation, since *supposita* ampliation happens when 'a term [is] interpreted for both actual and non-existent supposita'.[44] Formally, we shall restrict the term '*supposita* ampliation' so that it only applies to those cases of ampliation which do not fall under tense ampliation, that is, when the non-existents in question are not past or future existents. This means that we must only address those predicates which ampliate to past, present, and future non-existents.

Additionally, we stipulate that we antecedently know which predicates are the ones that ampliate in this way. These are not defined by any syntactic property, but can only be collected by ostension. Lambert says that

> But in order to recognize which verbs ampliate and which ones do not, it is important to know that an action can be related to a substance in two different ways: in one way as regards that in which it is and of which it is stated [...]—in the other way as regards that of which it is stated although it is not in it.[45]

We let S be the set of all predicates which ampliate by means of *supposita*.

---

44) Lambert of Auxerre, 'Properties of Terms', 138; idem, *Logica* (*Summa Lamberti*), VIII, 229: 'Terminum [tenet] pro suppositis actu et non existentibus.'

45) Lambert of Auxerre, 'Properties of Terms', 117; idem, *Logica* (*Summa Lamberti*), VIII, 214: 'Ad cognoscendum autem que verba ampliant et que non, sciendum quod ad substantiam actus potest comparari duplicitier: uno modo quantum ad illud in quo est et de quo enunciatur[...] alio modo tamquam ad id de quo enunciatur non tamen in ipso est.'

Definition 5.7 (*Supposita* ampliation). If $R \in S$, then $\mathrm{Amps}(Q,t)=V'(Q)$.

When we say 'a chimaera is thought of' or 'a man is praised', this is true if there is *any* chimaera or *any* man, past, present, or future, existing or not, which is thought of or is praised.

This concludes the presentation of our formal model.

6. Applying the Formal Model

In this section we investigate the formal properties of the model presented in the previous section, with particular attention to showing that it satisfies the rules put down in §3.

We consider Rule 3.1 first, which says that the supposition of a term in a present-tensed, non-ampliated, non-restricted proposition will be the term's *appellata* if it has appellata, and non-existents otherwise. The truth conditions that we gave for categorical propositions may seem on first consideration not to respect this rule, since no mention is made of *appellata*, but it is straightforward to prove lemmas outlining the interaction of categorical propositions with propositions about objects actually existing:

Lemma 6.1. If $\mathrm{Ap}(Q,t)\neq\emptyset$ and $t \models Q\mathrm{ua}S$, then $t \models \forall x(Qx \to Sx)$

Proof. Let $a \in \mathrm{Ap}(Q,t)$ be arbitrary. Since $a \in \mathrm{Ap}(Q,t)$, $t \models \hat{a}$. From $t \models Q\mathrm{ua}S$, we know that $V(S,t) \subseteq V(Q,t)$. It follows then that $a \in \mathrm{Ap}(S,t)$\$. Since a was arbitrary, this means that $\mathrm{Ap}(Q,t) \subseteq \mathrm{Ap}(S,t)$, which is sufficient to show our conclusion.

Similar lemmas for the other three types of categorical statements are easily proven. We can also prove the converses for particular categoricals:

Lemma 6.2. If $t \models \exists x(Sx \wedge Qx)$ then $t \models S\mathrm{pa}Q$.

Proof. Assume that $t \models \exists x(Sx \wedge Qx)$. It follows that $\mathrm{Ap}(s,t) \cap \mathrm{Ap}(Q,t) \neq \emptyset$. Since $\mathrm{Ap}(Q,t) \subset V(Q,t)$ and $\mathrm{Ap}(S,t) \subset V(Q,t)$, it follows that $V(Q,t) \cap V(S,t) \neq \emptyset$, and so $t \models S\mathrm{pa}Q$. Note that from this result, the soundness of the conversion rule for particular affirmative statements can be derived; if $t \models S\mathrm{pa}Q$, then $t \models Q\mathrm{pa}S$.

Lemma VI.3. If $t \models \exists x(Sx \wedge {\sim}Qx)$, then $t \models S\mathrm{pn}Q$.

Proof. Assume that $t \models \exists x(Sx \wedge {\sim}Qx)$. Then $\mathrm{Ap}(S,t) \cap \mathrm{Ap}({\sim}Q,t) \neq \emptyset$. By the definition of the appellation of infinite predicates, we know that

$\mathrm{Ap}(S,t) \cap \{a : t \models \hat{a}\} \setminus \mathrm{Ap}(Q,t) \neq \emptyset$.

This implies that there is some a such that $a \in \mathrm{Ap}(S,t)$ and $a \notin \mathrm{Ap}(Q,t)$, so $\mathrm{Ap}(S,t) \not\subseteq \mathrm{Ap}(Q,t)$. It then follows that $V(S,t) \not\subseteq V(Q,t)$ since $\mathrm{Ap}(S,t) \subseteq V(S,t)$, and $b \in \mathrm{Ap}(S,t)$ and $b \in V(Q,t)$ implies $b \in \mathrm{Ap}(Q,t)$.

That the converses for universal categoricals are not provable is easily demonstrable.

Lemma 6.4. $t \models \forall x(Sx \rightarrow Qx)$ does not imply $t \models S\mathrm{ua}Q$.

Proof. Let M be such that $O=\{a,b\}$ and there is a $t \in T$ such that $t \in E(a)$ but $t \notin E(b)$. Let $d \in V(S,t)$ for all $d \in O$ and $a \in V(Q,t)$. Then $t \models \forall x(Qx)$ and hence $t \models \forall x(Sx \rightarrow Qx)$. However, $V(S,t) \not\subseteq V(Q,t)$, so $t \not\models S\mathrm{ua}Q$.

A similar proof can be given for the universal negative categorical propositions.

In order to discuss Rules 3.2 and 3.3, we need to introduce a few more definitions:

Definition 6.5 (Substantial term). Q is a *substantial term* iff $\forall a \in O$, if $\exists t \in E(a)$ and $t \models Qa$, then $\forall t' \in E(a)$, $t' \models Qa$.

Definition 6.6 (Accidental term). Q is an *accidental term* iff $\exists a \in O$ such that $\exists t, t' \in E(a)$ where $t \models Qa$ and $t' \not\models Qa$.

Next we need to formalize the different ways in which propositions with tensed verbs, such as 'will', or modal verbs, such as 'is able', can be interpreted. Lambert says:

> We have to say that 'A white thing will be Socrates' has two interpretations; for it can be interpreted in this sense: That which will be white will be Socrates; or in this sense: What is white will be Socrates.[46]

The distinction is between 'there is something that exists now and which will be white in the future and will be Socrates in the future' and 'there is something which exists now and is white now, and will be Socrates in the future'. Since modal terms such as 'can', 'is able', and 'possibly' are all analysed in temporal terms, this means that the same things hold for modal sentences such as 'A white thing can be Socrates'.

---

[46] Lambert of Auxerre, 'Properties of Terms', 133; idem, *Logica* (*Summa Lamberti*), VIII, 225: 'Dicendum quod: 'album erit Socrates' habet duas acceptiones: potest enim accipi sub hoc sensu: id quod erit album erit Sortes; vel sub isto: quod est album erit Sortes.'

Lambert's distinction here is similar to, but not quite the same as, the distinction that other authors make between the divided and composite interpretations of modal and temporal statements.[47] The divided interpretation of the modal statement 'A white thing can be black' is 'there is something which is now white and which will be black in the future', which is distinguished from the composite interpretation, which is 'it will be the case that there is something which is both white and black'. The statement is true under the divided interpretation but false under the composite interpretation.

This gives us three ways that past and future tensed statements can be interpreted. Let *t* indicate the present moment and *a* be some object:

$$\mathrm{t} \models \hat{a} \quad \text{and} \quad \exists t'<t, \exists t'<t, t' \models Wa \quad \text{and} \quad t' \models a{=}\mathrm{Soc} \tag{1}$$
$$\mathrm{t} \models \hat{a} \quad \text{and} \quad t \models Wa \quad \text{and} \quad \exists t>t', t' \models \models a{=}\mathrm{Soc} \tag{2}$$
$$\exists t>t', t' \models Wa \wedge a{=}\mathrm{Soc} \tag{3}$$

(To obtain future tense analogues, just change the direction of the >.) The first of these corresponds to the divided sense and to Lambert's first interpretation; the second corresponds to Lambert's second interpretation; and the third to the composite sense. It is the composite sense which corresponds to the interpretation of the temporal operators given in §4. However, this sense is too narrow to capture what we intend to express with tensed quantificational sentences. In general, we want to be able to make statements of types one and two.

It turns out that for accidental terms, Rules 3.2 and 3.3 cause the first two distinctions to collapse. This will be clear when we give formal expressions of the truth conditions given informally in those rules, below. Before we do so, we first distinguish types 1 and 2 from type 3 by formalizing the latter as *P*(*Q*•*S*) where • is any of the four categorical connectives, and abusing notation to formalize the former as *QP*•*S* (we trust that this will not be confusing since we never use *P* as a predicate variable); we call tenses in sentences of this second type 'embedded tenses' or 'embedded modalities'. To take an example, we read formulas of the form *P*(*Q*pa*S*) as 'it was the case that some *Q* is *S*' and ones of the form *QP*pa*S* as 'it is the case that some *Q* was *S*', and similarly for the other connectives. Now we can give the formal versions of the rules:

---

47) See, e.g., Knuuttila (1982), 347-348, 354-357.

Definition 6.7 (Rule 3.2). If Q is an unrestricted accidental term and $S\notin S$, then

$$\begin{array}{lll} t\models QPuaS & \text{iff} & \{\mathrm{Ap}(Q,t)\cup\mathrm{Ampt}^{\mathrm{P}}(Q,t)\}\subseteq\mathrm{Ampt}^{\mathrm{P}}(S,t) \\ t\models QPpaS & \text{iff} & \{\mathrm{Ap}(Q,t)\cup\mathrm{Ampt}^{\mathrm{P}}(Q,t)\}\cap\mathrm{Ampt}^{\mathrm{P}}(S,t)\neq\varnothing \\ t\models QPunS & \text{iff} & \{\mathrm{Ap}(Q,t)\cup\mathrm{Ampt}^{\mathrm{P}}(Q,t)\}\cap\mathrm{Ampt}^{\mathrm{P}}(S,t)=\varnothing \\ t\models QPpnS & \text{iff} & \{\mathrm{Ap}(Q,t)\cup\mathrm{Ampt}^{\mathrm{P}}(Q,t)\}\not\subseteq\mathrm{Ampt}^{\mathrm{P}}(S,t) \end{array}$$

and

$$\begin{array}{lll} t\models QHuaS & \text{iff} & \{\mathrm{Ap}(Q,t)\cup\mathrm{Ampt}^{\mathrm{P}}(Q,t)\}\subseteq\mathrm{Ampt}^{\mathrm{H}}(S,t) \\ t\models QHpaS & \text{iff} & \{\mathrm{Ap}(Q,t)\cup\mathrm{Ampt}^{\mathrm{P}}(Q,t)\}\cap\mathrm{Ampt}^{\mathrm{H}}(S,t)\neq\varnothing \\ t\models QHunS & \text{iff} & \{\mathrm{Ap}(Q,t)\cup\mathrm{Ampt}^{\mathrm{P}}(Q,t)\}\cap\mathrm{Ampt}^{\mathrm{H}}(S,t)=\varnothing \\ t\models QHpnS & \text{iff} & \{\mathrm{Ap}(Q,t)\cup\mathrm{Ampt}^{\mathrm{P}}(Q,t)\}\not\subseteq\mathrm{Ampt}^{\mathrm{H}}(S,t) \end{array}$$

If Q is an unrestricted substantial term and $S\notin S$, then

$$\begin{array}{lll} t\models QPuaS & \text{iff} & \mathrm{Ampt}^{\mathrm{P}}(Q,t)\subseteq\mathrm{Ampt}^{\mathrm{P}}(S,t) \\ t\models QPpaS & \text{iff} & \mathrm{Ampt}^{\mathrm{P}}(Q,t)\cap\mathrm{Ampt}^{\mathrm{P}}(S,t)\neq\varnothing \\ t\models QPunS & \text{iff} & \mathrm{Ampt}^{\mathrm{P}}(Q,t)\cap\mathrm{Ampt}^{\mathrm{P}}(S,t)=\varnothing \\ t\models QPpnS & \text{iff} & \mathrm{Ampt}^{\mathrm{P}}(Q,t)\not\subseteq\mathrm{Ampt}^{\mathrm{P}}(S,t) \end{array}$$

and

$$\begin{array}{lll} t\models QHuaS & \text{iff} & \mathrm{Ampt}^{\mathrm{P}}(Q,t)\subseteq\mathrm{Ampt}^{\mathrm{H}}(S,t) \\ t\models QHpaS & \text{iff} & \mathrm{Ampt}^{\mathrm{P}}(Q,t)\cap\mathrm{Ampt}^{\mathrm{H}}(S,t)\neq\varnothing \\ t\models QHunS & \text{iff} & \mathrm{Ampt}^{\mathrm{P}}(Q,t)\cap\mathrm{Ampt}^{\mathrm{H}}(S,t)=\varnothing \\ t\models QHpnS & \text{iff} & \mathrm{Ampt}^{\mathrm{P}}(Q,t)\not\subseteq\mathrm{Ampt}^{\mathrm{H}}(S,t) \end{array}$$

Definition 6.8 (Rule 3.3). The formalization of Rule 3.3. can be obtained by replacing P with F and H with G throughout.

Note that it follows from these rules that sentences with substantial terms as their subject terms can only be interpreted in the first of the two interpretations that Lambert gives.

We make just one more remark before concluding our application of the formal model to Lambert's theory of supposition. In §4 we noted that the temporal operators H and G can be defined as $\sim P\sim$ and $\sim F\sim$, respectively, and that is why we focused only on P and F throughout the current section and the preceding one. While it is clear that this interdefinability holds for sentences interpreted in the third way (the composite sense), it is by no means obvious that the same is true when we use P and F as in the two rules. In fact, as we have defined the truth conditions for sentences of the form $QP\bullet S$ and $QF\bullet S$, the following holds:

Lemma 6.9. $t \models QP\bullet S$ iff $t \models Q{\sim}P{\sim}\bullet S$ and $t \models QF\bullet S$ iff $t \models Q{\sim}F{\sim}\bullet S$.

Proof. We prove just the case of $t \models QP\text{pa}S$ iff $t \models Q{\sim}P{\sim}\text{pa}S$, where Q is accidental and unrestricted, and leave the other cases as exercises for the reader.

($\leftarrow$) Suppose $t \models Q{\sim}P{\sim}\text{pa}S$. Since the categorical propositions respect the relationships in the square of opposition, ~pa can be replaced with un. If $\{\text{Ap}(Q,t)\cup\text{Ampt}^{\text{P}}(Q,t)\}\cap\text{Ampt}^{\text{P}}(S,t)=\varnothing$, then $t \models QP\text{un}S$. Since $t \models Q{\sim}P\text{un}S$, it follows by modus tollens that $\{\text{Ap}(Q,t)\cup\text{Ampt}^{\text{P}}(Q,t)\}\cap\text{Ampt}^{\text{P}}(S,t)\neq\varnothing$, and hence $t \models QP\text{pa}S$.

($\rightarrow$) As all of the implications involved in the proof of the other direction are equivalences, this case is symmetrical.

7. Conclusions

There is one interesting issue which is not generally addressed in medieval theories of supposition and which, because of its potential applicability to modern philosophical problems, warrants further investigation, and that is the issue of iterated tenses. Syntactically, nothing prevents us from nesting temporal operators, e.g., PFP(Qua$S$), $QGHF$ua$S$, etc. Three questions immediately arise from this: First, what sense can we give to the interpretation of these strings of temporal operators? Second, how must we modify the definitions in order to allow for iterated temporal operators? Third, what strings of temporal operators result in the same semantic outcome, that is, when can iterated temporal operators be reduced to a single temporal operator?

We briefly comment on the first question. When the iterated tenses are being used in their usual fashion, e.g., $PFP\varphi$, then the answer to the first question is straight-forward: the formula is read from left to right as normally: 'it was the case that it will be the case that it was the case that φ'. When used in the special way that we introduced above, e.g., QHFua$S$, the most natural reading of the sentence is to attach the first tense to the copula and the remaining tenses to the predicate, e.g., 'All Q's were-always will-be S'. When there are more than two temporal operators, this natural reading becomes more stilted, but even so we can still make sense of things like $QGHF$ua$S$ 'All Q's will-always-be were-always will-be S'. Tenses iterated in the first sense are well understood in both linear and branching time structures. This means that future investigation should focus on nested temporal operators used in the second fashion, since they are the ones that make use of the truth conditions based on ampliation and appellation.

The brief excursus into modern logic in the previous section shows the depth and breadth of Lambert's theory of supposition, and opens up the possibility of applying this theory to modern philosophical problems, such as questions about reference to non-existent entities and issues with combining quantification and modality. We have shown how from a relatively basic theory of supposition a very interesting and distinctive modal and temporal logic can be extracted. What this points to is that that rise of supposition theory over the course of the twelfth to fourteenth centuries was not just the rise of supposition theory, but the rise of well-defined and widely applicable modal and temporal logic. Since many of problems considered by the medieval logicians in their supposition theories still arise in philosophical discussions today, we have shown yet another reason why it is important for contemporary philosophical logicians to be familiar with the developments of their medieval predecessors, as there is much to be gained from a close investigation of their theories.

Bibliography

Primary Literature

Guillelmus de Shireswood: William of Sherwood, *Introductiones in Logicam*, in: William of Sherwood, *Introduction to logic*. Translated with an introduction and notes by N. Kretzmann (Minneapolis 1966)

——, *Introductiones in Logicam. Einführung in die Logik*, ed. H. Brands and C. Kann (Hamburg 1995)

Lambertus Autissiodorensis, *Logica* (*Summa Lamberti*), ed. F. Alessio (Firenze 1971)

Lambertus Autissiodorensis: Lambert of Auxerre, 'Properties of Terms', transl. N. Kretzmann, E. Stump, in: N. Kretzmann, E. Stump, eds., The *Cambridge Translations of Medieval Philosophical Texts*, 1: Logic and the Philosophy of Language (Cambridge 1988, 102-165)

Lambertus Autissiodorensis: Lambert of Lagny, *Appellationes*. ed. A. de Libera, in: id., 'Le *Tractatus de appellatione* de Lambert de Lagny (*Summa Lamberti* VIII)', in: *Archives d'histoire doctrinale et littéraire du moyen âge* 48 (1982), 227-285

Petrus Hispanus: Peter of Spain, *Tractatus, called afterwards Summulae logicales*. First Critical Edition from the Manuscripts with an Introduction, by L.M. de Rijk (Philosophical Texts and Studies, XXII; Assen 1972)

——, *Language in dispute*, transl. F.P. Dinneen, S.J. (Amsterdam-Philadelphia 1990)

Roger Bacon, *Summule dialectices*, ed. R. Steele (Opera hactenus inedita Rogeri Baconi, fasc. 15; Oxford 1940)

——, *Summulae dialectices*, ed. A. de Libera, in: id., 'Les *Summulae dialectices* de Roger Bacon. I-II. De termino. De enuntiatione', in: *Archives d' histoire doctrinale et littéraire du moyen-âge* 53 (1986), 139-289

——, *Summulae dialectices*, ed. A. de Libera, in: id., 'Les *Summulae dialectices* de Roger Bacon. III. De argumentatione, in: *Archives d'histoire doctrinale et littéraire du moyen âge* 54 (1988), 171-278

William of Sherwood *vide s.n.* Guillelmus de Shireswood

Secondary Literature

Chagrov, A. and M. Zakharyaschev (1997), *Modal Logic* (Oxford 1997)

Fitting, M., and R.L. Mendelsohn (1998), *First-order Modal Logic* (Synthese Historical Library, 277; Dordrecht 1998)

Knuuttila, S. (1982), 'Modal Logic', in: N. Kretzmann, A. Kenny and J. Pinborg, eds., E. Stump, ass. ed., *The Cambridge History of Later Medieval Philosophy From the Rediscovery of Aristotle to the Desintegration of Scholasticism 1100-1600* (Cambridge-London-New York-New Rochelle-Melbourne-Sydney 1982, 342-357)

—— (1993), *Modalities in Medieval Philosophy* (London-New York 1993)

Libera, A. de: see Lambertus Autissiodorensis; Roger Bacon

Prior, A.N. (1957), *Time and Modality* (Oxford 1957)

Rijk, L.M. de (1976), 'Some Thirteenth Century Tracts on the Game of Obligation, III', in: *Vivarium* 14 (1976), 26-49

The Expressive Power of Medieval Logic

Terry Parsons
University of California, Los Angeles

Abstract

This paper is about the development of logic in the Aristotelian tradition, from Aristotle to the mid-fourteenth century.

I will compare four systems of logic with regard to their expressive power.

1. Aristotle's own logic, based mostly on chapters 1-2 and 4-7 of his *Prior Analytics*.

2. An expanded version of Aristotle's logic that one finds, e.g., in Sherwood's *Introduction to Logic* and Peter of Spain's *Tractatus*

3-5. Versions of the logic of later supposition theorists such as William Ockham, John Buridan, and Paul of Venice. Version 4 is the logic without relatives (anaphoric pronouns); version 5 adds relatives.

I am ignoring modals, conditionals that are not ut nunc, infinitizing negation, exclusives and exceptives, all exponibles, all insolubles, and terms with simple or material supposition, ampliation and restriction, and many other things.

Keywords

supposition, reduction, exposition, expository syllogism, logical structure vs. truth conditions

1. Aristotle's Logic

Aristotle's logic is limited to the four forms of proposition in the traditional square of opposition plus some elementary principles about singular propositions.

Four forms of proposition.

Every A is B	*No A is B*
Some A is B	*Some A is not B*

Diagonally opposite propositions are contradictories; the universal propositions at the top are contraries. Regarding singular propositions, these are contradictories:

a is B ↔ *a is not B*

Logical inferences include:

- Conversions: *No A is B* ≈ *No B is A*

  Some A is B ≈ *Some B is A*

  Every A is B ⇒ *Some B is A*

- 19 forms of syllogism, such as: *Every B is C*

  Every A is B

  ∴ *Every A is C*

This logic is equivalent to a very restricted portion of the monadic predicate calculus. Syllogisms are decidable by means of simple mechanical tests, such as Venn diagrams.

Singular Propositions: Aristotle uses singular propositions of the form '*n is B*' or '*n is not B*' but he does not develop their logic completely. For example, he does not discuss the following argument, though he *uses* it at *PA*1.6 (28 a 25):

	Every A is B
	n is A
∴	*n is B*

Aristotle used three principles to establish his principles of conversion and syllogisms. They are:

Reductio	
Exposition:	*Some A is B*, so *c is A*, and *c is B*
Expository syllogism:	*c is A; c is B*, so *Some A is B*
	c is A; c is not B, so *Some A is not B*

It will be useful to employ these to justify the argument above, which I will call 'Aristotle's shortcut'.

1.	Every A is B	<premise>
2.	n is A	<premise>
3.	n is not B	<assumption for reductio>
4.	Some A is not B	Expository syllogism from 2, 3
5.	n is B	Reductio; line 4 contradicts line 1

Aristotle's own proof of the principle of conversion for universal negatives goes like this; to show:

No B is A
∴ No A is B

1.	No B is A	<premise>
2.	Some A is B	<assumption for reductio>
3.	c is A	Exposition from 2
4.	c is B	Exposition from 2
5.	Some B is A	Expository syllogism from 3,4
6.	No A is B	Reductio: line 5 contradicts line 1

Reduction of Syllogisms: Aristotle uses the conversion principles plus the first figure universal syllogisms (BARBARA and CELARENT) to prove all of the remaining forms of syllogism. Interestingly, he could also have proved the first figure universal syllogisms using only his principles of Reduction, Exposition, and Expository Syllogism. For example, here is a proof of Barbara:

1.	Every B is C	
2.	Every A is B	
3.	Some A is not C	<assumption for reductio>
4.	a is A	Exposition from 3
5.	a is not C	Exposition from 3
6.	a is B	Aristotle's shortcut from lines 2, 4
7.	Some B is not C	Expository syllogism from 6,5
8.	Every A is B	Reductio: line 7 contradicts line 1

2. The Thirteenth-Century Logic of William Sherwood and Peter of Spain

In some thirteenth-century writings Aristotle's notation is enhanced so as to make it more expressive. New sentence forms are available and new principles of inference are added.

Expansion of Aristotle's notation:

- Quantifying signs are allowed to accompany predicate terms

  No animal is every donkey

- Negation signs may occur before and after any terms (with their quantifier signs)

  Not every animal is not some donkey

- There are arbitrary verbs in addition to the copula:

  Some donkey <u>*runs*</u>
  Some donkey <u>*sees*</u> *every horse*

New principles of inference:

- Quantifier exchange laws, such as:
  '*every*' ≈ '*not some not*'
- The principle that any proposition of the form '*not P*' contradicts '*P*'.
- The principle that singular terms commute with common terms (together with their quantifier signs) and with negations.

Also allowed are conjunctions and disjunctions of propositions, and conditionals (both pure and 'ut nunc').

Ignoring pure conditionals, this system of logic is equivalent to a fragment of the monadic predicate calculus with identity. This is a richer symbolism than the monadic predicate calculus without identity. For example, quantifier signs occur within the scopes of other quantifier signs, and these embeddings cannot be eliminated; an example is '*No animal is every donkey*'.

This system of logic is still decidable, but it is harder to test for validity; for example, Venn diagrams are inadequate. So it is more sophisticated than the 'Aristotelian logic' taught in twentieth-century texts.

Sherwood also invoked some principles of supposition theory, such as 'from determinate supposition to merely confused supposition'; his illustration of this principle is:

A man is seen by every man
∴ Every man sees a man

The example may be confusing because of the unnecessary use of 'man' twice; the same principle is illustrated (in word order that is grammatical in Latin) by:

A donkey every man sees
∴ *Every man a donkey sees*

(Case inflections on the nouns make clear that '*man*' is the subject of '*sees*' and '*donkey*' is the direct object in both propositions.) I think that these sorts of examples did not actually require any new kinds of proof technique. For example, the argument just given could be proved using the thirteenth-century expansions of Aristotle's principles, as follows.

1.	A donkey every man sees	
2.	not every man a donkey sees	<assumption for reductio>
3.	some man not a donkey sees	Quantifier exchange from 2
4.	some man every donkey doesn't see	Quantifier exchange from 3
5.	m is a man	Exposition from 4
6.	m every donkey doesn't see	Exposition from 4
7.	every donkey m doesn't see	Permutation of singular term
8.	d is a donkey	Exposition from 1
9.	d every man sees	Exposition from 1
10.	every man sees d	Permutation of singular term
11.	m sees d	Aristotle's shortcut, lines 5, 10
12.	d m doesn't see	Aristotle's shortcut, lines 7,8
13.	m doesn't see d	Permutation of singular term
14.	Every man a donkey sees	Reductio; 11,13 contradict

Although this argument and many more like it can be derived using basic techniques available to medieval logicians, this was probably not widely known. So some of the rules that were stated in terms of modes of supposition were clearly useful.

3. The Fourteenth-Century Logic of Burley, Ockham, Buridan et al.

By the early fourteenth century writers were using a much expanded notation.

The inherited stock of propositional forms was expanded by adding:

- Propositions in which singular terms occur wherever common terms (with their signs) occur; e.g., singular terms now appear in predicate position.
- There are relational common terms:
  - participles of transitive verbs
    - *Socrates is a man seeing* <i.e., Socrates is a seer of a man>
  - terms which are objects of genitives
    - *Of-Socrates a donkey is running*
- There are relative clauses
  - *Every man who is white is running*

There are additional logical techniques. For example, there is an analogue of Leibniz's Law:

$$\begin{array}{ll} & a\ [\text{is}]\ \phi \\ & a \text{ is } b \\ \therefore & b\ [\text{is}]\ \phi \end{array}$$

(This was not needed previously because the second premise was not expressible until singular terms were allowed to appear in predicate position.)

Also, there was a theory of 'causes of (modes of) supposition' that gives an algorithm for determining whether a common term is determinate, distributed, or merely confused, and this classification sanctions certain descents and ascents.

These modes also allow the statement of some general and useful principles, such as 'From an inferior to a superior for terms with determinate or merely confused supposition' and 'From a superior to an inferior for terms with distributive supposition.'

Aristotle's techniques of Exposition and Expository Syllogism are now commonly used; the latter, but not the former, is explicitly discussed.

These vastly expanded the power of the logic. In order to say how much, a certain amount of formalization is useful.

4. A modern Symbolization of Medieval logic

The point of this section is to provide a modern symbolization of medieval logic. We begin by asking: What kind of *logical structure* do medieval logicians attribute to the propositions that they deal with? This question is equivalent to asking what kind of *grammatical structure* that is relevant to logic do medieval logicians attribute to the propositions that they deal with? I phrase the question in this way because it appears to me that the logical notation used by fourteenth-century logicians is somewhat regimented ordinary Latin, and so the only structure available to them is grammatical structure. I want to construct a logical notation that codifies only the grammatical structure available to medieval logicians.

- To begin, medieval logicians concentrated on sentences in which quantifier signs (like 'every', 'some', 'no') immediately precede the term that they are grammatically construed with. I'll group these signs and their terms together with parentheses:

  (Some horse) (every donkey) sees.

- Next, each term has a unique grammatical role in the proposition. For example, in the sentence above, the verb 'sees' provides two grammatical roles: subject and object. I will indicate these positions by placing a variable before the verb for its subject place and a variable after the verb for its direct object place:

  x sees y

Then I'll index each term with a variable to indicate which term plays which role. This gives a notation like this:

> (Some horse x) (every donkey y) x sees y.
> (Some horse y) (every donkey x) x sees y.

Sometimes inflections on a term tell you which role it has, and sometimes not; if not, the proposition has two readings, and the proposition stands in logical relations only with respect to a reading of it, which will be indicated with variables.

Another grammatical relation needing annotating is that between a genitive term and the term that it relates to (the possessor term and the possessed term). Variables can also be used here:

(of-every farmer x)(some donkey-of-x y) y runs

The term 'donkey' has its own grammatical role (it is the subject of 'runs', indicated by its variable 'y') and it also licenses another term to be its 'possessor'.

Likewise for participles of transitive verbs:

(every horse x)(some donkey y)(a seeing-y z) x is z

Lastly, a relative pronoun occupies a grammatical role in its relative clause, and the whole relative clause combines with a common term to make a complex common term, such as:

{horse which_y (some donkey x) x sees y}

Complex common terms occur wherever other common terms appear, as in:

(Every {horse which_y (some donkey x) x sees y} z)(some man u) u sees z.

I call these forms with variables *annotated propositions*. This resulting notation resembles that of symbolic logic with 'restricted' quantifiers such as ($\forall$Px), ($\exists$Qy). In fact, it is possible to assign suppositions to terms and to use such assignments in a contemporary way to give truth-conditions for annotated propositions which are the truth-conditions that medieval logicians attributed to them. These truth-conditions determine logical relations that medieval logicians attributed to them. This remains true when one allows conjunctions, disjunctions, and ut nunc conditionals of categorical propositions. This project is a long one, and I do not plan to carry it through and justify it here. (It will be available in book form in *Articulating Medieval Logic*, forthcoming from Oxford University Press.) I call the resulting system 'codified medieval logic'. I want to ask the question here: what is the expressive power of medieval logic, so understood?

Codified medieval logic has approximately the expressive power of the full predicate calculus with restricted quantifiers and with identity, but with a special constraint: predicate calculus sentences are limited to those in which each (restricted) quantifier binds exactly one occurrence of a variable. For example, in a formula of the form '($\forall$Px)ϕ' the subformula 'ϕ' contains exactly one free occurrence of the variable 'x'. This restriction is a reflection of the fact that each term of medieval logic has a unique grammatical role, and variables are used in codified medieval logic solely to encode these grammatical roles.

What is this codified symbolism like? On the one hand, it is quite rich, allowing the expression of things like:

> *Belonging-to-every man which some woman sees, some running donkey sees a horse which is not sitting*

On the other hand, this system of logic is (I think) decidable—that is, there is a mechanical way to determine for each sentence in the notation whether it is logically true or not. This is because every sentence that is satisfiable is satisfiable in a finite domain. It is not *easily* decidable, like classical syllogistic, but it is decidable.

One could, of course, expand this notation by allowing a restricted quantifier to bind more than one occurrence of its variable. But that would be to give up what is distinctive of medieval logic, which is that logical relations are based entirely on grammatical structure.

5. Buridan, Ockham with Relatives (i.e., Terms with Antecedents)

One asks naturally what happens if the notation of codified medieval logic is expanded with the addition of what medieval logicians call 'relatives', that is, of pronouns with antecedents. That depends on how this is done.

The theory of relatives occurs as early as Peter of Spain. The common theory of relatives, shared by several researchers, is that relatives have modes of supposition of their own, where the mode of supposition of a relative is stipulated to be the same as that of its antecedent. And a relative is said to 'supposit for' whatever its antecedent is 'verified for'. For example, in '*Some donkey is grey and it is sitting*' the relative '*it*' has determinate supposition (because its antecedent, '*donkey*' does), and it supposits for donkeys which are grey—which are the things that '*donkey*' is verified for in '*Some donkey is grey*'.

Without going into details, this account does not work well. For example, taken literally it would sanction this descent:

> *Every donkey sees itself* ⇒
> Every donkey sees this donkey & every donkey sees that donkey & . . . (for all donkeys that see themselves)

However, when the stated account is (supposedly) illustrated in texts, the authors never actually show a descent under a relative. Instead, they show a descent under the antecedent of the relative under discussion, and the descended term becomes the new antecedent of the relative, which itself is unchanged. For example, the semantics of '*itself*' is illustrated by this descent which is supposedly under '*itself*' but is actually under 'donkey':

Every donkey sees itself ⇒
This donkey sees itself & that donkey sees itself & . . . (for all donkeys).

The relative itself remains unchanged in descent, but its antecedent is now '*this donkey*' or '*that donkey*' instead of '*every donkey*'. Then (every writer agrees that) when the antecedent of a relative is a singular term, the relative may be replaced by it. So it turns out that 'Every donkey sees itself' is equivalent to 'This donkey sees this donkey & that donkey sees that donkey & . . . (for all donkeys)'. This seems to me to be a completely accurate account.

I think that this system of logic is not decidable. One can formulate a sentence that is satisfiable in an infinite domain, but not in a finite domain:

No thing exceeds itself, and
every thing is exceeded by some thing, and
every thing is exceeded by every thing which exceeds a thing which exceeds it (= the first thing).

In symbols:

(No thing x) (itself^{x} y) x exceeds y
and
(Every thing x) (some thing y) y exceeds x
and
(Every thing x) (every {thing which$_{y}$ (a {thing which$_{z}$ (it^{x} u) z exceeds u}v) y exceeds v} w) w exceeds x

I would *like* to claim that given any sentence of the first-order predicate calculus with restricted (nonvacuous) quantifiers and with identity there is a sentence of medieval logic with an equivalent semantics. Is this so? The answer is at present uncertain. The problem again is grammar: there are no natural subjects or direct objects in predicate calculus formulas, so most predicate calculus formulas correspond to ungrammatical sentences of natural language.

For example, given a statement of symmetry in the predicate calculus:

$$\forall x \forall y (xRy \rightarrow yRx)$$

one can try to convert it into codified medieval logic by writing:

(every thing x)(every thing y) if xRy then (it^x u)(it^y v)uRv

But this has an 'if' between the 'everything' and the rest of the sentence that the 'everything' is supposed to be grammatically part of. This is not grammatically well-formed.

A slightly artificial solution would be to introduce the locution 'is such that' into the regimented grammar. It takes a single subject and it is followed by a grammatical sentence. Then instead of:

$\forall x \forall y(xRy \to yRx)$

one can write:

(every thing x) x is such that (every thing y) y is such that if (it^x u)(it^y v)uRv then (it^y v)(it^x u)vRu

That is, without the variables:

Everything is such that everything is such that if it (the first thing) Rs it (the second thing) then it (the second thing) Rs it (the first thing).

But this 'such that' locution was not used by medieval logicians in this way.

The exact relation between contemporary and medieval logic is still not fully understood.

INDEXES

Index Locorum

Adam Wodeham, *Lectura super Sententias*, in: ms Vatican, *Biblioteca Apostolica Vaticana*, lat. 955
I, 33.1.2, ff. 177v-178r 269
I, 33.1.2, f. 177v 268
I, 33.1.3.3, f. 192r 273
I, 33.3.1, f. 186r 271
I, 33.3.2, ff. 1r-208v 268
I, 33.3.2, f. 187v 272
I, 33.3.2, f. 188r 271, 272
I, 33.3.2, f. 188r-v 272
Adam Wodeham, *Super quattuor libros Sententiarum*, abbreviated by Henry Totting of Oyta (1512)
I, 33 268
Aegidius Romanus, *Expositio super libros Elenchorum*, ed. Venetiis (1496)
f. 9vb 215
ff. 10vb-11ra 343
f. 15ra 218
f. 16rb 218
f. 17rb 218, 220, 221
ff. 54rb-55va 224
f. 55rb 223
f. 87vb 219
Aegidius Romanus, *Quaestio quid sit medium in demonstratione*, ed. Pinborg (1984)
III, 255-268 209
Alanus ab Insulis, *Regulae caelestis iuris*, ed. N.M. Häring (1981)
217-226 133
Albertus de Saxonia, *Quaestiones Circa Logicam*, ed. Fritzgerald (2003)
qu. XXIV, pp. 311-318 417
qu. XXIV, p. 315, ll. 12-15 417
Albert of Saxony, *Perutilis Logica*, ed. Venice (1522)
II, 2, f. 11ra 338
II, 2, f. 11va 387
II, 4, f. 11ra 392
II, 4, f. 11va 397
II, 10, f. 15vb 356
Albert de Saxonia, *Sophismata*, in: Pozzi (1987)
316 366
Albertus Magnus, *Liber Elenchorum*, ed. Borgnet (1890)
I, t. 3, cc. 1-3, 557-560 218
I, t. 3, c. 3, 560 218
I, t. 3, c. 4, 565 225
I, t. 3, c. 6, 571 215
I, t. 3, c. 20, 596 207
I, t. 4, c. 1, 602-603 214
I, t. 4, c. 1, 603 208
I, t. 4, c. 2, 604 208
I, t. 4, c. 7, 612 208
I, t. 5, c. 1, 616 208
II, t. 1, c. 7, 681 218
II, t. 1, c. 7, 684 224
II, t. 2, c. 7, 681-683 218
II, t. 3, c. 3, 561 217
II, t. 3, c. 3, 563 217
II, t. 3, c. 4, 563 220
Alexander of Aphrodisias, *In Aristotelis Analytica Priora*, ed. Wallies (1882)
32, 32-33,15 100
99, 19-100, 26 100
Ammonius, *In Aristotelis De Interpretatione*, ed. Busse (1897)
186, 15 94
214, 25 106
215, 14-16 106
Anonymus *Commentum (...) in primum et quartum tractatus Petri Hyspani*, ed. Hagennaw (1495)
sig. n 7 r-v 394
sig. o 4 v 392
sig. q 1 r 390, 392
sig. q 1 v 392
sig. q 4 r 392
sig. q 5 v 385, 386
sig. q 7 r 390, 394

sig. q 7 r-v 390
sig. r 1 v 392, 405
sig. r 2 r 402
Anonymus Aurelianensis III, *Commentarium in Analytica Priora*, in: ms Orléans, *Bibliothèque municipale*, 283
188A 64
Anonymus, *Ars Meliduna*, in: ms. Oxford, *Bodleian Library, Digby*, 174
f. 217vb 66
f. 218va 70
f. 225ra 70
f. 227va 70
Anonymus, *Tractatus de dissimilitudine argumentationum*, ed. De Rijk (1962-1967), II/2
489, l. 34 218
Anonymus, *Ars Emmerana*, ed. De Rijk (1962-1967), II/2
154 170
Anonymus, *Logica 'Ut dicit'*, ed. De Rijk (1962-1967), II/2 174
383 170
388 217
409 192
446. 192
447 193
Anonymus, *Dialectica Monacensis*, ed. De Rijk (1962-1967), II/ 2 57, 71, 72, 74, 83, 172, 175, 184
VI 110
409 83
446-448 83
456 83
469 170
582 172
584-586 83
585 217, 218
591 217, 218
605-638 76
606 110
606 107, 166
606, ll. 23-26 107
607-608 177
608, ll. 14-19 83
616, ll. 20-30 110
Anonymus, *Tractatus de proprietatibus sermonum*, ed. De Rijk (1962-1967), II/2
709-710 175
713 176
723 176
Anonymus, *Tractatus 'Cum sit nostra'*, ed. De Rijk (1962-1967), II/2 174, 190
446 182
447 132, 183
448 137
Anonymus, *Glose in Aristotelis Sophisticos Elenchos*, ed. De Rijk (1962-1967), I
215 217
Anonymus, *Glossae super Porphyrium*, ed. Geyer (1919)
17, 12-28 93
Anonymus, *Glossae super Praedicamenta*, ed. Geyer (1919)
122, 29 93
Anonymus, *Summe Metenses*, ed. De Rijk (1962-1967) 175
455 132, 177
456 177
Anonymi: Incertorum Auctorum *Quaestiones super Sophisticos Elenchos*, Anonymus SF, ed. Ebbesen (1977)
q. 13, n. 16, 369 242
q. 47, 102 243
q. 48, 102 242
q. 48, 106-107 238, 243, 247
q. 69, 158-162 213
q. 80, 179 218
q. 80, 183 218
q. 82, 189, ll. 34 f. 218
q. 83 218
q. 85, 196 224
Anonymi: Incertorum Auctorum *Quaestiones super Sophisticos Elenchos*, Anonymus C, ed. Ebbesen (1977)
q. 835, 350-351 207
q. 839, 368 218
q. 840, 371 218
Anonymus Monacensis, *Commentum in Sophisticos Elenchos*, in: ms Admont, *Stiftsbibliothek*, 241
f. 17vb 67
Anonymus Monacensis, *Commentum in Sophisticos Elenchos*, in: ms Munich, *Bayerische Staatsbibliothek, clm*, 14246
f. 8rb 67

Anonymus Tabarroneus, *Sophismata* (1995)
315-318 221
Anonymus, *Introductiones Parisienses*, ed. De Rijk (1962-1967), II/1 74, 184
Anonymus, *Summule antiquorum*, ed. De Rijk (1968) 175, 180, 184, 185, 187
13 179
Anonymus, *Commentary on the Prior Analytics*, in: ms. Cambridge, *Peterhouse* 206
f. 100va 62
Anonymus, *Commentarium in Boethii De topicis differentiis*, ed. Hansen (2005)
67 63
87-88 63
Anonymus, *Compendium logicae Porretanum*, eds. Ebbesen *et al.* (1983)
I, 20, 9 61
I, 23, 10-11 61
III, 29, 52 61
V, I. 7 116
V, I. 8 116
Anonymus, *Lectura Tractatuum*, ed. De Rijk (1969)
143 198
146 198, 199
147 196
Anonymus, *Fallaciae Londinenses*, ed. De Rijk (1962-67), II/ 2
669 217, 218
Anonymus, *Fallacie Vindobonenses*, ed. De Rijk (1962-1967), I
527, ll. 25-27 218
Anonymus, *Fallacie Parvipontane*, ed. De Rijk (1962-1967), I
565 137
569 137
Anonymus Parisiensis, *Compendium Sophisticorum Elenchorum*, ed. Ebbesen (1990)
288 218
Anonymus, *Notabilia in libros Topicorum, II*, in: ms *Bibliotheca Apostolica Vaticana, Ottob. Lat.*, 318
f. 258rb 239
Anonymus, *Summa Sophisticorum Elenchorum*, ed. De Rijk (1962-1967)
I, 356 216
Anonymus, *Summa Zwettlensis*, ed. Häring (1977)
9 121
30, 18 130
30-33 133
62, 127 124
62, 128 125, 126
62, 129 125, 126
62, 130 125
62, 136 126
62, 137 126
63, 130-131 127
63, 132 124
64, 134 125, 134
64, 135 126, 129
66, 145 130
67, 149 124
Anonymus, *Fallacie Magistri Willelmi*, ed. De Rijk (1962-1967), II/2 57
691 218
Anonymus, *Tractatus de univocatione Monacensis*, ed. De Rijk (1962-1967), II/ 2 173
ff. 121r-141r 71, 76
Anonymus, *Tractatus Monacensis, Fallaciae*, in: ms München, *Bayerische Staatsbibliothek, clm*, 14763
f. 123vb 71
Anonymus, *Commentary on Perihermeneias*, in: ms. Orléans, *Bibliothèque Municipale* 266
f. 261b 32
Aristoteles Latinus, *De sophisticis elenchis*, transl. Boethii, ed. Minio-Paluello (1975)
11, ll. 8-16 (V, 166 b 29-37) 216
15, l. 25-16 l. 2 (V, 168 a 37-168 b 4) 217
15, ll. 20-21 (V, 168 a 35) 216
47, ll. 9-13 (XXIV, 179 b 38-39) 216
Aristoteles, *Metaphysica*
VII, 6, 1031 b 25-26 37
VII, 6, 1032 a 1 99
VII, 6, 1031 a 24-29 216
VII, 4 92
VII, 6 98
VII, 17-VIII, 1-6 42
XIII, 3, 1077 b 31-1078 a 5 99

Aristoteles, *De interpretatione*
I, 16 a 3-8 102
I, 16 a 6-7 17
III, 16 b 6-7 34
III, 16 b 8-9 34
III, 16 b 24-25 29
VII, 3-5 37
VII, 4 92
VII, 6 98
VII, 6, 1031 a 24-29 216
VII, 6, 1032 a 1 99
VII, 17- VIII, 1-6 42
VIII 17, 23
X, 21 a 25-8 94
Aristoteles, *Physica*
IV, 10-14 19
Aristoteles, *Analytica Posteriora*
I, 4 216
I, 22 92
I, 13, 32 b 24-32 99
I, 15, 34 a 34 - b 6 99
II, 1-2 99
Aristoteles, *Praedicamenta (Categoriae)*
I, 1 a 12-15 92
V 99
V, 3 b 10-12 17
V, 3 a 21-4 92
III, 1 b 10-11 216
V, 3 b 13-14 216
V, 3a 21-24 92
VII, 7 a 35-39 21
VIII 98
X, 10 a 27-b 11 92
Aristoteles, *Analytca Priora*
I, 6, 29 a 4-10 102
I, 9, 30 b 5-6 102
I, 13, 32 b 24-32 99
I, 15, 34 a 34-b 6 99
Aristoteles, *Sophistici Elenchi*
IV, 166 b 10-19 150
V, 166 b 29-37 216
V, 168 a 35 216
VI, 168 b 1-4 217
VI, 168 b 28-30 217
XXIV, 179 b 38-39 216
Aristoteles, *Topica*
I, 7, 103 a 9-10, 216
V, 4, 133 b 31-36 216
VI, 6, 144 a 20-1 92
Augustine of Ferrara, *Quaestiones super librum Praedicamentorum Aristotelis*, ed. Andrews (2000)
44-45 263
Augustinus Hunaeus, *Logices fundamentum*, ed. Antwerp (1568)
7 467
Augustinus Nifus, *Dialectica ludrica*, ed. Venetiis (1521)
ff. 86v-87r 465
Avicenna, *Al-ʿIbāra*, ed. M. al-Ḳuḍayrī (1970)
15 100
77, 8-84, 13 90
82, 16-18 94
103, 5 102
109, 2-110, 1 94
110, 2 ff 94
110, 7-14 94
23, 1 111
16 ff. 102
Avicenna, *Al-Ilāhiyyāt*, ed. Anawati (1960)
29, 5-6 102
31, 5 102
195, 8 109
196, 6-197, 5 106
201, 7-13 86
202, 2-203, 8 106
202, 3-8 87
204, 16-205 87
205, 2 86
207, 5-12 86
208, 5-9 87
236, 6-8 86
241, 5-16 86
Avicenna, *Al-Ishārāt*, ed. Dunya (1971)
1010-1012 87
III, 2, 1, 271, 8-12 .93
Avicenna, *Al-Madkhal*, eds. Al-Ahwānī *et al.* (1952)
15, 1-6 85
28, 13-29, 6 87
36, 8 86
48, 15 86
74, 11-75, 21 87
Avicenna, *Al-Nafs*, eds. Anawati and S. Zayid (1975)
32, 7 ff. 94
147, 1 ff. 94
148, 14-15 89
184, 9-10 89
208, 3-209, 7 89
Avicenna, *Al-Najāt*, ed. Kurdi (1938)
451 87
Avicenna, *Al-Ṭabiyyāt*, ed. Zāyid (1965-1983)
vol. 2.1, 13, 1-12 87

Avicenna, *Al-Qiyās*, ed. Zāyid (1964)
20, 11-21, 5 91
26, 18-27, 2 96
76, 2-4 100
27, 4-5 96
82, 15-83, 1 95
83, 10-11 93, 109
84, 16-85, 2 93
Avicenna, *Kitāb al-Burhān*, eds. Alfifi and Madjur (1956)
71, 15 93
Avicenna, *Kitāb al-Najāt*, ed. Al-Kurdi (1963)
451 87
6, 2-3 94
Avicenna, *Manṭiq al-Mašriqiyyīn*, eds. Al-Ḵaṭīb *et al.* (1910)
64, 2-4 93
71, 26-72, 6 96
Bartholmaeus Arnoldi of Usingen, *Summa compendiaria*, ed. Basle (1507)
f. 6r 473
Bartholomaeus Arnoldi of Usingen, *Exercitium veteris artis*, ed. Erfurt (1514)
f. S3r 471
Boethius, *De syllogismis categoricis* 31
Boethius, *In Categorias Aristotelis libri quattuor*, in: Opera omnia, ed. Migne (1860)
175D-176A 216
Boethius, *In Porphyrium*, editio prima, ed. Brandt (1906)
81-82 192
Boethius de Dacia, *Quaestiones super libros Physicorum*, ed. Sajó (1954)
I, 12, 152-154 72
Boethius de Dacia, *Quaestiones super libros Topicorum*, eds. Pinborg and Green Pedersen (1976)
32, ll. 34-42 220
Boethius de Dacia (pseudo), *Quaestiones in libros Analyticorum Posteriorum*, in: Pinborg (1971)
I, q. 26, p. 53 250
Bonaventura, *Commentaria in quatuor libros Sententiarum*, ed. Collegium S. Bonaventurae (1882)
I. d. 5, a. 1, q. 1, ad 2-3 265
I, d. 21, a. 1, q. 1 254, 265
Christian August Crusius, *Wege zur Gewissheit* (...), ed. Leipzig (1747)
1080 478
Christian Weise, *Doctrina logica* (...), ed. Leipzig-Frankfurt (1690)
440ff. 476
Crathorn, *In I Sententias*, ed. Hoffmann (1988)
q. 13 50
q. 13, p. 392, 16-23 50
q. 13, p. 392, 24-393, 8 51
q. 13, p. 394, 4-9 52
q. 13, p. 394, 18-26 52
David Hume, *An Enquiry Concerning Human Understanding*, ed. Selby Bigge (1981)
§ 7, pp. 134-147 54
§ 8, pp. 148-164 54
Everardus Yprensis, *Dialogus Ratii et Everardi*, ed. Häring (1953)
143-147 121
252 134
253 134
254 135
257 140
269 136
270 135, 137
270, 28 128
273 f. 126
Dominicus Soto, *Summule*, ed. Burgos (1528)
f. xvi va 392
f. xix rb 391
f. xix rb-va 391
f. xx rb 396
f. xx rb-vb 396
f. xxiii va-vb 395
f. xxiii vb 399, 400
f. xxiiii ra-rb 399
f. xxiv ra 394, 396, 399
f. xxiiii rb 391, 392, 399, 400
f. xxv rb-va. 404
f. xxv vb 396, 402, 403
f. xxv va-vb 406
f. xxviii va-xxix ra 400
Franciscus of Prato, *Tractatus de suppositionibus*, ed. Amerini (1999/2000)
I, 487-488, ll.46-69 339
II, 492, ll. 120-123 348
VI, 501, 4-7 339
VI, 501, 7-9 339

VII, 510, 171-184 340
VII, 504-505, 4-5 340
VII, 504-505, 18-20 340
VII, 504-505, 36-81 340
VII, 508, 122-134 340
VIII-IX, 511-518 347
Appendix 2, 549-550 347
Gerard Frilden, *Exercitium veteris artis*, ed. Rostock (1507)
f. v3va 471
Georgius Bruxellensis, *Expositio Georgii super summulis magistri Petri Hyspani*, ed. Parisius (1491)
sig. A 1vb-sig. A 2 ra 394
sig. A 5va 401
sig. A 7rb 392
Georgius Rovegnatinus, *Ex divo Thoma suppositionum collectula* (ed. post 1504)
I, f a ii^v 331, 343
III, ff. b i^r-b iii^v 343
III, ff. a iv^v-b i^r 344
Gilbertus Porretanus, *Expositio in Contra Euticen*, ed. Häring (1966)
293, 49 f. 130
296, 31-33 122
296, 31-297, 66 137
301, 77-82 122
345, 40-346, 47 122
349, 50-51 122
Gilbertus Porretanus, *Expositio in Utrum Pater*, ed. Häring (1966)
17, 37 128
166, 78 ff. 128
Gilbertus Porretanus, *Expositio super librum Boecii De trinitate*, ed. Häring (1966)
67, 50-54 122
86, 31-37 137
176, 17-19 120
Giraldus Odonis, *Tractatus de suppositionibus*, ed. Brown (1975)
1.11-1.12 346
1.2-1.21, 14 346
Girolamo Savonarola, *Compendium logicae*, eds. Garfagnini and Garin (1982) 328
10. 12, p. 144, ll. 19-22 342
10. 13, p. 144, l. 24-145, l. 7 342
10. 13, p. 145, ll. 8-24 344
Gregorius Ariminensis, *Lectura super primum et secundum Sententiarum*, ed. Trapp *et al.* (1979-1984) 357
Gualterus Burleus, *Middle Commentary on Aristotle's Peihermeneias*, ed. Brown (1973)
1. 819, 94 356
Gualterus Burleus *De puritate artis logicae tractatus longior*, ed. Boehner (1955)
1.1.1, p. 2, l. 14 - p. 3, l. 25 338
I.1.1, p. 2, ll. 17-18 306
1.1.2, p. 2, ll. 17-18 306
I.1.3 103
I.1.3, p. 7, ll. 6-10 305
1.1.3, p. 7, l 17 - 10, l. 34 341
I.1.3, pp. 6-10 308
I.1.4, p. 21 244, 246
I.4, pp. 28-33 276
I.5-6, pp. 45-47 276
I.1.3, p. 2, ll. 6-10 305
Gualterus Burleus, *Expositio in libros octo Physicorum Aristotelis*. Prologus, ed. Venetiis (1501)
f. 8vb 322
Gualterus Burleus, *Expositio super Praedicamenta Aristotelis: de sufficientia praedicamentorum*, ed. Venetiis (1509)
f. 19va-b 311
f. 21ra 311
f. 24rb 311
f. 34rb 311
Gualterus Burleus, *Quaestiones in librum Perihermeneias*, ed. Brown (1974)
1, 88, 215 356
Gualterus Burleus, *Tractatus de suppositionibus*, ed. Brown (1972)
2.2-2.3, 35-36 338
35-36 308
Guillelmus de Heytesbury, *De sensu diviso et composito*, ed. Venetiis (1494)
f. 4vb 356
Guillelmus de Heytesbury, *Sophismata*, ed. Pironet (1994)
sophism 16, p. 3 356
Guillelmus de Ockham, *In librum secundum Sententiarum, Reportatio*, eds. Gál and Wood (1981)
q. 1, p. 20, ll. 10-14 358
q. 1, p. 22, ll. 10-16 33
q. 15, p. 376, ll. 5-9 358

Guillelmus de Ockham, *Quaestiones in librum quartum Sententiarum. Reportatio*, eds. Wood and Gál (1984)
IV, q. 16, pp. 358-359 379
Guillelmus de Ockham, *Quodlibeta septem*, ed. Wey (1980)
VI, q. 29, p. 696 33
Guillelmus de Ockham, *Summa logicae*, ed. Gál (1974)
I, 4, pp. 15-16 235
I, 8, ll. 8-32 105
I, 8, ll. 55 ff. 105
I. 13 102
I, 15, pp. 50-51 318
I, 33 103
I, 63 360, 383
1, 63, p. 193, ll. 2-3 338
I, 63, p. 193 360, 383
I, 63, p. 194, ll. 16-32 360
1, 63, p. 194, ll. 33-35 336
I, 64, p. 195, ll. 5-6 103
1, 64, p. 195, ll. 4-7 336
I, 64, p. 195-197 371
I, 64, pp. 195-196, ll. 4-37 105
1, 64, p. 196, ll. 31-32 104, 336
1, 64, p. 196, ll. 33-37 337
I, 64, p. 196, ll. 38-39 336
I, 64, p. 196 371
I, 64, p. 196 371
I, 64, p. 197 373
I, 65 111
1, 65, pp. 197-198, ll. 3-10 337
I, 65, pp. 197-198 337, 374
I, 65, p. 197 379
I, 65, p. 199, ll. 43-57 337
I, 65, p. 198 379, 380
I, 65, p. 198, ll. 17-21 361
I, 70 84
I, 70, p. 210, ll. 21-25 361
I, 70, p. 210 361, 395
I, 70, p. 211 396
I, 72, pp. 218-219, ll. 113-130 363
I, 72, p. 219, ll. 135-139 368
I, 73, p. 227, ll. 24-27 394
I, 73, pp. 231-233 394
I, 74, p. 196 164
I, 75, p. 231 396
I, 76, p. 235, l. 69 396
I, 76, p. 177, ll. 233-238 276
II, 5, p. 266, ll. 192-204 390
II, 2, pp. 249-250, ll. 8-21 360
II, 2-4, pp. 249-266 264
III, 1.4, pp. 370-371 268
III, 3, 32, pp. 708-714 398
III, 4, 4, p. 763 375
Guillelmus de Ockham, *Tractatus de praedestinatione et praescientia Dei* [...], eds. Boehner-Brown (1978)
q. 1, 508, l. 42 108
Guillelmus de Ockham, *Scriptum in Librum primum Sententiarum. Ordinatio*, eds. Etzkorn and Kelley (1978)
d. 27, q. 3, p. 255, ll. 3-8 359
Guillelmus de Ockham, *Expositio in librum Praedicamentorum Aristotelis*, ed. Gál (1978)
c. 8. 1, pp. 164-168 318
c. 9, 13. p. 201, ll. 30-34 355
Guillelmus de Ockham, *Quaestiones in libros Physicorum Aristotelis*, ed. Brown (1984)
q. 3, p. 402, ll. 53-60 358
Guillelmus de Ockham, *Expositio in librum Perihermeneias Aristotelis*, ed. Gambatese and Brown (1978)
Prooe., 9, p. 366, 75-86 359
c. II, p. 383 35
c. II, p. 389, 14-19 35
c. III, p. 389 35
Guillelmus de Shireswode, *Introductiones ad Logicam*, eds. Brands and Kann (1995)
V, 132 65, 235
V, 134 173
V, 136 213
V, 136-138 84
V, 138 244, 248
V, 140 189
V, 140-142 247
V, 142 245
V, 144 194
VI, 196 218
Henricus de Gandavo (adscr.), *Syncategoreumata*, eds. Braakhuis et al. (2011)
58, 1772-1774 47
Hervaeus Natalis, *Quodlibet*, ed. Venice (1513)
I, q. 2, ff. 4vb-7va 335
III, q. 1, ff. 67vb-70ra 348
Hervaeus Natalis, *Scriptum super IV Sententiarum*, ed. Paris (1647)
prol., q. 7, f. 21^{b}B 334
1.5.1, f. 51^{b}A-C 334
1.23.1, ff. 112^{b}B-113^{a}C 334

Hieronymus de Sancto Marco, *Compendium*, ed. Cologne (1507)
sig. B 3 r 394
sig. C 1 v 399
sig. C 2 v 395
Hume, *see s.v.* David Hume.
Jacopo Zabarella, *De natura logicae*, ed. Cologne (1597)
62 f. 471
Joachim Ehrenfried Pfeiffer, *Elementa hermeneuticae* (...), ed. Jena (1743)
108-111 476
132-134 476
Joachim Georg Darjes, *Introductio in artem inveniendi*, ed. Jena (1742)
257 477
267 477
Joachim Georg Darjes, *Weg zur Wahrheit*, Frankfurt/Oder (1776)
159 476, 477
Johann Clauberg, *Logica vetus et nova quadripartita*, ed. Amsterdam (1691)
852 f. 475
Johann Conrad Dannhauer, *Idea boni disputatoris*, ed. Strassburg (1629/2009)
53-65 474
54 ff. 475
Johann Conrad Dannhauer, *Idea boni interpretis* [...], ed. Strasburg (1630/2004)
4 470
11 472
24 f. 472
101 f. 473
118 474
196 473
Johann Peter Reusch, *Systema logicum*, ed. Jena (1732)
266 476
Johannes Arboreus, *Compendiaria in dialecticen elementa introducttio*, ed. Lyons (s.a.)
48 465
51 464
Johannes Buridanus, *Quaestiones De anima (secundum tertiam lecturam)*, ed. Patar (1991)
III, 9, Appendix V, p. 826 382
Johannes Buridanus, *De fallaciis*, ed. Ebbesen (1976)
7.3.4, p. 156 373
Johannes Buridanus, *Quaestiones in Aristotelis Metaphysicam*, ed. Paris (1518)
IV, q. vi 40
IV, q. xii 43
IV, q. xv, f. 25va 417
IV, q. xv, f. 25vb 415
VII, q. i 42
VII, q. iii 38
Johannes Buridanus, *Tractatus de consequentiis*, ed. Hubien (1976)
I, 5, pp. 25-26 264
III,1.4.1, p. 86 268
III, II, 3, p. 102 377
Johannes Buridanus, *Questiones longe super Perihermeneias*, ed. Van der Lecq (1983)
I, 9, p. 44, ll. 12-14 414
Johannes Buridanus, *Summulae. De practica sophismatum*, ed. Pironet (2004)
II, 6, p. 42- p. 46 389
III, 5, p. 53 398
VI, 1, p. 116, 3-8 419
VI, 1, p. 116, 9-11 419
VI, 1, p. 119, 10-13 420
VI, 2, p. 119, 17-18 420
VI, 2, p. 120, 6-9 421
VI, 2, p. 120, 14-20 421
VI, 4, p. 124, 3-5 432
VII, 1, p. 128, 22-p. 129, 8, 414
VII, 1, p. 129, 1-2 414
VII, 2, p. 130, 14-17 414
VII, 2, p. 130, 17-19 405
VII, 3, p. 131, 11-16 418
VII, 4, p. 131, 17- p. 133, 10 417
VIII, 2, p. 144, 22-28 412
Johannes Buridanus, *Summulae. De suppositionibus*, ed. Van der Lecq (1998)
4.1.1, p. 7 383
4.1.2. 104, 372
4.1.4. 108
4.1.2, p. 9, 20-p. 10, 2 339
4.3.1, p. 37 371, 372
4.3.2, p. 38 372
4.3.2, p. 38-44 338
4.3.4, p. 47, 10-12 416

4.3.4, p. 47, 25-27 416
4.3.5, p. 49-50 395
4.3.6, p. 50 396
4.3.6, p. 51 402
4.3.8.4, p. 68 377
4.4.1, p. 70 392
4.4.4, p. 72 388
4.5.1 108
4.5.3, p. 84 376
4.6.1. 110

Johannes Dacus, *Summa Gramatica*, ed. Otto (1955)
112 165
180 165
371 248

Johannes Dorp, *Commentary in: Johannes Buridanus, Compendium logicae*, ed. Venetiis (1499)
sig. b 1vb 390
sig. b 4ra 386, 394
sig. b 4ra-rb 394
sig. b 6va 394
sig. c 1rb 394
sig. c 3ra, 394
sig. h 6ra 397
sig. h 6vb 397
sig. i 1ra 397
sig. i 1ra-rb 394
sig. i 2ra 392
sig. i 3ra-rb 390
sig. i 3va 401

Johannes Duns Scotus (pseudo), *Quaestiones super libros Posteriorum*, eds. Wadding-Vivès (1891)
I, q. 23, p. 257 222

Johannes Duns Scotus, *Lectura in librum primum Sententiarum*, ed. Commissio Scotistica (1960)
I, d. iv, q. unica, n. 4, 408 236
I, d. xxi, q. unica, n. 13, 299 237, 238

Johannes Duns Scotus, *Ordinatio*, dist. iv-x, ed. commsssio Scotistica (1956)
I, d. iv, q. unica, n. 11, p. 5 254

Johannes Duns Scotus, *Lectura* I, ed. Commissio Scotistica (1960)
I, 2.2.1-4, 136 266
I, 2.2.1-4, 280-281 266
I, d. 21, q. unica, n. 13, 299 237, 238

Johannes Duns Scotus, *Ordinatio* I, ed. Commissio Scotistica (1959)
I, d. 21, q. unica, n. 30, 337, n. 2 226, 244

Johannes Duns Scotus, *Ordinatio* I, eds. Balić *et al.*, vol. 2 (1950)
I.2.2, qu. 1-4, n. 415, p. 363 267
I.5.1, qu unica, nn. 32-33, 282-289 266
I.8.1, qu. 4, nn. 218-222, pp. 274-277 266

Johannes Duns Scotus, *Quaestiones in librum Porphyrii Isagoge*, eds. Andrews *et al.* (1999)
q. 24, n. 12, 153 248
qq. 9-11, n. 7, 45 247
qq. 9-11, n. 25 90, 239

Johannes Duns Scotus, *Quaestiones super librum Elenchorum Aristotelis*, eds. Andrews *et al.* (2004)
q. 15, n. 7, 333 237
q. 15, n. 11, 334-335 253
q. 15, n. 15, 335-336 236
q. 15, n. 17, 336 215
q. 41, n. 11, 461 253
q. 41, n. 15, 462 253

Johannes Duns Scotus, *Quaestiones super Praedicamenta Aristotelis*, eds. Andrews *et al.* (1999)
q. 8, n. 2, p. 314 241
q. 8, n. 19, p. 319 234
q. 8, n. 24, p. 321 242, 249
q. 11, n. 28, p. 360 238
q. 12, n. 13, p. 355 244
q. 12, n. 13, p. 358 236, 238, 241, 244, 251
q. 12, n. 16, p. 359 238
q. 12, n. 15, p. 359 238
q. 12, n. 16, p. 359 238
q. 12, n. 23, p. 360 238, 266
q. 12, n. 28, p. 362 222
q. 12, n. 30, p. 362 244
q. 12, n. 31, p. 372 238
q. 13, n. 3, p. 365 251
q. 13, n. 14, p. 368 242
q. 13, n. 15, p. 368-p. 369 250
q. 13, n. 17, p. 369 238, 260
q. 13, n. 31, p. 372 250
q. 13, n. 33, p. 373 237, 250, 254
q. 13, n. 37, p. 374 242, 244, 250
q. 13, n. 46, p. 377 242
q. 15, n. 17, p. 369 238
q. 15, n. 31, p. 372 238

Johannes Duns Scotus, *Quaestiones in duos libros Perihermenias*, ed. Andrews *et al.* (2004)
I, qq. 5-8, n. 44, p. 83 251
I, q. 6, p. 167 ff. 252
I, q. 9, p. 97 ff. 237
I, qq. 9-11, n. 24, p. 103 251

I, qq. 9-11, n. 25, p. 104 239
I, qq. 9-11, n. 44, p. 111 251
I, q. 13 234
I, q. 13, p. 131 237
I, q. 13, n. 14, p. 126 240
I, q. 13, n. 28, p. 130 240
I, q. 13, n. 29, p. 130 247
I, q. 13, n. 29-n. 33, p. 130-p. 132 240
I, q. 13, n. 32, p. 131 237
Johannes Duns Scotus, *Reportata Parisiensia*, ed. Wadding (1639)
I, d. 42, q. 4, n. 20, 414b 235
Johannes Duns Scotus. *Notabilia in libros Topicorum*, II, in: ms Vatican, *Bibliotheca Apostolica Vaticana*, Ottob. 318
f. 258ra-rb 246
f. 257va 252
Johannes Duns Scotus. *Ordinatio* I, dist. xi-xxv, ed. Commissio Scotistica (1959)
I, d. 21, q. unica, n. 13, p. 299 237, 238, 254
Johannes Eckius, *In Summulas Petri Hispani*, ed. Augustae Vindelicorum (1516)
f. lxxxix rb 399
f. lxxxix va 391
f. lxxxix vb 390
f. xc rb 392
Johannes Pagus, *Appellationes*, ed. De Libera (1984)
176 62
193 66
224 62
193-255 184
Johannes Philoponus, *In Analytica Priora*, ed. Wallies (1905)
43, 8-18 96
Johannes Scharfiuss, *Manuale Logicum* ed. Leipzig-Frakfurt (1701)
319-322 476
Johannes Sharpe, *Quaestio super universalia*, ed. Conti (1990)
199-207 261
Johannes Wyclif, *De ente praedicamentali*, ed. Beer (1891)
ch. 5, 38-39 323
ch. 7, 61 312
Johannes Wyclif, *De actibus animae*, ed. Dziewicki (1902)
pars II, ch. 4, 122-123 312
pars II, ch. 4, 127 312
Johannes Wyclif, *Logicae continuatio*, ed. Dziewicki (1893-1899)
tr. 3, ch. 2, 26-27 325
tr. 3, ch. 2, 27 315
tr. 3, ch. 2, 28 316
tr. 3, ch. 2, 35-37 317
tr. 3, ch. 2, 38 317
tr. 3, ch. 2, 39 318
tr. 3, ch. 2, 40-42 319
tr. 3, ch. 2, 42 320
tr. 3, ch. 2, 43 320
tr. 3, ch. 2, 48 321
tr. 3, ch. 3, 62 322
tr. 3, ch. 3, 64 322
tr. 3, ch. 3, 67 322
Johannes Wyclif, *Tractatus de logica*, ed. Dziewicki (1893-1899)
ch. 1, 2 308
ch. 1, 2-3 308
ch. 1, 3 308
ch. 1, 5 309
ch. 1, 5-6 309
ch. 1, 7 309, 310
ch. 3, 11 312
ch. 5, 14 309
ch. 5, 15 324
ch. 7, 112 f. 46
ch. 12, 39 307
ch. 12, 39-40 307
ch. 12, 40-41 313
ch. 12, 41 314
ch. 12, 41-42 314
ch. 12, 42 317
Johannes Wyclif, *Tractatus de universalibus*, ed. Mueller (1985)
ch. 1, 15-16 309, 312
ch. 1, 35-36 319
ch. 2, 56 324
ch. 2, 64 312
ch. 3, 70 323
ch. 3, 74 323
ch. 4, 86-87 312
ch. 4, 90-92 312
ch. 6, 116-124 309
ch. 11, 239-240 319
ch. 11, 240 313
Johannes Wyclif, *De materia et forma*, ed. Müller (1902)
IV, 185-186 309
Juan Caramuel y Lobkowitz, *Praecursor logicus complectens*

Grammaticam audacem, Frankfurt (1654)
11 479
Juan de Celaya, *Magne Suppositiones*, ed. Parisiis (1526)
sig. B 1 rb 394, 399
Lambertus de Lagny, *Appellationes*, ed. De Libera (1981)
254-255 65, 66
Lambertus Autissiodorensis (Lambert of), *Logica* (*Summa Lamberti*), ed. Alessio (1971)
VII, 175, ll. 7-13 218
VII, 181 224
VIII, 205 488
VIII, 206 488
VIII, 208 494
VIII, 209 490
VIII, 210 490
VIII, 211 491
VIII, 212 492
VIII, 213 491
VIII, 214 403
VIII, 219 502
VIII, 223 495
VIII, 225 505
VIII, 226 493
VIII, 227 493
VIII, 228 494
VIII, 229 494, 497, 503
Marsilius ab Inghen, *Suppositiones*, ed. Bos (1980)
52 387, 392
56 397
58 397
Matthew of Orleans, *Sophistaria*, ed. Spruyt (2001)
II, 22 47
II, 23 47
Nicolaus Drukken de Dacia, *Tractatus de suppositionibus*, ed. Ebbesen (1997)
1-5, p. 235, ll. 1-19 345
4-5, p. 235, ll. 11-16 345
Paul of Pergula, *Logica*, ed. Brown (1961)
48 402
Paulus Venetus, *De universalibus*, ed. Conti (1990)
201 261, 262
203 263
Paulus Venetus, *Logica Magna (Tractatus de suppositionibus)*, ed. Perreiah (1971)
1.2 346
1.9 346
2.1-2.5 346
2.8-2.14 345
2.20 346
3.1 346
3.8 ff. 346
3.8, p. 88 397
Paulus Venetus, *Logica Parva*, ed. Perreiah (2002)
71 388
Petrus Abaelardus, *Ethica sive Scito teipsum*, ed. Luscombe (1971)
6, 24-28 47
Petrus Abaelardus, *Dialectica*, ed. De Rijk (1970)
65, 24-31 93
Petrus Abaelardus, *Glossae super Porphyrium*, ed. Geyer (1919)
17, 12-28 93
Petrus Abaelardus, *Glossae super Peri Hermenias*, ed. Geyer (1919)
360, 23-34 93
Petrus Abaelardus, *Logica 'Ingredientibus'*, ed. Geyer (1919-1927)
60.13 264
Petrus Abaelardus, *Glossae super Praedicamenta*, ed. Geyer (1919)
122, 29 93
Petrus Abaelardus, *Theologia christiana*, ed. Buytaert (1969)
IV, 52-53 265
Petrus de Alvernia, *Sophisma* V, in: Marmo (1999)
V, p. 95, n. 37 250
Petrus Hispanus, *Tractatus*, ed. De Rijk (1972) 196, 200, 201
lxxvii-lxxxiii 83
lxxviii-lxxix 83
I, § 8 (4-5) 170
II, § 19 186
V, § 5-7 (81, 5 - 82, 12) 340
VI, § 4 340
VI, § 8-9 107
VI, § 17 107
VI, § 79 236

VI, § 80 185, 237
VI, § 81 186
VI, § 81-82 186
VI, § 82 (14-22) 366
VII, § 135 218
VII, § 101 218
VII, § 106 218
VII, § 107 218
VII, §§ 146-147 217
IX, § 194 186
X, §§ 197-198 187

Petrus Hispanus, *Textus et copulata omnium tractatuum Petri Hispani*, ed. Cologne (1493)
36 14

Petrus Ramus, *Scholae in artes liberales*, Basle (1569)
159 471

Petrus Tartaretus, *Expositio* [...] *in Summulas Petri Hispani*, ed. Venice (1514)
f. 51v 473

Philippus Melanchthon, *De corrigendis adolescentiae studiis*, ed. Stupperich (1961)
34 ff. 466

Philippus Melanchthon, *Erotemata dialectices*, ed. Brettschneider (1846)
col. 750 465, 466, 467
col. 751 468, 469

Plato, *Laches*
188 D 97

Plato, *Phaedrus*
265 E 97

Porphyrius, *Isagoge*, ed. Minio-Paluello (1966)
82, 85, 192

Priscianus, *Institutiones Grammatice*, ed. Hertz (1855-1859)
II, 55, 6 171
XVII, 3, 23 61

Proclus, *Elementatio*, ed. Dodds (1963)
prop. 2 26
prop. 24, expositio 26
prop. 28, p. 32, 19-21 24
prop. 99, expositio 26, 27
prop. 99, p. 88, 20-23 26

Radulphus Brito, *Quaestiones in Elenchos*, q. 23, in: ms Bruxelles, *Bibliothèque Royale*, 3540-47
q. 23, f. 505ra 248, 253
q. 35 218
q. 23, f. 504va-vb 238

Radolphus Brito, *Quaestiones super sophistcos elenchos*, ed. Ebbesen (1975-1976)
q. 35 224

Radulphus Brito, *Quaestiones in libros Analyticorum Posteriorum*, in Pinborg (1976)
I, q. 46, p. 272-275 238

Radulphus Brito, *Quaestiones super Analytica Priora*, ed. Pinborg (1984)
I, q. 46, III, 73 192, 306

Radulphus Brito, *Quaestiones super Priscianum minorem*, ed. Pinborg (1980)
I, q. 18 28
I, q. 19 28
I, q. 21 28
I, q. 22 28

Ricardus Brinkley, *Logica* II (*De universalibus*), ed. Cesalli (2008)
§ 2 290
§ 10 292
§ 33, 310, ll. 319-325 345

Ricardus Brinkley, *Tractatus de suppositionibus*, eds. Cesalli and Lonfat, this volume, pp. 295-300

Ricardus Campsall, *Logica*, ed. Synan (1955) pp. 205-300
1.50 346
I. 51, ll. 198-199, 200-203 346

Robertus Holkot, *Commentary on the fourth book of the Sentences*, ed. http://www. abelard.paris-sorbonne. fr/retrholkot.htm
f. a2v 357

Robertus Holkot, *Quodlibetal Dispute*, in: Courtenay (1971)
p. 7 357

Robertus Kilwardby, *Commentum in Analytica Priora*, in: ms Cambridge, *Peterhouse* 205
f. 88vb-89ra 67

Robertus Kilwardby, *Commentum in Sophisticos Elenchos*, in: ms Cambridge, *Peterhouse* 205
f. 335rb 67

Robertus Kilwardby, *Commentum in Sophisticos Elenchos*, in: ms Paris,

Bibliothèque nationale de France, nal. 16619.
f. 62vb 67
Robertus Kilwardby, *De ortu scientiarum*, ed. Judy (1973),
151, n. 438 224
Rogerus Bacon, *Summulae dialectices*, in: De Libera (1986) 174
2.1, p. 266 244
2.2, p. 288 246
Rogerus Roseth, *Lectura super Sententias*, 3-5, ed. Hallamaa (2005)
3.1 (67-68) 261
Sigerus de Courtraco, *Ars priorum*, ed. Wallerand (1913)
12 204
Simon Faversham, *Quaestiones super libro Porphyrii*, ed. Mazzarella (1957)
32, l. 18 211
33, ll. 27-35 209
33, ll. 31-38 210
47, l. 9 211
52-53, ll. 32-18 210
53, ll. 1-3 211
53, ll. 10-11 211
55, ll. 23 ff. 210
63 213
65, ll. 24-25 209
46 ff. 210
Simon Faversham, *Quaestiones super libro Praedicamentorum*, ed. Mazzarella (1957)
93, ll. 27 ff. 209
95, ll. 24-40 211
97, l. 20 211
107, ll.1-2 211
109, ll. 11-20 211
112 213
123-5, ll. 28-30 218
132, ll. 15-22 209
142, ll. 13-15 209
Simon Faversham, *Quaestiones super libro Perihermeneias*, ed. Mazzarella (1957)
153, ll. 27-30 209
155, ll. 9-29 209
163, ll. 30-35 208
Simon Faversham, *Quaestiones super libro Analyticorum*, ed. Mazzarella (1957)
97, n. 40 250
Simon Faversham, *Quaestiones super libro Elencorum. Quaestiones novae*, ed. Ebbesen *et al.* (1984)
15 224
51 215
69 212
76 212
124, ll. 62-78 211
134-135 212
140-141 213
145 207
148 224
148-149 212
149 218
151, ll. 26-27 218
154 211
155 223
169 208
173 214
Simplicius (pseudo), *In libros Aristotelis De Anima*, ed. Hayduck (1882)
127, 26-32 100
278, 20-29 100
Thomas Aquinas, *Questio disputata De unione Verbi*, ed. Marietti (1931)
2 ad arg. 4 330
3 ad arg. 5 335
Thomas Aquinas, *Expositio libri Peryermenias*, ed. Leonina (1989)
I.2, 10-11, 95-112 329
5 ff. 471
Thomas Aquinas, *Expositio libri Posteriorum Analyticorum*, ed. Gauthier (1989)
I. 4, l. 9, n. 3 209
I. 4, l. 9, 3-4 209
Thomas Aquinas, *Scriptum super IV Sententiarum*, ed. Paris (1647)
I. 1 3 72
1.4.1.3 ad arg. 4 332
3.1.2.4, ad arg. 6 334, 336
3.6.1.2, ad arg. 4 330
Thomas Aquinas, *Quaestiones disputatae de potentia Dei*, ed. Taurini-Romae (1931)
9.4 329
9.4 ad arg. 6 330
9.4 ad arg. 18 335
9.6 329
Thomas Aquinas, *Summa theologiae*, ed. Romae (1888-1906)
I, 29, 2 209

1, 36,.3 331
1, 36, 4 ad arg. 4-6 333
1, 39, 4 331, 336
1, 39, 4 ad arg. 2 335, 336
1, 39, 4 ad arg. 3 331
1, 39, 5 ad arg. 5 330
1, 39, 5 330
III, 2, 2 209
III, 2, 3 209
III, 3, 6 ad arg. 3 331, 334
III, 4, 3 335
III, 4, 3 ad arg. 2 335
III, 16, 7 332
III, 17, 2 335

Thomas Aquinas, *Summa contra Gentiles*, ed. Romae-Taurini (1961)
4. 49, 339, no. 3847 330
4.40, 329, no. 3785 330

Thomas Aquinas, *Expositio libri Metaphysicorum*, ed. Taurini-Romae (1950)
5.9 330
7.5 330

Thomas Aquinas (pseudo), *De fallaciis*, ed. Taurini-Romae (1954) 429
234, n. 680 218
233-234, n. 677 217

Thomas Bradwardine, *Insolubilia*, ed. Read (2010)
6.3 355

Vincentius Ferrer, *Tractatus de suppositionibus*, ed. Trentman (1977)
III, 100-101 343
III, 100-103 343
VII, 163 343
X, 181-184 342

William Penbygull, *De universalibus*, ed. Conti (1982)
196-197 310

Wittgenstein, Ludwig, *Wittgenstein's Lectures*, eds. King and Lee (1989)
2 478

Wittgenstein, Ludwig, *Tractatus*
3.3. 478

Index Nominum

Adam Wodeham: 260, 262, 268, 269, 271-273
Aegidius Romanus: 215, 218-221, 223, 224; *see also* 'Guillelmus Arnaldi'
Alain de Lille *see s.n.* Alanus ab Insulis
Al-Afifi, A.: 93
Al-Ahwani, A.: 85-87
Alanus ab Insulis (Alain de Lille): 121, 124, 133, 141, 275, 284
Albertus de Saxonia: 338, 349, 356, 366, 374, 392, 397, 417, 434, 443, 445, 447, 448, 451-455, 458, 460-462
Albertus Magnus (Albert the Great): 198, 205, 208, 215, 217, 218, 220, 224, 225, 227, 465
Alessio, F.: 218, 224, 486, 488, 489, 515
Alexander (ps.): 218
Alexander of Aphrodisias: 100, 248
Al-Katib: 83, 93, 96
Al-Kudayri: 90, 94, 100-102, 111
Al-Kurdi, A.: 87, 94
Amerini, F.: 327, 338-340, 347, 348
Ammonius: 91, 94, 106
Anawati, G.: 86, 87, 89, 94, 96, 102, 106, 109
Andrés Limos: 429
Andrew Sunesen: 65, 66, 469
Andrews, R.: 51, 57, 207, 212, 215, 219, 228, 233-254, 263
Angel Stanyol: 429
Anonymus Aurelianensis III: 38, 63, 64
Anonymus C: 207, 218, 224
Anonymus Monacensis: 67
Anonymus Parisiensis: 218
Anonymus Pragensis: 205, 208, 213, 214, 218, 221, 223-225
Anonymus SF: 206, 213, 218, 224, 238, 242, 243
Anonymus Tabbaroneus: 221
Anscombe, G.: 8
Antonio Andrés: 429
Antonio Vallarono: 429
Aragona: 429, 435, 442
Ardeshir, M.: 100
Aristoteles, Aristotelian: x, xi, xiii, 4, 13-23, 28-32, 34-42, 44, 49, 50, 53, 55-58, 64, 66, 77, 78, 85, 86, 88, 91, 92, 94, 97-103, 106, 108-115, 149, 150, 165, 167, 168, 170, 173, 197, 207, 211, 212, 215-218, 226-229, 233, 234, 236-239, 241, 242, 244-247, 249, 250, 252-254, 256-259, 263, 268, 270, 273, 274, 284, 296, 298, 300, 302, 311, 318, 321, 322, 325, 328, 333, 364, 356, 358, 359, 369, 407, 408, 414, 429, 434, 435, 439, 445, 449, 465, 466, 470-472, 477, 481, 510-516
Arnulphus: 63
Ashworth, E.J.: 235, 287, 341, 391, 394, 394, 401, 403, 409, 465
Augustinus de Ferrara: 263, 264, 270
Augustinus Hunaeus: 467
Augustinus Niphus: 465
Averroes: 72, 101,
Avicenna: 81-115, 210, 212, 300
Avignon: 46, 435

Bäck, A.: 21, 35, 57, 81, 85-87, 89, 92, 93, 95, 97, 98, 105, 106, 109, 113
Bakker, P.J.: 37, 41, 42
Barcelona: 429, 430
Bartholomeus Arnoldi of Usingen: 471, 472
Bartholomeus magister: 427-430, 432-436, 439
Bartolomé López: 435, 436
Bartolomé Rodríguez de Caracena: 435
Beer, R.: 312, 323
Beltrán de Heredia: 435, 437, 445, 462
Benedictus de Undis: 429, 433
Berger, H.: 338, 374
Bérubé, C.: 287
Beth, E.W.: 7
Boehner, Ph.: 14, 15, 56, 84, 102-106, 108, 111, 112, 164, 235, 244, 246, 256, 268, 273, 276, 301, 305, 306, 308, 318, 336-338, 341, 348-350, 352, 360, 361, 363, 368, 369, 371, 373-375, 379, 383, 384, 390, 391, 394-396, 398, 408
Boer, R.J. de: 82

Boethius de Dacia: 72, 220, 250
Boethius, Anicius Manlius Severinus: 32, 63, 72, 119, 120, 122, 137, 141, 142, 192, 195, 203, 216, 220, 224, 249, 250, 266, 471
Bonaventura: 265, 269
Borgnet, A.: 207, 208, 214, 215, 217, 218, 220, 224, 225
Bos, E.P.: xii, 58, 72, 76, 233, 387, 392, 397, 408
Boulnois, O.: 235
Bover, J.M.: 431
Braakhuis, H.A.G.: xii, 47, 71, 83, 177, 235
Brands, H.: 65, 83, 84, 170, 188, 189, 193, 194, 203, 218, 235, 243-245, 247, 248, 256, 280, 301, 340, 342, 487, 509
Brettschneider, C.G.: 465-469
Brown, S.: 35, 56, 108, 112, 113, 256, 308, 325, 328, 338, 346, 348-350, 355, 356, 358, 359, 398
Brumberg-Chaumont, J.: 179, 183, 187, 195, 196, 213, 287
Bueno, A.: 217
Busse, A.: 94, 106
Buytaert, E.: 265

Carnap, R.: 6, 8
Casas Homs, J.: 430
Castile, Castilian: 429, 431, 432
Castrillo, C.: 439
Catalan: 429
Catalani, L.: 120
Cerdo (magister): 429, 430, 436, 437, 442
Cesalli, L.: 275, 291, 295, 301, 302, 309, 345, 355
Chagrov, A.: 498
Chartres: 71
Chenu, M.-D.: 66
Christian August Crusius: 478
Christian Weise: 475, 476
Christian Wolff: 450
Chrysippus: 17
Chrysostomus Javelli: 465
Clemens VI (Petrus Rogerius, Pierre Roger): 46
Codoñer, C.: 430-432
Conrad Dannhauer: 464, 470-475
Conti, A.: 261-263, 304, 310-312, 314, 316, 323
Copi, I.: 217
Courtenay, W.: 294, 357
Cousin, V.: 27
Crathorn: 50-54, 56
Cross, R.: 267

Dahlstrom, D.O.: 234
Davidson, D.: 383, 384, 473
Diego de Navalmarcuende: 438
Diego García de Castro: 438
Dinneen, F.P.: 487
Dodds, E.R.: 24, 27
Dominicus Soto (Domingo de Soto): 385, 387, 391, 392, 384-396, 399, 400, 402-408, 465
Donati, S.: 208
Dunya, S.: 87, 93
Dutilh Novaes, C.: 83, 148, 163, 167, 345, 350, 352, 353, 355, 359, 368, 370, 388, 395, 409, 412, 469
Dziewicki, M.H.: 46, 305, 307-310, 312-322, 324, 325

Ebbesen, S.: 60, 61, 63-69, 74, 76, 101, 107, 114, 121, 150, 157, 167, 175, 195, 204, 206-209, 211-215, 218, 220, 221, 223, 224, 235, 238, 241-243, 248, 256-258, 302, 345, 373, 469
Eco, U.: 353
Elamrani-Jamal, A.: 82
Elbe: 71
Emden, A.B.: 275
England, G.: 51
Etienne Gaudet: 285
Etzkorn, G.: 47, 56, 359
Everardus of Ypres: 119, 111, 130, 134, 134-137, 139-142

Fernando de Córdoba: 429
Fernando de Enzinas: 445
Fitting, M.: 498, 499, 510
Fitzgerald, M.J.: 295, 301
Franciscus de Prato: 327-329, 338-340, 344, 347, 348
Franquera (magister): 427, 429-431, 436, 438-461
Frege, G.: 6, 8, 9, 83, 378

Gál, G.: 33, 275, 276, 295, 301, 318, 379
Gambatese, A.: 35, 358
García de Castello: 429, 430, 436-438, 440, 442-445
Garfagnini, G.C.: 342, 344, 345
Garin, E.: 342, 344, 345, 348
Gaskin, R.: 295
Gasparus Lax: 445
Gauthier, R.: 69, 196, 206
Geach, P.: 365
Gelber, H.: 217, 261

Georgius Bruxellensis: 392, 394, 401
Georgius Rovegnatinus: 327, 331, 335, 343, 344, 348, 479
Geraldus Odonis: xiii, xiv, 43, 345, 346
Geyer, B.: 93, 264
Gilbertus Porretanus (of Poitiers): xii, 61, 72, 119-124, 128, 130, 132, 133, 135, 137, 141
Giles of Rome: *vide s.n.* Aegidius Romanus
Gilson, E.: 105
Girolamo Savonarola: 327-329, 341-345, 347, 348
Goichon, A.-M.: 87
Goodman, N.: 6
Goubier, F.: 179, 202, 235, 256, 258, 275, 282, 284, 294
Gratiadeus Asculanus (of Ascoli): 328
Green Pedersen, N.J.: 220
Gregorius Ariminensis (Gregory of Rimini): 357, 366
Grellard, C.: 382
Gualterus Burleus: xii, 3, 83, 84, 101, 103, 108, 112, 113, 164, 167, 244, 246, 276, 277, 291, 305, 306, 308, 310, 311, 313, 322, 327-329, 338, 340, 341, 344-349, 355, 356, 379, 427, 434, 445-447, 451-462, 516
Gualterus Chatton: 338
Guillelmus Arnaldi: 69
Guillelmus de Heytesbury: 284, 356, 429
Guillelmus de Montibus: 69
Guillelmus de Ockham, Ockhamist: xiii, 4, 32-35, 50, 52, 53, 56-58, 83-85, 101-108, 148, 163, 164, 166, 217, 235, 262, 267, 268, 276, 277, 281, 282, 284, 301, 305, 306, 318, 325, 327-329, 336-342, 345-368, 371-383, 385, 386, 388-391, 393-398, 404, 427, 442, 444-446, 450, 455, 458, 511, 516, 519
Guillelmus de Shireswood: 65, 75, 83, 105, 169-171, 174, 184, 186-190-196, 198, 201, 203, 234, 235, 237, 243-245, 247, 248, 254, 255, 285, 291, 301, 302, 306, 338, 340, 342, 349, 487, 511, 514, 515
Guillelmus Penbygull: 310, 314
Guillermo de Osma: 429, 430, 435, 439
Guthrie, W.K.C.: 16, 18, 21, 22

Hallamaa, O.: 261
Hansen, H.: 63, 66
Häring, N.: 72, 120-122, 124-131, 133 140
Hayduck, M.: 100
Henricus de Gandavo: 47
Henricus Totting of Oyta: 268
Hervaeus Natalis: ix, xiv, 28, 45, 327-329, 334, 340, 348
Hieronymus de Sancto Marco: 328, 394, 399, 404, 405, 408
Hieronymus Pardo: 387, 445
Hoenen, M.J.F.M.: 294
Hoffmann, F.: 50-52, 56, 57
Hubien, H.: 264, 268, 377
Huelsen, R.: 150
Huesca: 429
Hume, David: 54

Irwin, T.: 97
Iwakuma, Y.: 64, 69-71, 74

Jacobi, K.: 5, 93, 120, 133, 134
Jacopo Zabarella: 471
Jamblichus: 25
Joachim Georg Darjes: 476, 477
Joannes Roderici de la Franqueira: 437
Johann Conrad Dannhauer: 464, 470-476
Johann Jacob Stübelius: 476
Johann Philipp Ebel: 472
Johannes Arboreus: 464, 465
Johannes Buridanus: xiii, xiv, 3, 23, 36-43, 46, 83, 85, 104, 108, 110, 112, 116, 167, 262, 264, 267, 268, 273, 302, 338, 339, 349, 354, 365-369, 371-378, 380-390, 392, 394-398, 401, 405, 408-422, 511, 516, 519
Johannes Clauberg: 475
Johannes Dacus: 165, 248
Johannes de Sancta Cruce: 442
Johannes Dorp: 386, 387, 390, 392, 394, 397, 398, 401, 405
Johannes Duns Scotus: 205, 207, 208, 212, 215, 218, 222, 227, 233-242, 244, 246-260, 265, 267, 284
Johannes Eckius: 390-392, 399
Johannes Mair: 445
Johannes Pagus: 62, 66, 78, 184, 285, 286, 288
Johannes Philoponus: 96
Johannes Sharpe: 261
Johannes Tarteys: 314
Johannes Wyclif: 46, 47, 261, 291, 304-326
Johannes XXI: 69
Jolivet, J.: 120, 133
Juan Caramuel y Lubkowitz: 479
Juan de Celaya: 387, 394, 399
Juan de la Cerda de Quintanapalla: 437 439
Juan de Medina: 429
Juan de Pastrana: 429-432, 439
Juan de Santa Cruz: 429, 430, 439
Juan Rodríguez de Franqueira: 437-440
Judy, A.G.: 224

Kahn, C.: 17, 22, 23
Kaluza, Z.: 228, 259, 285, 286, 295
Kann, C.: 65, 83, 84, 170, 188, 189, 193, 194, 203, 218, 235, 243-245, 247, 248, 256, 280, 301, 340, 342, 487, 509
Karger, E.: 337, 352, 353, 361-364, 370, 386, 393, 395
Kaufmann, M.: 85, 103, 105, 106
Kelley, F.: 359
Kenny, A.: 312
King, P.: 478
Kintsch, W.: 111
Klima, G.: 104, 110, 365, 373, 407, 414, 416, 418-422
Kneale, M.: 171
Kneale, W.: 172
Kneepkens, C.H.: xiv, 32, 61, 62, 65, 120, 121, 123, 130, 134, 140, 149, 170
Knudsen, C.: 107
Knuuttila, S.:260, 261, 264, 497, 506
Kretzmann, N: 473, 486, 487, 495
Kripke, S.: 487, 495, 496

Lambert of Auxerre, *vide s.n.* Lambertus Autissiodoriensis
Lambert de Lagny *vide s.n.* Lambertus Autissiodoriensis
Lambertus Autissiodoriensis: 65-67, 76, 83, 202, 218, 224, 243, 285, 302, 485-495, 497, 502, 503, 505-507, 509
Lecq, R. van der: 104, 108, 110, 338, 339, 365, 371, 372, 376, 377, 383, 386-388, 392, 395, 396, 402, 404, 415, 416
Lee, D.: 478
Leibniz, G.W.: 261, 516
Lewis, F.A.: 217
Libera, A. de: 61, 62, 65-67, 121, 174, 184, 206, 235, 244, 245, 255, 284, 285, 288, 328, 486, 487
Lilao, O.: 439
Lloyd, A.C.: 218
Lohr, C.: 213
Lonfat, J.: 275, 295
Lop Otín, M.J.: 432, 439
Louis VII: 70
Loux, M.J.: 48, 337
Lozano Guillén, C.: 430
Luscombe, D.: 47, 48

Madkour, I.: 114
Maierù, A.: 208, 217, 261, 264, 268, 274, 306, 328, 386, 397, 399
Maioli, B.: 120, 121, 130
Mallorca: 431, 432
Marcos Rodríguez, F.: 437
Marenbon, J.: 93, 120, 137
Maritain, J.: 55
Marmo, C.: 206-208, 215, 228, 233-236, 239, 250, 255, 257, 287, 293, 353, 448
Marsilius of Inghen: 37, 164, 386-388, 392, 397, 408
Martín de Espinosa: 438
Martin, C.: 119, 233
Master of Abstractions: 147, 148, 163
Mattheus Aurelianensis: 47
Matthew of Orléans: *see s.n.* Mattheus Aurelianensis
Matthews, G.: 148, 163, 167, 217, 387, 394
Mazzarella, P.: 208-213
McCord Adams, M.: 337
Meier-Oeser, S.: 331, 464, 470
Mendelsohn, R.L.: 498-499
Mill, J.S.: 354
Minio-Paluello, L.: 216, 217
Moody, E.A.: 5, 30, 106, 313, 352
Mueller, I.J.: 309, 313, 315, 319, 323, 324
Murè, D.: 205

Nebrija: 430
Neumann, J. von: 8, 9
Nicolaus Metensis (Nicholas of Metz): 71, 177
Nicolaus Parisiensis (Nicholas of Paris): 71, 177, 185
Nicolás Eymerich: 429
Nicolás Surrina: 429
Nicolaus Drukken de Dacia: 345
Nielsen, L.: 61, 120, 130, 134
Normore, C.: 164, 337

Ockhamist, *see s.n.* Guilelmus de Ockham
Ors, A. d': 69, 152, 427
Otto, A.: 165, 166, 246, 248, 252
Owen, G.E.L.: 98
Oxford: 70, 184, 261, 268, 275, 310, 346
Oxonian tradition: 184

Panaccio, C.: 280, 337, 352, 371, 375, 378, 380
Paris, Parisian: 46, 51, 68, 71, 101, 131, 178, 184, 185, 242, 244, 245, 284, 285, 445, 487
Park, S.C.: 206
Parmenides: 27
Parsons, T.: 328, 392

Patar, B.: 382
Paulus Pergulensis: 397
Paulus Venetus (Paul of Venice): 260-264, 345, 346, 347, 397, 511
Pedro de Castrovol: 429
Pedro de Padilla: 438
Pedro Fajardo: 429, 430, 439
Pérez-Ilzarbe, P.: 387
Perini-Santos, E.: 337, 353, 363, 367, 378, 411, 415
Perler, D.: 233, 235
Perreiah, A.: 345-347, 349, 397
Perry, J.: 423
Peter of Poitiers, *see s.n.* Petrus Pictavensis
Peter of Spain, vide *s.n.* Petrus Hispanus
Petrus Abaelardus: x-xiii, 28, 32, 35, 45-48, 57, 93, 137, 264-267
Petrus de Alvernia: 108, 250
Petrus Fonseca: 474
Petrus Helias: 61, 170, 430
Petrus Hispanus (non-papa): 83
Petrus Hispanus: xi, xiii, 69, 77, 78, 83, 105, 107, 110, 170, 173, 174, 178, 180, 185-187, 190, 196-198, 200-204, 217, 218, 227, 229, 234, 236, 243, 254, 257-259, 306, 328, 336, 338, 340, 349, 354, 365, 366, 368, 369, 372, 427-429, 432-435, 445, 447-449, 454, 455, 460, 462, 487, 509, 511, 514, 519
Petrus Hurtado de Mendoza: 474
Petrus Lombardus: 128, 357
Petrus Pictavienss: 72, 119, 123
Petrus Ramus: 471
Petrus Rogerius: 46. *see also s.n.* Clemens VI
Petrus Tartaretus: 473
Petrus Thomae: 264
Petrus Viennensis (Peter of Vienna): 121
Philippus Melanchthon: 465-470, 474
Philosophus: *vide etiam s.n.* Aristoteles
Pinborg, J.: ix, xii, xiii, 28, 61, 65, 69, 121, 157, 184, 195, 206, 209, 215, 220, 238, 250, 256-259, 469
Pini, G.: 233, 235, 267
Pironet, F.: 387, 389, 410, 412, 414, 415, 417-422
Plato, Platonic: x, xii, 15-19, 23, 27, 42, 89, 97-99, 103, 106, 114, 122, 143, 151, 152, 195, 197, 200, 210, 228, 245, 268, 278
Pontevedra: 440
Porphyrius: 82, 85, 101, 192, 193, 195, 206, 211, 224, 234
Pozzi, L.: 356, 370
Prantl, C.: 81
Priest, G.: 107, 162, 166, 167, 386, 388, 404
Prior, A.: 417, 418, 424, 498
Priscianus: 28, 60, 61, 76, 131, 134, 165, 170, 171, 182
Proclus: 23, 24, 26, 27

Quine, W.V.O.: 6, 8, 156, 158, 473, 498

Radulphus Brito: 28, 206, 218, 238, 248, 250, 253
Radulphus Strode: 397-399, 401, 402, 429
Rashed, R.: 111
Raymundus Lullus: 431
Reach, K.: 8
Read, S.: 83, 101, 102, 107, 156, 162, 164, 165, 167, 168, 337, 355, 386, 388, 402, 404, 412
Richardus Sophista: 147
Richardus Billingham: 427, 429, 431, 442-447, 451-454, 462
Richardus Brinkley: 3, 269, 276-293, 295, 301, 345
Richardus Campsall: 346
Richardus, Master of Abstractions: 147, 148, 163
Rijk, L. M. de: *passim*
Robertus Alyngton: 314
Robertus Grosseteste, *vide s.n.* Robertus Lincolniensis
Robertus Holkot: 357
Robertus Kilwardby: 67, 75, 165, 218, 224
Robertus Kilwardby (ps.): 165, 166
Robertus Lincolniensis: 75, 289
Rode, C.: 338
Rogerus Bacon: 70, 75, 174, 184, 234, 235, 241, 243, 245, 246, 255, 487
Rogerus Roseth: 260
Rogerus Whelpdale: 314
Rosier-Catach, I.: 121, 122, 130, 206, 235, 236, 287, 293
Ross, W.D.: 92, 112, 216
Russell, B.: 354, 378

Sajó, G.: 72
Salamanca: 427, 429, 430-444, 461, 462
Salvador de Terradis: 429
Scholz, O.R.: 473
Schoot, H.: 206, 336
Schulthess, P.: 103
Schupp, F.: 439
Schweiss, K.: 134

Scott, T.K.: 85, 162, 388
Seine: 71
Shank, M.H.: 268
Sigerus de Cortraco: 219
Simon de Faversham: 44, 205, 207-215, 218, 220, 223, 224, 250
Simplicius (ps.): 100
Sirridge, M.: 147
Spade, P.V.: 4, 83-85, 101, 108, 112, 114, 115, 148, 154, 162-165, 167, 235, 268, 295, 322, 326, 337, 338, 351, 352, 357, 370, 374, 375, 378, 380, 375-389, 395-397, 401, 404, 408
Spain: 429
Spiazzi, R.-M.: 218
Spruyt, J.: xiii, 1, 46, 47, 143, 304
Steele, R.: 487
Stephan Langton: 61, 65, 66, 469
Street, T.: 81, 83, 90, 93
Streveler, P.O.: 147
Stump, E.: 486, 495
Sundholm, B.G.: 3
Sweeney, L.: 336
Synan, E.A.: 346

Tabarroni, A.: 37, 38, 168, 196, 218
Tarski, A.: 6-8, 380
Thijssen, J.M.M.H.: 294
Thom, P.: 109
Thomas Aquinas (ps.): 217
Thomas Aquinas: xii, 35, 72, 196, 197, 200, 216, 209, 210, 211, 217, 227, 229, 327-336, 338, 344, 350, 429, 471, 473
Thomas Bradwardine: 355, 357, 366
Thomas Maulevelt: 51
Thomsen, W.R.: 314
Toledo: 440
Toulouse: 78, 196
Trentman, J.: 342, 343
Truhlář, J.: 295

Uckelman, S.L.: 485, 488

Valente, L.: 61, 72, 119-123, 129, 130, 133
Valladolid: 429
Vendler, Z.: 19
Verdelho, T.: 432
Vincentius Ferrer: 342, 343, 349, 429
Vivès, L.: 222
Vos, A.: 233

Wadding, L.: 224, 235
Wallies, M.: 96, 100
Walter Burley *see s.n.* Gualterus Burleus
Weber, S.: 431
Weidemann, H.: 386
Wey, J.C.: 33, 338
Wheeler, M.: 97
Wilks, I.: 93
William (of) Heytesbury: *see s.n.* Guillelmus Hentisberi
William Arnaldi, *see s.n.* Guillelmus Arnaldi; *see also s.n.* Aegidius Romanus
William of Ockham: *see s.n.* Guillelmus de Ockham
William of Sherwood: *see s.n.* Guillelmus de Sheriswode.
Williams, C.J.F.: 217
Wilson, C.: 47
Wittenberg: 466
Wittgenstein, L.: 478, 479
Wood, R.: 56, 275, 276, 295, 358, 379

Yrjönsuuri, M.: 412

Zakharyaschev, M.: 498
Zaragoza: 429, 439, 442
Zayid, S.: 87, 89, 91, 93-96, 100, 109
Zimmermann, F.W.: 94
Zupko, J.: 41

Index Rerum

acceptance (*acceptio*) of a term in a context: 6, 8, 30, 144, 180, 185, 186, 198, 208, 223, 224, 236-240, 242, 243, 249, 300, 321, 339, 343, 356, 365, 392, 405, 447-451, 455, 456, 458, 467, 475, 477, 479, 488, 492, 505
accidents: 46, 49, 71, 85, 87-89, 91, 92, 96, 98-101, 195, 311, 324; concrete accidents 312, 313; *per se* accidents: 87, 92
algorithm: 111, 163, 515; algorithmetic hermeneutics: 166
ambiguity: 8, 22, 23, 44, 47, 91, 107, 153, 170, 248, 371, 375, 376, 380; ambiguity in mental language: 377
ampliation (*ampliatio*): vi, 4, 67, 70, 81, 82, 107, 110, 114, 159, 162, 175, 179, 187, 285, 286, 287, 289, 293, 296, 297, 306, 356, 388, 391, 394, 400, 442, 446, 474, 485, 487, 491-495, 499-504, 509, 511; ampliation to imaginary and impossible objects: 391, 400; ampliative power: 286
anomaly: 172
Antichrist: 110, 153, 179, 180, 251, 283, 288, 289, 290, 296, 297, 399, 400, 494
apophantics as opposed to onomastics: 21, 28, 31
appellation (*appellatio*)*:* 4, 19-21, 44-46, 60, 64, 66, 67, 69-76, 95, 103, 108, 110, 123, 125, 126-128, 131, 134, 136, 138, 168, 175, 177-184, 187, 188, 191, 203, 228, 252, 268, 275, 284-290, 293, 296, 302, 303, 376-378, 394, 408, 436, 446, 449, 485, 487, 492, 493, 495, 499, 501 502, 504, 509 *appellatio communis*: 125; *appellatio rationis*: 377; *appellatio formae:* 46; *appellata* and *supposita* as extensionally equivalent: 103; appellation does not concern pronouns: 188; appellation in the early sense of present existence: 95; appellation may be restricted or ampliated: 69; appellation of discrete terms: 203; appellation or supposition of singular terms: 175; quidditative appellation: 21
ascent (*ascensus*): 114, 162, 167, 294, 347, 362, 363, 385-387, 389-391, 393-400, 402, 403, 404, 406-408, 410, 516

Barcan formula: 499, 500
being: consequential or dispositional being: 158; *see also s.v. 'ens'*
Bible: 437, 469; examples of biblical sentences: 468, 469
biconditional: 411, 413

Caesar (as name of indivividual of the past): 110, 180, 181, 203, 243, 251
category: *passim*; categorial line: 319; category of being x, 101
categorization: 13, 20, 21, 31, 36, 50, 55, 193
cause (*causa*), causation: *passim*; causation in Hume 54, 56; tension between Proclean causation and participation: 23, 25; First Cause: 23; superior cause: 26; transcendent cause: 26; *causa apparentiae*: 60, 71, 73, 215, 218; the formas cause of an individual: 138, 312; *variatio medii* as cause of a fallecy: 220, 253; intention of the *usus loquendi* as a cause for figurative or improper supposition: 277; cause eof existence: 317; causes of truth: 362
chimera: 109, 180, 361, 365, 390-392, 399, 400
cognition: 31, 33, 34, 43-45, 49, 50, 53, 166, 179, 207, 347, 362, 469; psychological act of cognition: 347; reliability of cognition: 43, 49
common nature: 209, 235, 240, 250, 256, 262, 267, 268, 270, 306-312, 314, 317, 320, 321, 324, 330, 388; *see also s.v.* quiddity
complexe significabile: 354, 368
concept: *passim*; complex concept: 381, 382; first-order concept: 381; second-order concept: 288, 289, 380-384; conceptual ambiguity: 23; conceptual approach: 35; concept of the copula: 33; to produce a concept of a human by means of his

humanity: 134; concepts as unified individuals: 95; *see also s.v.* syncategorematic concept
conditionals: 147, 501, 511, 514, 518: universally quantified conditionals: 501; *see also s.v.* biconditional
connotation: 20, 30, 31, 34, 38-41, 43, 45-47, 66, 132, 446, 467; connotation as opposed to denotation: 31; connotation or secondary significate: 450
consignification: 34: temporal consignification: 449; consignificative function to the copula *'est'*: 34
constantia: 397, 399, 400
context: *passim*; extra-linguistic context: 289; contextual approach: x, 29, 35, 36, 119, 122, 127; Aristotle's argumentative contextuality mirrors the medieval contextual approach: 36
copula: passim; mental copula: 416; vocal copula: 416; monadic ('copula-less') anatomy: 29
copulatio (conjunction): 4, 66, 233, 241, 245, 251, 252, 446, 449

deictic force: 170; deictic pronoun: 170; deictic expression: 286, 287
demonstration; scientific demonstration: 61, 64, 92, 93, 209, 480; *medium demonstrationis*, 20, 21; *demonstratio ad intellectum*: 288, 290, 293; *demonstratio ad sensum:* 288
denominatio: 121, 269
denotation: 81, 119, 205, 206, 353, 354, 370
descent (*descensus*): 24, 66, 82, 84, 95, 100, 102, 104, 108-111, 148, 150, 152, 156, 158, 159, 161-162, 167, 173-175, 178, 208, 244, 250, 253-255, 278, 279, 294, 295, 314, 347, 356, 362-364, 385-408, 444, 451, 459-461, 474, 490, 491, 516, 519, 520; '*copulatim*' and '*disiunctim*' forms of descent: 460
discourse: mental discourse: 372-374, 381, 411, 418-423; spoken or written discourse: 377, 380; theological discourse: 121
disquotation: 421, 423
distribution: 148, 150, 152, 154, 219, 267, 392, 403, 404, 406, 407, 409, 492; distribution and distributive supposition: 392; *see also s.v.* suppositio
divine essence: 266-268, 330, 331; divine nature: 124, 128, 130, 330-332

ens: see also s.v. 'being'; *ens transcendens*: 316, 317
entity: non-existent entities: 509; past, future and possible entities: 354; *see also s.v.* 'existence'.
equivalence: equivalence principle: 280, 285, 293: equivalence between propositions: 385, 389, 395, 404; equivalence between *modi intelligendi:* 27; formal equivalence: 408; referential equivalence: 30, 31; logical equivalence: 395-400
equivocation: 150, 154, 158, 239, 242, 243, 248, 302, 320, 330, 332, 375, 379; equivocation follows signification and not supposition: 239; *see also s.v.* 'fallacy'
est as a copula: 32; *'est'* as syncategorematic term: 33-35; vocal copula '*est*' is ambiguous: 416
eucharist: 39, 41, 42, 57
existence: possible existence: 93; presupposition of existence: 362; existential import: 109, 203, 393, 500, 501; *see also s.v.* entity.
expository syllogism: 261, 270, 271, 511-513, 516
expression: types of equiform expressions: 6

fallacia (fallacy): xi, 14, 53, 55, 69, 71, 149-151, 159, 205-207, 215, 218, 305, 325, 315, 313, 316, 388; fallacies: root of terminism: 4, 55; *fallacia accidentis:* 156, 211, 216-222, 225, 248, 254; *fallacia consequentis*: 153; *fallacia figure dictionis*: v, 148, 168, 169, 248, 253, 333; *fallacia in dictione*: 151, 207, 208, 214; *fallaciae extra dictionem*: 151, 207, 208, 214, 216, 226; *fallacia equivocationis:* 320; *fallacia compositionis et divisionis:* 213; *fallacia medii:* 218
fictitious objects: 100
form: metaphysical form: 196-202; relationship between the individual and the form: 189; singular form: 138, 140, 178, 195, 270; *forma per modum inherentis substantie:* 449

grammar, grammarians: 4, 60, 61, 132, 135, 150, 165-168, 171, 182, 183, 195, 237, 469, 474, 520, 521; Priscianic grammar: 60; the influence of grammar on the theory of the properties of terms: 172; Aristotelian-style grammarians: 167; speculative grammar: 27, 166; *see also s.v.* modism

hermeneutics: general hermeneutics: 130, 464, 469-473, 477; principle of hermeneutic charity: 290, 467, 473; algorithmic hermeneutics: 163, 166; twelfth-century hermeneutics: 130

identity (*identitas*): 48, 193, 211, 214, 215, 217-220, 261, 264, 265, 267, 270, 280, 317-321, 325, 405, 407, 514, 518, 410; *identitas essentialis*: 266; *identitas formalis*: 266; *identitas suppositiva*: 266; identity between *albedo* and *esse albedinem:* 38; referential identity: 19, 36, 37, 42; accidental or partial identity: 321; syntactic identity of sentences: 149, the identity of the *suppositum*, the *appellatum* and the significate in the case of discrete terms: 186, 205; identity of expression and identity of meaning 220; identity theory of predication: 260, 262, 264, 267; real identity and formal distinction: 263; intensional identity: 265; identity between the particular and its quiddity: 37; identity of logical terms and their metaphysical principles: 207
imagination: 89, 175
impositio (Imposition), impositor.: 101, 129, 135, 178, 181, 215, 234, 294, 298, 300, 339, 390, 448, 450-456, 454, 455, 458, 460, 478; *primaria impositio*: 448, 450, 452, 454; *impositor*: 215, 235
inflection: 165, 515, 517
intellection: 39, 43, 44, 187, 190, 200, 208, 278; primary and secondary intellection of an extramental thing: 44; intellection of the individual 198; *see also s.v.* modus intelligendi
intention: intention of the author: 278; intention of the mind: 339-341, 346, 347, 366; intention of the producer of the sentence token 418; intention of the speaker: 287, 293, 411-423; term of first intention: 88, 102-104, 309, 320, 453; term of second intention: 44, 99, 102, 103, 247, 309, 314, 453, 454; intentionality debate: xiii, 43; intentional mental verbs like 'to know', 'to love': 378
intentionalistae: 44
intensionality: 31, 43-45, 49, 56, 104; 264, 265, 279, 315, 332, 353, 354, 369, 362, 367; intensional operator: 110; intensional relation: 292; intensional interpretation: 353, 363, 354; intensional terms: 367

Jesuits: 474

language: *passim*; mental language: 102, 104, 325, 338, 349, 371-385, 452-453, 455, 458, 485; natural language: 14, 214, 215, 219, 220, 224, 485, 520; spoken language: 368, 376, 372, 411, 420; written language: 337, 348, 372, 383, 377, 380, 411, 420, 423, 452, 455, 458, 472, 485;
see also s.v. trinitarian language
logic: *passim*; modal logic: 97, 395, 497, 498; English and Continental traditions in medieval logic: 235; protestant logic tradition: 464, 466, 469; Stoic logic: 31; modern mathematical logic: 6; philosophical logic: 205, 486. 509; quantification theory of terminist logic: 388; temporal logic, 485-487, 499, 519; differentiation of logic into tensed and untensed: 485

metaphysics: xii, 17-19, 22, 23, 30, 31, 37, 48, 61, 101, 111, 120, 191, 193, 196-226, 253, 267, 275, 276, 278, 283, 290-293, 300, 309, 311, 319, 321, 323, 324; metaphysical counterpart of predictability: 192; metaphysical implications of the semantic properties: 171; metaphysics of modality: 48; metaphysical presuppositions of the theory of predicables: 174; metaphysical structure of the things signified: 189; Aristotelian metaphysics: 53; metaphysics of form according to the Porretans: 61
modality: 48, 106, 330, 487, 495, 487, 496, 498, 500, 501, 509; modal contexts: 96; modal expression 'can': 156; modal or intensional operator: 157, 159, 497; embedded modalities: 504; metaphysics of modality: 48; *see also s.v.* 'logica'
moderni: 110, 400
modus. *modus essendi*: accidental or coincidental modes of being: 17, 20, 21, 27, 35; *modus intelligendi*: 236, 238, 255; *modus praedicandi*: 131, 209, 243, 255; *modus significandi*: 77, 114, 226, 236, 237; *modus supponendi*: 150, 152, 239, 255

modism/modists: 101, 107, 108, 111, 165, 166, 177, 206, 207, 209, 213-215, 233, 234, 246-249, 241, 248, 220, 251, 255; modistic framework: 239; modistic terminology: 215; modists' semantic views in Scotus' philosophy of language: 255

Neo-platonism: xii, 23, 25, 28
natura universalis or *communis: see s.v.* common nature
nomen (noun, name): proper name: 7, 8, 63, 97, 123, 126, 128, 129, 131, 132, 137, 169, 151, 192-194, 196, 243, 251, 313, 489, 498; empty names: 187, 195, 361; non-empty proper names: 196, 501; *qualitas nominis*: 120; *nomen concretivum*: 136; *nomen discretivum*: 125, 132; appellative noun or: appellative term: 19, 20, 44, 45, 56, 123, 128, 131, 134-136, 178; *see also s.v.* pronoun
nominalism: 36, 55, 83, 104, 106, 320, 412

onomastics: 21, 28, 31
ontology/ontological: 8, 13, 19, 21, 22, 42, 43, 46, 52, 56, 83, 111, 119, 120, 122, 123, 140, 165, 209 212, 221-223, 225, 226, 268, 279, 292, 293, 302, 304, 305, 308, 309, 318, 323, 326, 328, 329, 335, 341, 347, 348, 354, 355, 412; ontology and the eucharist: 43; ontology and nominalism: 83; ontology of modern physics and Aristotelian metaphysics of substance: 111; ontology of types and tokens: 165; Platonic ontology: 268; ontology and material supposition: 293, 336; ontological entailment: 25; ontological features of signified things: 309; ontological status of what is denoted: 348, 355; ontological level: 348; ontological presuppositions: 279, 347; ontological properties for semantic features: 323; Aristotelian ontology: 19

paradox: Liar paradox: 267
parallellism paradigm: 49, 55
paralogism: 97, 212, 215, 215, 224, 260, 262, 277, 270, 271, 272, 307; the Trinitarian Barbara paralogism: 267, 270
paratactic approach: 380
Philosophus: *see the index of names s.v.* 'Aristoteles'
phoenix: 109, 155, 390, 391
Platonism: *see index of names s.v.* 'Plato'
pragmatics: 292, 293, 404; pragmatic suspension of the *regula appellationum*: 293
Preachers (Dominican order): 327, 328, 331, 334, 342, 431, 479, 547
predicability: 169, 183, 186, 187, 190, 192, 193, 195, 197, 201; predicability is a logical and a metaphysical property: 193
predication: *passim*; the scope of predicates of second intention: 454; real predicate: 321, 324; predication by essence: 319, 321; predication *per inhaerentiam*: 265; predication *quale* and predication *quid*: 154; predication *secundum habitudinem*: 320; causal predication: 319; formal predication: 262, 263, 266, 269, 270, 319, 321; habitudinal predication: 319, 321; identity theory of predication: 260, 262, 264, 267, 270; aspect theory of predication: 89, 96; predication of substantial forms: 318; accidental predication: 22, 217, 318, 320, 321; predicative combination of things qua understood: 347; *see also s.v.* trinitarian identical predication; *see also s.v. modus praedicandi; see also s.v.* trinitarian identical predication
pronoun: 170, 171, 177, 178, 185, 188, 191, 193, 194, 273, 279, 286, 287, 306, 308, 313, 315, 320, 360, 459, 490, 511, 518, 419; demonstrative pronoun: 173, 286-288; anaphoric pronoun: 511; deictic pronoun: 156; relative pronoun: 518; personal pronoun: 459
proposition/*propositio*: *passim*; dyadic analysis of medieval *propositio*: 32; cognitive and semantic process of producing a proposition: 286; affirmative proposition: 260, 314, 352, 361-365, 460; singular propositions: 170, 173, 178, 203, 244, 254, 270, 362, 363, 385, 389, 390, 396, 401, 403, 405, 511, 512; disjunctive proposition and a proposition with a disjoint predicate: 279, 401; false proposition: 282-284, 290, 352, 354, 361, 374, 377, 404, 407, 473; hypothetical (molecular) proposition: 362; mental proposition: 283, 289, 373-381, 383, 422-426; negative proposition: 314, 362-364, 390, 415; predicational versus quantified proposition: 93; real

proposition: 323; proposition-token, 411, 412, 418-420, 423, 424; proposition-type: 412, 501; scientific proposition: 416, 417; self-reflexive proposition: 412; sophismatic proposition: 149-151

quantification theory: x, 162, 388, 486; quantifiers: 84, 93, 94, 238, 239, 244, 249, 389, 490, 499, 511, 512, 514, 515, 517-520; quantified modal-temporal framework: 487
quiddity: quiddities in themselves: 85-88, 93-106; quiddities in the mind: 86, 88, 89, 94, 98, 101-104; necessary quiddities: 89, 94, 98, 101, 102-104; Avicenna's threefold distinction of quiddity: 81, 84-86; the referential identity between the particular quiddity and the individual: 42; existence of a quiddity: 36; quiddity only applies to things designated by substantive terms: 37

razor: William of Ockham's razor: 50
reduction: 53, 342, 396, 404, 414, 418, 435, 511, 513, 515; reduction of supposition to signification: 342; reduction of enthymemes: 435; reduction of syllogisms: 513; Aristotle's reductive principle: 49;
reference: vi, ix, 3, 31, 46, 81-84, 93, 96, 97, 100-108, 110, 111, 122, 123, 130-132, 135, 154, 155, 157, 159, 162-166, 170, 175, 176, 181, 182, 189, 190, 196, 198, 200, 211, 238, 240, 241, 244, 249, 256, 265, 270, 292, 305, 359,387, 388, 392, 395, 397, 417, 492; self-reference: 17; self-reference of linguistic entities by self-instantiation: 155; ambiguous reference: 81; empty reference: 292; reference to non-existent entities 509
referentiality: 41, 43, 49; referentiality *annex* epistemology: 43
relation, relationship: intentional relation: 43, 292; extensional relation: 292; rational relation: 33; real relation: 33, 319; linguistic correlation between language and thought: 15; intransitive relationship between causation and participation: 25; subordinate relationship: 66; metaphysical relaion: 279; inferential relation: 363; accessibility relation: 496, 497; logical relation: 497
restrictio (restriction): 70, 81, 82, 107, 110, 134, 137, 175, 179, 180, 187, 204, 213, 239, 240, 251, 285, 287, 388, 473, 485, 487, 491-495, 499, 501, 502, 513, 518; *restrictio naturalis*: 493; extension or restriction of the range of possible referents: 213
science/scientific: vii, ix, xiii, 68, 92-95, 100, 109, 120, 140, 468, 472; Aristotelian science: 92; *see also s.v.* demonstration
scope, scope distinction: x, xiii, 14, 22, 26, 27, 53, 108, 110, 167, 173, 174, 201, 239, 240, 282, 286, 287, 332, 355, 388, 454, 456, 464, 476, 493, 498, 524
semantics: semantic stratification: 31, 45, 47, 48; interaction between the semantic theory and metaphysical issues: 191
sense: sense in the author's mind and the reader's attention: 122; opaque sense: 156, 157, 160; sense perception: 89; composite (compound) sense: 93, 149, 157, 158, 246, 252, 356, 506, 507; divided sense: 156-158, 212, 246, 252, 356
significate: 17, 18, 41, 44, 46, 169, 172, 181-183, 186, 187, 189, 193-198, 291, 298, 372, 381, 388, 400, 452, 456, 459, 477; significates of singular expressions: 84, 98, 300, 310; immediate significate of a spoken word: 372; rule of multiple significate (RMS): 17-19, 31; significate and *suppositum*: 180, 186, 187, 200
significatio: change of meaning (*significatio*) by virtue of context: 215; signification of vocal terms: 476; modistic view of signification: 236; natural signification: 356, 376-378, 383; transferred signification: 450
signum materialitatis: 453
speculative grammar: 27, 166; *see also s.v.* modism
square of oppositions: 90, 100, 304, 501, 508, 511
state of affairs: 17, 323, 354, 486
status: triplex status naturae: 81, 84, 85, 100
subordination: 63, 66, 93, 97, 104, 245, 294, 296, 327, 339, 344, 382, 393, 420, 432, 450; subordination of supposition to signification: 327, 339
subsistentia (subsistence): 17, 27, 50, 51, 131, 135
sun: 390, 391
suppositio:
Supposition and other logical notions: association of theories of supposition to

theories of reference: 353; extensional character of Ockham's supposition theory: 359; division of supposition by combining a *semantical* with an *ontological* condition: 336; ontic foundation of the division principle of supposition: 276; ontologically based distinction between personal and simple supposition: 279; supposition divides simple supposition into supposition of the subject and supposition of the predicate: 186, 270; hermeneutical approach to supposition: 477; interconnection existing between supposition and theory of interpretation: 333; link between supposition and verification of propositions: 250, 365; supposition requires both *acceptio* and verification: 392; relationship between supposition and signification: 327, 328, 335, 337, 339; co-suppositum: 367; disjoint supposition: 385, 401-403; suppositional descent: 385, 393; *see also s.v.* descent

divisions of supposition, first into '*propria*' and '*impropria*', and the latter into '*antonomastica*', '*methonomastica*' and '*sinecdochica*': 451; autonomastic and methononimic (methonomastic) ssupposition: 245; relation between simple and personal supposition: 293; the relationship between *suppositio simplex* and *personalis*: 241; personal and simple supposition in mental language: 378; interaction of restriction and ampliation with supposition: 494

kinds of supposition: empty supposition: 361-363, 366; *suppositio abstractiva*: 310; *suppositio accidentalis*: 174, 185, 197, 200-202, 447, 448, 489; *suppositio antonomastic:* 245, *suppositio collectiva:* 476; *suppositio comparata generalis*: 457; *suppositio comparata specialis*: 458; *suppositio concretiva*: 135; *suppositio confusa distributiva:* 84, 244, 277, 307, 314, 324, 392, 451, 460, 461; *suppositio confusa tantum*: 84, 136, 385, 386, 389, 393, 396, 401-404, 515, 516; *suppositio determinata*: 74, 161, 208, 243, 250, 361, 459, 314, 365, 391, 395, 397, 402-407, 490, 515, 519; *suppositio discreta*: 155-190, 229, 391, 490, 453; *suppositio discretiva*: 132, 133; *suppositio essentialis*: 65, 261, 261; *suppositio formalis*: 7, 83, 104, 188, 243, 343, 306, 308, 340, 342, 344; 453, 454, *suppositio materialis*: 6, 7, 81, 83, 84, 102, 104, 111, 147, 150, 151, 153, 156, 157, 159, 161-166, 193, 247, 277, 279, 280, 282, 290, 293, 327, 334-339, 341, 342, 346, 350, 371-374, 378, 380, 382, 412, 417, 511; *suppositio metonomastica*: 245, 451; *suppositio naturalis*: xii, 186, 200-202, 229, 259, 275, 285, 286, 293, 294, 331, 334-336, 338, 339, 447, 448, 451, 488, 489; *suppositio personalis*: 65, 82-85, 95, 102-105, 107-110, 132, 147, 151. 162-164, 166, 169, 173, 175, 179, 180-184, 187, 189-192, 196-198, 204, 239, 241-243, 245-250, 253-255, 161, 279, 281, 290, 291, 293, 294, 298, 300, 306, 307, 308, 310, 322, 331-333, 336-337, 340-342, 344, 345, 347, 361, 373-379, 379-381, 385-389, 393-397, 402, 408, 450, 452, 454, 456, 457, 458-460, 489, 490, 492; *suppositio propria* and *impropria*: 287, 339, 472, 432, 451; *suppositio simplex:* 83, 84, 86, 101, 103-105, 109, 119, 139, 141, 147, 150, 151, 155, 164, 166, 167, 169, 171, 172, 174, 182, 183, 186, 187, 189-192, 194, 196-202, 241, 242, 245-247, 249, 252-255, 261, 279-280, 290, 291, 294, 298, 305-307, 314, 320, 322, 327, 331, 332, 336, 337, 339, 340-342, 345-349, 361, 372-376, 378-384, 404, 427, 443-445, 454-457, 459, 460, 467, 466; *suppositio simplex respectiva* or *comparativa*: 457, *suppositio vaga*: 244, 249, 252

suppositum: suppositum as 'what is denoted' by a signifying term: 206, 354; *suppositum* is or is claimed to be the bearer of a certain form: 60, 61, 329, 330, 334, 335, 340; *suppositum per accidens*: 207; *suppositum per se*: 207, 250; in Priscian *suppositum* means 'grammatical subject': 47, 60, 62; *suppositum-appositum* analysis: 61, 196; *see also s.v.* significate

syncategorematic concept: 33-35; syncategorematic terms: 4, 33-35, 105, 153, 177, 258, 339, 395, 444, 460, 461; *see also s.v.* concept

syntactic components: 238, 251; syntactic rules of instantiations for quantifiers: 83, 84; syntactical analysis of propositions: 388; syntctic identity of sentences: 149;

surface syntactic transormations: 165; syntactical detemination of supposition 279

terms: accidental terms: 90, 108, 308, 310, 321, 325, 334, 459, 504; common terms, that have a triple relation: 241; concrete terms: 88, 89, 95, 98, 99, 102-105, 107, 120, 309-311, 313, 323; discrete terms: 169-202, 459, 491, 492; semantic change in discrete terms: 176; marginal status of discrete terms: 173; mental terms: 373, 378, 420, 455; paronymous terms: 91-93, 107, 108, 111; *quaternio terminorum*: 214, 215, 220, 475; singular terms: 88, 161, 169, 202, 223, 390, 399, 400, 456, 459, 514-516, 520; aggregate singular terms: 446, 448; transcendental terms *ens* and *unum*: 207; *see also s.v.* intention
terminism: 13, 14, 28-30, 36, 44, 49, 50, 53, 55, 56, 169, 171, 243, 238, 427; historiography of terminism: 157; terminist logic: x, 3, 13-15, 28, 30, 31, 49, 52, 53, 55, 107, 133, 141, 169, 170, 171, 174, 206, 207, 234, 236, 237, 243, 388, 478
theology, theological: 13, 14, 74, 66, 69, 72, 119-121, 130, 133, 135, 136, 143, 255, 262, 266, 267, 327, 329, 332, 358, 408, 470, 474; theological truths: 320; theological propositions: 121, 124, 130; theologically true singular propositions: 270; theological theory of supposition: 65; predication in theology: 130
time: time connotation: 34, 45, 56; time of utterance of a proposition: 256, 411, 413, 423; determination of the extension of the time about which a proposition is denoted to be true: 416; interval of time containing the totality of the utterance event: 413
token: token of the concept: 383; sentence-tokens: 6, 15, 63, 165, 174, 282, 310, 341, 377, 378, 380, 383, 411-413, 418-420, 423; *see also s.v.* type
transumptio or *improprietas*: 121, 137, 140
trinity: 124, 129, 255, 260, 265, 267, 270; trinitarian paralogism: 262; trinitarian identical predication: 273; trinitarian language: 123, 132
truth: truth claims: 388; truth conditions: 81, 84, 85, 90, 93, 270, 305, 309, 347, 367, 387-390, 393, 404, 413, 417, 418, 485, 496, 497, 499, 501, 502, 504, 506, 508, 509, 511, 518; truth conditions based on ampliation and appellation: 509; truth conditions for quantification: 487; truth conditions for simple present-tensed predications: 501, 502; cause of truth: 422; truth-conditions of copulative sentences: 315; truth values: 304, 305, 332, 411, 412; bearer of truth-values are assertions: 412
type: spoken and written types and tokens of expressions: 165; different types of language (oral and written): 452

unicorn: 361
use versus mention: xviii, 6; the actual situation of use of words is an imaginary situation described by the word: 422
usus loquendi: 277, 451, 452
utterance; utterance situation: 421; the tense of the verb plus the time of the utterance: 414; the time of evaluation of the utterance: 415

verb: verb expressing a possibility: 287; verb signifying an act of the soul: 376, 377
vetula: 382, 384
virtus: de virtute sermonis: 206, 212, 219, 220, 251, 394, 362, 363, 412, 415, 451, 452

Manuscripts Mentioned

Admont, *Stiftbibliothek*, 241: 67, 76
Assisi, *Biblioteca communale*, 662: 46
Barcelona, *Biblioteca de Catalunya*, 71: 428
Barcelona, *Biblioteca de Catalunya*, 768: 428
Barcelona, *Ripoll*, 141: 428
Barcelona, *Ripoll*, 166: 428
Brugge, *Stedelijke Bibliotheek*, 497 (referred to in the contribution by *siglum* B): 147, 154; 157, 160
Bruxelles, *Bibliothèque Royale*, 3540-3547: 238, 248, 253
Cambridge, *Corpus Christi College*, 378: 428, 435
Cambridge, *Corpus Christi College*, E 293 B: 147
Cambridge, *Jesus College*, Q.B. 17: 69
Cambridge, *Peterhouse*, 205: 67, 76
Cambridge, *Peterhouse*, 206: 62, 76
Carpentras, *Bibliothèque Inguimbertine*, 292 (referred to in the contribution by *siglum* C): 38-41, 43
Erfurt, *Universitätsbibliothek*, Amplon., F. 322: 40
Gdansk (Danzig), *Municipal Library*, 2181: 428, 439
Kobenhavn, *Der Kongelige Bibliotek*, Fragm. 1075: 147
Leipzig, *Universitätsbibliothek*, 1360: 295, 301
London, *British Library, Harley*, 3243: 295, 301
London, *British Museum, Royal*, 12 F VII: 147
London, *British Museum, Royal*, 8 A VI: 178
Madrid, *Biblioteca Nacional*, 1070: 434
Madrid, *Biblioteca Nacional*, 2166: 428
Madrid, *Biblioteca Nacional*, 9748: 430
München, *Bayerische Staatsbibliothek*, *clm* 14. 246: 67
München, *Bayerische Staatsbibliothek*, *clm* 14. 763: 71
Orléans, *Bibliothèque municipale*, 266: 32, 56
Orléans, *Bibliothèque municipale*, 283: 64
Oxford, *Bodleian Library, Lat. misc. e* 100: 428, 439, 442
Oxford, *Bodleian Library, Digby* 2: 147
Oxford, *Bodleian Library, Digby* 24: 70, 147
Oxford, *Bodleian Library, Digby* 174: 66, 70, 76
Oxford, *Corpus Christi College*, E 293B: 147
Pamplona, *Archivo y Biblioteca Capitular*, 6: 428
Paris, *Bibliothèque Nationale de France, lat.*, 6433: 428, 433, 434
Paris, *Bibliothèque Nationale de France, lat.*, 14069: 147
Paris, *Bibliothèque Nationale de France, lat.*, 14.716 (referred to in the contribution by *siglum* P): 38-41, 43
Paris, *Bibliothèque Nationale de France, lat.*, 16.619: 67
Paris, *Bibliothèque Nationale de France, Nouv. acq. lat.*, 258: 428
Prague, *Státní Knihovna* CSR, 396 8 (III. A. 11): 295, 301
Praha, *Knihovna Metropolitni Kapitaly*, L. 6: 208, 213; 214, 219, 222, 224
Roma, *Biblioteca Casanatense*, 5445: 428, 439, 441, 442
Salamanca, *Biblioteca Universitaria*, 1735: 428, 436, 439, 441-444
Salamanca, *Biblioteca Universitaria*, 1882: 428-431, 434
Salamanca, *Biblioteca Universitaria*, 2002: 428
Salamanca, *Biblioteca Universitaria*, 2080: 428, 434
Salamanca, *Biblioteca Universitaria*, 2107: 428, 430
Segovia, *Archivo de la Catedral*, B-283 (olim 30): 428, 436, 438, 440
Segovia, *Archivo de la Catedral*, B-293 (olim 31): 428, 434, 437, 439, 440, 442, 443
Segovia, *Archivo de la Catedral*, B-355 (olim 33): 428, 429, 433
Sevilla, *Biblioteca Capitular y Colombina*, 7-7-7: 428, 433-435, 457

Sevilla, *Biblioteca Capitular y Colombina*, 5-1-14: 428
Sevilla, *Biblioteca Capitular y Colombina*, 7-2-22: 430
Sevilla, *Biblioteca Capitular y Colombina*, 7-3-13: 428
Toledo, *Archivo y Biblioteca Capitular*, 94-27: 428, 441, 444, 430, 441, 443, 445
Toledo, *Biblioteca Capitular*, 94-28: 428
Tortosa, *Biblioteca Capitular*, 25: 428
Vatican, *Bibliotheca Apostolica, Ottob. Lat.*, 318: 239, 252
Vatican, *Bibliotheca Apostolica, lat.* 955: 268, 269, 271-273
Zaragoza, *Biblioteca del Cabildo Metropolitano*, 15-82 (now lost): 428

www.ingramcontent.com/pod-product-compliance
Lightning Source LLC
LaVergne TN
LVHW102225300926
849160LV00005B/23
9789004259836